Vitamin B12

Advances and Insights

Vitamin B12

Advances and Insights
Advanced Materials

Edited by

Rima Obeid

Aarhus Institute of Advanced Studies
University of Aarhus
School of Engineering and Applied Sciences
Aarhus C, Denmark

CRC Press
Taylor & Francis Group
Boca Raton London New York

CRC Press is an imprint of the
Taylor & Francis Group, an **informa** business

A SCIENCE PUBLISHERS BOOK

CRC Press
Taylor & Francis Group
6000 Broken Sound Parkway NW, Suite 300
Boca Raton, FL 33487-2742

First issued in paperback 2021

ISBN-13: 978-0-367-78239-9 (pbk)
ISBN-13: 978-1-4987-0699-5 (hbk)

Library of Congress Cataloging-in-Publication Data

Names: Obeid, Rima, editor.
Title: Vitamin B12 : advances and insights / edited by Rima Obeid.
Other titles: Vitamin B12 (Obeid)
Description: Boca Raton, FL : Taylor & Francis Group, [2016] | Includes bibliographical references and index.
Identifiers: LCCN 2016039101| ISBN 9781498706995 (hardback : alk. paper) | ISBN 9781498707008 (e-book : alk. paper)
Subjects: | MESH: Vitamin B 12 Deficiency | Vitamin B 12--metabolism
Classification: LCC QP772.V52 | NLM WD 120 | DDC 612.3/99--dc23
LC record available at https://lccn.loc.gov/2016039101

Visit the Taylor & Francis Web site at
http://www.taylorandfrancis.com

and the CRC Press Web site at
http://www.crcpress.com

Dedication

To my son, Jean-Paul

Preface

Cobalamin(s) (vitamin B12) have been known for 100 years. Key milestones in the study of cobalamin have been through over 10 decades of trial and error, research and discoveries. Remarkable discoveries in the field have saved many lives and were awarded 2 Nobel Prizes; Minot and Murphy (Nobel Prize in Physiology or Medicine 1934) and Dorothy Hodgkin (Nobel Prize in Chemistry 1964). Still, cobalamin constitutes an amazing area of research with many undiscovered facets.

The health relevance of cobalamin became evident long before discovering its chemical entity. Liver extracts containing few micrograms of the healing factor, cobalamin, were used in the 1920s up to the early 1930s as a life-saving medication against fatal pernicious anemia. The purification and production of large quantities of the 'liver factor' were real challenges, but the biggest challenge was for patients to eat these extracts as an alternative to death. Now that cobalamin has become available as over-the-counter supplements, or as injections containing a few micrograms to milligrams, its relevance to health and disease has gained more importance over the time. Cobalamin's 'lifting effect' has been experienced by millions of patients and doctors. Today, the impact of cobalamin on human health has changed from 'treating a severe disease' to 'prevention of a yet not-manifested condition'. The meaning of cobalamin has now taken new dimensions on a population level after implementing modern laboratory diagnosis tests. Using modern biomarkers has shown that subclinical cobalamin deficiency affects many individuals in critical life phases.

'Vitamin B12: advances and insights' is an extract of knowledge of experienced scientists who have been working on nutritional, structural, chemical, and clinical aspects of the vitamin. This book has introduced an innovative and unclassical approach by addressing 'gaps in knowledge' that surround the topic. These gaps are identified by scientists who are very close to the cobalamin epi-center and intended to provide a direction for future research.

The book is an in-depth study on the vitamin from basic science to modern health challenges. Early knowledge on cobalamin in the light of recent scientific discoveries (Chapter 1); Dietary requirements and nutritional supply (Chapter 2); Cobalamin uptake and intracellular processing (Chapter 3);

Congenital cobalamin disorders (Chapter 4); Acquired causes of cobalamin deficiency and clinical consequences (Chapter 5); Cobalamin deficiency: a public health problem in developing countries (Chapter 6); The role of cobalamin in the nervous system, its relevance to brain aging, and potential mechanisms surrounding this area (Chapters 7 and 8); Cobalamin deficiency biomarkers and diagnosis (Chapter 9); Cobalamin deficiency in critical age phases such as pregnancy, lactation and early life (Chapters 10 and 11); Cobalamin deficiency in the era of folic acid fortification (Chapter 12); Cobalamin unexplained extreme values in clinical practice (Chapter 13); the role of Cobalamin in drug transport and development (Chapter 14).

The target audience for this book are experts and researchers looking for in-depth knowledge in the above mentioned areas of cobalamin science; health care providers who take part in diagnosis, treatment, and prevention of deficiency conditions; policy makers who can influence implementation of diagnosis tools or nutritional policies on a country and population levels; and stakeholders and pharmaceutical companies who are interested in producing diagnosis tools, supplements, fortified foods or other pharmaceutical products that use cobalamin as a drug carrier.

This book is by no mean a complete documentation of what is going on around the topic. However, it constitutes an attempt to grasp the current knowledge on a few areas related to cobalamin and to provide insights into unexplored questions and issues.

Contents

1

Milestones in the Discovery of Pernicious Anemia and its Treatment

Jörn Schneede

OVERVIEW

Pernicious anemia (PA) is a serious form of vitamin B12-deficiency. Vitamin B12 belongs together with heme and chlorophyll to the tetrapyrrole family (Figure 1) (Yin and Bauer 2013). Vitamin B12 is an evolutionarily ancient (≈ 3.8 x 10^9 years old) cofactor that was responsible for energy production through fermentation of small organic molecules in the absence of exogenous electron acceptors in the prokaryotic anaerobic world (Figure 2) (Santander et al. 1997). In the course of evolution siroheme later allowed making use of inorganic electron acceptors, before oxygen production by chlorophyll made aerobic respiration through heme possible. Almost 1% of the genome of S. typhymurium is dedicated to vitamin B12 synthesis and transport (Roth et al. 1996). Though being one of the structurally most complex, non-polymeric biomolecules synthesized by nature, eukaryotic organisms do not produce B12 (Figure 2). As a consequence, this vitamin is essential for human metabolism, albeit only required in trace amounts (possibly as low as 1 µg/d, while the recommended daily allowances in adults are 2.4 µg/d) and functions as cofactor in only two enzymes, methionine synthase and (R)-methylmalonyl-CoA mutase (Helliwell et al. 2011). The remarkable discovery of vitamin B12 was only possible and proceeded by the endeavor to find effective

Department of Clinical Pharmacology. University of Umeå 901 85 Umeå, Sweden.
Email: jorn.schneede@medbio.umu.se

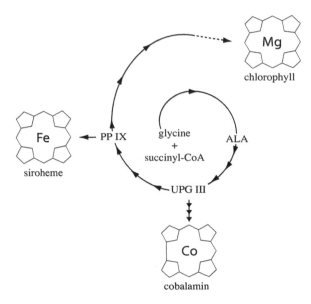

Figure 1. Schematic depiction of the evolutionary development of the tetrapyrrole biosynthetic pathways. Synthesis of tetrapyrrols starts with succinyl-CoA and glycine to form 5-aminolevulinic acid (ALA). Uroporphyrinogen III (UPG III) is used for synthesis of cobalamin (vitamin B12), while protoporphyrin IX (PP IX) is the starting point for siroheme and chlorophyll synthesis. Note the different metal ions bound to the corrin rings: Co, Fe and Mg. Compared to (siro-)heme- and chlorophyll, synthesis of cobalamin is far more complicated and complex, involving considerably more enzymatic steps (at least 25 enzymes uniquely involved), ring contraction, insertion of cobalt, modification of the tetrapyrrole ring and insertion of a nucleotide tail. Adapted and modified from (Yin and Bauer 2013).

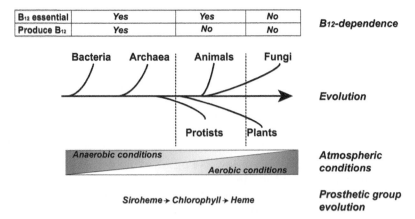

Figure 2. Distribution and dependency on vitamin B12 among different living forms during the evolutionary process and under different atmospheric conditions. Adapted and modified from (Roth et al. 1996).

treatment options for PA, one of the conditions causing severe vitamin B12-deficiency. The scientific progress was, however, slow and stretched over a period of almost 200 years from the first description of PA, to the evolution of theories about possible causes and ultimately the invention of effective treatments. The search for effective treatment options also resulted in the resolution of the pathogenesis of pernicious anemia and ultimately the discovery of B12. Through isolation of an unknown *extrinsic factor* from liver extracts that was accountable for clinical response in pernicious anemia patients it was eventually possibly to elucidate the chemical structure of vitamin B12. In parallel, efforts were started to map the production of vitamin B12 in certain bacteria. The elucidation of the chemical pathways of bacterial production of the vitamin finally made the complete synthesis of the vitamin in the laboratory possible. This enterprise has sometimes been called "Mount Everest of biosynthesis" and it was not before in 2013 that the complete anaerobic pathway of B12-synthesis had been charted (Moore et al. 2013). All in all, vitamin B12 research has resulted in two Nobel prizes-so far. Notwithstanding these achievements, synthesis of the vitamin in the laboratory is far too complicated and resource-demanding for commercial purposes and large-scale industrial production of vitamin B12 is still carried out by aerobic fermentation using *Pseudomonas denitrificans* (Xia et al. 2015) (Figure 3).

The course of history of vitamin B12 can arbitrarily be divided into different eras and stages (Figure 4). During this journey different therapeutic approaches for the treatment of vitamin B12-deficiency were developed, first oral therapy with raw liver, then oral or parenteral administration of liver extracts, followed by more refined liver concentrates that could be injected or taken by mouth with and without addition of intrinsic factor isolated from gastric juice. Later, crystalline B12 (cyanocobalamin) was isolated from the liver or produced by bacterial fermentation. With this advance, parenteral therapy with highly concentrated cobalamin preparations became feasible and affordable. In parallel, there was a continuous development of diagnostic tests for detection of vitamin B12-deficiency (Moridani and Ben-Poorat 2006). Parenteral therapy with intramuscular injections was soon considered the most reliable method of treating pernicious anemia (Bethell et al. 1959). However, recent clinical experience and health technology assessments from Sweden and other countries indicate that oral therapy with vitamin B12 tablets is both clinically feasible and more cost-effective than injections (Berlin et al. 1968b; Kolber and Houle 2014).

The following historical review will present a survey over the history of vitamin B12 and changing concepts for the treatment of vitamin B12-deficiency over the last two centuries.

Figure 3. Fermenter for pharmaceutical production of vitamin B12. Microbial fermentation still is the most commonly used method for industrial scale production of many vitamins, including vitamin B12 (Xia et al. 2015). The picture shows a fermenter for industrial scale production of vitamin B12 under both aerobic and anaerobic conditions. The tank volume of the depicted fermenter can be up to 6.000 liters. In large scale industrial production fermenters of 120.000 liters can be used yielding up to 198 mg/l of B12 (Xia et al. 2015). Reproduced with permission (INOXPA 2015).

1. History of Vitamin B12

1.1 The era before treatment of pernicious anemia was possible

1.1.1 Discovery of pernicious anemia as a first step in identification of vitamin B12

Pernicious anemia, an extreme form of vitamin B12-deficiency, was most likely first portrayed in 1822 by James Combe (1793–1860), a Scottish physician from Edinburgh (Combe 1824). He described the disease history of a 47-year-old man, Alexander Haynes, who suffered from a peculiar, rather

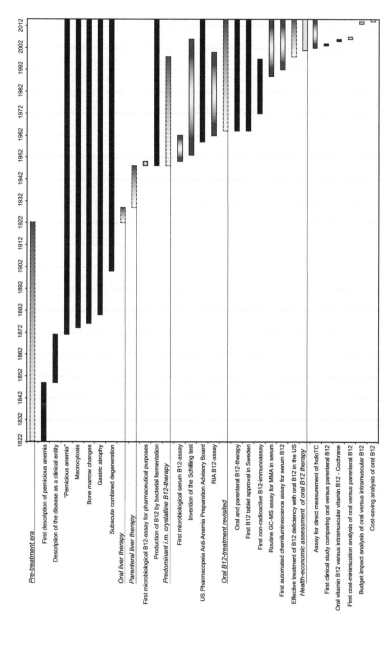

Figure 4. Time lines of different eras in the history of vitamin B12-deficiency and its treatment. The graph portrays an arbitrary division of the history of vitamin B12-deficiency into different eras and milestones of discoveries. Treatment-related accomplishments are depicted by central gradient bars with dashed-lined borders, major diagnostic achievements by central gradient bars with solid-lined borders and health-economic evaluations of oral vitamin B12 treatment are illustrated by grey bars with solid-lined borders. For about 100 years after the first description of pernicious anemia it was a virtually untreatable, fatal disease. For detailed references see Table 1.

rapidly progressing condition of severe anemia combined with gastrointestinal symptoms. Haynes finally deceased in a state of pulmonary edema within a period of less than seven months after the first contact with Combe. Combe recommended *"chalybeates,*[1] *tonics and a nourishing diet"* to treat the condition, with no success.

After the first description of PA as a clinical entity it would take over 100 years before an effective treatment strategies became available. During this period, PA was considered an untreatable and inevitably lethal ("pernicious") disease. The exact cause of the disease remained obscure for even many more years to come. This disease harvested many casualties. Before effective treatment became available, PA accounted for the death of more than 50,000 people per year in the US alone (Ahrens 1993; Jarcho and Brown 1977).

The constantly fatal outcome of this disease accompanied by lack of conclusive autopsy findings must indeed have been an agonizing experience for doctors at that time, but also spurred scientific efforts to find the cause and potential therapies of this disease. In 1855, Thomas Addison (1793–1860), a London physician, described more details about the disease from observations in 11 patients who were admitted to Guy's Hospital in London (Pearce 2004). Addison reported:

"a very remarkable form of general anaemia occurring without any discernable cause whatever." The patients had *"no previous loss of blood, no exhausting diarrhoea, no chlorosis, no purpura, no renal, splenic, miasmatic, glandular, strumous or malignant disease."*

Addison's observations of this "idiopathic anemia" initially received little attention in the scientific community. In 1872, however, Anton Biermer, a German internist working at that time in Zürich, gave a comprehensive description of this disorder during a meeting and used for the first time the expression "progressive pernicious anemia". He chose the phrase "pernicious anemia" (PA) as the disease had an insidious onset with slow progress and because it was deemed to be untreatable and lethal at that time. This time, the disorder received more interest worldwide, which resulted in a large number of publications during the years to follow (Cohnheim 1876; Ehrlich 1880; Eichhorst 1878; Fenwick 1880; Lichtheim 1887). As an acknowledgment of the contribution of both Addison and Biermer, PA is also called "Addison-Biermer disease" (Ewing 1901).

During the years to come the morphological characteristics of PA were identified and described in more detail. Cohnheim (1839–1884) observed increased cellularity in the bone marrow (Cohnheim 1876) and Paul Ehrlich (1854–1915) discovered the occurrence of megaloblasts in the peripheral blood of PA patients in 1880 (Ehrlich 1880). In 1900, Russell described spinal cord

---

[1] Water from a mineral spring with a high content of iron.

involvement in PA and coined the term "subacute combined degeneration of the spinal cord" (Russell et al. 1900). Moreover, gastric atrophy was detected in PA patients (Fenwick 1880). It was noted that hematologic abnormalities in PA patients resembled tropical sprue, which normally responded to a diet containing milk, meat, cod-liver oil and oranges. This may be the reason why similar treatments were used for patients with PA in the early days (Wills 1948).

Thus, in the period from 1876 to 1900, it became clear that PA was not solely a hematological disease, but also had gastrointestinal and nervous system components (Table 1).

Because of the treacherous start and slow progression of PA, this disease was often misinterpreted, and initial symptoms were often attributed to normal or premature ageing processes (Tobin and Cargnello 1993). Consequently, the disease was often identified at a late stage and many patients only had a short time to live after diagnosis (Combe 1824).

Nowadays it is difficult to imagine the seriousness of untreated pernicious anemia. However, thanks to William P Murphy, one of three laureates sharing the Nobel Prize in Physiology or Medicine in 1934, we are still able to watch a motion picture giving a vivid picture of the graveness of the disease in historic PA patients before and under treatment with liver extracts. The motion picture was presented during Murphy's Nobel Lecture, December 12, 1934 (Murphy 1934). This motion picture has been recently made available to the general public through the Blood journal's website (Kumar et al. 2006).

An overview over the milestones in the discovery of pernicious anemia and vitamin B12 is given in Table 1.

2. Early Treatment Approaches of PA

The first therapeutic approaches to a disease of an unknown cause and pathogenesis were highly experimental and arbitrary, and quite often obscure. Until the discovery that liver contains important nutrients and factors for hematopoiesis, the methods intended to treat pernicious anemia only had temporary effects and were on the whole unsuccessful (Combe 1824; Sinclair 2008). Around 1900 a renowned textbook of hematology considered iron being contraindicated for the treatment of certain forms of anemia where megalocytes with increased hemoglobin content appeared in the peripheral blood. Instead arsenic was supposed to exert almost specific effects and would result in increasing numbers of red cells and in stimulating the production of more uniformly distributed Hb (Ewing 1901). In the eighth edition of Sir William Osler's *"THE PRINCIPLES AND PRACTICE OF MEDICINE— DESIGNED FOR THE USE OF PRACTITIONERS AND STUDENTS OF MEDICINE"* from 1915, one of the most authoritative text books at that time, it was suggested—among other remedies—to try Fowler's Solution, sodium cacodylate or *Atoxyl*. Fowler's Solution was an arsenical preparation, which may well have accelerated the death of many PA patients. Also *Atoxyl*, a

Table 1. Historical track record of discovery and treatment of pernicious anemia that later lead to the discovery of vitamin B12.

Era/Discovery	Year	Discovery/Event	Ref.
Pre-treatment, disease finding era. Characterization of the pernicious anemia and early therapeutic approaches	1822	Unexplained cases of anemia, first description of PA	(Combe 1824)
	1849	PA identified as a clinical entity	(Addison 1849)
	1871	15 cases of PA described—PA gained general interest in the medical community for the first time	(Biermer 1872)
	1874	First blood count in PA—observed large size of cells	(Sørensen 1874)
	1876	Bone marrow examination—increased cellularity	(Cohnheim 1876)
	1878	Comprehensive monograph on progressive PA	(Eichhorst 1878)
	1880	Monograph on atrophy of the stomach	(Fenwick 1880)
	1880	Recognition of megaloblasts in the blood of PA patients	(Ehrlich 1880)
	1887	Neurological components of PA	(Lichtheim 1887)
	1900	Subacute combined degeneration of the spinal cord (SCDC)	(Russell et al. 1900)
Oral and parenteral liver treatment and discovery of the pathogenesis of pernicious anemia	1926	Liver therapy of pernicious anemia	(Minot and Murphy 1926)
	1929	Achylia gastrica (atrophic gastritis) associated with PA—intrinsic factor contained in gastric juice	(Castle 1929)
	1947	First (microbiological) assay for quantification of B12	(Shorb 1947b)
Isolation and characterization of B12; predominantly parenteral treatment with crystalline B12	1948	Isolation and crystallization of B12 from liver	(Fantes et al. 1950; Rickes et al. 1948)
	1949	Production of B12 by bacterial fermentation	(Stokstad et al. 1949)
	1952	Assay of cobalamin in human serum	(Ross 1952)
	1953	Schilling test for evaluation of intestinal absorption of B12	(Schilling 1953)
	1954	Complete structure of B12 resolved	(Brink et al. 1954)
Oral treatment revisited. First B12-tablet approved in Sweden and final elucidation of synthesis pathways of B12	1964	First B12 tablet approved by Swedish Medical Products Agency	(Ågren 1964)
	1965	B12-binding proteins described	(Hall and Finkler 1965)
	1973	Total synthesis of vitamin B12 achieved	(Woodward 1973)
	1994	Entire aerobic pathway of B12 synthesis resolved	(Battersby 1994)
	2013	Elucidation of the entire anaerobic pathway of B12 synthesis	(Moore et al. 2013)

precursor of sulfonamide antibiotics, contained arsenic (Riethmiller 2005), which—together with sodium cacodylate—was used for the treatment of syphilis (Nichols 1911). The following quote from Osler's textbook illustrates the prevailing ignorance of the medical community about the proper treatment of PA at that time (Osler 1915):

> *"There are five essentials: first, a diagnosis; secondly, rest in bed for weeks or even months, if possible (thirdly) in the open air; fourthly, all the good food the patient can take; the outlook depends largely on the stomach; fifthly, arsenic; Fowler's solution in increasing dosis beginning with m iii or v (0.2 to 0.3 c.c.) three times a day, and increasing to m i each week until the patient takes m xv (1 c.c.) three times a day. Other forms of arsenic may be tried, as the sodium cacodylate[2] or the atoxyl hypodermically. Atoxyl can be given in doses of gr. ss (0.032 gm.) every five days, and the amount is gradually increased. Accessories are oil inunctions; bone-marrow, which has the merit of a recommendation by Galen; in some cases iron seems to do good. Care should be taken of the mouth and teeth. Gastric lavage and irrigations of the colon are useful in some cases.*
>
> *Injections of blood serum and defibrinated blood have been given. The serum is given in small amounts, 10 to 20 c.c., usually into a vein; rabbit serum is perhaps the best. Defibrinated human blood should be given intravenously in large amounts, up to 500 c.c." (Osler 1915).*

Part of the short-lived effectiveness of arsenic for treatment of PA may be explained by liberation of B12 from the body's own cells through arsenic-induced cell-lysis (Dunlop 1973; Riedmann et al. 2015; Weber 1932).

3. Recovery from PA by Liver Treatment

3.1 The discovery and development of oral liver therapy

The First World War triggered research into blood substitutes and ways of improving recovery and hematopoiesis after massive blood loss (Sinclair 2008). This may also have stimulated George Whipple, who was an expert in liver diseases working at the University of California, to examine the liver's role in hematopoiesis. In 1920 he conducted a series of experiments in dogs that had been made anemic through venesectio and investigated the effects of various dietary treatments (Whipple et al. 1920). Interestingly, similar experiments carried out in pancreoectomized dogs resulted in the discovery of insulin by Fredrick Banting and Charles Best at the University of Toronto, Canada, during 1920–1924 (Banting et al. 1991). Whipple later moved to the University of Rochester, School of Medicine and Dentistry in New York State

---

[2] Chemical compound with the formula $(CH_3)_2AsO_2Na$ at that time used—together with Salvarsan and Atoxyl—for the treatment of syphilis

and continued his research on the effects of dietary regimes including liver, iron pills, arsenic and germanium dioxide for treatment of chronic anemia. Only liver, and especially raw, uncooked liver turned out to be effective in treatment of anemia (Robscheit-Robbins and Whipple 1925). The finding that raw liver was more effective than cooked liver was pure serendipity. Disobeying the instructions of Whipple, a laboratory technicians responsible for the dogs, fed the anemic animals raw liver instead of cooked and a more pronounce hematological effect was observed (Sinclair 2008). We now know that liver is rich in vitamin B12, folate, and other nutrients. Further, vitamin B12 is heat-stable while folate is not. Therefore, the chance finding that raw liver was more effective than cooked liver in restoring hemoglobin levels could indicate that apart from heat-stable vitamin B12 other, heat-labile hematopoietic factors such as folate contained in raw liver might have been responsible for the superior hematological effects in anemic dogs. Still, in 1923 George Minot and William Murphy, two physicians from Bostom, took notice of Whipple's discovery in dogs and decided to try raw liver for the treatment of patients with PA (Sinclair 2008). In 1926, Minot and Murphy presented their results of the first 45 patients who had been given a high protein diet that included 100–240 g of liver and 120 g of meat for between six weeks and two years at a meeting of the Association of American Physicians in Boston (Minot and Murphy 1926).

Interestingly, Minot had learnt a method of counting reticulocytes in the meanwhile that allowed him to study early hematological responses to liver treatment (Sinclair 2008). Minot and Murphy observed raised reticulocyte counts within four to ten days after starting the diet (Kumar et al. 2006). Other signs of hematological response such as increased hemoglobin levels and red cell counts and improvement of jaundice in addition to neurological recovery followed later during therapy.

3.2 Rather die than being treated with raw liver that tasted dreadful

Raw liver was assumed to contain a yet unidentified *extrinsic factor* responsible for the clinical effects. However, this diet tasted dreadful. David Hilbert (1862–1943), one of the greatest mathematicians of the first half of the twentieth century and director of the Mathematical Institute of Göttingen (Reid 1996) was diagnosed with pernicious anemia during autumn 1925. The disease had gone undetected for a long time because the first symptoms—taking into account his age of 65 y—had been interpreted as a merely age-related phenomenon. At the time of diagnosis, however, Hilbert was no longer able to leave his house because he was too weak to walk and he taught his students at home. The doctors gave him at best a few months or even weeks to live. A pharmacologist friend in Göttingen by chance read the paper of Minot and Murphy in JAMA from 1926 (Minot and Murphy 1926). By the intervention of Marianne Landau, daughter of the Nobel laureate Paul Ehrlich, contact was established with several Harvard professors in mathematics who finally could convince Minot to send experimental liver extracts from the United States despite the scarcity of the preparations. Until the arrival of the liver extracts, Hilbert had, however, to follow the original raw liver diet. A colleague of Hilbert, EU Condon, visiting Göttingen in the summer of 1926, heard Hilbert complain that he would rather die than eat that much raw liver. Yet, his condition improved almost immediately upon the liver therapy (Reid 1986).

3.3 Nobel Prize in Medicine in 1934 and development of oral and parenteral liver treatment

Vitamin B12 research resulted in two Nobel prizes. On December 10, 1934, the Caroline Institute awarded the Nobel Prize in Physiology or Medicine to three American investigators: George R. Minot and William P Murphy of the Harvard Medical School (Boston, MA) and George H Whipple of the University of Rochester School of Medicine and Dentistry (Rochester, NY), "in recognition of their discoveries respecting liver therapy in anaemias." On December 12, 1934, Murphy presented this motion picture (Murphy 2006) as part of his Nobel lecture (Murphy 1934). He introduced the movie with the following words:

> *"Rather than enlarge further upon the details and results of the treatment of pernicious anemia, I shall now present, with your permission, a motion picture which will illustrate many points more clearly than I could discuss them here."*

The motion picture, made at the Peter Bent Brigham Hospital in Boston emphasizes the superiority of parenteral to oral therapy with liver extract in the treatment of PA. The movie consists of two parts. In the first part hematologic and neurologic signs and symptoms in PA are illustrated and a synopsis of normal hematopoiesis as well as pathology seen in PA is given. Further, different treatment schemes with whole liver, oral liver extracts, and concentrated extracts for intramuscular injections are compared. The second part depicts the improvement in the peripheral smear with liver therapy, and the greater clinical effectiveness of parenteral therapy compared with oral treatment with liver or liver extracts. Even a cost-effectiveness analysis of the parenteral treatment is presented and the importance of maintenance therapy is highlighted. Oral liver therapy was used in clinical practice for a relatively short period of about seven years before it was substituted by parenteral therapy and at the time of the Nobel lecture oral liver therapy had been more or less replaced by intramuscular injections of liver extracts (Figure 4). Liver extraction methods were rather crude at that time (Gänsslen 1930) and preparations certainly contained other hematopoietic factors in addition to vitamin B12 (Okuda 1999). Ironically, improved purification of liver extracts may have removed these additional hematopoietic factors and contributed to the lower potency and greater batch variability of the preparations during the course of the years between 1940 and 1948 (Mollin 1950). Until that time, the "extrinsic factor" contained in liver and curing PA still remained undiscovered.

4. Isolation and Crystallization of Vitamin B12

The discovery of the unknown liver factor, i.e., vitamin B12, was delayed because no quantitative *in vitro* tests were available at that time to measure the potency of the different liver extracts. Thus, the only way to evaluate the effectiveness of the extracts was to test them on patients with PA, which was time-consuming (Rickes et al. 1948; Shive 2002; Vora 1956). The discovery of a microbiological assay for the measurement of vitamin B12 activity in 1947 (Shorb 1947a; Shorb 1947b) accelerated the isolation of the *extrinsic factor* contained in liver that was responsible for the alleviating the clinical symptoms. Vitamin B12 in the liver extract was able to enhance the growth of the bacteria, and the growth rate could be used as a measure of the amount of the unknown factor in the extract. Unfortunately, extraction and isolation of crystalline vitamin B12 from liver was highly inefficacious. One and a half tons of beef liver was needed to produce 1 gram of vitamin B12. Finally, in 1948 two independent groups, Folkers and co-workers at Merck, United States, and Lester-Smith and co-workers at Glaxo, England, succeeded in isolating the *extrinsic factor* in 2 different crystalline forms (hydroxyl and cyano-cobalamin) and they named it vitamin B12 (Wagner and Folkers 1963). Vitamin B12 received its name (B12) just after folate (vitamin B9) had been discovered and thus it was given the number 12 in the B-group. The gaps in the numbering of the B-vitamin complex are due to the fact that a number of substances initially mistaken for vitamins were gradually removed from the group of B-vitamins (Elliot 2008).

4.1 Second Nobel Prize in Chemistry in 1955 and the discovery of the chemical structure of vitamin B12

The final product of the extraction, vitamin B12, turned out to be odor- and tasteless, bright red needle-shaped crystals (Howard 2003). However, it took seven more years before the exact chemical structure was finally resolved through X-ray crystallography by Dorothy Hodgkin in 1955 (Brink et al. 1954; Hodgkin et al. 1955). About ten million calculations had been necessary to clarify the structure of this factor. For this achievement Dorothy Crowfoot Hodgkin received the Nobel Prize in 1964. During the 1950's studies on isolates of various bacteria and molds primarily used for antibiotic production, and rumen microorganisms revealed that many of these microbes were also able to synthesize vitamin B12 (Halbrook et al. 1950; Johnson et al. 1956).

Still, the complete laboratory chemical synthesis of vitamin B12 was not accomplished before 1972 (Woodward 1973). Total chemical synthesis requires more than 70 steps and is extremely resource demanding. Microorganisms are far more efficient at synthesis of this vitamin and bacterial fermentation remains the main source of production of vitamin B12 at industrial scale even today (Xia et al. 2015) (Figure 3).

5. Oral Treatment Revisited

5.1 Oral or parenteral treatment with liver extracts

Even though the exact cause of PA remained unknown for many years to come, involvement of the gastric ventricle in the pathogenesis of this disease was anticipated as early as in 1880 (Fenwick 1880). Gastric atrophy and achlorhydria were common findings in pernicious anemia patients. In 1926, it was clear that liver obviously contained an, at that time, unidentified *extrinsic* (food/liver) *factor* accountable for the clinical response, but for more effective oral treatment in addition an *intrinsic* (gastric) *factor* was needed and this factor was most likely contained in the gastric juice of healthy humans or animals (Castle 1929).

In 1927, Castle performed the first experiments that demonstrated the existence of an additional endogenous gastric substance involved in the pathogenesis of PA, which he called the *intrinsic factor* (Castle 1929).

Castle found that neither normal human gastric juice nor nearly raw hamburger meat alone could induce a reticulocyte response in PA patients. However, hamburger meat that had stayed in Castle's own stomach for 1 h before it was regurgitated and then fed to PA patients via a nasogastric tube triggered a reticulocyte response. Castle stated:

> *'that in contrast to the conditions within the stomach of the pernicious anaemia patient, there is found within the normal stomach during digestion of beef muscle some substance capable of promptly and markedly relieving the anaemia of these patients'* (Castle 1929).

Development of processed liver concentrates for oral administration followed. However, oral treatment demanded relatively high doses and was thus very expensive (Ungley 1955). It turned out that the potency of the liver extracts could be improved by adding the yet unidentified *intrinsic factor* (Castle 1953), which was supplied as liquefied stomach contents of a healthy normal person or desiccated hog's stomach (Glass and Boyd 1953). However, all of these preparations tasted dreadful and the period of exclusively oral treatment of PA with liver or liver extracts was soon replaced by more "palatable" parenteral regimes with injectable liver extracts (Schultzer 1934). Parenteral treatment had additional advantages as it was—at that time—considerably cheaper compared to oral therapy due to lower dose-demands and no *intrinsic factor* was needed (Ungley 1950a; Ungley 1950b). However, injectable liver extracts also had serious side effects, sometimes even fatal, and the potency unfortunately showed considerable variability between vendors as well as over time resulting in many relapses during maintenance therapy (Anonymous 1965; Mollin 1950). Moreover, the cleaner the liver extracts became the poorer was their potency as other potentially active factors such as folate and iron from crude liver extracts gradually disappeared (Conley and Krevans 1955). By the end of the 1940s pure crystalline cyanocobalamin (vitamin B12) was preferred over liver extracts for treatment of PA (Blackburn et al. 1952; Blackburn et al. 1955; Wagner and Folkers 1963).

Since then, parenteral supplementation with crystalline cyanocobalalmin has been the mainstay of treatment of most forms of vitamin B12 deficiency in the majority of countries world-wide (Stabler 2013). Adherence to this therapeutic tradition is most likely also a result of the 1959 US Pharmacopeia Anti-Anemia Preparations Advisory Board recommendation, which advised against the use of oral therapy for pernicious anemia, mostly because of its unpredictable efficacy (Bethell et al. 1959). Typically, arguments against oral cyanocobalamin therapy were based on findings of inadequately low serum cobalamin concentrations achieved in patients taking oral vitamin B12 doses of 100–250 µg/d without *intrinsic factor* (Glass and Boyd 1953). According to our current knowledge, when oral doses > 10 µg/d are used, only approximately 1.5% of the dose is expected to be absorbed through passive diffusion, thus explaining the relatively low efficacy of doses < 500 µg/d in treatment of PA.

5.2 Oral treatment with crystalline vitamin B12 with and without intrinsic factor

Parallel with the parenteral use of crystalline vitamin B12, investigators searched for oral alternatives to liver extracts and experimented with oral use of small amounts of crystalline vitamin B12 in combination with *intrinsic factor* from various sources to improve bioavailability (Blackburn et al. 1955). By the end of the 1950s more efficient, large-scale industrial production of vitamin B12 by microbial fermentation was accomplished, securing the supply of inexpensive cyanocobalamin (Mervyn and Smith 1964). This opened up for the use of higher oral doses of cyanocobalamin without *intrinsic factor* despite poor bioavailability (Waife et al. 1963). During the 1950's a large number of dose-finding studies of oral therapy with crystalline B12 were performed with and without addition of intrinsic factor and clinical responses were monitored thoroughly (Brody et al. 1959; Chalmers and Hall 1954; Chalmers and Shinton 1958; Conley and Krevans 1955; Doscherholmen and Hagen 1957; Gaffney et al. 1959; McIntyre et al. 1960; Reisner et al. 1955; Ross et al. 1954; Schwartz et al. 1959; Shinton 1961; Spies et al. 1949; Ungley 1950a; Ungley 1950b; Ungley 1950c; Ungley and Childs 1950; Waife et al. 1963). The required oral doses varied considerably between the studies and ranged from repeated daily doses of 50–100 µg (Brody et al. 1959) to 1000 µg per week (Reisner et al. 1955) to at highest 3000 µg/d (Ungley 1950a). Consistently, it was found that oral doses needed to be 30–60 times higher than parenteral doses in patients with pernicious anemia (Chalmers and Hall 1954; Spies et al. 1949). In Sweden, during the late 1950's and early 1960's basic research was carried out studying the feasibility of oral treatment of pernicious anemia with tablets containing very high doses of cyanocobalamin without an intrinsic factor (Berlin et al. 1965; Berlin et al. 1966; Berlin et al. 1968b; Berlin et al. 1958; Berlin et al. 1961). Little by little, this research provided convincing evidence from long-term follow-up of pernicious anemia patients, showing that oral treatment with vitamin B12-tablets was indeed possible and reliable (Berlin et al. 1965; Berlin et al. 1968b). Vitamin B12 tablets were approved by the Swedish Medical Products Agency in 1964 (Ågren 1964). Since then, vitamin B12-substitution with tablets has gradually replaced parenteral therapy in Sweden, where vitamin B12 tablets constitute more than 80% of vitamin B12-prescription drugs (Nilsson et al. 2005).

6. The Pharmacology of Oral Vitamin B12

To understand the feasibility of oral vitamin B12-treatment it is necessary to recognize the clinical pharmacology of vitamin B12, which is quite complex. Most of the work on vitamin B12 pharmacokinetics was carried out during the 1960s (Adams et al. 1971; Ardeman et al. 1964; Boddy et al. 1968; Gottlieb et al. 1965; Herbert 1968; Hertz et al. 1964; Heyssel et al. 1966; Skouby 1966). Physiological losses of vitamin B12 through renal and biliary elimination routes are minimal and daily losses in healthy subjects account for only 0.1–0.2% of the total body reserves of 3–5 mg, and merely this portions needs to be replenished (Table 3) (Combs 2008). Therefore, the daily cobalamin requirements in order to maintain normal vitamin status in healthy subjects are extremely low. The recommended daily allowance (RDA) for adults is 2.4 µg/d as set by the US Institute of Medicine in 1998.

6.1 Pharmacokinetics

Absorption, transport and cellular uptake as well as retention in the body depend on a number of transporters, binding proteins and receptors that all have a high specificity for vitamin B12 (Table 2).

The free binding capacity of most of the binding proteins is adapted to the physiologically low vitamin B12-supply and demands. Therefore, the free binding capacity of B12 binders is generally low as regards to both active intestinal absorption and transport in the blood. In line with this, the maximum capacity of active intrinsic factor-mediated absorption of vitamin B12 is only 2.5 to 3.0 µg per serving (Heyssel et al. 1966). It takes about 4–6 hours before maximum active absorption capacity is completely restored (Heyssel et al. 1966). Further, the unsaturated vitamin B12 binding capacity in human plasma

Table 2. Main proteins involved in vitamin B12 homeostasis and transport.*

	Binding proteins or transporters			Receptor proteins		
	IF	HC	TCII	Cubilin/ amnion-less	Megalin	TC-receptor CD320
Main function:						
Intestinal absorption	x			x		
Blood transport		x	x			
Cellular uptake			x		x	x
Entero-hepatic circulation	x		x	x		
Renal tubular re-absorption				x	x	x
Biliary elimination		x				

* The table is based on data from the following references (Banerjee et al. 2009; Birn 2006; Fyfe et al. 2004; Grasbeck 2006; Herbert 1994; Kanazawa et al. 1983; Quadros and Sequeira 2013; Schjonsby 1989). IF = intrinsic factor; HC = haptocorrin; TCII = transcobalamin II; TC-receptor = transcobalamin receptor

ranges from 230 to 1380 pmol/l (Herbert 1968), mostly constituted of apo-transcobalamin II (apo TC-II) (Markle 1996; Obeid et al. 2006; Teplitsky et al. 2003). This corresponds to a total binding capacity of only 3 to 5 μg of newly absorbed cobalamin (Gottlieb et al. 1965). Vitamin B12 unbound to transporters or plasma proteins is subject to glomerular filtration and rapid renal excretion (Herbert 1968). This limits maximum (active) absorption and body retention when vitamin B12 is supplemented in pharmacological doses (Table 3). Further, the bioavailability and total body retention of vitamin B12 not only depends on the route of administration, oral or parenteral, the capacity of vitamin B12 binding proteins, but also on the formulation of vitamin B12 preparations (i.e., cyano- and hydroxocobalamin) (Adams et al. 1971; Boddy et al. 1968; Hertz et al. 1964; Skouby 1966) (Table 3). At high doses oral bioavailability or retention of *i.m.* vitamin B12 is very similar in healthy subjects and patients suffering from PA (Table 3, Figure 4).

6.1.1 Oral absorption

Gastric dysfunction such as chronic atrophic gastritis is a major cause of reduced oral up-take of cobalamin from food sources (food cobalamin malabsorption) (Nielsen et al. 2012). Interestingly, reduced availability of the intrinsic factor due to gastric atrophy is not a rate-limiting factor in this process. The stomach appears to have a large reserve capacity for *intrinsic factor* secretion and daily production of *intrinsic factor* suffices for uptake of 100–150 μg/d (Ardeman et al. 1964). Only extreme forms of atrophic gastritis and selective destruction of

Table 3. Bioavailability and retention of cobalamin.

	Healthy subject		Pernicious anemia patient	
Daily loss*	≈ 1μg/d		≈ 2 μg/d	
RDA	2.4 μg/d		1000 μg/d p.o. 1000 μg/90 days i.m.	
Bioavailability of a single oral dose μg (% of the dose)				
0.5 μg	0.38 μg (75%)		0.006 μg (1.2%)	
1–2 μg	0.5–1 μg (50%)		0.012–0.024 μg (1.2%)	
10 μg	1.6 μg (16%)		0.12 μg (1.2%)	
50 μg	2.0 μg (4%)		0.6 μg (1.2%)	
500 μg	10 μg (2%)		6 μg (1.2%)	
1000 μg	≈ 14 μg (1.4%)		≈ 12 (1.2%)	
Retention of a single i.m. dose (% of the dose)	CN-Cbl	OH-Cbl	CN-Cbl	OH-Cbl
3 μg	100%	100%	100%	100%
10 μg	97%	98%	98%	98%
25 μg	95%	96%	96%	98%
40 μg	93%	94%	94%	96%
100 μg	55%	90%	60%	94%
500 μg	20%	50%	30%	55%
1000 μg	15%	30%	20%	35%

* Mainly represented by vitamin excreted in urine and bile. RDA, Recommended Dietary Allowance. This table is a compilation of data from the following studies (Ardeman et al. 1964; Berlin et al. 1968a; Heyssel et al. 1966; Scott 1997).

gastric parietal cells by autoantibodies in pernicious anemia can reduce *intrinsic factor* production to a significant level. Nevertheless, atrophic gastritis per se may limit the bioavailability of oral vitamin B-12 through other mechanisms, such as impaired release of the vitamin from food proteins due to impaired acid secretion and reduced digestion by pepsin (Selhub et al. 2000). These restraints do of course not apply for crystalline vitamin B12 tablets. The rate-limiting factor of oral bioavailability of low doses of food vitamin B12 is primarily saturation of ileal receptors, which recognize the cobalamin-*intrinsic factor* complex (Heyssel et al. 1966). The active transport of the cobalamin-*intrinsic factor*-complex is easily saturable. The maximum amount of vitamin B12 that can be absorbed from a single meal is about 2 μg (Scott 1997; Watanabe 2007) and the fractional absorption decreases as oral doses are increased. About 50% of a single oral dose of 1 μg is retained, 20% of a 5 μg dose, and only 5% of a 25 mg dose (Table 3). To improve oral bioavailability of vitamin B12 repeated daily dosing at least 4–6 hours apart without concomitant food intake would appear advantageous (Berlin et al. 1968a; Brody et al. 1959; Heyssel et al. 1966). At oral doses of 500–1000 μg and above a constant fraction of cobalamin, approximately 1.5 of the dose, is absorbed by simple diffusion from the lumen to the intestinal epithelium independent of *intrinsic factor* (Berlin et al. 1968a). In pernicious anemia active uptake of vitamin B12 through ileal receptors does not occur due to lack of *intrinsic factor*, but passive absorption by simple diffusion is more than adequate to meet the daily requirements for patients without *intrinsic factor* when daily oral dosages of 1000 µg are ingested (Berlin et al. 1968a). When using low oral doses of vitamin B12 a saturation of the plasma total vitamin B12 binding capacity is normally not achieved. However, at oral doses exceeding 1–10 mg significant urinary excretion of newly absorbed vitamin B12 is observed (Berlin et al. 1968a).

At increasing oral doses the absolute and relative difference in vitamin B12 uptake between healthy subjects and patients with pernicious anemia becomes narrower (Berlin et al. 1968a) (Figure 5).

6.2 Pharmacodynamics

Vitamin B12 is essential for cell growth and replication and participates in transmethylation reactions during the synthesis of methionine, choline, creatinine and nucleic acids. Vitamin B12 supplements reverse the hematopoietic and, if started timely, neurological symptoms of vitamin B12

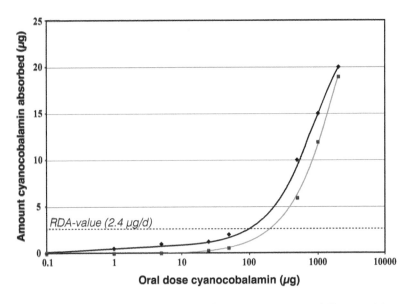

Figure 5. Schematic representation of estimated vitamin B12-uptake at different oral doses (per serving) in healthy subjects (black color, black diamond symbols) and cyanocobalamin tablets patients with pernicious anemia (grey color, solid grey square symbols) based on data from literature (Doscherholmen and Hagen 1957; Gaffney et al. 1959). The figure illustrates the dual mechanism of active (saturable) and passive absorption of oral cyanocobalamin (Doscherholmen and Hagen 1957) and that similar absolute amounts of cyanocobalamin are absorbed in healthy subjects and patients suffering from PA when very high oral doses are used.

deficiency (Stabler 2013). Vitamin B12 does not exert direct pharmacodynamic activity, but acts as a co-factor in two different forms, methylcobalamin and 5-deoxyadenosincobalamin, for the enzymes methionine synthase and methylmalonyl-CoA mutase, respectively (Banerjee 2006). In addition, vitamin B12 has repeatedly been proposed as a carrier molecule for up-take and targeting of drugs to certain tissues (Clardy-James et al. 2013).

6.3 Side effects

Generally, both oral vitamin B12 and vitamin B12-injections are well tolerated. However, injection site reactions including pain, erythema, pruritus, induration, swelling, and necrosis can occur (2013) causing patient discomfort and inconvenience (van Walraven et al. 2001a; van Walraven et al. 2001b). Oral supplementation reduces the risk of injection complications such as infections or cyst formations and nerve injuries are avoided. In addition, the risk of allergic reactions is lower with oral administration (Bilwani et al. 2005). Finally, oral treatment is preferred to intramuscular injections in patients on anticoagulation therapy as there is increased risk of hematoma formation when i.m. injections have to be performed regularly (Kim and Hyung 2011).

7. Diagnostic Achievements in the Course of Discovery of PA

Milestones in the development of diagnostic procedures for vitamin B12-deficiency are depicted by central gradient bars with solid-lined borders in Figure 4. Development of assays for identification and quantification of the putative *extrinsic factor* were crucial for final isolation of vitamin B12 from liver extracts, and measuring its concentration in these extracts (Vora 1956) and later diagnostically in blood samples. Before that, the potency of liver extracts assumed to contain the vital *extrinsic factor* that cured anemia had to be tested on the basis of the hematological response in affected patients, which was very time-consuming and not without risks (Shive 2002; Vora 1956). However, in 1947/48 important observations of Shorb of a linear relationship between the amount of a presumed growth factor for *Lactobacillus lactis* Dorner in liver extracts and the potency of the same factor for the treatment of patients suffering from pernicious anemia resulted in the first microbial vitamin B12-assay (Shorb 1947a; Shorb 1947b; Shorb 1948). Having a quantitative *in vitro* assay that allowed testing the biological activity of the extracts greatly accelerated the successful isolation of vitamin B12 (Ahrens 1993).

Soon, the same principle of microbial assays was found to be useful for measuring of vitamin B12 concentrations in serum and other body fluids. Measuring blood concentrations of vitamin B12 soon became the first step in diagnosing PA (Ross 1952).

However, even microbial assays were laborious had many limitations such as multiple steps and long incubation times, difficulties to automate the assay, risk of microbial contamination, and suppression of growth by antibiotics and cytotoxic drugs and were relatively soon abandoned from the clinical routine laboratory (O'Sullivan et al. 1992). Today, microbiological assays for vitamin B12 are adapted to microtiter plate format and carried out by robotic workstations and are mainly used for scientific purposes (Molloy and Scott 1997; O'Broin and Kelleher 1992; Taneja et al. 2007).

In the 1960s, radioimmunoassays, RIAs, were introduced using [57]Co-cyanocobalamin and *intrinsic factor* and R-binder as binders (Moridani and Ben-Poorat 2006; O'Sullivan et al. 1992). RIA tests had the drawback of demanding manual sample pretreatment, radiation exposure risk and costs associated with disposal of the radioimmunoassay components (Kuemmerle et al. 1992; Moridani and Ben-Poorat 2006). By 2007, the manufacturer of the most commonly used RIA-assay, the Bio-Rad RIA, discontinued the production (Yetley et al. 2011). The first automated, non-isotopic chemiluminescence assays for measuring vitamin B12 in serum were developed in the early 1990s (Kuemmerle et al. 1992). Current automated routine assays are mostly based on these non-isotopic procedures using chemiluminescence or more recently electroluminescence detection and exhibit relatively good measuring agreement between the methods (Karmi et al. 2011; Vogeser and Lorenzl 2007).

In parallel, alternative or complementary tests to serum vitamin B12 assays were developed, including more sensitive and specific functional markers of vitamin B12 status such as methylmalonic acid and homocysteine that increase in blood and urine of people with deficiency. By the end of the 1980s and through the 1990s different methods for determination of the functional vitamin B12-marker methylmalonic acid (MMA) became available, including GC-MS, HPLC, capillary electrophoresis, LC-MS and LC-MS-MS techniques (Stabler et al. 1986; Schneede and Ueland 1993; Schneede and Ueland 1995; Windelberg et al. 2005; Lakso et al. 2008). Together with total homocysteine, MMA nowadays represents one of the cornerstones of vitamin B12 deficiency diagnostic tests (Langan and Zawistoski 2011; Remacha et al. 2014; Risch et al. 2015).

During the same period, the concept of holotranscobalamin (holoTC) as a measure of biologically active cobalamin was developed (Carmel 1985; Herbert et al. 1990; Herzlich and Herbert 1988; Lindemans et al. 1983; Remacha et al. 2014). Vitamin B12 in serum is carried by two binding proteins, transcobalamin and haptocorrin. Although the fraction of transcobalamin-bound vitamin B12 (holoTC) in relation to total vitamin B12 is small ($\approx$ 20–30 %), holoTC is considered representing newly absorbed vitamin B12 and being responsible for delivering cobalamin to cells through a receptor mediated up-take. HoloTC is considered being the functionally active fraction of the vitamin (Carmel 2011; Nexö and Hoffmann-Lucke 2011). The first commercial RIA-assay for determination of holoTC became available in 2002 (Ulleland et al. 2002) and has now been replaced by an automated sandwich microparticle enzyme immunoassay that can be run on standard analytical platforms (Brady et al. 2008). The clinical utility of the holo-TC test has been evaluated during the recent years and its place in vitamin B12-diagnostics is still under debate (Herrmann and Obeid 2013; Remacha et al. 2014; Risch et al. 2015).

Another landmark in revealing the cause of vitamin B12-deficiency through diagnostic test was the Schilling test. The Schilling test was introduced by Schilling in 1953 and was designed to assess the ability of the patient to absorb small oral doses of radioactively labeled vitamin B12 (Schilling 1953). This test remained the mainstay of diagnostic tests for detection and differentiation of potential causes of vitamin B12 deficiency for more than five decades until it was abandoned due to high costs, lack of sensitivity under certain conditions and terminated production of cobalt radioisotopes and labeled cobalamin forms (Moridani and Ben-Poorat 2006; Palmer et al. 2012; Yetley et al. 2011). More recently, a non-radioactive vitamin B12 absorption test (CobaSorb) has been developed, but also this test has several limitations and has not achieved widespread adoption so far (Hardlei et al. 2010; Hvas et al. 2011; Hvas et al. 2007).

8. Health-economic Assessments of Oral B12 Treatment

Vitamin B12-treatment was from the very beginning not just a matter of clinical effectiveness, but also cost-effectiveness (Kumar et al. 2006). Already in 1934 Murphy addressed in his Nobel lecture cost-aspects of liver-treatment and potential savings with liver extracts for intramuscular use compared to peroral treatment (Murphy 1934). The first modern health-economic assessment of oral versus parenteral B12 supplementation that also included a sensitivity analysis was performed in 2001 (Figure 4) (van Walraven et al. 2001a), grey bars with solid-lined borders in Figure 4. Sensitivity analysis indicated that the number of injection-associated physician visits that could be avoided by switching patients to oral therapy had major impact on the cost-effectiveness. Later, publications on cost minimization analyses (Vidal-Alaball et al. 2006) and cost-saving analyses (Houle et al. 2014) as well as budget impact analyses (Masucci and Goeree 2013) performed in the UK, Canada and Spain followed.

Nowadays, the switch to oral vitamin B12 supplementation with tablets is generally considered feasible and a cost-effective alternative to parenteral treatment (Kolber and Houle 2014; Kwong et al. 2005).

9. Summary

Vitamin B12 is an archaic vitamin in many aspects. As other vitamins, vitamin B12 is vital for all higher organisms and functions as cofactor. Humans require dietary supply of these organic micronutrients, but microorganisms and many plants synthesize de novo the cofactors they need. Vitamin B12 is also an ancient molecule, as it was first synthesized by prokaryotic cells. Conceivably, the history of vitamin B12 deficiency and PA in humans has to be considerably shorter than the history of the vitamin itself, and can arbitrarily be divided into different epochs. It is almost 200 years since the first description of PA and the pre-treatment era stretches over 100 years before effective treatment options were developed (Figure 4). Oral treatment with raw or slightly cooked liver or liver extracts followed, but was soon abandoned due to high costs and unacceptable taste. The epoch of oral liver therapy was succeeded by a period of intramuscular administration of liver extracts, which lasted for about 20 years. With the isolation and characterization of the *extrinsic factor* (B12) from liver and the advent of large-scale, cost-effective production of vitamin B12 by bacterial fermentation (Xia et al. 2015) it was possible to use pure crystalline vitamin B12 for intramuscular administration, which still is the predominant treatment option of vitamin B12 deficiency world-wide today. The introduction and approval of high-dose vitamin B12 tablets in Sweden in the early 1960's ushered in the renaissance of oral treatment of vitamin B12-deficiency. The need to diagnose PA and other causes of vitamin B12 deficiency triggered the development of a range of diagnostic tests, some of them now constitute the basis of diagnostic strategies in clinical routine. During the last decade several health-economic assessments have confirmed the cost-effectiveness of oral vitamin B12 treatment over parenteral therapy. Industrial production by bacterial fermentation and the return of oral treatment vitamin B12 deficiency underscore the archaic nature of this precious co-factor. Almost 200 years after discovering PA and after approximately 70 years of efforts to find the cause of PA and to understand the synthesis of vitamin B12, large parts of the puzzle now seem to have been solved, but in fact many questions on vitamin B12 deficiency are still unanswered and warrant further investigations (Gräsbeck 2013).

Keywords: Vitamin B12 deficiency, cobalamin, history, management, diagnostics, supplementation, homocysteine, methylmalonic acid, holo-transcobalamin, health economics.

Abbreviations

PA	:	Pernicious anaemia
HC	:	Haptocorrin
TCII	:	Transcobalamin II
HoloTC	:	holotranscobalamin
RDA	:	Recommended Daily Allowance

References

Adams JF, Ross SK, Mervyn L, Boddy K and King P. 1971. Absorption of cyanocobalamin, coenzyme B 12, methylcobalamin, and hydroxocobalamin at different dose levels. Scandinavian journal of Gastroenterology. 6(3): 249–252.

Addison T. 1849. Chronic suprarenal insufficiency, usually due to tuberculosis of suprarenal Capsule. London Medical Gazette. 43: 517–518.

Ågren A. 1964. Farmacevtiska specialiteten Behepan tabletter 1 mg, file number 7399. Stockholm: Kungliga Medicinalstyrelsen (Swedish Medical Products Agency). p 1–20.

Ahrens RA. 1993. Mary Shaw Shorb (1907–1990). J Nutr. 123(5): 791–796.

Anonymous. 1965. Research News. Section II Development of the institute's research activities. The institute and research on pernicious anemia. Research News Ann Arbor: Office of Research Administration, the University of Michigan. Ann Arbor. pp. 15–19.

Ardeman S, Chanarin I and Doyle JC. 1964. Studies on secretion of gastric intrinsic factor in man. British Medical Journal. 2(5409): 600–603.

Banerjee R. 2006. B12 Trafficking in Mammals: A Case for Coenzyme Escort Service. ACS Chemical Biology. 1(3): 149–159.

Banerjee R, Gherasim C and Padovani D. 2009. The tinker, tailor, soldier in intracellular B12 trafficking. Current Opinion in Chemical Biology. 13(4): 484–491.

Banting FG, Best CH, Collip JB, Campbell WR and Fletcher AA. 1991. Pancreatic extracts in the treatment of diabetes mellitus: preliminary report. 1922. CMAJ: Canadian Medical Association Journal. 145(10): 1281–1286.

Battersby AR. 1994. How nature builds the pigments of life: the conquest of vitamin B12. Science (New York, NY). 264(5165): 1551–1557.

Berlin H, Berlin R and Brante G. 1965. [Peroral Treatment of Pernicious Anemia with High Doses of Vitamin B12 without Intrinsic Factor]. Lakartidningen. 62: 773–781.

Berlin H, Berlin R and Brante G. 1966. Crude or refined intrinsic factor in preparations for the oral treatment of pernicious anaemia. Scandinavian Journal of Haematology. 3(3): 236–244.

Berlin H, Berlin R and Brante G. 1968a. Oral treatment of pernicious anemia with high doses of vitamin B12 without intrinsic factor. Acta Medica Scandinavica. 184(4): 247–258.

Berlin H, Berlin R and Brante G. 1968b. Oral treatment of pernicious anemia with high doses of vitamin B12 without intrinsic factor. Acta Medica Scandinavica. 184: 247–258.

Berlin H, Berlin R, Brante G and Sjoberg SG. 1958. Studies on intrinsic factor and pernicious anemia. I. Oral uptake of vitamin B12 in pernicious anemia with increasing doses of an intrinsic factor concentrate. Scand J Clin Lab Invest. 10(3): 278–282.

Berlin R, Berlin H and Brante G. 1961. The absorption of IF-bound and free B12 in various clinical conditions. Second European Symposium on Vitamin B12 and Intrinsic Factor. Hamburg: Enke Verlag Stuttgart.

Bethell FH, Castle WB, Conley CL and London IM. 1959. Present status of treatment of pernicious anemia. J Am Med Assoc. 171: 2092–2094.

Biermer AM. 1872. Über eine eigenthümliche Form von progressiver, perniciöser Anaemie. Correspondenz-Blatt für Schweizer Aerzte. 2(1): 15–17.

Bilwani F, Adil SN, Sheikh U, Humera A and Khurshid M. 2005. Anaphylactic reaction after intramuscular injection of cyanocobalamin (vitamin B12): a case report. JPMA The Journal of the Pakistan Medical Association. 55(5): 217–219.

Birn H. 2006. The kidney in vitamin B12 and folate homeostasis: characterization of receptors for tubular uptake of vitamins and carrier proteins. American Journal of Physiology Renal Physiology. 291(1): F22-36-F22-36.

Blackburn EK, Burke J, Roseman C and Wayne EJ. 1952. Comparison of liver extract and vitamin B12 (cyanocobalamin) in maintenance treatment of pernicious anaemia. British Medical Journal. 2(4778): 245–248.

Blackburn EK, Cohen H and Wilson GM. 1955. Oral treatment of pernicious anaemia with a combined vitamin B12 and intrinsic factor preparation. British Medical Journal. 2(4937): 461–463.

Boddy K, King P, Mervyn L, Macleod A and Adams JF. 1968. Retention of cyanocobalamin, hydroxocobalamin, and coenzyme B12 after parenteral administration. Lancet 2(7570): 710–712.

Brady J, Wilson L, McGregor L, Valente E and Orning L. 2008. Active B12: a rapid, automated assay for holotranscobalamin on the Abbott AxSYM analyzer. Clinical Chemistry. 54(3): 567–573.

Brink C, Hodgkin DC, Lindsey J, Pickworth J, Robertson JR and White JG. 1954. X-ray crystallographic evidence on the structure of vitamin B12. Nature. 174(4443): 1169–1171.

Brody EA, Estren S and Wasserman LR. 1959. Treatment of pernicious anemia by oral administration of vitamin B12 without added intrinsic factor. The New England Journal of Medicine. 260(8): 361–367.

Carmel R. 1985. The distribution of endogenous cobalamin among cobalamin-binding proteins in the blood in normal and abnormal states. Am J Clin Nutr. 41(4): 713–719.

Carmel R. 2011. Biomarkers of cobalamin (vitamin B-12) status in the epidemiologic setting: a critical overview of context, applications, and performance characteristics of cobalamin, methylmalonic acid, and holotranscobalamin II. Am J Clin Nutr. 94(1): 348s–358s.

Castle WB. 1929. The Aetiological Relationship of Achylia Gastrica to Pernicious Anaemia. Proc R Soc Med. 22(9): 1214–1216.

Castle WB. 1953. Development of knowledge concerning the gastric intrinsic factor and its relation to pernicious anemia. The New England Journal of Medicine. 249(15): 603–614.

Chalmers JN and Hall ZM. 1954. Treatment of pernicious anaemia with oral vitamin B12 without known source of intrinsic factor. British Medical Journal. 1(4872): 1179–1181.

Chalmers JN and Shinton NK. 1958. Absorption of orally administered vitamin B12 in pernicious anaemia. Lancet. 2(7060): 1298–1302.

Clardy-James S, Chepurny OG, Leech CA, Holz GG and Doyle RP. 2013. Synthesis, characterization and pharmacodynamics of vitamin-B12 -conjugated glucagon-like Peptide-1. Chem Med Chem. 8(4): 582–586.

Cohnheim JF. 1876. Erkrankungen des Knochenmarkes bei perniziöser Anämie. Virchows Archiv für Pathologische Anatomie und Physiologie und für Klinische Medizin. 68: 291–293.

Combe JS. 1824. History of a case of anaemia. Transactions of the Medico-Chirurgical Society of Edinburgh. 1(August 2.): 194–204.

Combs GF. 2008. Chapter 17: Vitamin B12. The vitamins: fundamental aspects in nutrition and health. 3rd ed. ed. Oxford: Academic. pp. 384–385.

Conley CL and Krevans JR. 1955. New developments in the diagnosis and treatment of pernicious anemia. Annals of Internal Medicine. 43(4): 758–766.

Doscherholmen A and Hagen PS. 1957. A dual mechanism of vitamin B12 plasma absorption. The Journal of clinical investigation 36(11): 1551–1557.

Dunlop DM. 1973. Medicines in our time. The Rock Carling Fellowship. London: Nuffield Provincial Hospitals Trust. pp. 1–12.

Ehrlich P. 1880. Über Regeneration und Degeneration der rothen Blutscheiben bei Anämien. Berliner Klinische Wochenschrift. 117: 405.

Eichhorst H. 1878. Die progressive perniziöse Anämie: eine klinische und kritische Untersuchung. Leipzig: Veit & Comp. xi, 375 p., 373 leaves of plates.

Elliot CM. 2008. Vitamin B : new research. New York: Nova Biomedical Books. xiv, 234 p.

Ewing J. 1901. Clinical pathology of the blood; a treatise on the general principles and special applications of hematology: Philadelphia, Lea Brothers & Co. 489 p.

Fantes KH, Page JE, Parker LFJ and Smith EL. 1950. Crystalline Anti-Pernicious Anaemia Factor from Liver. Proceedings of the Royal Society of London B: Biological Sciences. 136(885): 592–609.

Fenwick S. 1880. On Atrophy of the Stomach and on the nervous affections of the digestive organs. J. & A. Churchill: London. 242 p.

Fyfe JC, Madsen M, Hojrup P, Christensen EI, Tanner SM, de la Chapelle A, He Q and Moestrup SK. 2004. The functional cobalamin (vitamin B12)-intrinsic factor receptor is a novel complex of cubilin and amnionless. Blood. 103(5): 1573–1579.

Gaffney GW, Watkin DM and Chow BF. 1959. Vitamin B12 absorption: relationship between oral administration and urinary excretion of cobalt 60-labeled cyanocobalamin following a parenteral dose; study of doses of 2 to 250 mu g in 148 apparently health men 20 to 92 years old. J Lab Clin Med. 53(4): 525–534.

Gänsslen M. 1930. Ein Hochwirksamer, Injizierbarer Leberextrakt. Klin Wochenschr. 9(45): 2099–2102.

Glass GBJ and Boyd LJ. 1953. Oral Treatment of Pernicious Anemia with Small Doses of Vitamin B12 Combined with Mucinous Materials Derived from the Hog Stomach. Blood. 8(10): 867–892.

Gottlieb C, Lau K-S, Wasserman LR and Herbert V. 1965. Rapid Charcoal Assay for Intrinsic Factor (IF), Gastric Juice Unsaturated B12 Binding Capacity, Antibody to IF, and Serum Unsaturated B12 Binding Capacity. Blood. 25(6): 875–884.

Grasbeck R. 2006. Imerslund-Grasbeck syndrome (selective vitamin B(12) malabsorption with proteinuria). Orphanet Journal of Rare Diseases. 1: 17.

Gräsbeck R. 2013. Hooked to vitamin B12 since 1955: a historical perspective. Biochimie. 95(5): 970–975.

Halbrook ER, Cords F, Winter AR and Sutton TS. 1950. Vitamin B12 production by microorganisms isolated from poultry house litter and droppings. J Nutr. 41(4): 555–563.

Hall CA and Finkler AE. 1965. The Dynamics of Transcobalamin Ii. A Vitamin B12 Binding Substance in Plasma. J Lab Clin Med. 65: 459–468.

Hardlei TF, Mørkbak AL, Bor MV, Bailey LB, Hvas A-M and Nexo E. 2010. Assessment of vitamin B(12) absorption based on the accumulation of orally administered cyanocobalamin on transcobalamin. Clinical Chemistry. 56(3): 432–436.

Helliwell KE, Wheeler GL, Leptos KC, Goldstein RE and Smith AG. 2011. Insights into the evolution of vitamin B12 auxotrophy from sequenced algal genomes. Mol Biol Evol. 28(10): 2921–2933.

Herbert V. 1968. Diagnostic and Prognostic Values of Measurement of Serum Vitamin B12-Binding Proteins. Blood. 32(2): 305–312.

Herbert V. 1994. Staging vitamin B-12 (cobalamin) status in vegetarians. Am J Clin Nutr. 59(5 Suppl): 1213s–1222s.

Herbert V, Fong W, Gulle V and Stopler T. 1990. Low holotranscobalamin II is the earliest serum marker for subnormal vitamin B12 (cobalamin) absorption in patients with AIDS. American Journal of Hematology. 34(2): 132–139.

Herrmann W and Obeid R. 2013. Utility and limitations of biochemical markers of vitamin B12 deficiency. European Journal of Clinical Investigation. 43(3): 231–237.

Hertz H, Kristensen HP and Hoff-Jorgensen E. 1964. Studies on vitamin B12 retention. Comparison of retention following intramuscular injection of cyanocobalamin and hydroxocobalamin. Scand J Haematol. 1: 5–15.

Herzlich B and Herbert V. 1988. Depletion of serum holotranscobalamin II. An early sign of negative vitamin B12 balance. Laboratory investigation; A Journal of Technical Methods and Pathology. 58(3): 332–337.

Heyssel RM, Bozian RC, Darby WJ and Bell MC. 1966. Vitamin B12 turnover in man. The assimilation of vitamin B12 from natural foodstuff by man and estimates of minimal daily dietary requirements. Am J Clin Nutr. 18(3): 176–184.

Hodgkin DG, Pickworth J, Robertson JH, Trueblood KN, Prosen RJ and White JG. 1955. The crystal structure of the hexacarboxylic acid derived from B12 and the molecular structure of the vitamin. Nature. 176(4477): 325–328.

Houle SK, Kolber MR and Chuck AW. 2014. Should vitamin B12 tablets be included in more Canadian drug formularies? An economic model of the cost-saving potential from increased utilisation of oral versus intramuscular vitamin B12 maintenance therapy for Alberta seniors. BMJ open. 4(5): e004501.

Howard JAK. 2003. Dorothy Hodgkin and her contributions to biochemistry. Nature Reviews Molecular Cell Biology. 4: 891–896.

Hvas A-M, Morkbak AL, Hardlei TF and Nexo E. 2011. The vitamin B12 absorption test, CobaSorb, identifies patients not requiring vitamin B12 injection therapy. Scandinavian Journal of Clinical and Laboratory Investigation. 71(5): 432–438.

Hvas A-M, Morkbak AL and Nexo E. 2007. Plasma holotranscobalamin compared with plasma cobalamins for assessment of vitamin B12 absorption; optimisation of a non-radioactive vitamin B12 absorption test (CobaSorb). Clinica Chimica Acta; International Journal of Clinical Chemistry. 376(1-2): 150–154.

INOXPA. 2015. INOXPA—Fermenter. In: http: //www.inoxpa.com/uploads/producte/Fermentador/Fermenter-INOXPA.jpg, editor. http: //wwwinoxpacom/uploads/producte/Fermentador/Fermenter-INOXPAjpg. http: //www.inoxpa.com/uploads/producte/Fermentador/Fermenter-INOXPA.jpg.

Institute of Medicine (US) Standing Committee on the Scientific Evaluation of Dietary Reference Intakes and its Panel on Folate, Other B Vitamins, and Choline. 1998. Dietary Reference

Intakes for Thiamin, Riboflavin, Niacin, Vitamin B6, Folate, Vitamin B12, Pantothenic Acid, Biotin, and Choline: The National Academies Press.

Jarcho S and Brown G. 1977. Medicine and Health Care. New York: Ayer Co Pub. 398 p.

Johnson RR, Bentley OG and Moxon AL. 1956. Synthesis *in vitro* and *in vivo* of Co60 containing vitamin B12-active substances by rumen microorganisms. J Biol Chem. 218(1): 379–390.

Kanazawa S, Herbert V, Herzlich B, Drivas G and Manusselis C. 1983. Removal of cobalamin analogue in bile by enterohepatic circulation of vitamin B12. Lancet. 1(8326 Pt 1): 707–708.

Karmi O, Zayed A, Baraghethi S, Qadi M and Ghanem R. 2011. Measurement of vitamin B12 concentration: A review on the available methods. The IIOAB Journal. 2(2): 23–32.

Kim HI and Hyung WJ. 2011. Oral vitamin B12 therapy after total gastrectomy. Annals of surgical oncology 18 Suppl 3: 199.

Kolber MR and Houle SK. 2014. Oral vitamin B12: a cost-effective alternative. Canadian family physician Medecin de famille canadien. 60(2): 111–112.

Kuemmerle SC, Boltinghouse GL, Delby SM, Lane TL and Simondsen RP. 1992. Automated assay of vitamin B-12 by the Abbott IMx analyzer. Clinical Chemistry. 38(10): 2073–2077.

Kumar N, Boes CJ and Samuels MA. 2006. Liver therapy in anemia: a motion picture by William P. Murphy. Blood. 107(12): 4970.

Kwong JC, Carr D, Dhalla IA, Tom-Kun D and Upshur RE. 2005. Oral vitamin B12 therapy in the primary care setting: a qualitative and quantitative study of patient perspectives. BMC family Practice. 6(1): 8.

Lakso HA, Appelblad P and Schneede J. 2008. Quantification of methylmalonic acid in human plasma with hydrophilic interaction liquid chromatography separation and mass spectrometric detection. Clinical Chemistry. 54(12): 2028–2035.

Langan RC and Zawistoski KJ. 2011. Update on vitamin B12 deficiency. American Family Physician. 83(12): 1425–1430.

Lichtheim L. 1887. Zur Kenntnis der perniziösen Anämie. Schweizerische Medizinische Wochenschrift 34.

Lindemans J, Schoester M and van Kapel J. 1983. Application of a simple immunoadsorption assay for the measurement of saturated and unsaturated transcobalamin II and R-binders. Clinica Chimica Acta; International Journal of Clinical Chemistry. 132(1): 53–61.

Markle HV. 1996. Cobalamin. Critical Reviews in Clinical Laboratory Sciences. 33(4): 247–356.

Masucci L and Goeree R. 2013. Vitamin B12 intramuscular injections versus oral supplements: a budget impact analysis. Ontario health technology assessment series. 13(24): 1–24.

McIntyre PA, Hahn R, Masters JM and Krevans JR. 1960. Treatment of pernicious anemia with orally administered cyanocobalamin (vitamin b12). Archives of Internal Medicine. 106(2): 280–292.

Mervyn L and Smith EL. 1964. The biochemistry of vitamin B12 fermentation. Progress in Industrial Microbiology. 5: 151–201.

Minot GR and Murphy WP. 1926. Treatment of pernicious anemia by a special diet. Jama. 87(7): 470–476.

Mollin DL. 1950. Treatment of pernicious anaemia with parenteral liver extract; a review of 51 patients between 1940 and 1948. Lancet 1(6615): 1064–1068.

Molloy AM and Scott JM. 1997. Microbiological assay for serum, plasma, and red cell folate using cryopreserved, microtiter plate method. Methods in Enzymology. 281: 43–53.

Moore SJ, Lawrence AD, Biedendieck R, Deery E, Frank S, Howard MJ, Rigby SEJ and Warren MJ. 2013. Elucidation of the anaerobic pathway for the corrin component of cobalamin (vitamin B12). Proceedings of the National Academy of Sciences. 110(37): 14906–14911.

Moridani M and Ben-Poorat S. 2006. Laboratory Investigation of Vitamin B12 Deficiency. Lab Medicine. 37(3): 166–174.

Murphy WP. 1934. Nobel Lecture: Pernicious Anemia. Stockholm: Nobelprize.org. Nobel Media AB 2014.

Murphy WP. 2006. Liver therapy in anemia: a motion picture by William P. Murphy. Supplemental materials for: Kumar et al, Blood, Volume 107, Issue 12,4970: http: //bloodjournal. hematologylibrary.org/content/107/12/4970.1/suppl/DC1. Washington: Blood. Journal of The American Society of Hematology.

Nexo E and Hoffmann-Lucke E. 2011. Holotranscobalamin, a marker of vitamin B-12 status: analytical aspects and clinical utility. Am J Clin Nutr. 94(1): 359s–365s.

Nichols HJ. 1911. Salvarsan and sodium cacodylate. Journal of the American Medical Association. LVI(7): 492–495.

Nielsen MJ, Rasmussen MR, Andersen CB, Nexo E and Moestrup SK. 2012. Vitamin B12 transport from food to the body's cells—a sophisticated, multistep pathway. Nat Rev Gastroenterol Hepatol. 9(6): 345–354.

Nilsson M, Norberg B, Hultdin J, Sandstrom H, Westman G and Lokk J. 2005. Medical intelligence in Sweden. Vitamin B12: oral compared with parenteral? Postgraduate Medical Journal. 81(953): 191–193.

O'Broin S and Kelleher B. 1992. Microbiological assay on microtitre plates of folate in serum and red cells. Journal of Clinical Pathology. 45(4): 344–347.

O'Sullivan JJ, Leeming RJ, Lynch SS and Pollock A. 1992. Radioimmunoassay that measures serum vitamin B12. Journal of Clinical Pathology. 45(4): 328–331.

Obeid R, Morkbak AL, Munz W, Nexo E and Herrmann W. 2006. The cobalamin-binding proteins transcobalamin and haptocorrin in maternal and cord blood sera at birth. Clinical Chemistry. 52(2): 263–269.

Okuda K. 1999. Discovery of vitamin B12 in the liver and its absorption factor in the stomach: a historical review. J Gastroenterol Hepatol. 14(4): 301–308.

Osler W. 1915. The principles and practice of medicine. New York: D. Appleton and Company. 1225 p.

Palmer WC, Crozier JA and Petrucelli OM. 2012. 79-year-old woman with forgetfulness. Mayo Clinic Proceedings. 87(4): 408–411.

Pearce JMS. 2004. Thomas Addison (1793–1860). Journal of the Royal Society of Medicine. 97(6): 297–300.

Quadros EV and Sequeira JM. 2013. Cellular Uptake of Cobalamin: Transcobalamin and the TCblR/CD320 Receptor. Biochimie. 95(5): 1008–1018.

Reid C. 1986. Hilbert-Courant. New York: Springer-Verlag. xiv, 547 p., 546 p. of plates p.

Reid C. 1996. Hilbert. New York: Copernicus. ix, 228 p. p.

Reisner EH, Jr., Weiner L, Schittone MT and Henck EA. 1955. Oral treatment of pernicious anemia with vitamin B12 without intrinsic factor. The New England Journal of Medicine. 253(12): 502–506.

Remacha AF, Sarda MP, Canals C, Queralto JM, Zapico E, Remacha J and Carrascosa C. 2014. Role of serum holotranscobalamin (holoTC) in the diagnosis of patients with low serum cobalamin. Comparison with methylmalonic acid and homocysteine. Annals of Hematology. 93(4): 565–569.

Rickes EL, Brink NG, Koniuszy FR, Wood TR and Folkers K. 1948. Crystalline Vitamin B12. Science (New York, NY). 107(2781): 396–397.

Riedmann C, Ma Y, Melikishvili M, Godfrey S, Zhang Z, Chen K, Rouchka E and Fondufe-Mittendorf Y. 2015. Inorganic Arsenic-induced cellular transformation is coupled with genome wide changes in chromatin structure, transcriptome and splicing patterns. BMC Genomics. 16(1): 212.

Riethmiller S. 2005. From Atoxyl to Salvarsan: searching for the magic bullet. Chemotherapy. 51(5): 234–242.

Risch M, Meier DW, Sakem B, Medina Escobar P, Risch C, Nydegger U and Risch L. 2015. Vitamin B12 and folate levels in healthy Swiss senior citizens: a prospective study evaluating reference intervals and decision limits. BMC Geriatrics. 15: 82.

Robscheit-Robbins FS and Whipple GH. 1925. BLOOD REGENERATION IN SEVERE ANEMIA: II. Favorable Influence of Liver, Heart and Skeletal Muscle in Diet. American Journal of Physiology—Legacy Content. 72(3): 408–418.

Ross GI. 1952. Vitamin B12 assay in body fluids using Euglena gracilis. Journal of Clinical Pathology. 5(3): 250–256.

Ross GI, Mollin DL, Cox EV and Ungley CC. 1954. Hematologic responses and concentration of vitamin B12 in serum and urine following oral administration of vitamin B12 without intrinsic factor. Blood. 9(5): 473–488.

Roth JR, Lawrence JG and Bobik TA. 1996. Cobalamin (coenzyme B12): synthesis and biological significance. Annu Rev Microbiol. 50: 137–181.

Russell JSR, Batten FE and Collier J. 1900. Subacute combined degeneration of the spinal cord. Brain. 23(1): 39–110.

Santander PJ, Roessner CA, Stolowich NJ, Holderman MT and Scott AI. 1997. How corrinoids are synthesized without oxygen: nature's first pathway to vitamin B12. Chem Biol. 4(9): 659–666.

Schilling RF. 1953. Intrinsic factor studies II. The effect of gastric juice on the urinary excretion of radioactivity after the oral administration of radioactive vitamin B12. The Journal of Laboratory and Clinical Medicine. 42(6): 860–866.

Schjonsby H. 1989. Vitamin B12 absorption and malabsorption. Gut. 30(12): 1686–1691.

Schneede J and Ueland PM. 1993. Automated assay of methylmalonic acid in serum and urine by derivatization with 1-pyrenyldiazomethane, liquid chromatography, and fluorescence detection. Clinical Chemistry. 39(3): 392–399.

Schneede J and Ueland PM. 1995. Application of capillary electrophoresis with laser-induced fluorescence detection for routine determination of methylmalonic acid in human serum. Analytical Chemistry. 67(5): 812–819.

Schultzer P. 1934. Intramuscular Injections of Liver Extract for Initial and Maintenance Treatment of Pernicious Anemia. Acta medica Scandinavica. 82(5-6): 393–418.

Schwartz M, Lous P and Meulengracht E. 1959. [Vitamin B12 absorption in pernicious anemia; studies on the treatment-induced deficiency of vitamin B12 absorption after protracted therapy with some new combination preparations]. Ugeskr Laeger 121(10): 353–358.

Scott JM. 1997. Bioavailability of vitamin B12. Eur J Clin Nutr. 51 Suppl 1: S49–53.

Selhub J, Bagley LC, Miller J and Rosenberg IH. 2000. B vitamins, homocysteine, and neurocognitive function in the elderly. Am J Clin Nutr. 71(2): 614S–620S.

Shinton NK. 1961. Oral Treatment of Pernicious Anaemia with Vitamin-B(12)-Peptide. British Medical Journal. 1(5239): 1579–1582.

Shive W. 2002. Karl August Folkers 1906–1997. Biographical Memoirs. Washington D.C.: National Academies Press. p 101–115.

Shorb MS. 1947a. Unidentified essential growth factors for Lactobacillus lactis found in refined liver extracts and in certain natural materials. J Bacteriol. 53(5): 669.

Shorb MS. 1947b. Unidentified growth factors for Lactobacillus lactis in refined liver extracts. J Biol Chem 169(2): 455.

Shorb MS. 1948. Activity of Vitamin B12 for the Growth of Lactobacillus lactis. Science (New York, NY). 107(2781): 397–398.

Sinclair L. 2008. Recognizing, treating and understanding pernicious anaemia. Journal of the Royal Society of Medicine. 101(5): 262–264.

Skouby AP. 1966. Retention and distribution of B12 activity, and requirement for B12, following parenteral administration of hydroxocobalamin (Vibeden). Acta Medica Scandinavica. 180(1): 95–105.

Sørensen ST. 1874. Tællinger af blodlegemer i 3 tilfælde af excessiv oligocythaemi. Hospitals Tidende. 1: 513–521.

Spies TD, Stone RE, and et al. 1949. Vitamin B12 by mouth in pernicious and nutritional macrocytic anaemia and sprue. Lancet. 2(6576): 454–456.

Stabler SP. 2013. Vitamin B12 Deficiency. New England Journal of Medicine. 368(2): 149–160.

Stabler SP, Marcell PD, Podell ER, Allen RH and Lindenbaum J. 1986. Assay of methylmalonic acid in the serum of patients with cobalamin deficiency using capillary gas chromatography-mass spectrometry. The Journal of Clinical Investigation. 77(5): 1606–1612.

Stokstad ELR, Jukes TH, Pierce J, Page AC and Franklin AL. 1949. THE MULTIPLE NATURE OF THE ANIMAL PROTEIN FACTOR. Journal of Biological Chemistry. 180(2): 647–654.

Taneja S, Bhandari N, Strand TA, Sommerfelt H, Refsum H, Ueland PM, Schneede J, Bahl R and Bhan MK. 2007. Cobalamin and folate status in infants and young children in a low-to-middle income community in India. Am J Clin Nutr. 86(5): 1302–1309.

Teplitsky V, Huminer D, Zoldan J, Pitlik S, Shohat M and Mittelman M. 2003. Hereditary partial transcobalamin II deficiency with neurologic, mental and hematologic abnormalities in children and adults. The Israel Medical Association Journal : IMAJ. 5(12): 868–872.

Tobin DJ and Cargnello JA. 1993. Partial reversal of canities in a 22-year-old normal chinese male. Archives of Dermatology. 129(6): 789–791.

Ulleland M, Eilertsen I, Quadros EV, Rothenberg SP, Fedosov SN, Sundrehagen E and Orning L. 2002. Direct assay for cobalamin bound to transcobalamin (holo-transcobalamin) in serum. Clinical Chemistry. 48(3): 526–532.

Ungley CC. 1950a. Absorption of vitamin B12 in pernicious anaemia. I. Oral administration without a source of intrinsic factor. British Medical Journal. 2(4685): 905–908.

Ungley CC. 1950b. Absorption of vitamin B12 in pernicious anemia. II. Oral administration with normal gastric juice. British Medical Journal. 2(4685): 908–911.

Ungley CC. 1950c. Absorption of vitamin B12 in pernicious anemia. IV. Administration into buccal cavity, into washed segment of intestine, or after partial sterilization of bowel. British Medical Journal. 2(4685): 915–919.

Ungley CC. 1955. The chemotherapeutic action of vitamin B12. In: Harris RS, Marian GF, and Thimann KV, editors. VITAMINS AND HORMONES. New York: Academic Press. pp. 139–213.

Ungley CC and Childs GA. 1950. Absorption of vitamin B12 in pernicious anemia. III. Failure of fresh milk or concentrated whey to function as Castle's intrinsic factor or to potentiate the action of orally administered vitamin B12. British Medical Journal. 2(4685): 911–915.

van Walraven C, Austin P and Naylor CD. 2001a. Vitamin B12 injections versus oral supplements. How much money could be saved by switching from injections to pills? Canadian Family Physician Medecin de Famille Canadien. 47: 79–86.

van Walraven CG, Austin P and Naylor CD. 2001b. Vitamin B12 injections versus oral supplements. How much money could be saved by switching from injections to pills? Canadian Family Physician. 47: 79–86.

Vidal-Alaball J, Butler CC and Potter CC. 2006. Comparing costs of intramuscular and oral vitamin B12 administration in primary care: a cost-minimization analysis. The European Journal of General Practice. 12(4): 169–173.

Vogeser M and Lorenzl S. 2007. Comparison of automated assays for the determination of vitamin B12 in serum. Clinical Biochemistry. 40(16-17): 1342–1345.

Vora VC. 1956. Vitamin B12—Its chemistry, production & assay. J Sci Industr Res. 15A: 552–561.

Wagner AF and Folkers K. 1963. Vitamin B12. In: Florkin M, and Stotz EH, editors. Comprehensive Biochemistry Water-Soluble Vitamins, Hormones, Antibiotics. 1 ed. New York: Elsevier. pp. 103–115.

Waife SO, Jansen CJ, Jr., Crabtree RE, Grinnan EL and Fouts PJ. 1963. Oral vitamin B12 without intrinsic factor in the treatment of pernicious anemia. Annals of internal medicine 58: 810–817.

Watanabe F. 2007. Vitamin B12 sources and bioavailability. Exp Biol Med (Maywood). 232(10): 1266–1274.

Weber FP. 1932. An old Case of Pernicious Anæmia. Proceedings of the Royal Society of Medicine. 25(6): 800–801.

Whipple GH, Robscheit FS and Hooper CW. 1920. BLOOD REGENERATION FOLLOWING SIMPLE ANEMIA: IV. Influence of Meat, Liver and Various Extractives, Alone or Combined with Standard Diets. American Journal of Physiology—Legacy Content. 53(2): 236–262.

Wills L. 1948. Pernicious anemia, nutritional macrocytic anemia, and tropical sprue. Blood. 3(1): 36–56.

Windelberg A, Arseth O, Kvalheim G and Ueland PM. 2005. Automated assay for the determination of methylmalonic acid, total homocysteine, and related amino acids in human serum or plasma by means of methylchloroformate derivatization and gas chromatography-mass spectrometry. Clinical Chemistry. 51(11): 2103–2109.

Woodward RB. 1973. The total synthesis of vitamin B 12. Pure Appl Chem. 33(1): 145–177.

Xia W, Chen W, Peng W-f and Li K-t. 2015. Industrial vitamin B12 production by Pseudomonas denitrificans using maltose syrup and corn steep liquor as the cost-effective fermentation substrates. Bioprocess Biosyst Eng. 38(6): 1065–1073.

Yetley EA, Pfeiffer CM, Phinney KW, Bailey RL, Blackmore S, Bock JL, Brody LC, Carmel R, Curtin LR, Durazo-Arvizu RA et al. 2011. Biomarkers of vitamin B-12 status in NHANES: a roundtable summary. Am J Clin Nutr. 94(1): 313s–321s.

Yin L and Bauer CE. 2013. Controlling the delicate balance of tetrapyrrole biosynthesis. Philosophical Transactions of the Royal Society of London B: Biological Sciences. 368(1622)

2

Nutritional and Biochemical Aspects of Cobalamin Throughout Life

Eva Greibe

1. Introduction

Cobalamin (vitamin B12) is important in all stages of life from birth to death. Since humans cannot synthesize cobalamin, we rely on nutritional sources for supply of the vitamin.

This chapter describes the general nutritional aspects of cobalamin including content in food, dietary sources, bioavailability, forms of the vitamin in food, and dietary requirements. Also, the biochemical aspects of cobalamin including its role as a methyl donor are presented. In relation, acquired cobalamin deficiency is discussed.

2. Nutritional Aspects of Cobalamin

All forms of cobalamin found in nature, including cobalamin analogues, are synthesized by microorganisms via a complex pathway. Several cobalamin-dependent enzymes are present in algae; but no species of plants has been found to have the enzymes needed for synthesis of the vitamin. Herbivorous animals obtain the vitamin by consuming cobalamin-producing bacteria present in roots and legumes, or plants contaminated with faeces, and they store it in tissues or excrete it in milk. Therefore, animal products are main

Department of Clinical Biochemistry, Aarhus University Hospital, Palle Juul-Jensens Boulevard 99, 8200 Aarhus N, Denmark.
Email: greibe@clin.au.dk

dietary sources of cobalamin for human, who cannot synthesize the vitamin. Unlike humans, ruminants are unique in that they can absorb cobalamin synthesized by their intestinal microbiota. It is a common misunderstanding that humans also can absorb cobalamin in this manner. Bacterial synthesis of cobalamin in humans occurs mostly in the terminal part of the gastrointestinal tract, where intrinsic factor mediated absorption of cobalamin is unlikely to take place. Humans therefore depend on cobalamin provided by consumption of animal-based foods such as eggs, milk products, meats, fish, seafood, and poultry.

2.1 Content of cobalamin in food items

The cobalamin content of different food items is listed in Table 1. Livers of ruminant animals contain the largest amounts of cobalamin, but also shellfish, fish, and fish roe have high contents. Usually, ruminant meats such as lamb and beef contain higher amounts of cobalamin compared with meats of

Table 1. Content of Cobalamin in Uncooked Foods.

Food	Cobalamin (µg/100 g)
Pasteurized milk	0.4–0.5
Cheese[1]	1.5–1.6
Infant formula (non-fortified)	0.9–1.0
Eggs	2.0
Beef	1.4–2.4
Veal	1.3
Beef liver	60–122
Chicken	0.3–0.5
Pork	0.6–0.8
Mutton	3.0–5.0
Lamb	1.2
Fish[2]	3.0–8.0
Fish roe[3]	18
Shellfish[4]	2–58
Fortified cereals	10–28
Soybeans	0
Rice, flour	0
Fruit, nuts, vegetables	0

[1]Hard cheese, brie, parmesan, mozzarella. [2]Salmon, trout, mackerel, and tuna. [3]Roe from atlantic cod, lumpfish, and rainbow trout. [4]Clam, scallop, mussel, shrimp, and oyster.

Note: All value ranges are indicated for raw foods unless otherwise stated. The data has been extracted from a national food composition data bank (The Danish National Food Institute, 2015). Food composition databases do rarely disclose how the levels are obtained, and presumably the presented data represents the outcome of different methods. Because of this, the listed values should be regarded as estimates.

monogastric animals such as pigs and poultry, because of a larger population of cobalamin synthesizing bacteria in the rumen of these animals. Meats from older animals tend to have a higher cobalamin contents that of younger animals; therefore generally the content of cobalamin in mutton and beef is higher than in lamb and veal (Table 1) (summarized by (Gille and Schmid 2015)). The cobalamin content in cow's milk is moderate, and fermentation of milk for production of yogurt and cheese decreases the original amount of cobalamin with up to 60% (Gille and Schmid 2015). High cobalamin contents in foods can be achieved by fortification of foods. Cereals, flour, and soy products are usually used as fortification vehicles.

Individuals who are following alternative lifestyles (e.g., vegans, macrobiotics) commonly use commercially available plant-based products that claim to contain cobalamin. Some plants and algae contain cobalamin-like compounds, the so called cobalamin analogues. Cobalamin analogues appear inactive in mammals and thus of no nutritional value (Dagnelie et al. 1991) (see paragraph on Forms of Cobalamin in Foods below). In general, bacterially fermented soy products and other fermented products do not contain measurable amounts of cobalamin (van den Berg et al. 1988). However, new attempts to enhance cobalamin content in fermented soy products have shown great promise (Gu et al. 2015), and may result in commercially available products in the future. Small concentrations of cobalamin have been found in malt syrup, sourdough bread, parsley, and shiitake mushroom (0.02–0.5 µg/100 g) (van den Berg et al. 1988). However, due to the low concentrations of cobalamin, and the uncertainties in methods analyzing small amounts, these foods are not considered of any significant nutritional value as sources of cobalamin.

The estimated amounts of cobalamin in food may vary with the method employed, and caution should be exercised when interpreting the values obtained. Microbiological assays based on the growth of cobalamin-dependent microorganisms may give false results if the sample contains other components that affect microbial growth, such as cobalamin analogues. Also incomplete release of cobalamin from the food, and chemical alternation of the vitamin during the extraction procedure may compromise the cobalamin measurements (reviewed by (Lildballe and Nexo 2013; Watanabe 2007)). Besides microbiological assays, cobalamin content in foods can also be measured by chemiluminiscence assays and high-performance liquid chromatography.

2.2 Bioavailability of cobalamin

Knowledge of the cobalamin content does not alone provide information on the nutritional value of a food item as a source of cobalamin. This is because a high cobalamin content does not automatically ensure a high bioavailability, and vice versa. A limited bioavailability is mainly due to the fact that cobalamin in animal products is bound to food proteins that must be degraded by digestive enzymes, possibly in combination with low pH, in order for cobalamin to be absorbed. The nature of the food matrix in question seems to be of important for the bioavailability. For instance, cow's milk contain high amounts of the cobalamin-binding proteins transcobalamin (Fedosov et al. 1996), and human milk contain high amounts of the cobalamin-binding protein haptocorrin (Greibe et al. 2013a). Both transcobalamin and haptocorrin have a very high affinity for cobalamin; however it is still not clear whether they hinder or promote the intestinal absorption of the vitamin. In addition, heat preparations of food have been found to reduce the content of cobalamin. Czerwonga et al found that heating of steak to a core temperature of 70°C reduced the content of cobalamin with 32 percent compared with raw meat (Czerwonga et al. 2014). Others have performed similar experiments with a loss in cobalamin content up to 68 percent depending on the study design, the temperature and cooking method, and the nature of the food (Heyssel et al. 1966; Nishioka et al. 2011; The Danish National Food Institute 2015).

Bioavailability of cobalamin has traditionally been assessed by oral administration of 100 g food item labelled with radioactive [57][Co]-cobalamin followed by whole body counting or measurement of radioactivity in fecal and urine excretions (summarized by (Allen 2010)). The limitations of such studies are lack of knowledge concerning the behaviour of endogenous cobalamin in food as compared with the added labelled vitamin that is not bound to food proteins. The methods also lack sensitivity because the provided dose may be differently distributes and stored in tissues in different individuals which cannot be tested. Few studies have used alternative more physiological approaches. One approach employed chicken eggs from hens fed on [57][Co]-cobalamin (Doscherholmen et al. 1975), and another one used fish fed with [57][Co]-cobalamin before sacrifice (Aimone-Gastin et al. 1997; Doscherholmen et al. 1981). Today, it is not advised to use radioactivity in human absorption studies, and no golden method exist for determining bioavailability of cobalamin from different food sources.

Based on previous radioactivity-based methods in humans and animals, the bioavailability of cobalamin is considered high from milk (65 percent) (Matte et al. 2012; Russell et al. 2001) and chicken (61–65 percent) (Doscherholmen et al. 1978) and comparable to that of free crystallized cobalamin dissolved in water (Doscherholmen et al. 1975; Russell et al. 2001). In contrast, the bioavailabilies of cobalamin from fish (30–41 percent) (Doscherholmen et al. 1981) and eggs (24–36 percent) (Doscherholmen et al. 1975) are lower. The bioavailability of fortified food items is not well studied. Since crystallized free cobalamin (not protein-bound) is used for fortifying foods, the bioavailability of cobalamin in fortified foods would be expected to be higher than natural protein-bound cobalamin from food.

A further issue to take into account, when discussing bioavailability, is that the intrinsic factor mediated intestinal absorption system is saturable and that only a certain amount of cobalamin can be absorbed via a receptor-mediated process after the meal. The point of saturation is currently debated by experts. Many believe it to be around 1–2 µg cobalamin per meal, while others claim it to be higher. The fact that the system is saturable means that the bioavailability increases with increasing intake of cobalamin per meal up to a certain degree and then decreases if cobalamin content is higher than the capacity of the uptake system. The rational thinking to compare how different foods can contribute to cobalamin status is to compare the effect of controlled diets (e.g., milk based diet vs. meat based diet vs. fish-based diet) on modern plasma cobalamin markers like holoTC. Obviously, there is a lot of work that needs to be done in the field.

2.3 Dietary sources of cobalamin

Dairy products, meats, fish, and seafood are considered to be the major dietary contributing sources of the vitamin in the western world. Brouwer-Brolsma et al. (2015) recently found that consumption of meat and dairy—predominantly milk—were the main predictors of having a high serum cobalamin in western elderly individuals, followed by fish and shellfish. Tucker et al. (2000) also found milk to have a high impact on serum cobalamin, and meat, poultry, and fish to play a minor role. The positive impact of dairy products on serum cobalamin is surprising when looking at the moderate content of cobalamin in milk (0.4–0.5 µg/100 g). However, the bioavailability of milk cobalamin is considered high (Matte et al. 2012; Russell et al. 2001) in addition to the relative high consumption of dairy products in the western society (Vogiatzoglou et al. 2009). Consumption of fortified foods, such as cereals, contributes to increasing cobalamin status and has been associated with high serum cobalamin (Tucker et al. 2000). Consumption of eggs, on the other hand, appears not to contribute to a higher serum cobalamin (Brouwer-Brolsma et al. 2015; Vogiatzoglou et al. 2009). This could be explained by the low content of the vitamin in eggs (approx. 2 µg/100 g) combined with a poor bioavailability (Doscherholmen et al. 1975).

2.4 Forms of cobalamin in foods

Cobalamin consists of a corrin ring surrounding a central cobalt atom (Figure 1). All compounds containing this corrin ring are designated corrinoids. Attached to the cobalt atom are a lower (5' position) and an upper (6' position) ligand. The lower ligand of cobalamin consists of phosphate moiety, ribose, and a 5,6-dimethylbenzimidazole base. The upper ligand can be occupied by different anionic substituents (denoted –R in Figure 1). Several forms of cobalamin exists including cyanocobalamin (–CN), hydroxocobalamin (–OH), aquocobalamin (–H_2O), methylcobalamin (–CH_3), sulphitocobalamin (–SO_3), glutathionylcobalamin (–GS), and 5'-deoxyadenosylcobalamin (-5'-deoxyadenosyl) (Fedosov 2012; Gimsing and Nexo 1983; Herbert 1996; Xia et al. 2004). Corrinoids with an altered lower ligand, or simply no lower ligand at all, or otherwise modified, are called cobalamin analogues and are defined as cobalamin-like structures with no known biological activity in man

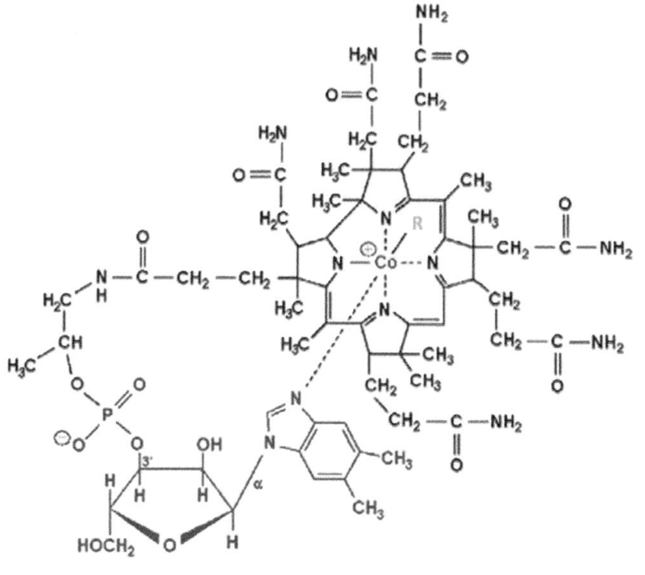

Figure 1. Structure of cobalamin. Cobalamin contains four reduced pyrrole rings (tetrapyrrole) forming a corrin ring system around a cobalt atom (red). The substitution group (upper ligand) designated R is indicated (green). Cleavage of the lower ligand (blue) converts cobalamin into the analogue cobinamide (black). The figure is adapted from (Brenner et al. 2010).

(Fedosov 2012; Gimsing and Nexo 1983; Herbert 1996). The origin of cobalamin analogues in the human body is unknown, but they are found in the circulation bound to the cobalamin-binding protein haptocorrin. Recently, cobalamin analogues have been detected in cord blood (Hardlei et al. 2013). This suggests that they may be produced in the body from circulating cobalamin. At this point, it is unknown if the amount of cobalamin analogues in the circulation is influenced by diet or different physiological conditions.

The principal forms of cobalamin found in foods are hydroxocobalamin, methylcobalamin, and 5′-deoxyadenosylcobalamin (Farquharson and Adams 1976). Methylcobalamin and 5′-deoxyadenosylcobalamin are light-sensitive and can be converted into hydroxocobalamin when exposed to light, and this may be part of the explanation to the high amounts of hydroxocobalamin in prepared and processed food items (Farquharson and Adams 1976). However, since only a small part of a solid food can be exposed to light and since protein-binding may shield the vitamin from photolysis, the forms of cobalamin present in foods can be a mixture of these three forms. Upon uptake of food cobalamin to the cell, it is converted to methylcobalamin and 5′-deoxyadenosylcobalamin that act as coenzymes in the conversions of homocysteine to methionine and methylmalonyl-CoA to succinyl-CoA, respectively.

Cyanocobalamin, the synthetic and most stable form of the vitamin, is found in fortified foods and nutritional supplementation. In most countries, food fortification with cyanocobalamin is discretionary and an individual choice of the food manufacturer, and only processed food products and non-alcoholic beverages may be fortified. It is not settled yet whether the bioavailability of cyanocobalamin is higher or lower than that of naturally occurring cobalamin forms such as hydroxocobalamin in their free form. Our newest data on animals suggest that free cyanocobalamin and hydroxocobalamin are handled differently in the body and homes for different tissues (Kornerup et al. 2015). However, it is currently not possible to speculate about physiological and therapeutic relevant differences. Obviously, further research is needed to establish the many missing informations in this field.

2.5 Dietary requirements of cobalamin throughout life

The Recommended Dietary Allowances (RDA) is generally defined as equal to the estimated average requirement (EAR) plus twice the CV to cover the needs of at least 97 percent of the individuals in the subgroup. For cobalamin, the RDA is 120 percent of the EAR. In this way, the RDA will assure that virtually everyone has an adequate intake. The RDA is classically based on scientific literature on the population for which the mean and standard deviation were determined. Thus, different populations (adults, children, pregnant women, etc.) have different RDAs. The RDAs for different age groups and populations are listed in Table 2. The RDA values were established by the Institute of Medicine (US) Standing Committee on the Scientific Evaluation of Dietary Reference Intakes and its Panel on Folate, Other B Vitamins, and Choline in 1998, and presented in the report Dietary Reference Intakes for Thiamin, Riboflavin, Niacin, Vitamin B6, Folate, Vitamin B12, Pantothenic Acid, Biotin, and Choline (1998). In the following, the underlying methods for establishing the RDAs in the report are given with focus on different populations.

Table 2. Recommended Dietary Allowance (RDA) or Adequate Intakes (AI).

Population group	Cobalamin (µg/day)
Adults and adolescence (age ≥ 14 years), men and women	2.4
Pregnant women	2.6
Lactating women	2.8
Children (age 9–13 years)	1.8
Children (age 4–8 years)	1.2
Children (age 1–3 years)	0.9
Infants (age 7–12 months)	0.5 (AI)
Infants (age 0–6 months)	0.4 (AI)

The RDA is 2.4 µg/day for adult men and women. As described above, the RDA is based on the EAR, which was estimated to be 2.0 µg/day for men and women aged 19 through 50 years. There was not sufficient data to distinguish between requirements of men and women within this group. The EAR was primarily based on the amount of cobalamin needed to maintain an adequate hematological status and concentration of serum cobalamin in patients with pernicious anemia, and in persons with low cobalamin intakes. The estimated EAR value was supported by data on the daily amount of cobalamin needed to maintain cobalamin body stores, and the estimation of dietary intake by healthy adults with adequate serum values of cobalamin and its biomarker methylmalonic acid (MMA).

For adults aged 51 years and older, the RDA was set to be the same as for younger adults (2.4 µg/day). The author finds this problematic. We know that 10–20% of the elderly in the western world show biomarkers indicative of cobalamin inadequacy (Koehler et al. 1996; Lindenbaum et al. 1994; Pennypacker et al. 1992), and also that different factors in the aging population that can affects cobalamin absorption and thereby the requirements for the vitamin. This includes atrophic gastritis, where a low stomach acid secretion can reduce the absorption of cobalamin from foods due to an inadequate release of the vitamin from the food matrix (Miceli et al. 2012). For this reason, the elderly may benefit from intakes of free cobalamin, as found in fortified foods and supplements, or from foods with a high bioavailability such as milk. In general, it would be adviserable to establish RDAs below and above 60 years of age. For further information on malabsorption in aging, see the paragraph on Cobalamin Malabsorption.

For pregnant and lactating women, the RDA is based on the values of normal non-pregnant/non-lactating adults but adjusted according to the assumed transport of cobalamin from mother to fetus/infant. The EAR during pregnancy is adjusted according to an estimated fetal deposition of 0.1–0.2 µg cobalamin/day and assumptions that maternal absorption of cobalamin is more efficient during pregnancy. It has, however, recently been shown that cobalamin absorption is unchanged during pregnancy (Greibe et al. 2011). The EAR during lactation has been adjusted according to the approximate secretion of 0.33 µg cobalamin/day into breast milk. This estimate was based on the average concentration of cobalamin in breast milk from nine Brazilian mothers with adequate cobalamin status at two months postpartum. Since the RDAs were established in 1998, it has been shown that measurement of cobalamin in breast milk without a prior removal of unsaturated haptocorrin provides inadequate results (Lildballe et al. 2009). It has also been shown that the content of cobalamin in breast milk greatly varies during the first 9 months of lactation (Greibe et al. 2013a). Recently it was shown that pregnant and lactating women have an increase in bioactive cobalamin (holotranscobalamin) upon oral intake of cobalamin supplements, which could suggest that the women could benefit from cobalamin intakes exceeding the current recommendation (Bae et al. 2015). Taken together, new knowledge on cobalamin metabolism during pregnancy and lactation suggest a need to re-establish the current RDAs for these subgroups.

For children and adolescence aged 1 through 18 years, no direct data was available on which to base the EAR. Therefore the EARs and RDAs for children and adolescence have been extrapolated down from adult values and rounded up. The RDAs for children and adolescence divided into different age groups are shown in Table 2.

For infants, adequate intake (AI) is given rather than RDA (Table 2). AI is established when data is insufficient to develop an RDA and set by experts at a level assumed to ensure nutritional adequacy based on what appear to be typical intakes of healthy-seeming people. In the case of infants, the AI reflects the observed average cobalamin intake of infants fed principally with human milk. The AI for infants aged 0 through 6 months was set to be 0.4 μg/day in 1998. The AI was based on the average concentration of cobalamin in milk collected from the above-mentioned Brazilian mothers and the estimate that an average woman produces 0.78 L breast milk/day. The adequacy of this intake was supported by evidence that it is above the intake levels that is associated with increased urinary excretion of MMA. The AI for infants aged 7 through 12 months was set to be 0.5 μg/day by adjusting the AI for infants aged 0 through 6 months according to body weight ratio.

A typical non-vegetarian western diet provides around 5–7 μg cobalamin per day, and a study suggest that doses of 6 μg cobalamin/day is needed to maintain serum biomarkers at a steady state levels (Bor et al. 2006). The fact that the RDAs were established almost 20 years ago, and that it for some populations are based on adjustments or extrapolations from the RDA determined for adults (infants, elderly, children, etc.), establishment of new RDAs needs to be considered.

3. Biochemical Aspects of Cobalamin

Cobalamin in transported to the cell bound to transcobalamin. The complex cobalamin-transcobalamin is hydrolysed in the cell lysosome. Cobalamin is then transferred to the cytosol where it is converted into methylcobalamin and to the mitochondria where it is converted to 5′-deoxyadenosylcobalamin. Cobalamin is a cofactor for only two reactions in mammalian cells. First, methylcobalamin is a co-factor for methionine synthase, the enzyme that converts homocysteine into methionine. The same reaction transfers methyl groups from 5-methyltetrahydrofolate to homocysteine. Second, adenosylcobalamin is a cofactor for the mitochondrial enzyme methylmalonyl-CoA mutase that converts methylmalonyl-CoA into succinyl-CoA. The excess of succinyl-CoA is converted to MMA. Cobalamin deficiency causes elevation of plasma concentrations of homocysteine (cytosolic origin) and MMA (mitochondrial origin). Cobalamin deficiency also causes low methionine and therefore, low S-adenosylmethionine, the main methyl donor in the cell. Plasma concentrations of MMA are considered sensitive and specific for cobalamin deficiency. Plasma concentrations of homocysteine are more affected by folate deficiency rather than by cobalamin deficiency. In addition, both MMA and homocysteine concentrations increase in plasma of people with renal insufficiency independent on cobalamin status. Therefore, using homocysteine and MMA as markers for cobalamin status though advantageous in many cases can have limitations in renal patients and elderly.

4. Acquired Cobalamin Deficiency—A Matter of Lifestyle?

There are two main forms of acquired cobalamin deficiency; the first is caused by nutritional factors and the second is caused by malabsorption. Both categories are outlined below. Cobalamin deficiency can also be caused by inborn errors in absorption, transport, or metabolism of the vitamin; but these conditions are rare and will not be discussed in this chapter.

4.1 Nutritional cobalamin deficiency

Cobalamin deficiency is endemic in some parts of the world such as India, Mexico, Guatamala, and Kenya (reviewed by (Stabler and Allen 2004)). Cobalamin deficiency caused by inadequate vitamin intake can occur if the diet contains low or no animal products. Strict vegetarians (vegans) are at high risk of cobalamin depletion because they consume a diet completely free of animal products. Lacto- and lacto-ovo vegetarians have a smaller risk of deficiency because of the presence of cobalamin in eggs and dairy products, but still they have a higher risk than omnivores to become deficient (Herrmann et al. 2003; Majchrzak et al. 2006). For this reason, vegetarians are encouraged to take daily cobalamin supplements in order to maintain their cobalamin status and health. It is generally believed, that in some vegan communities the only source of cobalamin is from contamination of food with microorganisms.

Nutritional cobalamin deficiency is also a major problem among individuals consuming a poor diet with low milk and meat intake, which is the case in several populations from developing countries, where animal-based foods are expensive and a luxury that only a few can afford with regularity. For this reason, the cobalamin status of many non-vegetarians in developing countries is only slightly better than that of lacto-ovo-vegetarians (Refsum et al. 2001). Poverty-imposed cobalamin deficiency is a worldwide problem affecting people in all stages of life. Infants born to cobalamin deplete mothers are at high risk of cobalamin deficiency, and the condition is worsened by months of exclusive breastfeeding (Casterline et al. 1997; Duggan et al. 2014). Cobalamin deficiency during infancy can cause apathy, megaloblastic anemia, stunting, and delayed developmental, and lead to poor cognitive and neuromotor performance when growing up (Dror and Allen 2008; Stabler 2013).

In the western society, cobalamin deficiency and depletion is common, particular among the elderly (Khodabandehloo et al. 2015; Lindenbaum et al. 1994; Pennypacker et al. 1992). This is partly due to an inadequate intake of animal products, simply due to loss of appetite and changes in dietary habits and partly to an age-related malabsorption for the vitamin (see below).

4.2 Cobalamin malabsorption

Cobalamin malabsorption causes deficiency and is especially common among the elderly ($\geq$ 65 years of age) (reviewed by (Carmel 1995; Wong 2015)). Malabsorption of cobalamin can occur as a result of a number of disorders that affect cobalamin release from food, its attachment to intrinsic factor, and uptake via the intrinsic factor receptor. Pernicious anaemia, the cobalamin malabsorption disease that originally led to discovery of the vitamin, has historically been considered the major cause of cobalamin deficiency, even though it is not a common condition. Pernicious anaemia is caused by lack of functional intrinsic factor, a glycoprotein secreted by gastric parietal cells that is required for cobalamin absorption from the terminal ileum. Lack of intrinsic factor can be caused by autoimmunity against the parietal cells that secrete intrinsic factor preventing production of intrinsic factor and secretion of hydrochloric acid. Pernicious anaemia can also be cause by autoimmunity against intrinsic factor itself where autoantibodies attach to intrinsic factor prevents binding to cobalamin in the intestine, or by preventing the absorption of the intrinsic factor-cobalamin complex by the ileal receptors (reviewed by (Rojas Hernandez and Oo 2015; Stabler 2013)).

Atrophic gastritis is another condition associated with cobalamin malabsorption (Miceli et al. 2012; Wong 2015). The disorder is characterized by a progressive reduction in the secretion of hydrochloric acid from the parietal cells. The condition leads to a gradual loss of gastric acid and is believed to reduce the release of protein-bound cobalamin contained in the food, thereby causing malabsorption of food cobalamin. Malabsorption of cobalamin from food is believed to be the main cause of deficiency among the elderly (Carmel 1995; Wong 2015) and explains why depletion occurs with aging. It has also been suggested that atrophic gastritis can lead to bacterial overgrowth (*Helicobacter pylori*) in the stomach and intestine and that this can reduce cobalamin absorption (Kountouras et al. 2007).

Other pathologies in the gastrointestinal tract such as pancreatic diseases and gastric reduction surgery may also cause or augment a deficient state (reviewed by (Carmel 1995; Schjonsby 1989)). Also parasitic infection with the fish tapeworm (*Diphyllobothrium latum*) can cause malabsorption because the worm is able to take up cobalamin from the small intestine of the host (Bjorkenheim 1966). Furthermore, medical treatments, such as proton-pump inhibitors, H2-receptor antagonists and the anti-diabetic drug metformin, have also been associated with low plasma cobalamin levels (Rozgony et al. 2010; Tomkin et al. 1971). However, new studies suggest that metformin treatment is not associated with functional cobalamin deficiency (elevated MMA or homocysteine) (Greibe et al. 2013b; Obeid et al. 2012). Taken together, disorders of cobalamin absorption are the most common factor affecting the bioavailability of the vitamin and potentially the nutritional requirements of people with such disorders. There are currently no specific recommendations on the nutritional requirements for patients with various conditions that cause cobalamin deficiency. However, such recommendations are needed and can help preventing a large number of deficiency cases when the vitamin can be supplemented.

4.3 Cobalamin dietary intervention studies

Dietary intervention, nutritional programs, fortification and intervention with low cobalamin doses to improve cobalamin status markers has been attempted in many populations and age groups with success. In Kenya, dietary intervention with milk and meat almost completely eliminated low plasma cobalamin concentrations (< 148 pmol/L) in school children after a two year period (McLean et al. 2007). A similar outcome was seen in young Indian adult subjected to a daily intake of milk for 14 days (Naik et al. 2013). In cobalamin-deficient women from Bangladesh, supplementation (250 µg/day) during pregnancy and lactation substantially improved maternal and infant cobalamin status and the content of cobalamin in breast milk (Siddiqua et al. 2015). Intervention with supplemental cobalamin also caused improvement in growth and weight gain in cobalamin deficient Indian children (Strand et al. 2015), and improved arterial function in vegetarians with subnormal cobalamin status (Kwok et al. 2012). In the elderly, intervention with cobalamin has not only been found to improve the hematological parameters, but has also been found to slow down cognitive and clinical decline (de Jager et al. 2012), and also brain atrophy rates in individuals with a high baseline of long chain omega 3 fatty acids (Jerneren et al. 2015).

Taken together, cobalamin deficiency is fairly easy to prevent through dietary changes, fortification and supplementation. Voluntary cobalamin supplementation and fortification of foods with cobalamin should be encouraged, especially in the third world and among the elderly, vegetarians and other high-risk groups.

5. Concluding Comments

Cobalamin deficiency is a world-wide public health problem affecting people in many geographical regions and of different life stages and life styles such as vegetarians, people living in developing countries, pregnant women, infants, children, and elderly. The current RDAs are not taking into consideration the development in the field within the last 20 years. There are no personalized recommendations for high risk groups that are known to benefit from higher intakes. The nutritional recommendations should aim at disease prevention in people liable to deficiency conditions. Therefore, I strongly recommend that the RDAs need to be re-evaluated based on the current knowledge and specific recommendations for the individual population subgroups need to be established.

Keywords: Cobalamin analogues, animal-based food, food content, nutrition, bioavailability, dietary sources, cobalamin forms, corrinoids, personalized recommendations, Recommended Dietary Allowance, recommendation, pregnancy, lactation, infancy, elderly, vegetarians, deficiency, malabsorption, pernicious anemia, atrophic gastritis

Abbreviations

AI	:	Average intakes
EAR	:	Estimated average requirement
RDA	:	Recommended dietary allowance
MMA	:	Methylmalonic acid

References

Aimone-Gastin I, Pierson H, Jeandel C, Bronowicki JP, Plenat F, Lambert D, Nabet-Belleville F and Gueant JL. 1997. Prospective evaluation of protein bound vitamin B12 (cobalamin) malabsorption in the elderly using trout flesh labelled *in vivo* with 57Co-cobalamin. Gut. 41: 475–479.

Allen LH. 2010. Bioavailability of vitamin B12. Int J Vitam Nutr Res. 80: 330–335.

Bae S, West AA, Yan J, Jiang X, Perry CA, Malysheva O, Stabler SP, Allen RH and Caudill MA. 2015. Vitamin B12 status differs among pregnant, lactating, and control women with equivalent nutrient intakes. J Nutr. 145: 1507–1514.

Bjorkenheim B. 1966. Optic neuropathy caused by vitamin-B12 deficiency in carriers of the fish tapeworm, Diphyllobothrium latum. Lancet. 1: 688–690.

Bor MV, Lydeking-Olsen E, Moller J and Nexo E. 2006. A daily intake of approximately 6 microg vitamin B12 appears to saturate all the vitamin B12-related variables in Danish postmenopausal women. Am J Clin Nutr. 83: 52–58.

Brenner M, Kim JG, Mahon SB, Lee J, Kreuter KA, Blackledge W, Mukai D, Patterson S, Mohammad O, Sharma VS and Boss GR. 2010. Intramuscular cobinamide sulfite in a rabbit model of sublethal cyanide toxicity. Ann Emerg Med. 55: 352–363.

Brouwer-Brolsma EM, Dhonukshe-Rutten RA, van Wijngaarden JP, Zwaluw NL, Velde N and de Groot LC. 2015. Dietary sources of vitamin B12 and their association with vitamin B12 status markers in healthy older adults in the B-PROOF study. Nutrients. 7: 7781–7797.

Carmel R. 1995. Malabsorption of food cobalamin. Baillieres Clin Haematol. 8: 639–655.

Casterline JE, Allen LH and Ruel MT. 1997. Vitamin B12 deficiency is very prevalent in lactating Guatemalan women and their infants at three months postpartum. J Nutr. 127: 1966–1972.

Czerwonga M, Szterk A and Waszkiewicz-Robak B. 2014. Vitamin B12 content in raw and cooked beef. Meat Sci. 96: 1371–1375.

Dagnelie PC, van Staveren WA and van den Berg H. 1991. Vitamin B12 from algae appears not to be bioavailable. Am J Clin Nutr. 53: 695–697.

de Jager CA, Oulhaj A, Jacoby R, Refsum H and Smith AD. 2012. Cognitive and clinical outcomes of homocysteine-lowering B-vitamin treatment in mild cognitive impairment: a randomized controlled trial. Int J Geriatr Psychiatry. 27: 592–600.

Doscherholmen A, McMahon J and Economon P. 1981. Vitamin B12 absorption from fish. Proc Soc Exp Biol Med. 167: 480–484.

Doscherholmen A, McMahon J and Ripley D. 1975. Vitamin B12 absorption from eggs. Proc Soc Exp Biol Med. 149: 987–990.

Doscherholmen A, McMahon J and Ripley D. 1978. Vitamin B12 assimilation from chicken meat. Am J Clin Nutr. 31: 825–830.

Dror DK and Allen LH. 2008. Effect of vitamin B12 deficiency on neurodevelopment in infants: current knowledge and possible mechanisms. Nutr Rev. 66: 250–255.

Duggan C, Srinivasan K, Thomas T, Samuel T, Rajendran R, Muthayya S, Finkelstein JL, Lukose A, Fawzi W, Allen LH, Bosch RJ and Kurpad AV. 2014. Vitamin B12 supplementation during pregnancy and early lactation increases maternal, breast milk, and infant measures of vitamin B12 status. J Nutr. 144: 758–764.

Farquharson J and Adams JF. 1976. The forms of vitamin B12 in foods. Br J Nutr. 36: 127–136.

Fedosov SN. 2012. Physiological and molecular aspects of cobalamin transport. Subcell Biochem. 56: 347–367.

Fedosov SN, Petersen TE and Nexo E. 1996. Transcobalamin from cow milk: isolation and physico-chemical properties. Biochim Biophys Acta. 1292: 113–119.

Gille D and Schmid A. 2015. Vitamin B12 in meat and dairy products. Nutr Rev. 73: 106–115.

Gimsing P and Nexo E. 1983. The forms of cobalamin in biological materials. pp. 7–30. In: The Cobalamins by Hall CA, Churchill Livingstone, Edinburgh London Melbourne and New York. Chapter 1 ed.

Greibe E, Andreasen BH, Lildballe DL, Morkbak AL, Hvas AM and Nexo E. 2011. Uptake of cobalamin and markers of cobalamin status: a longitudinal study of healthy pregnant women. Clin Chem Lab Med. 49: 1877–1882.

Greibe E, Lildballe DL, Streym S, Vestergaard P, Rejnmark L, Mosekilde L and Nexo E. 2013a. Cobalamin and haptocorrin in human milk and cobalamin-related variables in mother and child: a 9-mo longitudinal study. Am J Clin Nutr. 98: 389–395.

Greibe E, Trolle B, Bor MV, Lauszus FF and Nexo E. 2013b. Metformin lowers serum cobalamin without changing other markers of cobalamin status: a study on women with polycystic ovary syndrome. Nutrients. 5: 2475–2482.

Gu Q, Zhang C, Song D, Li P and Zhu X. 2015. Enhancing vitamin B12 content in soy-yogurt by Lactobacillus reuteri. Int J Food Microbiol. 206: 56–59.

Hardlei TF, Obeid R, Herrmann W and Nexo E. 2013. Cobalamin analogues in humans: a study on maternal and cord blood. PLoS One. 8: e61194.

Herbert V. 1996. Vitamin B12. In: Present Knowledge in Nutrition (7th edition) by Ziegler EE, Filer LJ, International Life Science Institute (ILSI), Washington DC, Chapter 20.

Herrmann W, Schorr H, Obeid R and Geisel J. 2003. Vitamin B12 status, particularly holotranscobalamin II and methylmalonic acid concentrations, and hyperhomocysteinemia in vegetarians. Am J Clin Nutr. 78: 131–136.

Heyssel RM, Bozian RC, Darby WJ and Bell MC. 1966. Vitamin B12 turnover in man. The assimilation of vitamin B12 from natural foodstuff by man and estimates of minimal daily dietary requirements. Am J Clin Nutr. 18: 176–184.

Jerneren F, Elshorbagy AK, Oulhaj A, Smith SM, Refsum H and Smith AD. 2015. Brain atrophy in cognitively impaired elderly: the importance of long-chain omega-3 fatty acids and B vitamin status in a randomized controlled trial. Am J Clin Nutr. 102: 215–221.

Khodabandehloo N, Vakili M, Hashemian Z and Zare ZH. 2015. Determining functional vitamin B12 deficiency in the elderly. Iran Red Crescent Med J. 17: e13138.

Koehler KM, Romero LJ, Stauber PM, Pareo-Tubbeh SL, Liang HC, Baumgartner RN, Garry PJ, Allen RH and Stabler SP. 1996. Vitamin supplementation and other variables affecting serum homocysteine and methylmalonic acid concentrations in elderly men and women. J Am Coll Nutr. 15: 364–376.

Kornerup LS, Juul CB, Fedosov SN, Heegaard CW, Greibe E and Nexo E. 2015. Absorption and retention of free and milk protein-bound cyano- and hydroxocobalamins. An experimental study in rats. Biochimie.

Kountouras J, Gavalas E, Boziki M and Zavos C. 2007. Helicobacter pylori may be involved in cognitive impairment and dementia development through induction of atrophic gastritis, vitamin B12 folate deficiency, and hyperhomocysteinemia sequence. Am J Clin Nutr. 86: 805–806.

Kwok T, Chook P, Qiao M, Tam L, Poon YK, Ahuja AT, Woo J, Celermajer DS and Woo KS. 2012. Vitamin B12 supplementation improves arterial function in vegetarians with subnormal vitamin B12 status. J Nutr Health Aging. 16: 569–573.

Lildballe DL, Hardlei TF, Allen LH and Nexo E. 2009. High concentrations of haptocorrin interfere with routine measurement of cobalamins in human serum and milk. A problem and its solution. Clin Chem Lab Med. 47: 182–187.

Lildballe DL and Nexo E. 2013. Analysis of cobalamins (vitamin B12) in human samples: An overview of methodology. In: B Vitamins and Folate: Chemistry, Analysis, Function and Effects, Chapter 26.

Lindenbaum J, Rosenberg IH, Wilson PW, Stabler SP and Allen RH. 1994. Prevalence of cobalamin deficiency in the Framingham elderly population. Am J Clin Nutr. 60: 2–11.

Majchrzak D, Singer I, Manner M, Rust P, Genser D, Wagner KH and Elmadfa I. 2006. B-vitamin status and concentrations of homocysteine in Austrian omnivores, vegetarians and vegans. Ann Nutr Metab. 50: 485–491.

Matte JJ, Guay F and Girard CL. 2012. Bioavailability of vitamin B12 in cows' milk. Br J Nutr. 107: 61–66.

McLean ED, Allen LH, Neumann CG, Peerson JM, Siekmann JH, Murphy SP, Bwibo NO and Demment MW. 2007. Low plasma vitamin B12 in Kenyan school children is highly prevalent and improved by supplemental animal source foods. J Nutr. 137: 676–682.

Miceli E, Lenti MV, Padula D, Luinetti O, Vattiato C, Monti CM, Di SM and Corazza GR. 2012. Common features of patients with autoimmune atrophic gastritis. Clin Gastroenterol Hepatol. 10: 812–814.

Naik S, Bhide V, Babhulkar A, Mahalle N, Parab S, Thakre R and Kulkarni M. 2013. Daily milk intake improves vitamin B12 status in young vegetarian Indians: an intervention trial. Nutr J. 12: 136.

Nishioka M, Kanosue F, Yabuta Y and Watanabe F. 2011. Loss of vitamin B(12) in fish (round herring) meats during various cooking treatments. J Nutr Sci Vitaminol (Tokyo). 57: 432–436.

Obeid R, Jung J, Falk J, Herrmann W, Geisel J, Friesenhahn-Ochs B, Lammert F, Fassbender K and Kostopoulos P. 2012. Serum vitamin B12 not reflecting vitamin B12 status in patients with type 2 diabetes. Biochimie.

Pennypacker LC, Allen RH, Kelly JP, Matthews LM, Grigsby J, Kaye K, Lindenbaum J and Stabler SP. 1992. High prevalence of cobalamin deficiency in elderly outpatients. J Am Geriatr Soc. 40: 1197–1204.

Refsum H, Yajnik CS, Gadkari M, Schneede J, Vollset SE, Orning L, Guttormsen AB, Joglekar A, Sayyad MG, Ulvik A and Ueland PM. 2001. Hyperhomocysteinemia and elevated methylmalonic acid indicate a high prevalence of cobalamin deficiency in Asian Indians. Am J Clin Nutr. 74: 233–241.

Rojas Hernandez CM and Oo TH. 2015. Advances in mechanisms, diagnosis, and treatment of pernicious anemia. Discov Med. 19: 159–168.

Rozgony NR, Fang C, Kuczmarski MF and Bob H. 2010. Vitamin B(12) deficiency is linked with long-term use of proton pump inhibitors in institutionalized older adults: could a cyanocobalamin nasal spray be beneficial? J Nutr Elder. 29: 87–99.

Russell RM, Baik H and Kehayias JJ. 2001. Older men and women efficiently absorb vitamin B12 from milk and fortified bread. J Nutr. 131: 291–293.

Schjonsby H. 1989. Vitamin B12 absorption and malabsorption. Gut. 30: 1686–1691.

Siddiqua TJ, Ahmad SM, Ahsan KB, Rashid M, Roy A, Rahman SM, Shahab-Ferdows S, Hampel D, Ahmed T, Allen LH and Raqib R. 2015. Vitamin B12 supplementation during pregnancy and postpartum improves B12 status of both mothers and infants but vaccine response in mothers only: a randomized clinical trial in Bangladesh. Eur J Nutr.

Stabler SP. 2013. Clinical practice. Vitamin B12 deficiency. N Engl J Med. 368: 149–160.

Stabler SP and Allen RH. 2004. Vitamin B12 deficiency as a worldwide problem. Annu Rev Nutr. 24: 299–326.

Strand TA, Taneja S, Kumar T, Manger MS, Refsum H, Yajnik CS and Bhandari N. 2015. Vitamin B12, folic acid, and growth in 6- to 30-month-old children: a randomized controlled trial. Pediatrics. 135: e918–e926.

The Danish National Food Institute, DF. 2015. The Danish National Food Institute, DTU Food, "Food Composition Databank - Version 7.01".

Tomkin GH, Hadden DR, Weaver JA and Montgomery DA. 1971. Vitamin-B12 status of patients on long-term metformin therapy. Br Med J. 2: 685–687.

Tucker KL, Rich S, Rosenberg I, Jacques P, Dallal G, Wilson PW and Selhub J. 2000. Plasma vitamin B12 concentrations relate to intake source in the Framingham offspring study. Am J Clin Nutr. 71: 514–522.

van den Berg H, Daqnelie PC and van Staveren WA. 1988. Vitamin B12 and seaweed. Lancet. 1: 242–243.

Vogiatzoglou A, Smith AD, Nurk E, Berstad P, Drevon CA, Ueland PM, Vollset SE, Tell GS and Refsum H. 2009. Dietary sources of vitamin B12 and their association with plasma vitamin B12 concentrations in the general population: the Hordaland Homocysteine Study. Am J Clin Nutr. 89: 1078–1087.

Watanabe F. 2007. Vitamin B12 sources and bioavailability. Exp Biol Med (Maywood). 232: 1266–1274.

Wong CW. 2015. Vitamin B12 deficiency in the elderly: is it worth screening? Hong Kong Med J. 21: 155–164.

Xia L, Cregan AG, Berben LA and Brasch NE. 2004. Studies on the formation of glutathionylcobalamin: any free intracellular aquacobalamin is likely to be rapidly and irreversibly converted to glutathionylcobalamin. Inorg Chem. 43: 6848–6857.

3

Intracellular Processing and Utilization of Cobalamins

Luciana Hannibal[1,*] and *Donald W Jacobsen*[2]

1. Introduction

Humans and other members of the Eukaryota domain are unable to synthesize cobalamin (Cbl; B12) With the exception of plants and fungi, most members in Eukaryota have an absolute requirement for this essential micronutrient.

The complex pathway for the biosynthesis of Cbl is found in a relatively few species of the Archaea and Bacteria domains (Martens et al. 2002) and its elucidation has been matter of intense research (Croft et al. 2005; Martens et al. 2002; Moore et al. 2013; Raux et al. 1998; Roth et al. 1996; Warren 2006; Warren et al. 2002). The exclusive biosynthesis performed by a few primitive microorganisms and the occurrence of rare, spontaneous and stereospecific photochemical A/D precorrin-to-corrin cycloisomerization observed during the artificial synthesis of vitamin B12 has led to the proposal that Cbl may be a primordial biomolecule, bridging life's transition from chemistry to biology (Eschenmoser 2011). Uncovering the pathway of cobalamin utilization in mammals has not been less intense. Since the elucidation of its structure by Dorothy Hodgkin (Hodgkin et al. 1955; Lenhert and Hodgkin 1961), the intracellular maps of cobalamin utilization continue to be drawn until these days (Gherasim et al. 2013b). Mutations in the genes that encode the enzymes or proteins involved in Cbl processing, trafficking and biosynthesis are defined as Cbl complementation groups (*cblA-cblG* and *mut*) (Rosenblatt and Whitehead 1999; Watkins and Rosenblatt 2011) (Figure 1). These inborn

[1] Laboratory of Clinical Biochemistry and Metabolism, Department of Pediatrics and Adolescence, Medical Center, University of Freiburg, Mathildenstr. 1, 79106-Freiburg, Germany.
[2] Department of Cellular and Molecular Medicine, Lerner Research Institute, Cleveland Clinic, 9500 Euclid Ave., Cleveland, Ohio 44195, USA.
* Corresponding author: luciana.hannibal@uniklinik-freiburg.de

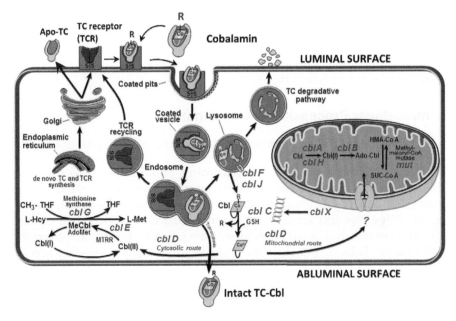

Figure 1. Intracellular events of processing, trafficking and delivery to acceptor proteins. Cobalamin processing begins upon its release from the lysosomal compartment by proteins CblF and CblJ. In the cytosol, the dedicated chaperone CblC process the upper axial ligand with concomitant generation of cob(II)alamin. Cob(II)alamin is directed to two possible compartments, the cytosol or the mitochondria by adapter protein CblD via protein-protein interactions with CblC. Cytosolic cob(II)alamin is further reduced by CblE enabling its entrance in the catalytic reaction of MS (cblG). Mitochondrial cob(II)alamin or cob(III)alamin (if oxidized) enters the catalytic cycle of mut (MCM) preceded by the actions of CblA and CblB.

errors of Cbl metabolism lead to either isolated or combined homocystinuria and methylmalonic aciduria. Cellular Cbl utilization begins with the import of circulating Cbl bound to transcobalamin (holo-TC, TC•XCbl, where 'X' represents the β-axial ligand of Cbl). The transcobalamin receptor (TCblR) captures holo-TC from circulation and internalizes the complex by absorptive endocytosis (Quadros et al. 2009; Quadros and Sequeira 2013). Once internalized, holo-TC dissociates from its receptor in acidic endosomes and TCblR is recycled back to the cell surface (Amagasaki et al. 1990). Holo-TC is degraded in lysosomal vesicles by proteolysis and the XCbl is exported to the cytosol via the lysosomal transport system involving LMBRD1, the cblF gene product, and ABCD4, the cblJ gene product (Coelho et al. 2012; Gailus et al. 2010a; Gailus et al. 2010b; Rutsch et al. 2011). XCbl then binds to the *cblC* gene product MMACHC (for *methylmalonic aciduria type C with homocystinuria*; hereafter referred to as CblC). The β-axial ligand "X-group" is enzymatically eliminated by CblC. The *cblD* gene product MMADHC (for *methylmalonic aciduria type D and homocystinuria*; hereafter referred to as CblD) is thought to direct CblC-bound Cbl to the mitochondria for AdoCbl synthesis (via *cblA* and *cblB*), or to cytosolic methionine synthase for MeCbl synthesis (via *cblE* and *cblG*) (Figure 1). The following sections present an overview of the key known players in intracellular Cbl processing and trafficking and the synthesis and utilization of the two essential cofactors, methylcobalamin (MeCbl) and adenosylcobalamin (AdoCbl).

2. Intracellular Processing of Cobalamins

A. CblC-mediated decyanation and dealkylation of B12

The routes of incorporation of incoming cobalamins into the two Cbl-dependent enzymes remained elusive for many years. Much of the current knowledge on Cbl processing (defined as the removal of the upper, β-axial ligand with either concerted or subsequent reduction of the cobalt center) arose from *ex vivo* studies with fibroblasts from patients carrying inborn errors of Cbl metabolism. A study by Chu et al suggested that even dietary MeCbl and AdoCbl must undergo processing of their upper axial ligand prior to their incorporation into B12-dependent methionine synthase (MS) and B12-dependent methylmalonyl-CoA mutase (MCM), respectively (Chu et al. 1993).

The first case report of functional Cbl deficiency caused by an inborn error of metabolism was provided by Harvey Mudd et al. more than 45 years ago (Mudd et al. 1969). The patient under study belonged to the *cblC* complementation group and presented with combined homocystinuria and methylmalonic aciduria (Mudd et al. 1969). Cultured patient fibroblasts displayed slightly reduced uptake of Cbl with respect to the normal fibroblasts, efflux of Cbl at long incubation times, and impaired biosynthesis of both MeCbl and AdoCbl (Mudd et al. 1969). This was the first evidence that the gene responsible for the *cblC* complementation group was required for a step prior to both cofactors biosynthesis. The number of patients presenting with both early and late onset *cblC* disease in 2009 was greater than 360 (Lerner-Ellis et al. 2009).

1. Cloning of the cblC (MMACHC) gene

Early studies with crude cell extracts showed that *cblC* fibroblasts possessed reduced Cbl β-transferase activity and/or Cbl reductase activity (Pezacka et al. 1988; Pezacka et al. 1990; Pezacka 1993; Pezacka and Rosenblatt 1994). These preliminary findings pointed to CblC as the protein responsible for processing the dietary Cbls. These findings also evidenced the participation of the most abundant intracellular thiol, glutathione, in the step preceding cofactor biosynthesis (Pezacka et al. 1988; Pezacka et al. 1990; Pezacka 1993; Pezacka and Rosenblatt 1994). It was not until 2006 that the gene responsible for the *cblC* phenotype was identified and characterized (Lerner-Ellis et al. 2006). Gene sequence alignment and motif searches revealed that the CblC protein is not a member of any previously identified gene family (Lerner-Ellis et al. 2006). Although it is well conserved among mammals, its C-terminal end does not seem to be conserved in eukaryotes outside mammalia, and no homologs are found in prokaryotes. The CblC protein has two motifs that are similar to motifs present in bacterial genes with Cbl-related functions: a) a Cbl-binding motif 52% identical to the corresponding motif of MCM of *Streptomyces avermitilis*, and b) a TonB motif ~40–50% identical to various TonB proteins from Gram negative bacteria (Lerner-Ellis et al. 2006). The *cblC* gene appeared to be expressed in most tissues. High mRNA levels were detected in fetal liver, lower levels were detected in spleen, lymph node, thymus and bone marrow, and no message was detected in peripheral blood leukocytes. In addition, the cellular *cblC* phenotype was complemented in two immortalized *cblC* fibroblast cell lines infected with wild-type *cblC* cDNA. Function of both MS and MCM was restored to control levels, or above, in infected *cblC* fibroblasts. Moreover, the conversion of CNCbl into AdoCbl and MeCbl was more effective in the complemented cell lines (Lerner-Ellis et al. 2006).

2. Expression and characterization of the cblC (MMACHC) gene product

The human CblC protein has been expressed and characterized by three independent groups (Deme et al. 2012; Froese et al. 2009; Kim et al. 2008; Plesa et al. 2011). The expression and characterization of bovine (Jeong et al. 2011) and *C. elegans* (Li et al. 2014a; Park and Kim 2015) variants of CblC have also been reported. Human recombinant CblC exists predominantly as a monomer with a molecular weight of 29 kDa. Studies by another group showed evidence that CblC can form dimers (Froese et al. 2012), consistent with phage display assays showing self-assembly regions in the protein (Deme et al. 2012). The relevance of oligomerization in CblC remains to be investigated. CblC does not bind any chromophoric ligands except Cbl upon reconstitution *in vitro*. Binding of Cbl to CblC leads to the base-off configuration of the micronutrient, that is, the detachment of the dimethylbenzimidazole moiety from the cobalt center (Figure 2). The binding affinity of CblC for Cbl is in the low micromolar range, with reported K_d values between 1 and 15 µM (Table 1). Binding of glutathione (GSH) to the bovine counterpart of CblC increases its affinity for Cbls (Jeong and Kim 2011), and protects bound aquacobalamin from catalyzing the oxidation of GSH to form GSSH (Jeong et al. 2011).

An interesting feature of CblC is its stability. Froese et al. demonstrated that CblC is naturally thermolabile ($T_m = 39°C$) and that some of the most frequent mutations that occur in humans exacerbate this property (Froese et al. 2010a) as well as its ability to bind Cbls (Froese et al. 2009). Further, the same study showed that CblC is most stable when bound to AdoCbl and MeCbl, with increased polydispersity and hydrodynamic radii (Froese et al. 2009). This may reflect conformational changes, as noted for bacterial apo-BtuF versus cyanocobalamin (CNCbl)-BtuF (James et al. 2009). Studies with the bovine isoform of CblC revealed that GSH stabilizes CblC, suggesting that intracellular redox control could play a role in the regulation of the protein's lifetime (Jeong

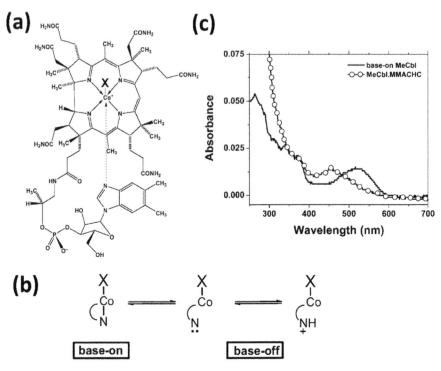

Figure 2. Binding of CblC to Cbl induces the base-off configuration. (a) Structure of cobalamin. (b) The base-on to base-off transition of Cbl is a pH-dependent process that involves protonation of the dimethylbenzimidazole moiety leading to rupture of the axial Co-N bond. (c) Incubation of stoichiometric amounts of MeCbl with CblC results in a base-off CblC•MeCbl complex. The UV-visible spectra of free MeCbl and CblC•MeCbl (EPPS buffer (40 mM, pH 7.6) supplemented with 150 mM NaCl and 10% glycerol) are shown as line and line-circled traces, respectively.

Table 1. Binding affinity of CblC for several naturally occurring cobalamins.

CblC	Binding affinity, K_d (µM)					
	CNCbl	**HOCbl**	**MeCbl**	**AdoCbl**	**GSCbl**	**Reference**
H. sapiens	15.5 ± 1.1	ND	ND	ND	ND	(Kim et al. 2008)
H. sapiens	18.1 ± 6.3	3 ± 0.4	2.1 ± 0.3	0.9 ± 0.1	ND	(Froese et al. 2010a)
H. sapiens	6.4 ± 0.04	9.8 ± 0.06	1.42 ± 0.02	1.68 ± 0.02	ND	(Plesa et al. 2011)
B. taurus	16.1 ± 1.0	16.7 ± 1.5	8.6 ± 2.2	11.7 ± 2.7	ND	(Jeong et al. 2011)
C. elegans	50.2 ± 9.5	ND	16.8 ± 6.1	22.0 ± 7.5	26.5 ± 5.5	(Park and Kim 2015)

ND: Not determined

et al. 2011; Jeong and Kim 2011; Park et al. 2012). Koutmos (Koutmos et al. 2011) and Froese (Froese et al. 2012) have independently obtained high-resolution X-ray crystal structures of CblC. The work by Koutmos et al. revealed that CblC possesses an N-terminal flavodoxin nitroreductase domain, which can use

FMN or FAD to catalyze the reductive decyanation of CNCbl (Koutmos et al. 2011). CblC possesses a large cavity for binding B12 in its base-off configuration (Figure 3, panel a), a binding mode thought to facilitate the reductive removal of CNCbl's cyanide group at the β-axial position. Unlike other Cbl-dependent enzymes, the base-off Cbl binding by CblC does not involve the coordination

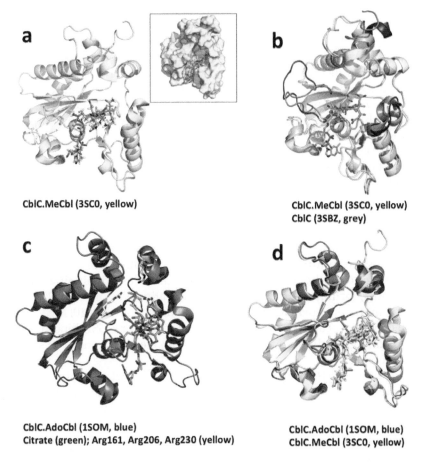

a

CblC.MeCbl (3SC0, yellow)

b

CblC.MeCbl (3SC0, yellow)
CblC (3SBZ, grey)

c

CblC.AdoCbl (1SOM, blue)
Citrate (green); Arg161, Arg206, Arg230 (yellow)

d

CblC.AdoCbl (1SOM, blue)
CblC.MeCbl (3SC0, yellow)

Figure 3. The structures of apo-CblC, CblC•MeCbl and CblC• AdoCbl. (a) The structure of CblC. MeCbl (3SC0, (Koutmos et al. 2011)) reveals that the cofactor exists in its base-off configuration. The bound cobalamin is located within a spacious cavity. The inset shows a surface representation depicting significant solvent accessibility of the Cbl moiety. (b) Binding of MeCbl to apo.CblC (3SBZ, (Koutmos et al. 2011)) triggers conformational changes in three loop regions comprising amino acid residues 103–115, 196–203 and 231–238. Loop regions in black and orange or apo. CblC and CblC.MeCbl, respectively. (c) The structure of CblC.AdoCbl (3SOM (Froese et al. 2012)) shows that conserved amino acid residues Arg161, Arg206 and Arg230 (yellow sticks) necessary for GSH binding are in close contact with a citrate molecule where GSH is predicted to bind under physiological conditions. (d) Superposition of the structures of CblC. MeCbl and CblC.AdoCbl confirms that the Cbl-bound proteins share highly similar composition of secondary structure elements, including the abovementioned loops that seem only sensitive to Cbl binding.

of a His residue from the protein backbone (Koutmos et al. 2011). Binding of MeCbl to CblC induces measurable conformational changes in three different loop-structured domains around the B12 cavity (Figure 3, panel b). Froese et al. elucidated the first structure of CblC bound to AdoCbl (Froese et al. 2012). The overall fold of CblC does not differ markedly from that reported of MeCbl bound to CblC, but revealed a highly conserved dimerization cap for the β-axial 5′-adenosyl ligand, and an arginine-rich GSH-binding pocket up above the β-axial ligand position (Froese et al. 2012). Importantly, the arginine-rich pocket comprises residues Arg161, Arg206 and Arg230 (Figure 3, panel c, yellow sticks) all of which are sites for point mutations that occur in humans, leading to *cblC* disease. A citrate molecule from the solvent was identified in the predicted region for GSH binding during catalysis (Figure 3, panel c, green). Froese et al. further showed that recombinant mutant Arg206Gln was insoluble suggesting a structural role for this residue, and that mutants Arg161Gln and Arg230Gln abolished GSH binding and dealkylase activity, which is a strong indication that these amino acid residues are critical for GSH binding (Froese et al. 2012). The authors noted that FMN, and to a lesser extent Cbl, induces the dimerization of CblC, a previously unrecognized feature of the protein (Froese et al. 2012). Biophysical characterization of CblC variants that mimic these naturally occurring mutations showed that the Arg triad is essential to maintain the enzyme coupled toward Cbl processing (Gherasim et al. 2015). Mutations in Arg residues lead to futile redox cycling that generates superoxide and in turn, H_2O_2, contributing to oxidative stress (Gherasim et al. 2015). Importantly, patients with the *cblC* disorder exhibit alterations in redox status, including abnormal GSH-GSSG ratios (Pastore et al. 2014) and elevated markers of oxidative stress (Jorge-Finnigan et al. 2010; Mc Guire et al. 2009; Richard et al. 2009). Thus, apart from supporting GSH binding, these conserved Arg residues might ensure optimum electron transfer to support Cbl processing.

Residues Arg161 and Arg206 appear to support the interaction of CblC with MS (Fofou-Caillierez et al. 2013). While a crystal structure of the complex to fully demonstrate this is not yet available, molecular modeling and docking studies suggest that a loop of MS interacts directly with residues Arg161 and Arg206. This seems plausible as both residues exhibit substantial accessibility to the solvent (Figure 3, panel c), which would facilitate protein recognition events. The structures of CblC bound to MeCbl and AdoCbl are markedly similar (Figure 3, panel d), which suggests that major changes are only observed during the Cbl-binding event. Analysis of a clinical case with a deletion mutant showed that amino acid residue Gln131 is essential to support CblC enzyme activity (Backe et al. 2013). This amino acid residue establishes direct hydrogen bond contacts with the Cbl moiety, and its absence is predicted to alter Cbl positioning within its binding pocket thereby disrupting electron transfer reactions (Backe et al. 2013). Structural elucidation revealed a highly disordered C-terminus in CblC (Koutmos et al. 2011). Further, it was found that a C-terminus truncated form of CblC is predominantly expressed in murine tissues (Koutmos et al. 2011). In contrast, western blot analysis of human cell lines (HEK93 and fibroblasts MCH46) showed that CblC is expressed as a full-length protein with a molecular mass of 32 kDa (predicted value is 31.9 kDa) (Deme et al. 2012). Altogether, these findings reveal that the structural requirements that enable CblC function vary across species.

3. Mechanism of CblC enzyme activity

Regardless of the chemical nature of their β-axial ligands, all dietary Cbls must undergo enzymatic processing by the CblC chaperone prior to their incorporation into the final acceptor proteins MS and MCM (Chu et al. 1993). The reductive decyanation of CNCbl is catalyzed by the CblC chaperone in the presence of a flavoprotein reductase and NADPH (Kim et al. 2008) or with GSH (Li et al. 2014a). Surprisingly, CblC was first found not to catalyze the removal of methyl and adenosyl groups from MeCbl and AdoCbl, two major dietary forms (Kim et al. 2008). A reexamination of this work with cultured fibroblasts carrying the *cblC* mutation indicated that the CblC protein was indispensable for the dealkylation of natural and synthetic alkylcobalamins (Hannibal et al. 2009).

Three *cblC* mutant fibroblasts isolated from severely ill and genetically unrelated patients were used in an *ex vivo* study to assess Cbl dealkylation. The biochemical profile of the *cblC* mutant cell lines WG1801, WG2176 and WG3354 resembled that reported for other *cblC* cell lines, i.e., substantial export of homocysteine and methylmalonic acid into culture medium (indicative of functional Cbl deficiency) and poor or negligible utilization of CNCbl as a substrate for cofactor biosynthesis (Hannibal et al. 2009). *cblC* mutant fibroblasts were unable to utilize [^{57}Co]-Propyl-Cbl as a substrate for AdoCbl and MeCbl biosynthesis, which demonstrated that the *cblC* protein was essential for dealkylation (Hannibal et al. 2009). These findings were later confirmed by *in vitro* studies performed with human recombinant CblC (Kim et al. 2009). The CblC protein catalyzed the dealkylation of MeCbl and AdoCbl in the presence of excess reduced GSH (Kim et al. 2009). Demethylation of MeCbl was much faster than the removal of the 5'-adenosyl group from AdoCbl (Kim et al. 2009). The ability of CblC to dealkylate a series of MeCbl analogues previously described to undergo processing in *ex vivo* studies, namely, ethylCbl, propylCbl, butylCbl, pentylCbl and hexylCbl (Hannibal et al. 2009) was also assessed. CblC catalyzed the removal of the alkyl group in the β-axial position of all MeCbl analogs examined (Kim et al. 2009). However, the rate of dealkylation decreased with increasing alkyl chain length (Kim et al. 2009). Whether the later is a result of conformational alterations in CblC induced by the more bulky alkyl moieties or due to an unfavorable incorporation of the longer alkyl carbocations into GSH remains to be elucidated. CblC exhibits marked versatility in terms of catalysis, adopting features of the GSH S-transferase family (removal of alkyl groups) as well as of the MS enzymes (transfer of methyl groups to the sulfur acceptor homocysteine). The GSH-mediated reactions catalyzed by CblC are summarized in Figure 4. Dealkylation occurs through the formation of the supernucleophile cob(I)alamin which quickly oxidizes to the more stable form cob(II)alamin (Kim et al. 2009). The cob(I) alamin species was recently captured under aerobic conditions in a reaction catalyzed by *C. elegans* CblC, though this is thought to be an exception to the highly unstable nature of cob(I)alamin (Li et al. 2014b). Decyanation of CNCbl is driven by reductase partners such as MS reductase (MSR) and novel reductase 1 (NR1) as well as GSH under anaerobic conditions (Kim et al. 2008; Li et al. 2014a).

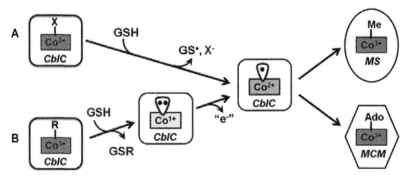

Figure 4. Mechanism of GSH-dependent catalysis by CblC. (A) Decyanation of CNCbl occurs via one-electron reductive elimination of cyanide with formation of glutathionyl radical and cob(II)alamin. Reduction of HOCbl to cob(II)alamin also proceeds via one-electron reduction by GSH (B) Dealkylation of MeCbl, AdoCbl and synthetic alkylcobalamins occurs via nucleophilic attack by GSH to yield cob(I)alamin and cob(II)alamin and the corresponding alkylated versions of glutathione (GSR). X: CN or HO/H$_2$O; R: alkyl group (Co-C bond). Reproduced from Li et al. (Li et al. 2014a), with permission.

4. Regulation of CblC enzyme activity

While our understanding of the reactions catalyzed by the CblC enzyme has deepened during the past 5 years, little knowledge exists on the factors that regulate its enzyme activity. Oxidized GSH (GSSG) appears to destabilize CblC via conformational changes as shown by thermal denaturation studies and an increased susceptibility to trypsin digestion (Park et al. 2012). It was found that GSSG binds to CblC with essentially the same affinity as GSH (Park et al. 2012). This finding suggests that subtle changes in the redox status of the cells that alter the ratio of GSH/GSSG might modulate CblC enzyme activity. *Ex-vivo* studies are warranted to further investigate this proposal. Some naturally occurring mutations identified in *cblC* patients have also been shown to alter CblC activity by reducing its thermal stability compared to the native protein (Froese et al. 2010a), decreasing enzymatic activity (Backe et al. 2013) and via futile redox cycling that deviates electron transfer from the Cbl processing pathway (Gherasim et al. 2015).

The expression of CblC is also subject of epigenetic (Loewy et al. 2009) and transcriptional regulation (Quintana et al. 2014; Yu et al. 2013). *Ex vivo* studies demonstrated that the methionione-dependent tumorigenic melanoma cell line MeWo-LC1 exhibited the same phenotype of cultured cells carrying the *cblC* inborn error of metabolism, yet, no mutations were identified in the *cblC* gene (Loewy et al. 2009). A fully methylated CpG island at the 5′-end of the *cblC* gene was identified in the MeWo-LC1 cell line, which led to low expression of CblC and, therefore, to methionine dependence (Loewy et al. 2009). Mutations in the global transcriptional co-regulator HCFC1 (*Homo sapiens* host cell factor C1) lead to severe reduction in the mRNA levels of the *cblC* gene, a condition that manifests as a late-onset form of the *cblC* disorder (Yu et al. 2013). Consensus HCFC1 binding sites were identified in the *cblC* gene, suggesting a conserved mechanism for regulation (Yu et al. 2013). This new case of combined homocystinuria and methylmalonic aciduria was named 'cblX', given that the mutation is linked to chromosome X (Yu et al. 2013). Disrupted control of *cblC* expression via mutation of the upstream co-regulator HCFC1 led to craniofacial abnormalities during the development of zebra fish, which was ascribed to deficient expression, and therefore reduced activity of the CblC protein required for Cbl processing (Quintana et al. 2014).

5. Proteomics of CblC-mediated B12 processing

One strategy to understand the biological relevance of the changes associated with functional Cbl deficiency is to examine the human proteome using the *cblC* genetic background as a model. Two such comprehensive studies have been carried out in cells (Hannibal et al. 2013; Hannibal et al. 2011) and plasma (Caterino et al. 2015) of *cblC* patients. Likewise, examining the proteome could shed light on the relationship between Cbl deficiency, the cognate functions of the CblC protein, and of seemingly unrelated metabolic pathways. The proteome of cells and plasma isolated from *cblC* patients share alterations in proteins involved in cellular detoxification, GSH metabolism, protein folding and cytoskeleton organization and assembly (Caterino et al. 2015; Hannibal et al. 2013; Hannibal et al. 2011). An important finding that emerged from these studies is the deregulation of the expression of proteins that are established pathological hallmarks of unrelated illnesses such as muscular dystrophy, Alzheimer's and Parkinson's diseases. Progressive cognitive decline, dementia, muscular dystonia and atrophy are common clinical manifestations of patients with *cblC* disease (Carrillo-Carrasco et al. 2012). Thus, it is reasonable to speculate that an absent or truncated CblC protein has effects beyond its catalytic reaction of processing dietary Cbls. Furthermore, the lack of functional CblC protein may impair the interacting partners of proteins downstream in the pathway of Cbl utilization, leading to widespread cellular stress. For example, the CblC protein interacts with MS in the cytosol (Fofou-Caillierez et al. 2013). It is presently unknown whether this interaction occurs solely for Cbl delivery or if it is serves additional functions in the cell.

B. CblD-mediated intracellular trafficking of B12

The molecular events that occur after processing of the β-axial ligand of Cbls by CblC but prior to delivery to MS and MCM are unknown. Mutations in the *cblD* (*MMADHC*) gene cause either combined or isolated homocystinuria and methylmalonic aciduria (Atkinson et al. 2014; Miousse et al. 2009; Stucki et al. 2012; Suormala et al. 2004) with heterogeneous clinical manifestations (Miousse et al. 2009). However, the *cblD* gene product CblD is not responsible for catalyzing decyanation or dealkylation of XCbl as cultured fibroblasts carrying mutations in this gene are able to process exogenously added radioactive XCbl normally (Suormala et al. 2004). CblD has been shown to direct the partition of newly processed Cbl between the cytosol and mitochondria (Gherasim et al. 2013a; Jusufi et al. 2014; Stucki et al. 2012; Suormala et al. 2004). Cultured fibroblasts from patients with *cblD* disease have an altered ratio of MeCbl/AdoCbl compared to normal individuals (Gherasim et al. 2013a; Suormala et al. 2004). Studies with purified CblD showed that the protein interacts with CblC, particularly under conditions that favor catalysis (Gherasim et al. 2013a). Thus, it is thought that the CblD protein functions as an adaptor that orchestrates the fate of newly made co(II)alamin for downstream delivery to MS and MCM.

1. Cloning of the cblD (MMADHC) gene

The *cblD* gene has been recently cloned and mapped to chromosome 2q23.2 (Coelho et al. 2008). The predicted gene product has sequence similarity with a bacterial ATP-binding cassette transporter, possesses a putative cobalamin binding motif (amino acid residues 81 to 86), and a putative mitochondrial targeting sequence (amino acid residues 1–12) (Coelho et al. 2008). The *cblD* gene transcribes into a single product of 891 base pairs, yielding a protein of 296 amino acid residues (32.8 kDa). A second initiation site has been identified at Met62, which would yield a cytosolic product. However, immunoprecipitation and western blot analyses identified the full-length protein as the only translated product (Coelho et al. 2008). Based on its predicted structure and mutational data from patients, it has been proposed that different mutations may determine distinct phenotypes of the disorder. For instance, mutations affecting the putative Cbl-binding site in the C-terminus of the protein, would lead to disrupted delivery of newly processed Cbls to both MS and MCM, resulting in combined homocystinuria and methylmalonic aciduria (a phenotype common to the *cblC* disorder). Mutations affecting the N-terminus of CblD have been identified in patients having isolated methylmalonic aciduria, consistent with disrupted delivery of Cbl to MCM. Impairments in the C-terminus of CblD lead to isolated homocystinuria. These sequence-based predictions have been partly confirmed experimentally (Deme et al. 2012; Jusufi et al. 2014; Plesa et al. 2011; Stucki et al. 2012).

2. Expression and characterization of the cblD gene product

Analysis of cell lysates prepared from HEK93 and MCH46 human cell lines demonstrated that CblD is expressed in full-length form, with a molecular mass equivalent to the protein lacking the mitochondrial leader sequence (CblD Δ1-12) (Deme et al. 2012; Stucki et al. 2012). Interestingly, another report showed that murine CblD is predominantly expressed as a truncated protein lacking a portion of its C-terminus (Koutmos et al. 2011). The human recombinant CblD protein was first cloned and expressed by Plesa et al. as a fusion with maltose binding protein (Plesa et al. 2011). Several other groups have reported CblD expression and isolation (Deme et al. 2012; Gherasim et al. 2013a; Jusufi et al. 2014). Human recombinant CblD is isolated as a monomeric polypeptide free of bound cofactors including Cbl (Deme et al. 2012; Gherasim et al. 2013a; Plesa et al. 2011), despite the presence of a putative Cbl binding site. Expression of CblD isoforms lacking the mitochondrial leader sequence (CblD Δ1-12) and the shorter product with predicted initiation site at Met62 (CblD Δ1-61) showed that both proteins are stable, monomeric and elongated according to circular dichroism studies. Prediction studies using algorithm PSI-PRED suggest that residues 26 to 109 (N-terminus) of the protein are highly unstructured and/or disordered. Algorithm DISOPRED identified three regions with substantial disorder, namely 28–48, 83–106 and 124–139 (Deme et al. 2012). Transfection studies performed on immortalized fibroblasts showed that synthesis of AdoCbl requires translocation of the CblD protein into mitochondria (residues 1-61) as well as an intact sequence downstream of the second presumptive starting site (Met62) (Stucki et al. 2012). Likewise, truncation of the first 115 amino acid residues yielded a protein that supports normal MeCbl synthesis (Stucki et al. 2012). Co-transfection studies with *cblC* and *cblD* suggested that the interaction between these proteins might regulate the intracellular flow into the MeCbl and AdoCbl synthesis pathways (Stucki et al. 2012). X-ray structural studies of a C-terminal fragment of cblD (cblD Δ1-108) sufficient for its interaction with cblC and to support the cytosolic Cbl trafficking pathway demonstrated that the protein is highly similar to CblC, which the authors referred to as 'molecular mimicry' (Yamada et al. 2015). CblD employs an α+β fold, reminiscent of the nitro-FMN reductase domains seen in CblC and the activation domain of MS (Yamada et al. 2015). These mammalian proteins share structural similarities with corrinoid-dependent reductive dehalogenases NpRdhA from *N. pacificus* (Payne et al. 2015), and PceA from *S. multivorans* (Bommer et al. 2014), and together they conform a new subclass within the nitroreductase superfamily (Yamada et al. 2015).

CblD possesses a putative Cbl binding domain, but it does not bind to Cbl. Similarly, the activation domain of MS does not bind Cbl, but instead, it binds AdoMet. While the equivalent of MS' AdoMet site is occupied by a loop (L7, Figure 5, panels a, b) in CblD, another cavity showing topological homology to the flavin binding site of CblC was identified (Yamada et al. 2015). In CblC, this flavin-binding site could only be observed in the crystal structure lacking Cbl (apo-CblC, PDB 3SBZ (Koutmos et al. 2011)). The authors proposed that the interaction of CblD with CblC and perhaps other biological partners may lead to reorganization of loop L7 (amino acid residues 226–245, Figure 5, panels a, b) in CblD to form a ligand-binding cavity (Yamada et al. 2015). This interesting proposal warrants further investigation. In the absence of interacting partners, CblD does not interact with Cbl, ATP, GTP, AdoMet, NADP+ or NADPH (Yamada et al. 2015). Importantly, the four known missense pathogenic mutations of the cblD disorder could be mapped in the protein structure. These are T182N, D246G, Y249C, L259P (Figure 5, panel c). Mutation D246G is located in loop 7. All of these mutations are located near a hairpin structure (hairpin β1′-β2′, Figure 5, panel a) proposed to be important for CblC-CblD recognition (Yamada et al. 2015).

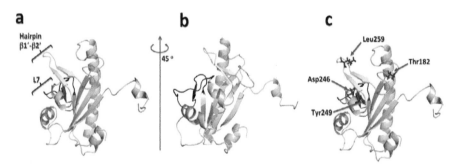

Figure 5. Structure of the C-terminal fragment (CblD Δ108, PDB 5CV0 (Yamada et al. 2015)) of human CblD. A deletion of the first 108 amino acid residues allowed crystallization of human CblD preserving its function for interactions with CblC. (a) Structure of human CblD D108 colored by secondary structure elements. CblD possesses several loops predicted to be flexible (grey). A Hairpin structure formed with sheets β1′-β2′ may be important for recognition with CblC. (b) Map of the four known missense mutations that cause cblD disorder in humans. Three of the mutations (D246G, Y249C and L259P) are located in or near hairpin β1′-β2′. Mutation T182N is located near sheet β2.

3. Interaction of CblD with CblC

An interaction between CblC and CblD has been described by several laboratories using *ex-vivo*, *in vitro* and *in silico* approaches (Deme et al. 2012; Gherasim et al. 2013a; Jusufi et al. 2014; Mah et al. 2013; Plesa et al. 2011; Stucki et al. 2012). Mah et al. demonstrated that CblC resides exclusively in the cytosol whereas CblD localizes both in cytosol and in mitochondria (Figure 6) (Mah et al. 2013). Analysis of MeCbl and AdoCbl synthesis in cultured fibroblasts from patients exhibiting each of the different phenotypes

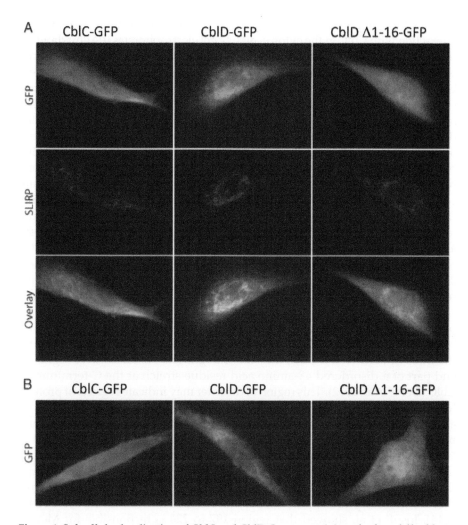

Figure 6. Subcellular localization of CblC and CblD. Immunostaining of cultured fibroblasts expressing green fluorescent protein fusions with CblC (CblC-GFP) or CblD (CblD-GFP). (A) Top panels show detection of the target proteins using an anti-GFP antibody. The middle row shows staining of mitochondria with marker anti-SLIRP antibody. An overlay of the top two rows shows that CblC resides in the cytosol whereas CblD localizes both to the cytosol and the mitochondrion. Deletion of the mitochondrial leading sequence of CblD disrupts its translocation to he organelle (third column). (B) Fibroblasts stained with anti-GFP antibody, without marking of the mitochondrion. Adapted from Mah et al. (Mah et al. 2013), with permission.

of the *cblD* disorder showed that complete absence of or presence of a truncated CblD protein leads to abnormal partition of the MeCbl and AdoCbl pathways (Gherasim et al. 2013a). Further, inability to synthesize one of the cofactors, leads to an overproduction of the other (Gherasim et al. 2013a). These data strongly suggest that CblD is essential for routing newly processed Cbl into the cofactor synthesis pathways, and that this function may be attained via its interaction with CblC in the cytosol. Native gel electrophoresis (PAGE) showed that CblC and CblD interact to form a 1:1 complex and that their interaction is strongest under conditions that favor dealkylation reactions, i.e., a Co-C bonded substrate and the presence of GSH (Gherasim et al. 2013a). *In vitro* studies with the recombinant proteins showed an interaction with relatively high affinity, and some specificity toward the β-axial ligand of the Cbl substrate. Bacterial two-hybrid systems and phage display revealed five distinct regions in CblD involved in its interaction with CblC, two near the center of the protein, comprising the homology domain of the ABC-transporter and three regions belonging in the C-terminus of CblD (Plesa et al. 2011). This finding concurs with another study where truncation variants of CblD suggest the existence of two domains that interact with either cytosolic or mitochondrial targets (Stucki et al. 2012). Another study employing a similar phage display approach identified four regions in CblD that would associate with CblC: region I (residues 34–57), region II (residues 221–238), region III (residues 236–250) and region IV (residues 261–280) (Deme et al. 2012). Importantly, regions I and II were mapped to the structure of CblC (PDB ID: 3SC0) and showed that binding in this region involves the core module and C-terminus of CblC, respectively, both of which undergo substantial mobility upon Cbl binding (Deme et al. 2012). While regions III and IV could not be mapped to the three dimensional model, they are thought to be extensions of region II and part of a disordered 48-amino acid residue stretch at the C-terminus of CblC (Deme et al. 2012). This region of disorder may indicate regions of protein interactions as observed in other proteins (Fong et al. 2009; Gunasekaran et al. 2003; Sugase et al. 2007). Overall, consensus exists that the N-terminal region of CblD is not important for its interaction with CblC (Deme et al. 2012; Gherasim et al. 2013a; Plesa et al. 2011; Stucki et al. 2012).

A study employing point mutagenesis of select amino acid residues within the different regions of CblD confirmed the presence of functional domains that are necessary to support MeCbl and AdoCbl biosynthesis (Jusufi et al. 2014). Jusufi et al. showed that region 197 to 226 is responsible for MeCbl synthesis, a truncation of more than 20 amino acids at the C-terminus leads to defects in AdoCbl synthesis, and region 259 to 266 is indispensable for both routes (Jusufi et al. 2014). Recreating these mutations in cultured fibroblast mimicked the three different phenotypes found in patients with the cblD disorder (Jusufi et al. 2014).

The CblD proteins harboring the four known pathogenic mutations in humans, namely T182N, D246G, Y249C, L259P have been expressed and their activity examined *in vitro* (Yamada et al. 2015). Binding of wild type CblD to CblC-cob(II)alamin accelerates the oxidation of cob(II)alamin to cob(III)alamin (Yamada et al. 2015), which is thought to facilitate release of Cbl from CblC. CblC has a lower binding affinity for cob(III)alamin compared to cob(II)alamin (Yamada et al. 2015). In contrast, all CblD pathogenic mutations were found to retard cob(II)alamin oxidation (Yamada et al. 2015). One proposed hypothesis is that wild type CblD weakens Cbl binding to CblC thus facilitating Cbl relay to acceptor proteins in the trafficking pathways. Mutations in the CblD gene impair this process by slowing the release of Cbl from CblC. In this scenario, CblC presumably operates as a transient trap for Cbl (Yamada et al. 2015).

In sum, much progress has been made in elucidating the properties of CblC and CblD and their association, both *in vitro* and *ex vivo*. Nonetheless, the molecular mechanism or signal that defines shifting Cbl trafficking into the cytosolic and mitochondrial compartments remains unresolved.

3. Intracellular Utilization of Cobalamins

Synthesis of MeCbl is performed by cytosolic methionine synthase (MS; complementation group *cblG*) (Gulati et al. 1996; Leclerc et al. 1998; Sillaots et al. 1992), and its intramolecular partner methionine synthase reductase (MSR; complementation group *cblE*) (Gulati et al. 1997; Leclerc et al. 1998; Watkins and Rosenblatt 1989). During the catalytic cycle of MS, the Cbl cofactor cycles between cob(I)alamin and MeCbl. The supernucleophile cob(I)alamin undergoes occasional oxidation (once for every 200 to 2,000 turns of the catalytic cycle) to an inactive cob(II)alamin-bound form of the enzyme, which is repaired by reductive methylation mediated by MSR and S-adenosylmethionine (SAM) (Banerjee and Matthews 1990; Matthews 2001). MSR, a dual flavoprotein oxidoreductase, reduces the cobalt center using NADPH as the source of electrons (Olteanu and Banerjee 2001).

The first step in the synthesis of AdoCbl is the *in situ* reduction of cob(II) alamin to cob(I)alamin by ATP-dependent cob(I)alamin adenosyltransferase (CblB; complementation group *cblB*) (Leal et al. 2003). A second protein from the *MMAA* gene (for methylmalonic aciduria, type A; complementation group *cblA*; hereafter referred to as the CblA protein) (Dobson et al. 2002a) protects methylmalonyl-CoA mutase (MCM; complementation group mut) from oxidative inactivation in the presence of nucleotides and increases its k_{cat} by 2-fold (Padovani et al. 2006). CblA forms a stable complex with MCM, and also interacts with ATR. Although the synthesis of AdoCbl is limited in *cblA* mutant cell lines (Lerner-Ellis et al. 2004), AdoCbl biosynthesis in extracts from these cells is comparable to that in control cell lines (Mahoney et al. 1975). It has therefore been speculated that CblA could also interact with the mutase and influence AdoCbl synthesis in the mitochondria (Banerjee 2006). The functions of the two Cbl-dependent enzymes and their chaperones are discussed in the sections below.

A. Methionine synthase

Methionine synthase is a tetramodular 140-kD enzyme that catalyzes the remethylation of homocysteine to yield the amino acid methionine, using N^5-methyltetrahydrofolate (CH$_3$-THF) as the methyl donor. This reaction regenerates tetrahydrofolate, an essential precursor for the biosynthesis of purines and pyrimidines, and involves the redox cycle of cob(I)alamin and MeCbl (formerly cob(III)alamin). MS is the only enzyme known to generate tetrahydrofolate from CH_3-THF. Occasional oxidation of cob(I)alamin to catalytically inert cob(II)alamin occurs every 200 to 2000 turns of the catalytic cycle (Drummond et al. 1993; Wolthers and Scrutton 2007). Inactive MS is repaired by electrons provided by MSR, a dual flavoenzyme that interacts with the C-terminal domain of MS to convert inert cob(III)alamin to the corresponding one- and two-electron reduced species. MS is one of four enzymes in the ubiquitous methionine cycle. The methionine formed by the remethylation of homocysteine is converted to SAM by methionine:ATP adenosyltransferase. SAM is used as a methyl donor substrate by over 200 different methyltransferase enzymes, catalyzing the methylation of DNA, RNA, proteins, lipids and numerous small molecules. S-adenosylhomocysteine (SAH) is then hydrolyzed to homocysteine and adenosine by SAH hydrolase to complete the methionine cycle.

1. Cloning of the methionine synthase (CblG) gene

Human MS possesses four discrete domains each binding homocysteine, CH_3-THF, Cbl and SAM. The identification and cloning of the MS gene was performed by three different research groups almost simultaneously (Chen et al. 1997; Leclerc et al. 1996; Li et al. 1996). The MS gene was localized to chromosome 1q43 and the open reading frame comprises 3798 nucleotides (Chen et al. 1997; Leclerc et al. 1996; Li et al. 1996). This yields a protein product of 1265 amino acids, with a predicted molecular mass of 140 kDa. Human MS shares 53 and 63% sequence identity with *E. coli* and *C. elegans* MS proteins, respectively (Chen et al. 1997). A common polymorphism was identified in human samples that results in a single-amino acid replacement (D919G) in the protein (Chen et al. 1997). Human MS mRNA was found in high levels in the pancreas, heart, skeletal muscle and kidney (Chen et al. 1997). Since the identification of the MS gene, several mutations have been mapped all belonging in the *cblG* complementation group, leading to homocystinuria (Fofou-Caillierez et al. 2013; Gulati et al. 1996; Harding et al. 1997; Morita et al. 1999; Watkins et al. 2002; Wilson et al. 1998).

2. Expression and characterization of methionine synthase

Full-length mammalian MS was first isolated from pig liver by Loughlin and colleagues (Loughlin et al. 1964) and later by Banerjee and colleagues (Banerjee et al. 1997; Chen et al. 1994). Human MS was first isolated from placenta by Kolhouse and colleagues (Utley et al. 1985). The native apo-protein could be reconstituted *in vitro* with various Cbls, only after reduction to cob(II)alamin (Kolhouse et al. 1991). Most Cbl analogues supported the reaction of MS after their reduction to cob(II)alamin (Kolhouse et al. 1991). This was critical evidence that MS exhibited specificity for the redox state and the coordination sphere of the incoming Cbl. The expression of recombinant human MS was a challenge for many years. Human MS is prone to proteolysis using conventional bacterial expression systems (Wolthers and Scrutton 2007). Scrutton and colleagues adopted the *Pichia pastoris* expression system with success (Wolthers and Scrutton 2009), with specific activity and yields comparable to the expression of MS in a baculovirus system (Yamada et al. 2006). In both systems, MS is recovered as an apo-protein that can be readily reconstituted with HOCbl, and fully activated in the presence of HOCbl, NADPH and MSR (Drummond et al. 1993; Wolthers and Scrutton 2009). The expression and crystallographic analysis of the activation domain of MS (Bandarian et al. 2002; Drennan et al. 1994) and MSR (Wolthers et al. 2007a; Wolthers et al. 2007b) has permitted a deeper understanding of structure-function relationships. The activation domain of MS has two modules that bind homocysteine and CH_3-THF, a third module that binds Cbl and a C-terminal module that binds SAM (Drummond et al. 1993). The activation domain is located in the C-terminal region of the gene and is separated from the rest of the protein by a linker of 38 amino acid residues (Wolthers et al. 2007b). Because MS switches between the catalytic and reactivation cycles, the conformational changes associated with this dual function have been investigated. Domain alternation has been proposed to modulate the function of MetH, the *E. coli* Cbl-dependent ortholog of MS, in particular, to facilitate the activation conformation (Bandarian et al. 2002), that is, the base-off configuration of Cbl and appropriate proximity to the SAM binding site to enable the methyl transfer step that regenerates MeCbl. A comparison of the two states of MS showed that reactivation involves a significant movement of the four-helix domain that caps Cbl in the resting and catalytic states (Figure 7). The structures showed that amino acid residues that connect the Cbl and cap domains enable the cap to adopt a new position, suitable for the reactivation cycle (Bandarian et al. 2002).

3. Mechanism of methionine synthase enzyme activity

Methionine synthase catalyzes the conversion of homocysteine to methionine using CH_3-THF as the methyl donor (Figure 8). Cobalamin mediates this methyl transfer reaction cycling between oxidation states cob(III)alamin and cob(I)alamin. Cob(I)alamin attacks the methyl group of the tertiary amine CH_3-THF, forming MeCbl and regenerating THF. Cob(I)alamin is easily oxidized by O_2 and reactive oxygen species to give the catalytically inert cob(II)alamin. Repair of cob(II)alamin to cob(I)alamin is catalyzed by the reductase domain of MS and is known as the reactivation cycle (Wolthers and Scrutton 2009) (Figure 8). Newly processed Cbl is presumably delivered as cob(II)alamin via a mechanisms that may involve a direct interaction with processing enzyme CblC (Fofou-Caillierez et al. 2013). Upon insertion into apo-MS, inert cob(II) alamin is reduced to cob(I)alamin by the reductase domain that transfers electrons from NADPH to the FAD and FMN subdomains and ultimately

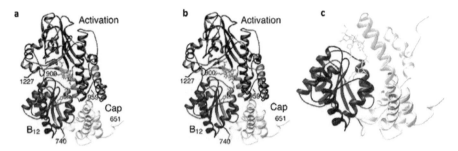

Figure 7. Structure of the activation domain of MS. The C-terminal fragment of methionine synthase (amino acid residues 651-1227) is shown in stereo ribbon representation. (a) The four-helix bundle that caps MeCbl in the resting state is depicted in yellow. The C-terminal activation domain (blue) contains the corrin ring and interacts with the Cbl domain (red). MeCbl is shown as gold sticks. (b) Position and movement of the cap domain that accompanies the interaction of the Cbl domain with the SAM activation domain. The position of the cap domain in the isolated fragment (649–896), corresponding to the resting state, is shown as silver ribbons. Transition to the activation conformation results in dramatic displacement (26.2 Å, and a 62.7 degree rotation) of this domain (compare silver versus yellow). Figures reproduced from Bandarian et al. (Bandarian et al. 2002), with permission.

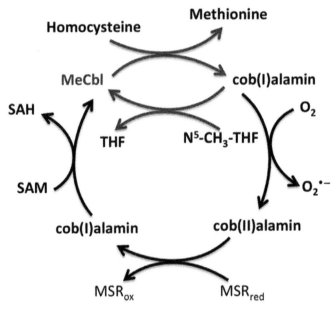

Figure 8. Methionine synthesis and reactivation of cob(II)alamin by MSR. MS catalyzes the methylation of homocysteine to form methionine, using N^5-CH$_3$-THF as a co-substrate for methyl transfer. This reaction regenerates THF, an essential precursor for nucleic acid biosynthesis. Every 200-2000 catalytic cycles, cob(I)alamin is oxidized to inert cob(II)alamin. MSR$_{red}$ reduces cob(II) alamin to cob(I)alamin, which abstracts a methyl group from S-adenosylmethionine (SAM), to re-enter the cycle as MeCbl. This reactivation cycle generates S-adenosylhomocysteine (SAH) and oxidized MSR (MSR$_{ox}$). The catalytic and reactivation cycles are depicted in grey and black, respectively. This figure was adapted from Bandarian et al. (Bandarian et al. 2002), with permission.

to Cbl in the activation domain (Matthews et al. 2008). Abstraction of the methyl group from CH_3-THF is driven by the highly nucleophilic cob(I) alamin, to form methionine and tetrahydrofolate, with transient formation of MeCbl (Matthews et al. 2008). The nature of the redox cycle demands that MS stabilizes six-coordinate MeCbl and a four-coordinate cob(I)alamin oxidation states. MS binds MeCbl in the base-off/His-on configuration, with the protein nitrogen ligand provided by amino acid residue His759. Histidine 759 forms H-bonds with Asp757 and Ser819, and this triad has been shown to be essential for stabilizing cob(I)alamin (Jarrett et al. 1996). This network of H-bonds supports heterolytic methyl transfer to and from substrates and inhibits unwanted hemolytic loss of the methyl group by photolysis (Jarrett et al. 1996). The sequence Asp-X-His-X-X-Gly which supports the base-off/His-on configuration upon binding of MeCbl to MS is also conserved in MCM and B12-dependent glutamate mutase, both of which employ this binding mode to facilitate catalysis (Drennan et al. 1994).

B. Methionine synthase reductase-mediated repair of inactivated methionine synthase

1. Cloning of the methionine synthase reductase (cblE) gene

Methionine synthase reductase is one of 77 flavoproteins present in the human proteome, of which 50 are associated with diseases (Lienhart et al. 2013). Mutations in the MSR gene (*cblE*) lead to a deficiency in methionine biosynthesis due to failure to support cob(I)alamin formation for subsequent methyl group abstraction from CH_3-THF. Along with nitric oxide synthase reductase and cytochrome P450 reductase, MSR is a dual flavoenzyme, employing both FAD and FMN as intramolecular electron carriers during catalysis (Haque et al. 2014). The MSR gene is localized to chromosome 5p15.2–15.3 (Leclerc et al. 1999; Leclerc et al. 1998). The MSR gene codes for 698 amino acids with a predicted molecular weight of 78 kDa. Nonetheless, analysis of the gene revealed the presence of two major transcription start sites that by alternative splicing would yield the translation of products of 698 and 725 amino acid residues (Froese et al. 2008). In either case, subcellular imaging studies showed that MSR is restricted to the cytosol, likely due to the lack of a mitochondrial leader sequence (Froese et al. 2008). The human MSR domain shares 38% identity with human cytochrome P450 reductase, and 43% identity with the putative MSR of *C. elegans* (Leclerc et al. 1998). MSR has two critical functions in methionine synthesis: (a) It stimulates Cbl uptake at the Cbl-binding site of MS and (b) it reduces cob(III)alamin and cob(II)alamin to cob(I)alamin so it can re-enter the catalytic cycle (Wolthers and Scrutton 2009). Fluorescence quenching of bound flavins showed that the role of the MSR domain involves specific protein-protein interactions with the MS Cbl-binding domain, which could not be reproduced by the reductase domain of the related dual flavoenzyme cytochrome P450 reductase (CPR)

(Wolthers and Scrutton 2009). Crystallization of the activation domain of human MS suggested formation of a putative dimer structure that could be of physiological relevance (Wolthers et al. 2007b).

2. Expression and characterization of methionine synthase reductase (MSR)

Isolated MSR is a monomeric protein of 78 kDa containing 1 molar equivalent each of FAD and FMN. In the presence of NADPH, the protein is self-sufficient in converting cob(III)alamin and cob(II)alamin into cob(I)alamin (Olteanu and Banerjee 2001). Human MSR contains two flavin-binding subdomains, termed NADPH/FAD–binding domain, which is the equivalent of bacterial ferredoxin-oxidoreductase (FNR) and a FMN-binding domain, related to the flavodoxins of prokaryotes. *In vitro*, MSR reduces non-native substrates such as cytochrome c and FeCN, (Olteanu and Banerjee 2001; Olteanu et al. 2002) a feature also observed by CPR and NOS reductase, and cob(II)alamin/cob(III) alamins (Wolthers and Scrutton 2009). *In vivo*, the flavins transfer electrons from the oxidation of NADPH to the Co-center following the sequence NADPH →FAD →FMN→cob(II)alamin (Olteanu and Banerjee 2001; Wolthers and Scrutton 2004). Studies by Wolthers and Scrutton showed that the FMN-domain of MSR interacts directly with the Cbl-containing activation domain of MS (Wolthers and Scrutton 2007). This interaction is weakened by high ionic strength, suggesting that electrostatic interactions are critical for MSR-MS interaction (Wolthers and Scrutton 2007). Cross-linking studies showed that the MSR domain exhibits mainly acidic residues in its interacting surface while the MS-activation domain presents basic surface residues at the predict protein-protein interface (Wolthers and Scrutton 2007). ITC studies showed that full length MSR binds to MS in a 2:1 stoichiometry, whereas the isolated FMN domain forms a 1:1 complex, suggesting that only 50% of the MSR forms a complex with MS (Wolthers and Scrutton 2007). One possibility is that MSR exists in two different conformations, where in one 'open' conformation, the FMN subdomain is able to interact with MS, whereas in the 'close' conformation, the FMN domain is closer to the NADPH-FAD binding domain (Wolthers and Scrutton 2007). This proposal is supported by crystallographic evidence of a flexible linker that would enable 'swing' between the two states of the FMN domain (Wolthers et al. 2007b).

3. Mechanism of methionine synthase reductase enzyme activity

The Cbl reductase activity of MSR involves electron transfer from the FMN (either FMN semiquinone or FMN hydroquinone) to the cobalt center of the Cbl residing the activation domain (Wolthers and Scrutton 2004). Reduction of cob(II)alamin by MSR coupled with methyl transfer from SAM restores the MeCbl-bound active state of MS. Compared to members of the same class, like cytochrome P450 reductase, inducible and neuronal NOS reductases, MSR is a rather sluggish reducing system, and thus, likely the rate-limiting step for methionine biosynthesis (Haque et al. 2014; Wolthers and Scrutton 2009). While the related dual flavoenzymes nitric oxide synthase reductase and cytochrome P450 reductase catalyze the reduction of Cbl *in vitro*, none of them support synthesis of MeCbl by MS (Wolthers and Scrutton 2009). This provided evidence that the effect of MSR on MeCbl biosynthesis extends beyond the mere reduction of Cbl (Wolthers and Scrutton 2009). In particular, addition of the isolated FNR subdomain of MSR to the activation domain of MS *in vitro* had the same stabilizing effect as adding MS (Wolthers and Scrutton 2009). Expression of the isolated subdomains of MSR showed that they conserve their thermodynamic properties, which suggests that FMN electronics is essential for methionine synthesis (Wolthers et al. 2003). Thus, in addition to reducing Cbl, MSR functions as a chaperone that stabilizes the activation domain of MS (Wolthers and Scrutton 2009).

C. Methylmalonyl-CoA mutase

1. Cloning of the mut gene

The conversion of propionyl-CoA to the Krebs cycle intermediate succinyl-CoA involves three enzymatic reactions: (1) carboxylation of propionyl-CoA to (2S)-methylmalonyl-CoA, (2) isomerization of (2S)-methylmalonyl-CoA to (2R)-methylmalonyl-CoA, and (3) rearrangement of (2R)-methylmalonyl-CoA to succinyl-CoA (Banerjee 1999). In mammals, the substrate propionyl-CoA is produced by catabolism of valine, isoleucine, methionine and threonine and also by the degradation of thymine, cholesterol and odd-chain fatty acids (Banerjee 1999). The product, succinyl-CoA is an important intermediate of the Krebs cycle and a precursor of heme biosynthesis (Hunter and Ferreira 2009). The human mut gene is located in chromosome 6p12–21.2 (Ledley et al. 1988a; Ledley et al. 1988b). It is subject to over 250 mutations that lead to methylmalonic aciduria (Froese and Gravel 2010). The open reading frame of human MCM comprises 750 amino acid residues, of which the initial 32 conform a mitochondrial leader sequence, targeting the enzyme to the mitochondrion (Andrews et al. 1993).

2. Expression and characterization of the mut gene product

Human MCM has been isolated as a homodimer of 150 kDa (Fenton et al. 1982; Kolhouse et al. 1980). Human recombinant MCM was first expressed in *S. cerevisiae*, and its enzymatic properties were undistinguishable from those of the native enzyme isolated from mammalian tissues (Andrews et al. 1993). MCM requires AdoCbl for its catalytic activity, and the assistance of two other proteins, CblA and CblB. The crystal structure of human MCM shows that the protein features two domains, namely a large TIM barrel located at the N-terminus binding to substrate methylmalonyl-CoA and a small AdoCbl-binding domain toward the C-terminus (Figure 9). These two domains are

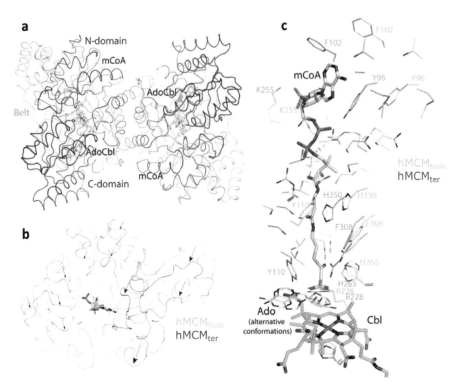

Figure 9. Structure of MCM and the two observed spatial configurations of AdoCbl in the presence of the substrate, MM-CoA. (a) Cartoon representation of homodimeric human MCM, highlighting N- and C-terminal domains (blue and magenta, respectively), the connecting linker (yellow) and bound AdoCbl located at the interface of N- and C-domains. (b) An overlay of the structures of holo-MCM and holo-MCM bound to substrate MM-CoA. The arrows show changes in secondary structure elements upon binding of MM-CoA. (c) Residues lining the substrate binding channel in holo-MCM (cyan) and holo-MCM bound to MM-CoA (pink). Substrate analogue malonyl-CoA (mCoA) and AdoCbl showing its 5'-adenosyl moiety in two conformations are shown in sticks. Adapted from Froese et al. (Froese et al. 2010b) with permission.

connected by a linker of 100 amino acid residues, and the active site forms at the N/C-domain interface (Froese et al. 2010b). A comparison of apo- and holo-MCM revealed significant conformational changes induced upon binding of Cbl (Froese et al. 2010b). Binding of AdoCbl features a base-off/His-on configuration, as seen in MS, and provokes substantial outward displacement of helical structures at the C-terminus of the protein whereas the N-terminus remains essentially unchanged (Froese et al. 2010b). These changes generate an induced-fit binding pocket that support the base-off/His-on AdoCbl binding mode, with the formation of new bonding interactions with His627, Leu674 and Ala675 and the burying of the dimethylbenzimidazole moiety in a glycine-rich hydrophobic cavity. In addition, exposure of the binding cleft to the solvent is minimized by a reorganization of flexible linker region comprising amino acids

500 to 505. The substrate binding channel maintains its solvent accessibility via a long crevice (Froese et al. 2010b). Analysis of the active in the structures of apo-MCM, holo-MCM and holo-MCM bound to methylmalonyl-CoA revealed that substrate binding influences the position of AdoCbl (Figure 9), especially of the 5'-adenosyl moiety which can adopt at least two different orientations, such to prime Co-Co homolysis (Froese et al. 2010b).

3. Mechanism of methylmalonyl-CoA mutase enzyme activity

Methylmalonyl-CoA mutase is an AdoCbl-dependent isomerase that catalyzes a carbon skeleton rearrangement reaction to convert substrate methylmalonyl-CoA into product succinyl-CoA (Banerjee 2003). The proposed first step of this reaction is the homolysis of the Co-C bond of AdoCbl to generate the radical species required for carbon atom rearrangement (Chowdhury and Banerjee 2000). In the case of MCM and related enzymes such as B12-dependent glutamate mutase, Co-C bond cleavage is coupled to formation of the next radical in the reaction pathway (Figure 10), which represents a substrate-centered radical (Padmakumar et al. 1997). Homolysis of the Co-C bond of AdoCbl is a thermodynamically uphill process. The rate for Co-C bond dissociation in base-on AdoCbl for the uncatalyzed reaction is 3.8 x 10^{-9} s^{-1}, whereas in the various AdoCbl-dependent enzymes the k_{cat} for the

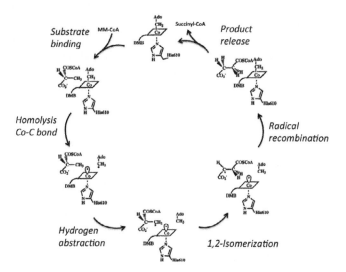

Figure 10. Mechanism of MCM catalysis. Binding of the substrate methylmalonyl-CoA (MM-CoA) produces a Michaelis-Menten enzyme-substrate complex. Substrate binding produces conformational changes at the active site that favor Co-C bond homolysis in AdoCbl, to generate the radical species required for presumably coupled hydrogen abstraction in the following step. Isomerization and radical recombination lead to product formation, succinyl-CoA, and the regeneration of AdoCbl. This figure was adapted from Chowdhury and Banerjee (Chowdhury and Banerjee 2000), with permission.

reaction is approximately $10^2 \, s^{-1}$ (Chowdhury and Banerjee 2000). The great extent of labilization induced by the isomerases has been mainly ascribed to destabilization and distorsion of the corrin ring (Brown and Brooks 1991; Kraeutler et al. 1994; Marzilli et al. 1979) and electronic trans effects (Geno and Halpern 1987; Ng and Rempel 1982). A comparison of the Co-C bond stretching frequencies by resonance Raman suggests major labilization occurs upon substrate binding. A study by Chowdhury and Banerjee concluded that substrate binding is an exothermic process, and that enthalpic and entropic factors contribute to lowering the activation energy for the hemolytic rupture of the Co-C bond in MCM (Banerjee 1999; Chowdhury and Banerjee 2000).

While the synthesis of MeCbl is self-reliant on methionine synthase and its dedicated reductase MSR, the biosynthesis of AdoCbl requires MCM and the assistance of two chaperones: cob(I)alamin adenosyltransferase *(MMAB for methylmalonic aciduria, type B; hereafter referred to as CblB)* and the GTPase protein MMAA, or CblA. Mutations in the MMAB and MMAA genes lead to methylmalonic aciduria due to failures in Cbl insertion into MCM or reduction to complete the redox cycle, as will be described in the following two sections.

D. Biosynthesis of adenosylcobalamin

1. Cloning of the cblB (MMAB) gene

The cblB *(MMAB)* locus encodes the enzyme ATP-dependent cob(I)alamin adenosyltransferase (CblB) and has been mapped to human chromosome 12 and comprises 1128 nucleotides (Leal et al. 2003). Human CblB shares 88% and 26% identity with bovine MMAB and the ATR PduO of *S. enterica*, respectively (Leal et al. 2003). A N-terminal stretch of 50–60 amino acids that is only present in the sequence of eukaryotic CblB harbors a mitochondrial leader sequence (Leal et al. 2003).

2. Expression of adenosyltransferase, the MMAB gene product

Human *MMAB* has been expressed in *E. coli* with and without the mitochondrial leader sequence (Leal et al. 2003). The CblB products exhibited the expected molecular masses, of 58 and 55 kDa, respectively (Leal et al. 2003). Enzymatic activity was detected in lysates of the soluble and insoluble fractions, being higher in the soluble fraction. Activity could only be observed in the presence of both ATP and Cbl (Leal et al. 2003). CblB carrying the mitochondrial leader sequence displayed lower activity than the product without it, both in bovine and human variants (Leal et al. 2003). Expression of the human CblB in a *S. enterica* strain deficient in ATP-dependent cob(I)alamin adenosyltransferase complemented its phenotype, demonstrating that the human enzyme was active, even in a foreign host (Leal et al. 2003). Human fibroblasts with the cblB disorder showed either absence of or the presence of smaller peptides (23–25 kDa) reactive toward the CblB antibody (Leal et al. 2003). The crystal structure of an archaeal CblB has been elucidated, namely, from *Thermoplasma acidophilum* (Saridakis et al. 2004). The protein is trimeric, both in solution and in crystalline form (Saridakis et al. 2004). This CblB has only 32% identity with its human counterpart, yet, 22 amino acid residues thought to be essential for catalytic activity are conserved in both species (Saridakis et al. 2004). These includes residues Arg-119, Arg-124, and Glu-126 in archaeal CblB corresponding to amino acid residues Arg-186, Arg-191, and Glu-193 in the human CblB, (Saridakis et al. 2004), which are mutated residues in *cblB* patients (Dobson et al. 2002b). The crystal structure of human CblB was elucidated by Schubert and Hill (Schubert and Hill 2006). The study revealed that 20 residues at the N-terminus of CblB become ordered upon binding to ATP, generating an unprecedented form of ATP-binding site that extends through a cleft likely to bind Cbl (Schubert and Hill 2006). This cleft has the appropriate dimensions to host base-off Cbl, which appears to be the binding mode of Cbl-CblB according to UV-visible and EPR studies (Stich et al. 2005; Yamanishi et al. 2005). These twenty conserved amino acid residues provide bonding contacts to support ATP binding as well as H-bonds for structural stability and binding to Cbl (Schubert and Hill 2006). Human CblB is trimeric, as seen in the archaeal counterpart and both these proteins share structural similarity with members of the ferritin family (Dobson et al. 2002b; Schubert and Hill 2006). While binding of ATP to CblB does not affect Cbl binding, the opposite is true. Binding of Cbl to CblB increases the affinity of the enzyme for ATP. Two critical residues, Arg190 and Arg186 are critical for the binding affinity of CblB for Cbl, and explained the phenotype of cblB patients carrying R190H and R186W mutations (Zhang et al. 2009).

3. Mechanism of cblB enzyme activity

The enzymatic activity of CblB involves a tetra-coordinated cob(II)alamin intermediate (St Maurice et al. 2008) and a rotary mechanism where the only vacant binding site in one of the monomers triggers AdoCbl ejection to client MCM from one of the two occupied active sites upon ATP binding (Padovani and Banerjee 2009b). Formation of a Co-C with the 5'-adenosyl moiety requires formation of the supernucleophile cob(I)alamin. This implies that cob(II) alamin must undergo reduction to cob(I)alamin in the CblB active site, and this reduction step is facilitated by formation of a four-coordinate cob(II) alamin intermediate (St Maurice et al. 2008). Indeed, this intermediate has been captured in the crystal structure of the CblB -like protein of *Lactobacillus reuteri*, along with the product AdoCbl partially occupied in the active site (St Maurice et al. 2008). Importantly, this base-off cob(II)alamin species only forms in the presence of bound ATP (Stich et al. 2004; Stich et al. 2005; Yamanishi et al. 2005). Structural analysis revealed that a 'close' conformation of the active site displaces the dimethyl-benzimidazole ligand destabilizing cob(II)alamin in its ground state thereby favoring a configuration analogous to cob(I)alamin, suitable for AdoCbl synthesis (St Maurice et al. 2008). The generation of a four-coordinate cob(II)alamin species makes it reduction to cob(I)alamin energetically favorable. Thus, similarly to the Circe effect, (Jencks 1975) where the favorable enthalpy associated with substrate binding pays for ground-state destabilization, binding of ATP to CblB favors the generation of an unstable four-coordinate cob(II)alamin intermediate. In this system, cob(I) alamin is generated in the absence of a dedicated reductase, which implies the occurrence of intramolecular reduction or disproportionation of cob(II) alamin (Yamada et al. 1968).

E. Regulation of MCM activity by the cblA gene product

1. Cloning of the cblA (MMAA) gene

The human gene that leads to the *cblA* disorder was first identified by Dobson and colleagues (Dobson et al. 2002b). Orthologous genes of *MMAA* are widespread from bacteria to eukaryotes. Mutations in the *MMAA* lead to methylmalonic aciduria, despite the presence of a normal gene for MCM. It has been documented that certain Cbl-dependent enzymes such as glycerol and diol dehydratases undergo inactivation by the substrate during catalysis brought about by irreversible breakage of the Co-C bond in AdoCbl to form 5'-deoxyadenosine and an alkylcobalamin-like species (Kajiura et al. 2001; Seifert et al. 2001; Toraya 2000). The modified Cbl remains bound to MCM without exchange of free intact cofactor until reactivation by a chaperone-like protein catalyzes the replacement with intact AdoCbl via an ATP-dependent reaction (Kajiura et al. 2001). In the case of MCM, this role is fulfilled by the MMAA chaperone, a.k.a. MeaB (Korotkova and Lidstrom 2004). In fact, mutations in MeaB lead to inactivation of MCM due to lack of AdoCbl (Korotkova and Lidstrom 2004).

2. Expression and characterization of CblA

Recombinant MeaB, a bacterial CblA ortholog, from *M. extorquens* has been expressed and purified in *E. coli* (Korotkova and Lidstrom 2004). The protein has a molecular mass of 35 kDa and size exclusion chromatography indicated both monomeric (Korotkova and Lidstrom 2004) and dimeric (Padovani et al. 2006) states in solution. MeaB hydrolyses GTP to produce GDP in the presence of $MgCl_2$, but does not hydrolyze ATP *in vitro* (Korotkova and Lidstrom 2004). Crystallographic analysis revealed that MeaB is homodimeric, with each subunit containing a α/β-core G domain typically observed in other members of the GTPase family (Figure 11) (Hubbard et al. 2007). MeaB possesses N- and C-terminal extensions not seen in other GTPases, which involved in protein-protein interactions and dimerization, respectively (Hubbard et al. 2007). Switch regions (loops, denominated switch I, II, and III), required for signal transduction upon GTP hydrolysis are located at the dimer interface rather than at the surface of the protein, suggesting that the interaction of MeaB with

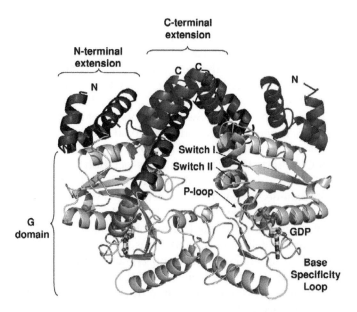

Figure 11. Crystal structure of MeaB, the bacterial ortholog of CblA. MeaB is homodimeric and presents distinctive N- and C-terminal extensions that enable protein-protein interactions and dimerization independent of GTP, respectively. Signal transduction switches I-III are located at the dimer interface, which suggest large molecular motions must occur to permit recognition of client MCM. The G-domain of MeaB resembles those observed in other members of the GTPase family. Reproduced from Hubbard et al. (Hubbard et al. 2007), with permission.

MCM must involve substantial conformational changes in MeaB (Figure 11) (Hubbard et al. 2007). Expression of variants of MeaB with mutations observed in patients with the *cblA* disorder led to the conclusion that these amino acid replacements predominantly affect protein stability, leading to failed folding (manifested as inclusion bodies), rupture of helical structures and the creation of a non-native disulfide bond (Hubbard et al. 2007).

3. Interaction of CblA with MCM

The first evidence of an interaction between CblA and MCM was perhaps provided by the observation that expression of *M. extorquens* MCM alone was not successful, but instead, co-expression of MeaB and MCM lead to the recovery of active MCM (Korotkova and Lidstrom 2004). Native PAGE confirmed the formation of a complex between MeaB and MCM, which was stabilized by the presence of AdoCbl or CNCbl, GTP and GDP, but not by ATP (Korotkova and Lidstrom 2004). Analysis of MCM activity *in vitro* showed that MeaB activates MCM almost 4-fold compared to control reactions, but this activation did not involve exchange of Cbl as it had been described for other Cbl-dependent enzymes (Korotkova and Lidstrom 2004). Studies by Padovani et al. showed that the two proteins interact with a dissociation constant of ranging from 34 to 524 nM, and that presence of the non-hydrolyzable GTP analogue GMP-PNP strengthened the interaction approximately 15-fold (Padovani et al. 2006). While the intrinsic GTPse activity of MeaB is low (0.04 min^{-1}), presence of MCM stimulates its activity by 100-fold. A later study revealed that the intrinsic GTPse activity of MeaB is kept low via an autoinhibitory loop denominated Switch II (Lofgren et al. 2013a). Analysis of binding energetics suggested large conformational rearrangements, involving an interacting surface of 4000–8600 Å^2 (Padovani et al. 2006). In the absence of MeaB, AdoCbl binds to MCM with a dissociation constant of 404 nM. Addition of MeaB increases the affinity of MCM for AdoCbl 2-fold (210 nM) (Padovani and Banerjee 2006). In the presence of GMP-PNP, no heat release could be detected for binding of AdoCbl to MCM in isothermal calorimetry measurements, suggesting that hydrolysis of GTP is a prerequisite for Cbl binding (Padovani and Banerjee 2006). Thus, *in vitro* studies demonstrate that MeaB has dual functions: (a) it supports the assembly of holo-MCM, a process that depends on GTP hydrolysis, and (b) it protects the mutase from oxidative inactivation during catalysis in a nucleotide-independent fashion (Padovani and Banerjee 2006).

The energy of GTP hydrolysis is utilized to release inactive Cbl from the active site of MCM during catalysis, which helped explain the inability of cblA patients to perform AdoCbl synthesis (Padovani and Banerjee 2009a). The crystal structure of MeaB in its apo and GMP-PNP-bound forms revealed that the third of its three mobile loops (Switches I to III), named 'Switch III', is essential for GTP-responsive communication between the active sites of MeaB and MCM (Lofgren et al. 2013b). Patients with a disrupted switch III fail to support AdoCbl biosynthesis, which leads to methylmalonic aciduria (Lofgren et al. 2013b). Additional crystallographic studies showed that the active site of MeaB is incomplete in the absence of client MCM, impairing the formation of the transition state in GTP hydrolysis (Lofgren et al. 2013a). Further, Switch II ensures that in the absence of MCM, MeaB remains in an inactive conformation, suppressing wasteful GTPase activity (Lofgren et al. 2013a).

The crystal structure of human CblA and MCM has been solved by Froese and colleagues (Froese et al. 2010b). Size exclusion chromatography showed the formation of high-molecular weight oligomers that could correspond to 2 CblA dimers: 1 MCM dimer (MW 352 kDa) and their corresponding 2 (observed: 740 kDa, calculated: 704 kDa) and 4 units (observed: 1410 kDa, calculated: 1408 kDa) of this core assembly (Froese et al. 2010b). While the function of the higher order species remains to be investigated, the authors speculate that formation of these large units may help to recruit apo-MCM toward MMAB for concerted transfer of newly synthesized AdoCbl to the CblA-MCM complex (Froese et al. 2010b). This supramolecular arrangement would minimize the leakage of AdoCbl into the mitochondrial milieu, allowing mammalian CblA to perform the gate-keeping function described for MeaB (Padovani and Banerjee 2009a). The crystal structure of human CblA and MCM is compatible with formation of a complex with 2:1 stoichiometry. Nucleotide binding sites and Switch motifs I and II from both subunits of the CblA dimer are juxtaposed in a central cavity such that they can be contacted, simultaneously, by one subunit of the MCM dimer (Froese et al. 2010b). This arrangement is different from that described for the MeaB-MCM complex of *M. extorquens*, featuring a 1:1 stoichiometry (Padovani et al. 2006). These species-specific differences are not unexpected. For example, the MCM of *M. extorquens* is a heterodimer, requires only 1 mole of AdoCbl, and its available surface is considerably smaller than that of the human counterpart (45,000 versus 53,000 Å^2) (Froese et al. 2010b). In its GTP-bound state, MeaB identifies and enables the transfer of AdoCbl from its source CblB to apo-MCM (gate keeping function) and surveys and removes inert oxidized Cbl from client MCM (editing function) (Padovani and Banerjee 2009a). In light of the structural differences between MeaB and human CblA, it would be interesting to assess whether human CblA performs the same gate-keeping and editing functions predicted for MeaB (Padovani and Banerjee 2009a).

4. Unresolved Issues of Intracellular Processing and Utilization of B12

A. Cellular import of cobalamin—new genes?

Cobalamins reach all cells in the body via the TC-TCblR interaction. The endocytosis pathway takes up the TC-TCblR complex and carries TC-Cbl to the lysosome. A case study showed that a homozygous missense mutation in gene ZFYVE20 encoding rabenosyn-5 (rbsn-5), a conserved multidomain protein involved in receptor-mediated endocytosis, causes transient Cbl deficiency of the combined phenotype and a series of clinical manifestations similar to those of established inborn errors of vitamin B12 metabolism (Stockler et al. 2014). The rabenosyn-5 protein may be critical to support holo-TC uptake via receptor-mediated absorptive endocytosis (Amagasaki et al. 1990; Jacobsen et al. 1980; Kishimoto et al. 1987; Takahashi et al. 1980). Fibroblasts isolated from the patient showed an abnormal ratio of MeCbl to AdoCbl biosynthesis, which suggests that rbsn-5 may also assist in Cbl trafficking downstream of cellular membrane-to-lysosome transit (Stockler et al. 2014). Thus, it is plausible that unexplained cases of functional Cbl deficiency may involve this and other genes not directly related to the classical *cbl* complementation groups.

B. Mitochondrial transport of cobalamin—who are the players?

Although essential Cbl synthesis and chaperone machineries reside in the mitochondrion, little is known about the transport system employed by newly processed Cbl to enter this organelle. Rosenberg and coworkers performed the first and only studies reporting that lysosome-free preparations of rat liver mitochondria take up [^{57}Co]-hydroxoCbl by a diffusion process that is dependent on mitochondrial swelling, rather than on energy or ion fluxes. (Fenton et al. 1976) This uptake system was saturable, unidirectional, and selective: [^{57}Co]-CNCbl was taken up less rapidly and to a lesser extent than [^{57}Co]-hydroxoCbl. Unlabeled MeCbl and AdoCbl inhibited [^{57}Co]-hydroxoCbl markedly, while CNCbl did not, suggesting some specificity of the putative mitochondrial transporter for the β-axial ligand of Cbl. (Fenton et al. 1976) Incubation of mitochondria with [^{57}Co]-hydroxoCbl, followed by Cbl extraction and chromatography, led to a ~80% recovery of [^{57}Co]-hydroxoCbl and ~20% recovery of a presumptive [^{57}Co]-sulfitoCbl species. The biological relevance of this finding is still unknown. (Fenton et al. 1976) A follow-up study by the same group reported that when free [^{57}Co]-hydroxoCbl is added to a suspension of intact rat liver mitochondria in the presence of ATP and a reducing system, AdoCbl synthesis could be observed. (Fenton and Rosenberg 1978) In contrast, when mitochondrial swelling was prevented, little or no synthesis of AdoCbl was observed. No synthesis of AdoCbl could be detected when [^{57}Co]-CNCbl was used as the source (Fenton and Rosenberg 1978). Intriguingly, this work also provided evidence that Cbl could be reduced with electrons derived from the respiratory chain, namely complexes I and II, in the presence of glutamate (Fenton and Rosenberg 1978). The authors found that newly synthesized [^{57}Co]-AdoCbl was present in the medium where mitochondria were suspended, which indicated that mechanisms exist for the export of Cbl from the mitochondrion (Fenton et al. 1976; Fenton and Rosenberg 1978). To date, the identity and exact mechanisms of Cbl transport across the mitochondrion continue to be a mystery. The discovery of a mitochondrial cblD isoform in humans points to a possible role of this protein in Cbl trafficking within the organelle, and perhaps, in facilitating its translocation across the mitochondrial membranes. These ideas await further investigation.

C. Regulation of cobalamin biosynthesis by oxidative stress

Mammalian MS and MCM are sensitive to O_2 and reactive oxygen species, namely N_2O and •NO, both *in vivo* and in vitro. (Ast et al. 1994; Brouwer et al. 1996; Christensen and Ueland 1993; Danishpajooh et al. 2001; Kambo et al. 2005; Kondo et al. 1981; Nicolaou et al. 1994; Nicolaou et al. 1996; Nicolaou et al. 1997) In humans, the use of N_2O, a general anesthetic, is associated with hematologic and neurologic abnormalities that mimic those seen in Cbl deficiency. Kondo and coworkers have shown that N_2O causes serious impairment of both MS and MCM activities in rats (Kondo et al. 1981). Of the two Cbl-dependent enzymes, MS was the most sensitive toward time and concentration-dependent inactivation by N_2O. Exposure to N_2O had multiple effects on Cbl metabolism. N_2O caused displacement of Cbl from MS, reduced synthesis of MeCbl, conversion of authentic Cbl to Cbl analogues, and an eventual decrease in MCM activity (Kondo et al. 1981). Although N_2O caused a rapid inhibition of MS activity, this process was reversed upon its removal. Studies performed by Christensen and Ueland showed that inhibition of

MS activity by N_2O increased homocysteine efflux and thereby the level of extracellular homocysteine in cultured cells and plasma. (Christensen and Ueland 1993) This was further evidence that MS inactivation is due to a failure to remethylate homocysteine. Similar effects have been observed after exposure to ˙NO; both Cbl-dependent enzymes are inactivated, in a concentration and time-dependent fashion, using both authentic ˙NO as well as nitric oxide donors. (Ast et al. 1994; Danishpajooh et al. 2001; Kambo et al. 2005; Nicolaou et al. 1994; Nicolaou et al. 1996; Nicolaou et al. 1997) *In vitro* and *in vivo* studies have shown that ˙NO is a more potent inhibitor than N_2O. (Danishpajooh et al. 2001; Kambo et al. 2005) It has been proposed that inhibition by ˙NO is caused by the formation of nitroxylcob(III)alamin (NOCbl), which cannot be recycled by MS or MCM. In support of that notion, model studies have shown that the following reactions exist and could be responsible for the observed inactivation by both NO and N_2O: (Zheng et al. 2002)

$$2Cbl\ (I)^- + 2˙NO + 2H^+ \leftrightarrow 2Cbl\ (II) + N_2O + H_2O$$
$$2Cbl\ (I)^- + N_2O + 2H^+ \leftrightarrow 2Cbl\ (II) + N_2 + H_2O$$
$$Cbl\ (II) + ˙NO \leftrightarrow NOCbl$$

Thus, conditions of oxidative stress that lead to oxidation of cob(I)alamin to cob(II)alamin or formation of NOCbl inactivate the two mammalian Cbl-dependent enzymes. In fact, the high reactivity of Cbl toward ROS, specifically with superoxide, has been demonstrated *in vitro* (Suarez-Moreira et al. 2009) and in cultured cells (Moreira et al. 2011). Repair mechanisms for the clearance of catalytically inert NOCbl and other Cbl analogues that may form upon scavenging of reactive oxygen species are yet to be described.

D. Inhibition of MS and MCM by cobalamin analogues

Stabler and coworkers provided some of the first evidence that Cbl analogues could efficiently inhibit both MS and MCM (Stabler et al. 1991). Synthesis of a series of Cbl derivatives featuring rational modifications at the side chains, corrin ring and nucleotide moiety showed that an intact, authentic Cbl is required to support the activity of MS and MCM *in vivo*. Modifications of the side chains of the B and C corrin rings (hydroxoCbl[c-lactam], hydroxoCbl[e-dimethylamide] and hydroxoCbl[e-methylamide]) resulted in the most potent inhibitors for both enzymes. For instance, administration of hydroxoCbl[c-lactam] to rats inhibited MCM and MS activities by 30% and 80%, respectively (Stabler et al. 1991). HydroxoCbl[c-lactam] has been successfully used to generate a model of B12-deficiency in cell culture (Sponne et al. 2000). One major topic of interest concerns the presence of substantial amounts of Cbl analogues in human cord blood (Carmel et al. 1988; Hardlei et al. 2013; Muir and Landon 1985). A comparison of Cbl analogues patterns and abundance in the cord serum of newborns and their mothers showed striking similarities, which suggests that Cbl analogues are derived from Cbl metabolism (Hardlei et al. 2013). A Cbl analog with HPLC elution properties matching those of dicyanocobinamide was two-fold more abundant in the newborns compared to their mothers (Hardlei et al. 2013). Although the presence of Cbl analogues has been well documented in various mammalian tissues (Allen and Stabler 2008; Gimsing 1995; Gimsing and Beck 1989; Hardlei et al. 2013; McLean et al. 1997), the low abundance of Cbl in cells and plasma has hampered unequivocal structural characterization and the elucidation of their physiological roles. Increased sensitivity of new mass spectrometry technologies will soon allow us to uncover the nature and potential functions of these Cbl analogues.

E. Transcytosis of cobalamins

Early studies showed that Cbl can be transported through the cell by transcytosis (Gueant et al. 1988). The transcellular transport of Cbl has been demonstrated in kidney (Ramanujam et al. 1991a; Ramanujam et al. 1994; Ramanujam et al. 1992; Strope et al. 2004) and in intestinal cells (Bose et al. 2007; Bose et al. 1997; Dan and Cutler 1994; Pons et al. 2000; Ramanujam et al. 1991b). This transport pathway has potentially important implications for tissue-to-tissue transfer of Cbl, supplying the micronutrient to tissues with limited access to circulation. This could contribute to multi-organ homeostasis, as suggested by a study where endothelial transport of Cbl appeared essential for Cbl access to the liver (Soda et al. 1985). Additional studies performed in a variety of cell types are required to assess the contribution of this pathway to global Cbl homeostasis, in particular, to help establish whether a relationship exists between intracellular and serum levels of the vitamin.

5. Summary and Conclusions

The identification of the genes involved in the complex pathways of Cbl processing, trafficking and coenzyme biosynthesis has enabled much progress in the field. In light of these discoveries and the likely primordial origin of B12, one could argue that Nature has deployed a complicated network of biochemical interactions and reactions to furnish the needs of two ancestrally ancient Cbl-dependent enzymes. Table 2 lists the proteins involved in intracellular Cbl pathways and their known functions. The structure-function relationships that ensure that demands for MeCbl and AdoCbl will be met are only now beginning to be understood. Even less is known about the protein-protein interactions involved in intra and extra-cellular communication of function and dysfunction events of Cbl metabolism. Apart from these established enzymatic functions, Cbl participates in intracellular signaling (Okada et al. 2011), apoptosis (Jorge-Finnigan et al. 2010; Orozco-Barrios et al. 2009; Richard et al. 2007), oxidative stress (Richard et al. 2009) and cytokine and growth factor-mediated regulation (Scalabrino 2009). We anticipate that the coming years will witness the discovery of new genes and modes of

Table 2. The cast of characters for cellular uptake, intracellular processing, trafficking and utilization of cobalamins (in order of appearance).

Gene or complementation group	Protein/enzyme	Clinical phenotype	References
TCN2, TC	TC Transcobalamin Cellular delivery of cobalamin	Combined homocystinuria and methylmalonic aciduria	(Hakami et al. 1971; Trakadis et al. 2014)
CD320	TCblR Cellular receptor for TC	Combined homocystinuria and methylmalonic aciduria	(Jiang et al. 2013; Quadros et al. 2010; Quadros et al. 2009)
Rbsn-5, ZFYVE20	Rabenosyn 5 Receptor-mediated endocytosis	Transient Cbl deficiency Combined homocystinuria and methylmalonic aciduria	(Stockler et al. 2014)
cblF, LMBD1	CblF Lysosomal Cbl efflux	Combined homocystinuria and methylmalonic aciduria	(Idriss and Jonas 1991; Laframboise et al. 1992; MacDonald et al. 1992; Rosenblatt et al. 1985; Rosenblatt et al. 1986; Shih et al. 1989; Vassiliadis et al. 1991; Watkins and Rosenblatt 1986)
cblJ, ABCD4	CblJ Lysosomal Cbl efflux	Combined homocystinuria and methylmalonic aciduria	(Coelho et al. 2012)
cblX, HCFC1	CblX Global transcriptional co-regulator HCFC1 Transcriptional regulation of MMACHC	Combined homocystinuria and methylmalonic aciduria	(Quintana et al. 2014; Yu et al. 2013)
cblC, MMACHC	Cbl Cytosolic chaperone, processing of newly internalized Cbl	Combined homocystinuria and methylmalonic aciduria	(Dillon et al. 1974; Lerner-Ellis et al. 2006; Linnell et al. 1976; Mahoney and Rosenberg 1971; Mahoney et al. 1971; Mellman et al. 1979a; Mellman et al. 1979b; Mudd et al. 1969; Mudd et al. 1970; Pezacka et al. 1990; Pezacka 1993;

Table 2. contd....

Table 2. contd....

Gene or complementation group	Protein/enzyme	Clinical phenotype	References
			Pezacka et al. 1992; Rosenberg et al. 1975; Watanabe et al. 1989; Watanabe et al. 1996; Willard et al. 1978; Willard and Rosenberg 1979)
cblD$_{cyto}$, MMADHC	CblD$_{cyto}$ Cytosolic and mitochondrial chaperone that orchestrates the fate of newly processed Cbl; functions downstream of the MMACHC protein	Isolated or combined homocystinuria and methylmalonic aciduria	(Dillon et al. 1974; Linnell et al. 1976; Mahoney and Rosenberg 1971; Mahoney et al. 1971; Mellman et al. 1979a; Mellman et al. 1979b; Mudd et al. 1969; Mudd et al. 1970; Pezacka et al. 1990; Pezacka 1993; Pezacka et al. 1992; Rosenberg et al. 1975; Watanabe et al. 1989; Watanabe et al. 1996; Willard et al. 1978; Willard and Rosenberg 1979)
cblG	MS Methionine synthase, synthesis of MeCbl, Met and regeneration of THF	Homocystinuria	(Gulati et al. 1996; Leclerc et al. 1996; Sillaots et al. 1992)
cblE	MSR Methionine synthase reductase, regeneration of cob(I)alamin	Homocystinuria	(Coelho et al. 2012; Gailus et al. 2010a; Gailus et al. 2010b; Gulati et al. 1997; Leclerc et al. 1998; Rutsch et al. 2009; Rutsch et al. 2011; Watkins and Rosenblatt 1989)

Table 2. contd.

Table 2. contd....

Gene or complementation group	Protein/enzyme	Clinical phenotype	References
$cblD_{mito}$, MMADHC	$CblD_{mito}$ Cytosolic and mitochondrial chaperone that orchestrates the fate of newly processed Cbl; functions downstream of the MMACHC protein	Isolated or combined homocystinuria and methylmalonic aciduria	(Dillon et al. 1974; Linnell et al. 1976; Mahoney and Rosenberg 1971; Mahoney et al. 1971; Mellman et al. 1979a; Mellman et al. 1979b; Mudd et al. 1969; Mudd et al. 1970; Pezacka et al. 1990; Pezacka 1993; Pezacka et al. 1992; Rosenberg et al. 1975; Watanabe et al. 1989; Watanabe et al. 1996; Willard et al. 1978; Willard and Rosenberg 1979)
cblA, MMAA	CblA Mitochondrial chaperone, gate-keeper and editor for AdoCbl insertion into apo-MCM	Methylmalonic aciduria	(Fenton and Rosenberg 1978; Gravel et al. 1975; Mahoney et al. 1975; Mahoney et al. 1971)
cblB, MMAB	CblB Mitochondrial cob(I)alamin ad-enosyltransferase	Methylmalonic aciduria	(Fenton and Rosenberg 1981; Gravel et al. 1975; Mahoney et al. 1975; Morrow et al. 1975)
mut	MCM Methylmalonyl-CoA mutase; synthesis of succinyl-CoA	Methylmalonic aciduria	(Acquaviva et al. 2005; Chang 2005; Kierstein et al. 2003; Sakamoto et al. 2007; Wilkemeyer et al. 1991)

regulation of intracellular Cbl metabolism as well as their interaction with other biochemical pathways.

Keywords: Vitamin B12, cobalamin, cobalamin deficiency, homocysteine, homocystinuria, methylmalonic aciduria, inborn errors of metabolism, cobalamin processing, cobalamin trafficking, methionine synthase, methylmalonyl-CoA mutase

Abbreviations

Hcy	:	homocysteine
MMA	:	methylmalonic acid
Cbl	:	cobalamin
MS	:	methionine synthase
MCM	:	methylmalonyl-CoA mutase
MMACHC	:	methylmalonic aciduria combined with homocystinuria type C
MMADHC	:	methylmalonic aciduria combined with homocystinuria type D

References

Acquaviva C, Benoist JF, Pereira S, Callebaut I, Koskas T, Porquet D and Elion J. 2005. Molecular basis of methylmalonyl-CoA mutase apoenzyme defect in 40 European patients affected by mut(o) and mut-forms of methylmalonic acidemia: identification of 29 novel mutations in the MUT gene. Hum Mutat. 25(2): 167–176.

Allen RH and Stabler SP. 2008. Identification and quantitation of cobalamin and cobalamin analogues in human feces. Am J Clin Nutr. 87(5): 1324–1335.

Amagasaki T, Green R and Jacobsen DW. 1990. Expression of transcobalamin II receptors by human leukemia K562 and HL-60 cells. Blood. 76(7): 1380–1386.

Andrews E, Jansen R, Crane AM, Cholin S, McDonnell D and Ledley FD. 1993. Expression of recombinant human methylmalonyl-CoA mutase: in primary mut fibroblasts and *Saccharomyces cerevisiae*. Biochem Med Metab Biol. 50(2): 135–144.

Ast T, Nicolaou A, Anderson MM, James C and Gibbons WA. 1994. Purification, properties and inhibition of rat liver cytosolic vitamin B12-dependent methionine synthase. Biochem Soc Trans. 22(2): 217S.

Atkinson C, Miousse IR, Watkins D, Rosenblatt DS and Raiman JA. 2014. Clinical, biochemical, and molecular presentation in a patient with the cblD-homocystinuria inborn error of cobalamin metabolism. JIMD Rep. 17: 77–81.

Backe PH, Ytre-Arne M, Rohr AK, Brodtkorb E, Fowler B, Rootwelt H, Bjoras M and Morkrid L. 2013. Novel deletion mutation identified in a patient with late-onset combined methylmalonic acidemia and homocystinuria, cblC type. JIMD Rep. 11: 79–85.

Bandarian V, Pattridge KA, Lennon BW, Huddler DP, Matthews RG and Ludwig ML. 2002. Domain alternation switches B(12)-dependent methionine synthase to the activation conformation. Nat Struct Biol. 9(1): 53–56.

Banerjee R. 1999. Chemistry and Biochemistry of B12. *In*: R Banerjee (ed.). John Wiley and Sons, New York, 921pp.

Banerjee R. 2003. Radical carbon skeleton rearrangements: catalysis by coenzyme B12-dependent mutases. Chem Rev. 103(6): 2083–2094.

Banerjee R. 2006. B12 trafficking in mammals: A for coenzyme escort service. ACS Chem Biol. 1(3): 149–159.

Banerjee R, Chen Z and Gulati S. 1997. Methionine synthase from pig liver. Methods Enzymol. 281: 189–196.

Banerjee RV and Matthews RG. 1990. Cobalamin-dependent methionine synthase. FASEB J. 4(5): 1450–1459.

Bommer M, Kunze C, Fesseler J, Schubert T, Diekert G and Dobbek H. 2014. Structural basis for organohalide respiration. Science. 346(6208): 455–458.

Bose S, Kalra S, Yammani RR, Ahuja R and Seetharam B. 2007. Plasma membrane delivery, endocytosis and turnover of transcobalamin receptor in polarized human intestinal epithelial cells. J Physiol. 581(Pt 2): 457–466.

Bose S, Seetharam S, Dahms NM and Seetharam B. 1997. Bipolar functional expression of transcobalamin II receptor in human intestinal epithelial Caco-2 cells. J Biol Chem. 272(6): 3538–3543.

Brouwer M, Chamulitrat W, Ferruzzi G, Sauls DL and Weinberg JB. 1996. Nitric oxide interactions with cobalamins: biochemical and functional consequences. Blood. 88(5): 1857–1864.

Brown KL and Brooks HB. 1991. Effects of axial ligation on the thermolysis of benzyl- and neopentylcobamides: analysis of the base-on effect. Inorg Chem. 30(18): 3420–3430.

Carmel R, Karnaze DS and Weiner JM. 1988. Neurologic abnormalities in cobalamin deficiency are associated with higher cobalamin "analogue" values than are hematologic abnormalities. J Lab Clin Med. 111(1): 57–62.

Carrillo-Carrasco N, Chandler RJ and Venditti CP. 2012. Combined methylmalonic acidemia and homocystinuria, cblC type. I. Clinical presentations, diagnosis and management. J Inherit Metab Dis. 35(1): 91–102.

Caterino M, Pastore A, Strozziero MG, Di Giovamberardino G, Imperlini E, Scolamiero E, Ingenito L, Boenzi S, Ceravolo F, Martinelli D et al. 2015. The proteome of cblC defect: *in vivo* elucidation of altered cellular pathways in humans. J Inherit Metab Dis. 38(5): 969–979.

Chang H. 2005. Gene symbol: MUT. Disease: Methylmalonic aciduria. Hum Genet. 117(2–3): 299.

Chen LH, Liu ML, Hwang HY, Chen LS, Korenberg J and Shane B. 1997. Human methionine synthase. cDNA cloning, gene localization, and expression. J Biol Chem. 272(6): 3628–3634.

Chen Z, Crippen K, Gulati S and Banerjee R. 1994. Purification and kinetic mechanism of a mammalian methionine synthase from pig liver. J Biol Chem. 269(44): 27193–27197.

Chowdhury S and Banerjee R. 2000. Thermodynamic and kinetic characterization of Co-C bond homolysis catalyzed by coenzyme B12-dependent methylmalonyl-CoA mutase. Biochemistry. 39(27): 7998–8006.

Christensen B and Ueland PM. 1993. Methionine synthase inactivation by nitrous oxide during methionine loading of normal human fibroblasts. Homocysteine remethylation as determinant of enzyme inactivation and homocysteine export. J Pharmacol Exp Ther. 267(3): 1298–1303.

Chu RC, Begley JA, Colligan PD and Hall CA. 1993. The methylcobalamin metabolism of cultured human fibroblasts. Metabolism. 42(3): 315–319.

Coelho D, Kim JC, Miousse IR, Fung S, du Moulin M, Buers I, Suormala T, Burda P, Frapolli M, Stucki M et al. 2012. Mutations in ABCD4 cause a new inborn error of vitamin B12 metabolism. Nat Genet. 44(10): 1152–1155.

Coelho D, Suormala T, Stucki M, Lerner-Ellis JP, Rosenblatt DS, Newbold RF, Baumgartner MR and Fowler B. 2008. Gene identification for the cblD defect of vitamin B12 metabolism. N Engl J Med. 358(14): 1454–1464.

Croft MT, Lawrence AD, Raux-Deery E, Warren MJ and Smith AG. 2005. Algae acquire vitamin B12 through a symbiotic relationship with bacteria. Nature. 438(7064): 90–93.

Dan N and Cutler DF. 1994. Transcytosis and processing of intrinsic factor-cobalamin in Caco-2 cells. J Biol Chem. 269(29): 18849–18855.

Danishpajooh IO, Gudi T, Chen Y, Kharitonov VG, Sharma VS and Boss GR. 2001. Nitric oxide inhibits methionine synthase activity *in vivo* and disrupts carbon flow through the folate pathway. J Biol Chem. 276(29): 27296–27303.

Deme JC, Miousse IR, Plesa M, Kim JC, Hancock MA, Mah W, Rosenblatt DS and Coulton JW. 2012. Structural features of recombinant MMADHC isoforms and their interactions with MMACHC, proteins of mammalian vitamin B12 metabolism. Mol Genet Metab. 107(3): 352–362.

Dillon MJ, England JM, Gompertz D, Goodey PA, Grant DB, Hussein HA, Linnell JC, Matthews DM, Mudd SH, Newns GH et al. 1974. Mental retardation, megaloblastic anaemia, methylmalonic aciduria and abnormal homocysteine metabolism due to an error in vitamin B12 metabolism. Clin Sci Mol Med. 47(1): 43–61.

Dobson CM, Wai T, Leclerc D, Kadir H, Narang M, Lerner-Ellis JP, Hudson TJ, Rosenblatt DS and Gravel RA. 2002a. Identification of the gene responsible for the cblB complementation group of vitamin B12-dependent methylmalonic aciduria. Hum Mol Genet. 11(26): 3361–3369.

Dobson CM, Wai T, Leclerc D, Wilson A, Wu X, Dore C, Hudson T, Rosenblatt DS and Gravel RA. 2002b. Identification of the gene responsible for the cblA complementation group of vitamin B12-responsive methylmalonic acidemia based on analysis of prokaryotic gene arrangements. Proc Natl Acad Sci U S A. 99(24): 15554–15559.

Drennan CL, Huang S, Drummond JT, Matthews RG and Lidwig ML. 1994. How a protein binds B12: A 3.0 A X-ray structure of B12-binding domains of methionine synthase. Science. 266(5191): 1669–1674.

Drummond JT, Huang S, Blumenthal RM and Matthews RG. 1993. Assignment of enzymatic function to specific protein regions of cobalamin-dependent methionine synthase from *Escherichia coli*. Biochemistry. 32(36): 9290–9295.

Eschenmoser A. 2011. Etiology of potentially primordial biomolecular structures: from vitamin B12 to the nucleic acids and an inquiry into the chemistry of life's origin: a retrospective. Angew Chem Int Ed Engl. 50(52): 12412–12472.

Fenton WA, Ambani LM and Rosenberg LE. 1976. Uptake of hydroxocobalamin by rat liver mitochondria. Binding to a mitochondrial protein. J Biol Chem. 251(21): 6616–6623.

Fenton WA, Hack AM, Willard HF, Gertler A and Rosenberg LE. 1982. Purification and properties of methylmalonyl coenzyme A mutase from human liver. Arch Biochem Biophys. 214(2): 815–823.

Fenton WA and Rosenberg LE. 1978. Mitochondrial metabolism of hydroxocobalamin: synthesis of adenosylcobalamin by intact rat liver mitochondria. Arch Biochem Biophys. 189(2): 441–447.

Fenton WA and Rosenberg LE. 1981. The defect in the cbl B class of human methylmalonic acidemia: deficiency of cob(I)alamin adenosyltransferase activity in extracts of cultured fibroblasts. Biochem Biophys Res Commun. 98(1): 283–289.

Fofou-Caillierez MB, Mrabet NT, Chery C, Dreumont N, Flayac J, Pupavac M, Paoli J, Alberto JM, Coelho D, Camadro JM et al. 2013. Interaction between methionine synthase isoforms and MMACHC: characterization in cblG-variant, cblG and cblC inherited causes of megaloblastic anaemia. Hum Mol Genet. 22(22): 4591–4601.

Fong JH, Shoemaker BA, Garbuzynskiy SO, Lobanov MY, Galzitskaya OV and Panchenko AR. 2009. Intrinsic disorder in protein interactions: insights from a comprehensive structural analysis. PLoS Comput Biol. 5(3): e1000316.

Froese DS and Gravel RA. 2010. Genetic disorders of vitamin B12 metabolism: eight complementation groups—eight genes. Expert Rev Mol Med. 12: e37.

Froese DS, Healy S, McDonald M, Kochan G, Oppermann U, Niesen FH and Gravel RA. 2010a. Thermolability of mutant MMACHC protein in the vitamin B12-responsive cblC disorder. Mol Genet Metab. 100(1): 29–36.

Froese DS, Kochan G, Muniz JR, Wu X, Gileadi C, Ugochukwu E, Krysztofinska E, Gravel RA, Oppermann U and Yue WW. 2010b. Structures of the human GTPase MMAA and vitamin B12-dependent methylmalonyl-CoA mutase and insight into their complex formation. J Biol Chem. 285(49): 38204–38213.

Froese DS, Krojer T, Wu X, Shrestha R, Kiyani W, von Delft F, Gravel RA, Oppermann U and Yue WW. 2012. Structure of MMACHC reveals an arginine-rich pocket and a domain-swapped dimer for its B12 processing function. Biochemistry. 51(25): 5083–5090.

Froese DS, Wu X, Zhang J, Dumas R, Schoel WM, Amrein M and Gravel RA. 2008. Restricted role for methionine synthase reductase defined by subcellular localization. Mol Genet Metab. 94(1): 68–77.

Froese DS, Zhang J, Healy S and Gravel RA. 2009. Mechanism of vitamin B12-responsiveness in cblC methylmalonic aciduria with homocystinuria. Mol Genet Metab. 98(4): 338–343.

Gailus S, Hohne W, Gasnier B, Nurnberg P, Fowler B and Rutsch F. 2010a. Insights into lysosomal cobalamin trafficking: lessons learned from cblF disease. J Mol Med (Berl). 88(5): 459–466.

Gailus S, Suormala T, Malerczyk-Aktas AG, Toliat MR, Wittkampf T, Stucki M, Nurnberg P, Fowler B, Hennermann JB and Rutsch F. 2010b. A novel mutation in LMBRD1 causes the cblF defect of vitamin B(12) metabolism in a Turkish patient. J Inherit Metab Dis. 33(1): 17–24.

Geno MK and Halpern J. 1987. Why does nature not use the porphyrin ligand in vitamin B12? J Am Chem Soc. 109(4): 1238–1240.

Gherasim C, Hannibal L, Rajagopalan D, Jacobsen DW and Banerjee R. 2013a. The C-terminal domain of CblD interacts with CblC and influences intracellular cobalamin partitioning. Biochimie. 95(5): 1023–1032.

Gherasim C, Lofgren M and Banerjee R. 2013b. Navigating the B12 road: assimilation, delivery, and disorders of cobalamin. J Biol Chem. 288(19): 13186–13193.

Gherasim C, Ruetz M, Li Z, Hudolin S and Banerjee R. 2015. Pathogenic mutations differentially affect the catalytic activities of the human B12-processing chaperone CblC and increase futile redox cycling. J Biol Chem. 290(18): 11393–11402.

Gimsing P. 1995. Cobalamin forms and analogues in plasma and myeloid cells during chronic myelogenous leukaemia related to clinical condition. Br J Haematol. 89(4): 812–819.

Gimsing P and Beck WS. 1989. Cobalamin analogues in plasma. An *in vitro* phenomenon? Scand J Clin Lab Invest Suppl. 194: 37–40.

Gravel RA, Mahoney MJ, Ruddle FH and Rosenberg LE. 1975. Genetic complementation in heterokaryons of human fibroblasts defective in cobalamin metabolism. Proc Natl Acad Sci U S A. 72(8): 3181–3185.

Gueant JL, Gerard A, Monin B, Champigneulle B, Gerard H and Nicolas JP. 1988. Radioautographic localisation of iodinated human intrinsic factor in the guinea pig ileum using electron microscopy. Gut. 29(10): 1370–1378.

Gulati S, Baker P, Li YN, Fowler B, Kruger W, Brody LC and Banerjee R. 1996. Defects in human methionine synthase in cblG patients. Hum Mol Genet. 5(12): 1859–1865.

Gulati S, Chen Z, Brody LC, Rosenblatt DS and Banerjee R. 1997. Defects in auxiliary redox proteins lead to functional methionine synthase deficiency. J Biol Chem. 272(31): 19171–19175.

Gunasekaran K, Tsai CJ, Kumar S, Zanuy D and Nussinov R. 2003. Extended disordered proteins: targeting function with less scaffold. Trends Biochem Sci. 28(2): 81–85.

Hakami N, Neiman PE, Canellos GP and Lazerson J. 1971. Neonatal megaloblastic anemia due to inherited transcobalamin II deficiency in two siblings. N Engl J Med. 285(21): 1163–1170.

Hannibal L, DiBello PM and Jacobsen DW. 2013. Proteomics of vitamin B12 processing. Clin Chem Lab Med. 51(3): 477–488.

Hannibal L, DiBello PM, Yu M, Miller A, Wang S, Willard B, Rosenblatt DS and Jacobsen DW. 2011. The MMACHC proteome: hallmarks of functional cobalamin deficiency in humans. Mol Genet Metab. 103(3): 226–239.

Hannibal L, Kim J, Brasch NE, Wang S, Rosenblatt DS, Banerjee R and Jacobsen DW. 2009. Processing of alkylcobalamins in mammalian cells: A role for the MMACHC (cblC) gene product. Mol Genet Metab. 97(4): 260–266.

Haque MM, Bayachou M, Tejero J, Kenney CT, Pearl NM, Im SC, Waskell L and Stuehr DJ. 2014. Distinct conformational behaviors of four mammalian dual-flavin reductases (cytochrome P450 reductase, methionine synthase reductase, neuronal nitric oxide synthase, endothelial nitric oxide synthase) determine their unique catalytic profiles. FEBS J. 281(23): 5325–5340.

Harding CO, Arnold G, Barness LA, Wolff JA and Rosenblatt DS. 1997. Functional methionine synthase deficiency due to cblG disorder: a report of two patients and a review. Am J Med Genet. 71(4): 384–390.

Hardlei TF, Obeid R, Herrmann W and Nexo E. 2013. Cobalamin analogues in humans: a study on maternal and cord blood. PLoS One. 8(4): e61194.

Hodgkin DC, Pickworth J, Robertson JH, Trueblood KN, Prosen RJ and White JG. 1955. Structure of vitamin B12. The crystal structure of the hexacarboxylic acid derived from B12 and the molecular structure of the vitamin. Nature. 176(4477): 325–328.

Hubbard PA, Padovani D, Labunska T, Mahlstedt SA, Banerjee R and Drennan CL. 2007. Crystal structure and mutagenesis of the metallochaperone MeaB: insight into the causes of methylmalonic aciduria. J Biol Chem. 282(43): 31308–31316.

Hunter GA and Ferreira GC. 2009. 5-aminolevulinate synthase: catalysis of the first step of heme biosynthesis. Cell Mol Biol (Noisy-le-grand). 55(1): 102–110.

Idriss JM and Jonas AJ. 1991. Vitamin B12 transport by rat liver lysosomal membrane vesicles. J Biol Chem. 266(15): 9438–9441.

Jacobsen DW, Montejano YD, Vitols KS and Huennekens FM. 1980. Adherence of L1210 murine leukemia cells to sephacryl-aminopropylcobalamin beads treated with transcobalamin-II. Blood. 55(1): 160–163.

James KJ, Hancock MA, Gagnon JN and Coulton JW. 2009. TonB interacts with BtuF, the *Escherichia coli* periplasmic binding protein for cyanocobalamin. Biochemistry. 48(39): 9212–9220.

Jarrett JT, Amaratunga M, Drennan CL, Scholten JD, Sands RH, Ludwig ML and Matthews RG. 1996. Mutations in the B12-binding region of methionine synthase: how the protein controls methylcobalamin reactivity. Biochemistry. 35(7): 2464–2475.

Jencks WP. 1975. Binding energy, specificity, and enzymic catalysis: the circe effect. Adv Enzymol Relat Areas Mol Biol. 43: 219–410.

Jeong J, Ha TS and Kim J. 2011. Protection of aquo/hydroxocobalamin from reduced glutathione by a B12 trafficking chaperone. BMB Rep. 44(3): 170–175.

Jeong J and Kim J. 2011. Glutathione increases the binding affinity of a bovine B12 trafficking chaperone bCblC for vitamin B12. Biochem Biophys Res Commun. 412(2): 360–365.

Jiang W, Nakayama Y, Sequeira JM and Quadros EV. 2013. Mapping the functional domains of TCblR/CD320, the receptor for cellular uptake of transcobalamin-bound cobalamin. FASEB J. 27(8): 2988–2994.

Jorge-Finnigan A, Gamez A, Perez B, Ugarte M and Richard E. 2010. Different altered pattern expression of genes related to apoptosis in isolated methylmalonic aciduria cblB type and combined with homocystinuria cblC type. Biochim Biophys Acta. 1802(11): 959–967.

Jusufi J, Suormala T, Burda P, Fowler B, Froese DS and Baumgartner MR. 2014. Characterization of functional domains of the cblD (MMADHC) gene product. J Inherit Metab Dis. 37(5): 841–849.

Kajiura H, Mori K, Tobimatsu T and Toraya T. 2001. Characterization and mechanism of action of a reactivating factor for adenosylcobalamin-dependent glycerol dehydratase. J Biol Chem. 276(39): 36514–36519.

Kambo A, Sharma VS, Casteel DE, Woods VL, Jr., Pilz RB and Boss GR. 2005. Nitric oxide inhibits mammalian methylmalonyl-CoA mutase. J Biol Chem. 280(11): 10073–10082.

Kierstein S, Peters U, Habermann FA, Fries R and Brenig B. 2003. Assignment of the methylmalonyl-CoA mutase gene (MUT) to porcine chromosome 7q13-->q14 by *in situ* hybridization and analysis of radiation hybrid panels. Cytogenet Genome Res. 101(1): 92F.

Kim J, Gherasim C and Banerjee R. 2008. Decyanation of vitamin B12 by a trafficking chaperone. Proc Natl Acad Sci U S A. 105(38): 14551–14554.

Kim J, Hannibal L, Gherasim C, Jacobsen DW and Banerjee R. 2009. A human vitamin B12 trafficking protein uses glutathione transferase activity for processing alkylcobalamins. J Biol Chem. 284(48): 33418–33424.

Kishimoto T, Tavassoli M, Green R and Jacobsen DW. 1987. Receptors for transferrin and transcobalamin II display segregated distribution on microvilli of leukemia L1210 cells. Biochem Biophys Res Commun. 146(3): 1102–1108.

Kolhouse JF, Utley C and Allen RH. 1980. Isolation and characterization of methylmalonyl-CoA mutase from human placenta. J Biol Chem. 255(7): 2708–2712.

Kolhouse JF, Utley C, Stabler SP and Allen RH. 1991. Mechanism of conversion of human apo- to holomethionine synthase by various forms of cobalamin. J Biol Chem. 266(34): 23010–23015.

Kondo H, Osborne ML, Kolhouse JF, Binder MJ, Podell ER, Utley CS, Abrams RS and Allen RH. 1981. Nitrous oxide has multiple deleterious effects on cobalamin metabolism and causes decreases in activities of both mammalian cobalamin-dependent enzymes in rats. J Clin Invest. 67(5): 1270–1283.

Korotkova N and Lidstrom ME. 2004. MeaB is a component of the methylmalonyl-CoA mutase complex required for protection of the enzyme from inactivation. J Biol Chem. 279(14): 13652–13658.

Koutmos M, Gherasim C, Smith JL and Banerjee R. 2011. Structural basis of multifunctionality in a vitamin B12-processing enzyme. J Biol Chem. 286(34): 29780–29787.

Kraeutler B, Konrat R, Stupperich E, Faerber G, Gruber K and Kratky C. 1994. Direct evidence for the conformational deformation of the corrin ring by the nucleotide base in vitamin B12: Synthesis and solution spectroscopic and crystal structure analysis of Co.beta.-cyanoimidazolylcobamide. Inorg Chem. 33(18): 4128–4139.

Laframboise R, Cooper BA and Rosenblatt DS. 1992. Malabsorption of vitamin B12 from the intestine in a child with cblF disease: evidence for lysosomal-mediated absorption. Blood. 80(1): 291–292.

Leal NA, Park SD, Kima PE and Bobik TA. 2003. Identification of the human and bovine ATP:Cob(I) alamin adenosyltransferase cDNAs based on complementation of a bacterial mutant. J Biol Chem. 278(11): 9227–9234.

Leclerc D, Campeau E, Goyette P, Adjalla CE, Christensen B, Ross M, Eydoux P, Rosenblatt DS, Rozen R and Gravel RA. 1996. Human methionine synthase: cDNA cloning and identification of mutations in patients of the cblG complementation group of folate/cobalamin disorders. Hum Mol Genet. 5(12): 1867–1874.

Leclerc D, Odievre M, Wu Q, Wilson A, Huizenga JJ, Rozen R, Scherer SW and Gravel RA. 1999. Molecular cloning, expression and physical mapping of the human methionine synthase reductase gene. Gene. 240(1): 75–88.

Leclerc D, Wilson A, Dumas R, Gafuik C, Song D, Watkins D, Heng HH, Rommens JM, Scherer SW, Rosenblatt DS et al. 1998. Cloning and mapping of a cDNA for methionine synthase reductase, a flavoprotein defective in patients with homocystinuria. Proc Natl Acad Sci U S A. 95(6): 3059–3064.

Ledley FD, Lumetta M, Nguyen PN, Kolhouse JF and Allen RH. 1988a. Molecular cloning of L-methylmalonyl-CoA mutase: gene transfer and analysis of mut cell lines. Proc Natl Acad Sci U S A. 85(10): 3518–3521.

Ledley FD, Lumetta MR, Zoghbi HY, VanTuinen P, Ledbetter SA and Ledbetter DH. 1988b. Mapping of human methylmalonyl CoA mutase (MUT) locus on chromosome 6. Am J Hum Genet. 42(6): 839–846.

Lenhert PG and Hodgkin DC. 1961. Structure of the 5,6-dimethylbenzimidazolylcobamide coenzyme. Nature. 192: 937–938.

Lerner-Ellis JP, Anastasio N, Liu J, Coelho D, Suormala T, Stucki M, Loewy AD, Gurd S, Grundberg E, Morel CF et al. 2009. Spectrum of mutations in MMACHC, allelic expression, and evidence for genotype-phenotype correlations. Hum Mutat. 30(7): 1072–1081.

Lerner-Ellis JP, Dobson CM, Wai T, Watkins D, Tirone JC, Leclerc D, Dore C, Lepage P, Gravel RA and Rosenblatt DS. 2004. Mutations in the MMAA gene in patients with the cblA disorder of vitamin B12 metabolism. Hum Mutat. 24(6): 509–516.

Lerner-Ellis JP, Tirone JC, Pawelek PD, Dore C, Atkinson JL, Watkins D, Morel CF, Fujiwara TM, Moras E, Hosack AR et al. 2006. Identification of the gene responsible for methylmalonic aciduria and homocystinuria, cblC type. Nat Genet. 38(1): 93–100.

Li YN, Gulati S, Baker PJ, Brody LC, Banerjee R and Kruger WD. 1996. Cloning, mapping and RNA analysis of the human methionine synthase gene. Hum Mol Genet. 5(12): 1851–1858.

Li Z, Gherasim C, Lesniak NA and Banerjee R. 2014a. Glutathione-dependent one-electron transfer reactions catalyzed by a B12 trafficking protein. J Biol Chem. 289(23): 16487–16497.

Li Z, Lesniak NA and Banerjee R. 2014b. Unusual aerobic stabilization of cob(I)alamin by a B12-trafficking protein allows chemoenzymatic synthesis of organocobalamins. J Am Chem Soc. 136(46): 16108–16111.

Lienhart W-D, Gudipati V and Macheroux P. 2013. The human flavoproteome. Arch Biochem Biophys. 535(2): 150–162.

Linnell JC, Matthews DM, Mudd SH, Uhlendorf BW and Wise IJ. 1976. Cobalamins in fibroblasts cultured from normal control subjects and patients with methylmalonic aciduria. Pediatr Res. 10(3): 179–183.

Loewy AD, Niles KM, Anastasio N, Watkins D, Lavoie J, Lerner-Ellis JP, Pastinen T, Trasler JM and Rosenblatt DS. 2009. Epigenetic modification of the gene for the vitamin B12 chaperone MMACHC can result in increased tumorigenicity and methionine dependence. Mol Genet Metab. 96(4): 261–267.

Lofgren M, Koutmos M and Banerjee R. 2013a. Autoinhibition and signaling by the switch II motif in the G-protein chaperone of a radical B12 enzyme. J Biol Chem. 288(43): 30980–30989.

Lofgren M, Padovani D, Koutmos M and Banerjee R. 2013b. A switch III motif relays signaling between a B12 enzyme and its G-protein chaperone. Nat Chem Biol. 9(9): 535–539.

Loughlin RE, Elford HL and Buchanan JM. 1964. Enzymatic synthesis of the methyl group of methionine. Vii. isolation of a cobalamin-containing transmethylase (5-Methyltetrahydro-Folate-Homocysteine) from mammalian liver. J Biol Chem. 239: 2888–2895.

MacDonald M, Wiltse H, Bever J and Rosenblatt DS. 1992. Clinical heterogeneity in two patients with cblF disease [Abstract]. Am J Hum Genet. 51: A353.

Mah W, Deme JC, Watkins D, Fung S, Janer A, Shoubridge EA, Rosenblatt DS and Coulton JW. 2013. Subcellular location of MMACHC and MMADHC, two human proteins central to intracellular vitamin B12 metabolism. Mol Genet Metab. 108(2): 112–118.

Mahoney MJ, Hart AC, Steen VD and Rosenberg LE. 1975. Methylmalonicacidemia: biochemical heterogeneity in defects of 5′-deoxyadenosylcobalamin synthesis. Proc Natl Acad Sci U S A. 72(7): 2799–2803.

Mahoney MJ and Rosenberg LE. 1971. Synthesis of cobalamin coenzymes by human cells in tissue culture. J Lab Clin Med. 78(2): 302–308.

Mahoney MJ, Rosenberg LE, Mudd SH and Uhlendorf BW. 1971. Defective metabolism of vitamin B12 in fibroblasts from children with methylmalonic aciduria. Biochem Biophys Res Commun. 44(2): 375–381.

Martens JH, Barg H, Warren MJ and Jahn D. 2002. Microbial production of vitamin B12. Appl Microbiol Biotechnol. 58(3): 275–285.

Marzilli LG, Toscano PJ, Randaccio L, Bresciani-Pahor N and Calligaris M. 1979. An unusually long cobalt-carbon bond. Molecular structure of trans-bis(dimethylglyoximato)(isopropyl)(pyridine)cobalt(III). Implications with regard to the conformational trigger mechanism of cobalt-carbon bond cleavage in coenzyme B12. J Am Chem Soc. 101(22): 6754–6756.

Matthews RG. 2001. Cobalamin-dependent methyltransferases. Acc Chem Res. 34(8): 681–689.

Matthews RG, Koutmos M and Datta S. 2008. Cobalamin-dependent and cobamide-dependent methyltransferases. Curr Opin Struct Biol. 18(6): 658–666.

Mc Guire PJ, Parikh A and Diaz GA. 2009. Profiling of oxidative stress in patients with inborn errors of metabolism. Mol Genet Metab. 98(1-2): 173–180.

McLean GR, Pathare PM, Wilbur DS, Morgan AC, Woodhouse CS, Schrader JW and Ziltener HJ. 1997. Cobalamin analogues modulate the growth of leukemia cells *in vitro*. Cancer Res. 57(18): 4015–4022.

Mellman I, Willard HF, Youngdahl-Turner P and Rosenberg LE. 1979a. Cobalamin coenzyme synthesis in normal and mutant human fibroblasts. Evidence for a processing enzyme activity deficient in cblC cells. J Biol Chem. 254(23): 11847–11853.

Mellman IS, Lin PF, Ruddle FH and Rosenberg LE. 1979b. Genetic control of cobalamin binding in normal and mutant cells: assignment of the gene for 5-methyltetrahydrofolate:L-homocysteine S-methyltransferase to human chromosome 1. Proc Natl Acad Sci U S A. 76(1): 405–409.

Miousse IR, Watkins D, Coelho D, Rupar T, Crombez EA, Vilain E, Bernstein JA, Cowan T, Lee-Messer C, Enns GM et al. 2009. Clinical and molecular heterogeneity in patients with the cblD inborn error of cobalamin metabolism. J Pediatr. 154(4): 551–556.

Moore SJ, Lawrence AD, Biedendieck R, Deery E, Frank S, Howard MJ, Rigby SE and Warren MJ. 2013. Elucidation of the anaerobic pathway for the corrin component of cobalamin (vitamin B12). Proc Natl Acad Sci U S A. 110(37): 14906–14911.

Moreira ES, Brasch NE and Yun J. 2011. Vitamin B12 protects against superoxide-induced cell injury in human aortic endothelial cells. Free Radic Biol Med. 51(4): 876–883.

Morita H, Kurihara H, Sugiyama T, Hamada C, Kurihara Y, Shindo T, Oh-hashi Y and Yazaki Y. 1999. Polymorphism of the methionine synthase gene : association with homocysteine metabolism and late-onset vascular diseases in the Japanese population. Arterioscler Thromb Vasc Biol. 19(2): 298–302.

Morrow G, 3rd, Mahoney MJ, Mathews C and Lebowitz J. 1975. Studies of methylmalonyl coenzyme A carbonylmutase activity in methylmalonic acidemia. I. Correlation of clinical, hepatic, and fibroblast data. Pediatr Res. 9(8): 641–644.

Mudd SH, Levy HL, Abeles RH and Jennedy JP, Jr. 1969. A derangement in B12 metabolism leading to homocystinemia, cystathioninemia and methylmalonic aciduria. Biochem Biophys Res Commun. 35(1): 121–126.

Mudd SH, Uhlendorf BW and Hinds KR. 1970. Deranged B12 metabolism: studies of fibroblasts grown in tissue culture. Biochem Med. 4(3): 215–239.

Muir M and Landon M. 1985. Endogenous origin of microbiologically-inactive cobalamins (cobalamin analogues) in the human fetus. Br J Haematol. 61(2): 303–306.

Ng FTT and Rempel GL. 1982. Ligand effects on transition metal-alkyl bond dissociation energies. J Am Chem Soc. 104(2): 621–623.

Nicolaou A, Ast T, Garcia CV, Anderson MM, Gibbons JM and Gibbons WA. 1994. *In vitro* NO and N2O inhibition of the branch point enzyme vitamin B12 dependent methionine synthase from rat brain synaptosomes. Biochem Soc Trans. 22(3): 296S.

Nicolaou A, Kenyon SH, Gibbons JM, Ast T and Gibbons WA. 1996. *In vitro* inactivation of mammalian methionine synthase by nitric oxide. Eur J Clin Invest. 26(2): 167–170.

Nicolaou A, Waterfield CJ, Kenyon SH and Gibbons WA. 1997. The inactivation of methionine synthase in isolated rat hepatocytes by sodium nitroprusside. Eur J Biochem. 244(3): 876–882.

Okada K, Tanaka H, Temporin K, Okamoto M, Kuroda Y, Moritomo H, Murase T and Yoshikawa H. 2011. Akt/mammalian target of rapamycin signaling pathway regulates neurite outgrowth in cerebellar granule neurons stimulated by methylcobalamin. Neurosci Lett. 495(3): 201–204.

Olteanu H and Banerjee R. 2001. Human methionine synthase reductase, a soluble P-450 reductase-like dual flavoprotein, is sufficient for NADPH-dependent methionine synthase activation. J Biol Chem. 276(38): 35558–35563.

Olteanu H, Munson T and Banerjee R. 2002. Differences in the efficiency of reductive activation of methionine synthase and exogenous electron acceptors between the common polymorphic variants of human methionine synthase reductase. Biochemistry. 41(45): 13378–13385.

Orozco-Barrios CE, Battaglia-Hsu SF, Arango-Rodriguez ML, Ayala-Davila J, Chery C, Alberto JM, Schroeder H, Daval JL, Martinez-Fong D and Gueant JL. 2009. Vitamin B12-impaired metabolism produces apoptosis and Parkinson phenotype in rats expressing the transcobalamin-oleosin chimera in substantia nigra. PLoS One. 4(12): e8268.

Padmakumar R, Padmakumar R and Banerjee R. 1997. Evidence that cobalt-carbon bond homolysis is coupled to hydrogen atom abstraction from substrate in methylmalonyl-CoA mutase. Biochemistry. 36(12): 3713–3718.

Padovani D and Banerjee R. 2006. Assembly and protection of the radical enzyme, methylmalonyl-CoA mutase, by its chaperone. Biochemistry. 45(30): 9300–9306.

Padovani D and Banerjee R. 2009a. A G-protein editor gates coenzyme B12 loading and is corrupted in methylmalonic aciduria. Proc Natl Acad Sci U S A. 106(51): 21567–21572.

Padovani D and Banerjee R. 2009b. A rotary mechanism for coenzyme B12 synthesis by adenosyltransferase. Biochemistry. 48(23): 5350–5357.

Padovani D, Labunska T and Banerjee R. 2006. Energetics of interaction between the G-protein chaperone, MeaB, and B12-dependent methylmalonyl-CoA mutase. J Biol Chem. 281(26): 17838–17844.

Park J, Jeong J and Kim J. 2012. Destabilization of a bovine B12 trafficking chaperone protein by oxidized form of glutathione. Biochem Biophys Res Commun. 420(3): 547–551.

Park J and Kim J. 2015. Characterization of a B12 trafficking chaperone protein from *caenorhabditis elegans*. Protein Pept Lett. 22(1): 31–38.

Pastore A, Martinelli D, Piemonte F, Tozzi G, Boenzi S, Di Giovamberardino G, Petrillo S, Bertini E and Dionisi-Vici C. 2014. Glutathione metabolism in cobalamin deficiency type C (cblC). J Inherit Metab Dis. 37(1): 125–129.

Payne KA, Quezada CP, Fisher K, Dunstan MS, Collins FA, Sjuts H, Levy C, Hay S, Rigby SE and Leys D. 2015. Reductive dehalogenase structure suggests a mechanism for B12-dependent dehalogenation. Nature. 517(7535): 513–516.

Pezacka E, Denison C, Green R and Jacobsen D. 1988. Biosynthesis of methylcobalamin: chemical model studies with thiol-cobalamin adducts and S-adenosylmethionine. J Cell Physiol. 107: 860a.

Pezacka E, Green R and Jacobsen DW. 1990. Glutathionylcobalamin as an intermediate in the formation of cobalamin coenzymes. Biochem Biophys Res Commun. 169(2): 443–450.

Pezacka EH. 1993. Identification and characterization of two enzymes involved in the intracellular metabolism of cobalamin. Cyanocobalamin beta-ligand transferase and microsomal cob(III) alamin reductase. Biochim Biophys Acta. 1157(2): 167–177.

Pezacka EH, Jacobsen DW, Luce K and Green R. 1992. Glial cells as a model for the role of cobalamin in the nervous system: impaired synthesis of cobalamin coenzymes in cultured human astrocytes following short-term cobalamin-deprivation. Biochem Biophys Res Commun. 184(2): 832–839.

Pezacka EH and Rosenblatt DS. 1994. Intracellular metabolism of cobalamin. Altered activities of β-axial-ligand transferase and microsomal cob(III)alamin reductase in cblC and cblD fibroblasts. Advances in Thomas Addison's Diseases. HR Bhatt, VHT James, GM Besser, GF Bottazzo and H Keen (eds.). Bristol. J. Endocrinol. pp. 315–323.

Plesa M, Kim J, Paquette SG, Gagnon H, Ng-Thow-Hing C, Gibbs BF, Hancock MA, Rosenblatt DS and Coulton JW. 2011. Interaction between MMACHC and MMADHC, two human proteins participating in intracellular vitamin B12 metabolism. Mol Genet Metab. 102(2): 139–148.

Pons L, Guy M, Lambert D, Hatier R and Gueant J. 2000. Transcytosis and coenzymatic conversion of [(57)Co]cobalamin bound to either endogenous transcobalamin II or exogenous intrinsic factor in caco-2 cells. Cell Physiol Biochem. 10(3): 135–148.

Quadros EV, Lai SC, Nakayama Y, Sequeira JM, Hannibal L, Wang S, Jacobsen DW, Fedosov S, Wright E, Gallagher RC et al. 2010. Positive newborn screen for methylmalonic aciduria identifies the first mutation in TCblR/CD320, the gene for cellular uptake of transcobalamin-bound vitamin B(12). Hum Mutat. 31(8): 924–929.

Quadros EV, Nakayama Y and Sequeira JM. 2009. The protein and the gene encoding the receptor for the cellular uptake of transcobalamin-bound cobalamin. Blood. 113(1): 186–192.

Quadros EV and Sequeira JM. 2013. Cellular uptake of cobalamin: transcobalamin and the TCblR/CD320 receptor. Biochimie. 95(5): 1008–1018.

Quintana AM, Geiger EA, Achilly N, Rosenblatt DS, Maclean KN, Stabler SP, Artinger KB, Appel B and Shaikh TH. 2014. Hcfc1b, a zebrafish ortholog of HCFC1, regulates craniofacial development by modulating mmachc expression. Dev Biol. 396(1): 94–106.

Ramanujam KS, Seetharam S, Dahms NM and Seetharam B. 1991a. Functional expression of intrinsic factor-cobalamin receptor by renal proximal tubular epithelial cells. J Biol Chem. 266(20): 13135–13140.

Ramanujam KS, Seetharam S, Dahms NM and Seetharam B. 1994. Effect of processing inhibitors on cobalamin (vitamin B12) transcytosis in polarized opossum kidney cells. Arch Biochem Biophys. 315(1): 8–15.

Ramanujam KS, Seetharam S, Ramasamy M and Seetharam B. 1991b. Expression of cobalamin transport proteins and cobalamin transcytosis by colon adenocarcinoma cells. Am J Physiol. 260(3 Pt 1): G416–422.

Ramanujam KS, Seetharam S and Seetharam B. 1992. Leupeptin and ammonium chloride inhibit intrinsic factor mediated transcytosis of [57Co]cobalamin across polarized renal epithelial cells. Biochem Biophys Res Commun. 182(2): 439–446.

Raux E, Lanois A, Rambach A, Warren MJ and Thermes C. 1998. Cobalamin (vitamin B12) biosynthesis: functional characterization of the Bacillus megaterium cbi genes required to convert uroporphyrinogen III into cobyrinic acid a,c-diamide. Biochem J. 335(Pt 1): 167–173.

Richard E, Alvarez-Barrientos A, Perez B, Desviat LR and Ugarte M. 2007. Methylmalonic acidaemia leads to increased production of reactive oxygen species and induction of apoptosis through the mitochondrial/caspase pathway. J Pathol. 213(4): 453–461.

Richard E, Jorge-Finnigan A, Garcia-Villoria J, Merinero B, Desviat LR, Gort L, Briones P, Leal F, Perez-Cerda C, Ribes A et al. 2009. Genetic and cellular studies of oxidative stress in methylmalonic aciduria (MMA) cobalamin deficiency type C (cblC) with homocystinuria (MMACHC). Hum Mutat. 30(11): 1558–1566.

Rosenberg LE, Patel L and Lilljeqvist AC. 1975. Absence of an intracellular cobalamin-binding protein in cultured fibroblasts from patients with defective synthesis of 5'-deoxyadenosylcobalamin and methylcobalamin. Proc Natl Acad Sci U S A. 72(11): 4617–4621.

Rosenblatt DS, Hosack A, Matiaszuk NV, Cooper BA and Laframboise R. 1985. Defect in vitamin B12 release from lysosomes: newly described inborn error of vitamin B12 metabolism. Science. 228(4705): 1319–1321.

Rosenblatt DS, Laframboise R, Pichette J, Langevin P, Cooper BA and Costa T. 1986. New disorder of vitamin B12 metabolism (cobalamin F) presenting as methylmalonic aciduria. Pediatrics. 78(1): 51–54.

Rosenblatt DS and Whitehead VM. 1999. Cobalamin and folate deficiency: acquired and hereditary disorders in children. Semin Hematol. 36(1): 19–34.

Roth JR, Lawrence JG and Bobik TA. 1996. Cobalamin (coenzyme B12): synthesis and biological significance. Annu Rev Microbiol. 50: 137–181.

Rutsch F, Gailus S, Miousse IR, Suormala T, Sagne C, Toliat MR, Nurnberg G, Wittkampf T, Buers I, Sharifi A et al. 2009. Identification of a putative lysosomal cobalamin exporter altered in the cblF defect of vitamin B12 metabolism. Nat Genet. 41(2): 234–239.

Rutsch F, Gailus S, Suormala T and Fowler B. 2011. LMBRD1: the gene for the cblF defect of vitamin B12 metabolism. J Inherit Metab Dis. 34(1): 121–126.

Sakamoto O, Ohura T, Matsubara Y, Takayanagi M and Tsuchiya S. 2007. Mutation and haplotype analyses of the MUT gene in Japanese patients with methylmalonic acidemia. J Hum Genet. 52(1): 48–55.

Saridakis V, Yakunin A, Xu X, Anandakumar P, Pennycooke M, Gu J, Cheung F, Lew JM, Sanishvili R, Joachimiak A et al. 2004. The structural basis for methylmalonic aciduria. The crystal structure of archaeal ATP:cobalamin adenosyltransferase. J Biol Chem. 279(22): 23646–23653.

Scalabrino G. 2009. The multi-faceted basis of vitamin B12 (cobalamin) neurotrophism in adult central nervous system: Lessons learned from its deficiency. Prog Neurobiol. 88(3): 203–220.

Schubert HL and Hill CP. 2006. Structure of ATP-bound human ATP:cobalamin adenosyltransferase. Biochemistry. 45(51): 15188–15196.

Seifert C, Bowien S, Gottschalk G and Daniel R. 2001. Identification and expression of the genes and purification and characterization of the gene products involved in reactivation of coenzyme B12-dependent glycerol dehydratase of *Citrobacter freundii*. Eur J Biochem. 268(8): 2369–2378.

Shih VE, Axel SM, Tewksbury JC, Watkins D, Cooper BA and Rosenblatt DS. 1989. Defective lysosomal release of vitamin B12 (cb1F): a hereditary cobalamin metabolic disorder associated with sudden death. Am J Med Genet. 33(4): 555–563.

Sillaots SL, Hall CA, Hurteloup V and Rosenblatt DS. 1992. Heterogeneity in cblG: differential retention of cobalamin on methionine synthase. Biochem Med Metab Biol. 47(3): 242–249.

Soda R, Tavassoli M and Jacobsen DW. 1985. Receptor distribution and the endothelial uptake of transcobalamin II in liver cell suspensions. Blood. 65(4): 795–802.

Sponne IE, Gaire D, Stabler SP, Droesch S, Barbe FM, Allen RH, Lambert DA and Nicolas JP. 2000. Inhibition of vitamin B12 metabolism by OH-cobalamin c-lactam in rat oligodendrocytes in culture: a model for studying neuropathy due to vitamin B12 deficiency. Neurosci Lett. 288(3): 191–194.

St Maurice M, Mera P, Park K, Brunold TC, Escalante-Semerena JC and Rayment I. 2008. Structural characterization of a human-type corrinoid adenosyltransferase confirms that coenzyme B12 is synthesized through a four-coordinate intermediate. Biochemistry. 47(21): 5755–5766.

Stabler SP, Brass EP, Marcell PD and Allen RH. 1991. Inhibition of cobalamin-dependent enzymes by cobalamin analogues in rats. J Clin Invest. 87(4): 1422–1430.

Stich TA, Buan NR and Brunold TC. 2004. Spectroscopic and computational studies of Co2+corrinoids: spectral and electronic properties of the biologically relevant base-on and base-off forms of Co2+cobalamin. J Am Chem Soc. 126(31): 9735–9749.

Stich TA, Yamanishi M, Banerjee R and Brunold TC. 2005. Spectroscopic evidence for the formation of a four-coordinate Co2+ cobalamin species upon binding to the human ATP:cobalamin adenosyltransferase. J Am Chem Soc. 127(21): 7660–7661.

Stockler S, Corvera S, Lambright D, Fogarty K, Nosova E, Leonard D, Steinfeld R, Ackerley C, Shyr C, Au N et al. 2014. Single point mutation in Rabenosyn-5 in a female with intractable seizures and evidence of defective endocytotic trafficking. Orphanet J Rare Dis. 9: 141.

Strope S, Rivi R, Metzger T, Manova K and Lacy E. 2004. Mouse amnionless, which is required for primitive streak assembly, mediates cell-surface localization and endocytic function of cubilin on visceral endoderm and kidney proximal tubules. Development. 131(19): 4787–4795.

Stucki M, Coelho D, Suormala T, Burda P, Fowler B and Baumgartner MR. 2012. Molecular mechanisms leading to three different phenotypes in the cblD defect of intracellular cobalamin metabolism. Hum Mol Genet. 21(6): 1410–1418.

Suarez-Moreira E, Yun J, Birch CS, Williams JH, McCaddon A and Brasch NE. 2009. Vitamin B12 and redox homeostasis: cob(II)alamin reacts with superoxide at rates approaching superoxide dismutase (SOD). J Am Chem Soc. 131(42): 15078–15079.

Sugase K, Dyson HJ and Wright PE. 2007. Mechanism of coupled folding and binding of an intrinsically disordered protein. Nature. 447(7147): 1021–1025.

Suormala T, Baumgartner MR, Coelho D, Zavadakova P, Kozich V, Koch HG, Berghauser M, Wraith JE, Burlina A, Sewell A et al. 2004. The cblD defect causes either isolated or combined deficiency of methylcobalamin and adenosylcobalamin synthesis. J Biol Chem. 279(41): 42742–42749.

Takahashi K, Tavassoli M and Jacobsen DW. 1980. Receptor binding and internalization of immobilized transcobalamin II by mouse leukaemia cells. Nature. 288(5792): 713–715.

Toraya T. 2000. Radical catalysis of B12 enzymes: structure, mechanism, inactivation, and reactivation of diol and glycerol dehydratases. Cell Mol Life Sci. 57(1): 106–127.

Trakadis YJ, Alfares A, Bodamer OA, Buyukavci M, Christodoulou J, Connor P, Glamuzina E, Gonzalez-Fernandez F, Bibi H, Echenne B et al. 2014. Update on transcobalamin deficiency: clinical presentation, treatment and outcome. J Inherit Metab Dis. 37(3): 461–473.

Utley CS, Marcell PD, Allen RH, Antony AC and Kolhouse JF. 1985. Isolation and characterization of methionine synthetase from human placenta. J Biol Chem. 260(25): 13656–13665.

Vassiliadis A, Rosenblatt DS, Cooper BA and Bergeron JJ. 1991. Lysosomal cobalamin accumulation in fibroblasts from a patient with an inborn error of cobalamin metabolism (cblF complementation group): visualization by electron microscope radioautography. Exp Cell Res. 195(2): 295–302.

Warren MJ. 2006. Finding the final pieces of the vitamin B12 biosynthetic jigsaw. Proc Natl Acad Sci U S A. 103(13): 4799–4800.

Warren MJ, Raux E, Schubert HL and Escalante-Semerena JC. 2002. The biosynthesis of adenosylcobalamin (vitamin B12). Nat Prod Rep. 19(4): 390–412.

Watanabe F, Nakano Y, Maruno S, Tachikake N, Tamura Y and Kitaoka S. 1989. NADH- and NADPH-linked aquacobalamin reductases occur in both mitochondrial and microsomal membranes of rat liver. Biochem Biophys Res Commun. 165(2): 675–679.

Watanabe F, Saido H, Yamaji R, Miyatake K, Isegawa Y, Ito A, Yubisui T, Rosenblatt DS and Nakano Y. 1996. Mitochondrial NADH- or NADPH-linked aquacobalamin reductase activity is low in human skin fibroblasts with defects in synthesis of cobalamin coenzymes. J Nutr. 126(12): 2947–2951.

Watkins D and Rosenblatt DS. 1986. Failure of lysosomal release of vitamin B12: a new complementation group causing methylmalonic aciduria (cblF). Am J Hum Genet. 39(3): 404–408.

Watkins D and Rosenblatt DS. 1989. Functional methionine synthase deficiency (cblE and cblG): clinical and biochemical heterogeneity. Am J Med Genet. 34(3): 427–434.

Watkins D and Rosenblatt DS. 2011. Inborn errors of cobalamin absorption and metabolism. Am J Med Genet C Semin Med Genet. 157C(1): 33–44.

Watkins D, Ru M, Hwang HY, Kim CD, Murray A, Philip NS, Kim W, Legakis H, Wai T, Hilton JF et al. 2002. Hyperhomocysteinemia due to methionine synthase deficiency, cblG: structure of the MTR gene, genotype diversity, and recognition of a common mutation, P1173L. Am J Hum Genet. 71(1): 143–153.

Wilkemeyer MF, Crane AM and Ledley FD. 1991. Differential diagnosis of mut and cbl methylmalonic aciduria by DNA-mediated gene transfer in primary fibroblasts. J Clin Invest. 87(3): 915–918.

Willard HF, Mellman IS and Rosenberg LE. 1978. Genetic complementation among inherited deficiencies of methylmalonyl-CoA mutase activity: evidence for a new class of human cobalamin mutant. Am J Hum Genet. 30(1): 1–13.

Willard HF and Rosenberg LE. 1979. Inborn errors of cobalamin metabolism: effect of cobalamin supplementation in culture on methylmalonyl CoA mutase activity in normal and mutant human fibroblasts. Biochem Genet. 17(1-2): 57–75.

Wilson A, Leclerc D, Saberi F, Campeau E, Hwang HY, Shane B, Phillips JA, 3rd, Rosenblatt DS and Gravel RA. 1998. Functionally null mutations in patients with the cblG-variant form of methionine synthase deficiency. Am J Hum Genet. 63(2): 409–414.

Wolthers KR, Basran J, Munro AW and Scrutton NS. 2003. Molecular dissection of human methionine synthase reductase: determination of the flavin redox potentials in full-length enzyme and isolated flavin-binding domains. Biochemistry. 42(13): 3911–3920.

Wolthers KR, Lou X, Toogood HS, Leys D and Scrutton NS. 2007a. Mechanism of coenzyme binding to human methionine synthase reductase revealed through the crystal structure of the FNR-like module and isothermal titration calorimetry. Biochemistry. 46(42): 11833–11844.

Wolthers KR, Toogood HS, Jowitt TA, Marshall KR, Leys D and Scrutton NS. 2007b. Crystal structure and solution characterization of the activation domain of human methionine synthase. FEBS J. 274(3): 738–750.

Wolthers KR and Scrutton NS. 2004. Electron transfer in human methionine synthase reductase studied by stopped-flow spectrophotometry. Biochemistry. 43(2): 490–500.

Wolthers KR and Scrutton NS. 2007. Protein interactions in the human methionine synthase-methionine synthase reductase complex and implications for the mechanism of enzyme reactivation. Biochemistry. 46(23): 6696–6709.

Wolthers KR and Scrutton NS. 2009. Cobalamin uptake and reactivation occurs through specific protein interactions in the methionine synthase-methionine synthase reductase complex. FEBS J. 276(7): 1942–1951.

Yamada K, Gherasim C, Banerjee R and Koutmos M. 2015. Structure of Human B12 trafficking protein CblD reveals molecular mimicry and identifies a new subfamily of Nitro-FMN reductases. J Biol Chem. 290(49): 29155–2916.

Yamada K, Gravel RA, Toraya T and Matthews RG. 2006. Human methionine synthase reductase is a molecular chaperone for human methionine synthase. Proc Natl Acad Sci U S A. 103(25): 9476–9481.

Yamada R, Shimizu S and Fukui S. 1968. Disproportionation of vitamin B12r under various mild conditions. Biochemistry. 7(5): 1713–1719.

Yamanishi M, Labunska T and Banerjee R. 2005. Mirror "base-off" conformation of coenzyme B12 in human adenosyltransferase and its downstream target, methylmalonyl-CoA mutase. J Am Chem Soc. 127(2): 526–527.

Yu HC, Sloan JL, Scharer G, Brebner A, Quintana AM, Achilly NP, Manoli I, Coughlin CR, 2nd, Geiger EA, Schneck U et al. 2013. An X-linked cobalamin disorder caused by mutations in transcriptional coregulator HCFC1. Am J Hum Genet. 93(3): 506–514.

Zhang J, Wu X, Padovani D, Schubert HL and Gravel RA. 2009. Ligand-binding by catalytically inactive mutants of the cblB complementation group defective in human ATP:cob(I)alamin adenosyltransferase. Mol Genet Metab. 98(3): 278–284.

Zheng D, Yan L and Birke RL. 2002. Electrochemical and spectral studies of the reactions of aquocobalamin with nitric oxide and nitrite ion. Inorg Chem. 41(9): 2548–2555.

4

Inherited Defects of Cobalamin Metabolism

*David Watkins** and *David S Rosenblatt*

1. Introduction

Inborn errors of cobalamin (vitamin B12) metabolism result in decreased synthesis of the cobalamin coenzyme derivatives adenosylcobalamin and methylcobalamin, either alone or in combination, and decreased function of one or both of the cobalamin-dependent enzymes methylmalonyl-CoA mutase and methionine synthase. Decreased methylmalonyl-CoA mutase activity results in accumulation of its substrate methylmalonic acid (MMA), and is associated with a tendency to development of life threatening metabolic acidotic crises and, in the long-term, moderate to severe intellectual handicap and end-stage renal failure. Decreased function of methionine synthase leads to increased serum homocysteine levels as well as decreased levels of methionine and its derivative S-adenosylmethionine, the major methyl group donor in cellular metabolism. Hyperhomocysteinemia has been associated with hypercoagulability and a variety of neurologic problems, while decreased methionine levels are associated with decreased central nervous system myelination. Inborn errors of cobalamin metabolism result in problems affecting multiple systems, including hematologic, neurologic, ophthalmologic, dermatologic, cardiovascular and renal.

The first inborn errors of cobalamin metabolism were reported in the late 1960s, with the description of patients with elevated serum and urine levels of MMA in the presence of serum vitamin B12 levels within the reference range, including cases in which MMA levels responded to therapy with cobalamin.

Department of Human Genetics, McGill University, Montreal, Quebec, Canada.
Email: david.watkins@mcgill.ca; david.rosenblatt@mcgill.ca
* Corresponding author

Subsequently additional patients with cobalamin-responsive combined methylmalonic aciduria and homocystinuria or isolated homocystinuria were described, and somatic cell complementation analysis was used to identify classes of patients with mutations at the same locus (Gravel et al. 1975; Willard et al. 1978; Watkins and Rosenblatt 1986; Watkins and Rosenblatt 1989). Currently, ten classes of inborn error of cellular cobalamin metabolism, designated *mut*, *cblA-cblG*, *cblJ* and *cblX*, are recognized (Table 1). These disorders are inherited as autosomal recessive traits with the exception of the *cblX* disorder, which is X-linked. Identification of the gene underlying each of these disorders has provided a means to study the reactions catalyzed by

Table 1. Inborn Errors of Cobalamin Transport and Metabolism.

Disorder	MIM Number	Biochemical Findings	Gene	Location	MIM Gene #
Transcobalamin Receptor Deficiency	613646	MMA + Hcy	CD320	19p13.2	606475
cblA	251100	MMA	MMAA	4q31.21	607481
cblB	251110	MMA	MMAB	12q24	607568
cblC	277400	MMA + Hcy	MMACHC	1p34.1	609831
cblD*	277410	MMA Hcy MMA + Hcy	MMADHC	2q23.2	611935
cblE	236270	Hcy	MTRR	5p15.31	602568
cblF	277380	MMA + Hcy	LMBRD1	6q13	612625
cblG	250940	MMA + Hcy	MTR	1q43	157570
cblJ	614857	MMA +Hcy	ABCD4	14q24.3	603214
cblX	309541	MMA + Hcy	HCFC1	Xq28	300019
Methylmalonyl-CoA Mutase Deficiency	251000	MMA	MUT	6p12.3	609058

MMA = methylmalonic acidemia (aciduria)
Hcy = hyperhomocysteinemia, homocystinuria
* patients with the *cblD* disorder can present with MMA, Hcy or MMA + Hcy, depending on the mutations present in the *MMADHC* gene.

their gene products, providing novel information on human cellular cobalamin metabolism.

   The biochemical characteristics of inborn errors of cobalamin metabolism reflect the step in cellular cobalamin metabolism that is affected (Figure 1). Exogenous cobalamin is internalized by carrier mediated endocytosis and converted to the active cobalamin derivatives adenosylcobalamin and methylcobalamin in the mitochondria and cytoplasm, respectively. Adenosylcobalamin is required for activity of the mitochondrial enzyme methylmalonyl-CoA mutase, which converts methylmalonyl-CoA to succinyl-

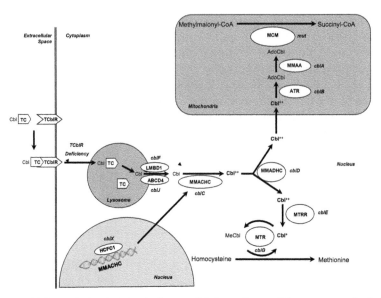

Figure 1. Cellular Cobalamin Metabolism and Inborn Errors. Inborn errors of cobalamin metabolism (*mut, cblA-cblG,* and *cblJ,* are shown beside the protein affected in each disorder. Abbreviations: ABCD4; product of the *ABCD4* gene, AdoCbl; - adenosylcobalamin, ATR; cobalamin adenosyltransferase (product of the *MMAB* gene), Cbl; cobalamin, cbl+; cob(I)alamin, cbl++; cob(II) alamin, HCFC1; product of the *HCFC1* gene, host factor 1, LMBD1; product of the *LMBRD1* gene, MCM; methylmalonyl-CoA mutase (product of the *MUT* gene), MeCbl; methylcobalamin, MMAA; product of the *MMAA* gene, MMACHC; product of the *MMACHC* gene, MMADHC; product of the *MMADHC* gene, MTR; methionine synthase (product of the *MTR* gene), MTRR; methionine synthase reductase (product of the *MTRR* gene), TC; transcobalamin, TCblR; transcobalamin receptor.

CoA; methylcobalamin is generated by the cytoplasmic enzyme methionine synthase, which catalyzes methylation of homocysteine to form methionine. Disorders that affect cellular uptake of cobalamin or early steps in cellular metabolism common to synthesis of both coenzyme derivatives result in accumulation of both MMA and homocysteine in blood (methylmalonic acidemia and hyperhomocysteinemia) and urine (methylmalonic aciduria and homocystinuria). Isolated methylmalonic aciduria occurs when synthesis of adenosylcobalamin alone is affected, while isolated homocystinuria occurs when synthesis of methylcobalamin is impaired. The *cblD* disorder represents a special case. Depending on the site and nature of the causal mutations, *cblD* patients may present with combined methylmalonic aciduria and homocystinuria, isolated methylmalonic aciduria, or isolated homocystinuria.

2. Diagnosis of Inborn Errors of Cobalamin Metabolism

The inborn errors have traditionally been diagnosed by studies of cultured patient skin fibroblasts. The function of methylmalonyl-CoA mutase is assessed by measuring incorporation of label from [^{14}C]propionate into trichloroacetic acid-precipitable cellular macromolecules in intact patient cells. Similarly, function of methionine synthase is assessed by measurement of incorporation of label from 5-[^{14}C]methyltetrahydrofolate into cellular macromolecules, or incorporation of label from [^{14}C]formate into cellular methionine. Ability to synthesize cobalamin coenzyme derivatives is determined by measurement of uptake of [^{57}Co]cyanocobalamin bound to human transcobalamin and its conversion to adenosylcobalamin and methylcobalamin. Assignment of the patient disorder to one of the known complementation classes is then achieved by somatic cell complementation analysis. With the identification of the causal genes for all of the known inborn errors of cobalamin, it is now possible to proceed directly to gene sequencing as the first step in diagnosis, and next generation sequencing panels that incorporate all known genes in the pathway are currently available. However, if no causal mutations can be identified or if the pathogenicity of novel sequence variants is not clear, biochemical studies of cultured patient fibroblasts remain an important means of establishing a diagnosis.

In the past, most patients with inborn errors of cobalamin metabolism came to medical attention when they developed symptoms of their disorder. In many cases there was a delay of variable length between development of symptoms and establishment of a diagnosis, resulting in delays in starting appropriate treatment. Neonatal screening for MMA in urine using chromatography was first performed in the 1980s and allowed for presymptomatic identification of patients with disorders characterized by methylmalonic aciduria with or without homocystinuria (Coulombe et al. 1981; Lemieux et al. 1988). More recently, screening for C3 acylcarnitine in newborn blood spots using tandem mass spectroscopy, followed by specific measurement of MMA and/or methylcitric acid in urine in patients with a positive result, has become common (Wilcken et al. 2003; Cheng et al. 2010). Techniques for tandem mass spectrometric measurement of homocysteine in newborn blood spots have been developed, with measurement of serum total homocysteine as second-tier measurement (Tortorelli et al. 2010; Turgeon et al. 2010). These have not been widely adopted to date but show promise (Wong et al. 2015) and will likely be used more frequently in the future.

Prenatal diagnosis has been performed for most of the disorders using cultured amniocytes or chorionic villus cells, measurements of metabolites in amniotic fluid or maternal blood and, as the underlying genes have been identified, sequencing of genomic DNA isolated from amniocytes or chorionic villus cells (Morel et al. 2005; Rosenblatt and Watkins 2010). Both cell studies and measurement of metabolites in amniotic fluid have given rise to false positive results, leading to suggestions that two independent techniques should always be used for prenatal diagnosis. Results of studies using amniocytes have been more reliable than those using chorionic villus cells (Morel et al. 2005). Prenatal diagnosis has been successfully carried out in a pregnancy at risk for the *mut* disorder using analysis of fetal cell-free DNA in maternal plasma (Gu et al. 2014).

Table 2 presents a scheme for diagnosis of infants or children with possible inborn errors of cobalamin metabolism based on identification of elevated methylmalonic acid and/or homocysteine levels. The scheme utilizes gene sequencing as the first-line strategy for identification of the patient's disorder; however, if the pathogenicity of sequence variants identified is unclear, additional somatic cell studies may be necessary to confirm the diagnosis. The next sections describe the different groups of inborn error of intracellular cobalamin metabolism.

3. Isolated Methylmalonic Aciduria

Methylmalonyl-CoA is generated in the mitochondria from propionyl-CoA formed during the breakdown of branched-chain amino acids, odd-chain fatty acids and other molecules. Conversion of methylmalonyl-CoA to succinyl-CoA depends on the activity of methylmalonyl-CoA mutase (the product of the *MUT* gene on chromosome 6p12.3) as well as cob(I)alamin adenosyltransferase (product of the *MMAB* gene on chromosome 12q24) and the G-protein chaperone MMAA (product of the *MMAA* gene on chromosome 4q31.21), which are required for synthesis of adenosylcobalamin, its transfer to methylmalonyl-CoA mutase and its stabilization. Mutations affecting activity of any of these proteins result in decreased ability to convert methylmalonyl-CoA to succinyl-CoA and accumulation of MMA in blood and urine, in the absence of elevated serum and urine homocysteine levels. Mutations in the *MUT* gene cause the *mut* disorder (MIM 251000); mutations in the *MMAA* gene result in the *cblA* disorder (MIM 251100); and mutations affecting the *MMAB* gene cause the *cblB* disorder (MIM 251110). Patients with the *mut* disorder have been subdivided into two classes: *mut⁰* patients have no detectable methylmalonyl-CoA mutase protein and no response of mutase activity in cultured fibroblasts to addition of hydroxocobalamin to the culture medium; *mut⁻* patients have detectable methylmalonyl-CoA mutase protein and respond to hydroxocobalamin supplementation by increasing methylmalonyl-CoA mutase activity.

Urine levels of MMA are elevated in patients with all three disorders, with values ranging from 10 to 10,000 or more nmol/mol creatinine (reference < 5 mmol/mol creatinine). Levels of metabolites of MMA (methylcitrate, 3-hydroxypropionate, lactate and propionylglycine) are also elevated (Fowler et al. 2008). Serum MMA levels are also elevated; mean levels of 70 µmol/L for relatively mildly affected patients and levels from 210–5452 µmol/L in patients severe disease have been reported (reference < 0.27 µmol/L) (Nizon et al. 2013; Niemi et al. 2015). Serum total homocysteine and cobalamin levels are within the reference range. In addition to *mut*, *cblA* and *cblB*, methylmalonic aciduria in the absence of homocystinuria also occurs in other genetic disorders. Mild methylmalonic aciduria occurs in patients with mutations affecting the *MCEE* gene, which encodes methylmalonyl-CoA

Table 2. Diagnosis of patients with methylmalonic aciduria and/or homocystinuria in infants and children.

A.   PATIENTS WITH ELEVATED SERUM MMA

Infant with elevated C3-carnitine on tandem mass spectrometry of newborn blood spot, and confirmation of elevated serum/urine MMA; or with elevated serum or urine MMA following symptomatic presentation.

Low or low-normal maternal serum vitamin B12:

Possible maternal cobalamin deficiency due to dietary insufficiency or undiagnosed pernicious anemia.

Elevated serum total homocysteine:

See C

Serum total homocysteine within reference range:

Isolated methylmalonic aciduria due to mutations in *MUT* (*mut⁰*, *mut⁻*), *MMAA* (*cblA*), *MMAB* (*cblB*) or *MMADHC* (*cblD* variant 2) genes.

Mild methylmalonic aciduria due to mutations in *MCEE* (methylmalonyl-CoA epimerase) gene.

Mild methylmalonic aciduria due to mutations in *ACSF3* gene; malonic aciduria may or may not be present as well.

Mild methylmalonic aciduria due to mutations in *SUCLG1*, *SUCLA2* (succinate-CoA ligase) genes; symptoms of mitochondrial disease will also be present.

B.   PATIENTS WITH ELEVATED SERUM HOMOCYSTEINE

Infant with elevated serum total homocysteine following symptomatic presentation, or with elevated homocysteine on newborn testing.

Elevated serum MMA or C3-carnitine, or elevated urine MMA:

See C

Elevated serum methionine:

Classical homocystinuria caused by mutations in *CBS* (cystathionine β-synthase) gene.

Low or low-normal methionine, presence of megaloblastic anemia:

Homocystinuria due to mutations in *MMADHC* (*cblD* variant 1), *MTRR* (*cblE*) or *MTR* (*cblG*) genes.

Homocystinuria due to mutations in the *MTHFD1* (methylenetetrahydrofolate dehydrogenase 1) gene.

Low or low-normal methionine, absence of megaloblastic anemia:

Homocystinuria due to mutations in the *MTHFR* (methylenetetrahydrofolate reductase) gene.

C.   PATIENTS WITH ELEVATED MMA AND HOMOCYSTEINE
Inborn errors of cellular cobalamin metabolism:

Caused by mutations at the *MMACHC* (*cblC*), *MMADHC* (*cblD*), *LMBRD1* (*cblF*), *ABCD4* (*cblJ*) genes; mutations in the *HCFC1* gene (*cblX*) if patient is male. Serum cobalamin levels normal.

Transcobalamin deficiency:

Caused by mutations in the *TCN2* gene. Serum cobalamin levels normal.

Inborn errors of intestinal cobalamin absorption:

Caused by mutations in the *GIF* (intrinsic factor deficiency), *CUBN* or *AMN* (Imerslund-Gräsbeck syndrome) genes. Presentation is typically later (after 1 year of age) after cobalamin stores present at birth are exhausted. Serum cobalamin levels decreased. Not detected by newborn screening.

Decreased cellular cobalamin uptake:

Caused by mutation in the *CD320* (transcobalamin receptor) gene. Serum cobalamin levels normal.

---

isomerase (racemase), the enzyme that catalyzes the step prior to mutase in propionyl-CoA metabolism (Dobson et al. 2006). Pathways for non-enzymic isomerization of methylmalonyl-CoA are believed to exist (Montgomery et al. 1983) and it is not clear that patients with MCEE mutations have any consistent clinical symptoms. Mutations in the *SUCLG1* and *SUCLA2* genes, which encode subunits of succinate-CoA ligase, the Krebs cycle enzyme that utilizes succinyl-CoA generated by methylmalonyl-CoA mutase, result in mild methylmalonic aciduria accompanied by symptoms of mitochondrial dysfunction (Ostergaard 2008). Mutations in the *ACSF3* gene, which encodes a member of the acyl-CoA synthetase family of unknown function, may also have methylmalonic aciduria with or without accompanying malonic aciduria (Alfares et al. 2011; Sloan et al. 2011).

Patients with isolated methylmalonic aciduria are typically well at birth. Severely affected individuals come to medical attention during the first hours or days of life. Most patients present before the end of the first year, although a few have come to medical attention later in childhood or even in adolescence. Patients are prone to potentially life threatening episodes of metabolic acidosis and hyperammonemia, which may be precipitated by febrile illnesses or increased protein intake. These crises can lead to multiorgan failure and death. Surviving patients often have a difficult clinical course with multiple hospital admissions due to episodes of metabolic decompensation. Patients typically present with lethargy, failure to thrive, recurrent vomiting, dehydration, respiratory distress and hypotonia. Laboratory testing shows elevated serum and urine concentrations of MMA as well as propionate and methylcitrate, ammonia, ketones, and glycine. There are also elevated levels of propionylcarnitine. Homocysteine levels are within the reference range. Neutropenia, thrombocytopenia, and/or anemia have been reported in approximately 50% of patients.

Long-term consequences of methylmalonic aciduria include moderate to severe intellectual handicap, chronic renal failure, growth and psychomotor retardation, mild microcephaly, and choreoathetosis and tremor. Bilateral globus pallidus infarcts are seen frequently, and small lacunar infarcts in the pars reticulata of the substantia nigra may also be present (Baker et al. 2015), but neurological impairment may be seen in patients with no evidence of basal ganglia injury (Ktena et al. 2015). The lesions represent specific sensitivity of cells in these particular structures to damage by high levels of MMA rather than interruption of vascular supply. Renal disease, characterized as chronic tubulo-interstitial nephritis affecting particularly the proximal tubules, can progress to end-stage renal failure requiring dialysis and kidney transplantation (Morath et al. 2013). The precise pathology underlying kidney damage in methylmalonic acidemia is unknown. High frequencies of joint hypermobility and pes planus have been noted (Ktena et al. 2015).

In general, patients with early-onset methylmalonic aciduria are more severely affected than those with later-onset disease. As well, studies have shown that mut^0 patients have an earlier age at onset, have a more severe clinical course, are more likely to develop end-stage renal disease, have a less favorable neurocognitive outcome, and are more likely to die than patients with other forms of methylmalonic aciduria; *cblB* patients tend to be more severely affected than *cblA* or *mut⁻* patients (Matsui et al. 1983; Hörster et al. 2007; Cosson et al. 2009; O'Shea et al. 2012). Although treatment outcomes have improved with time, mortality among *mut* patients remains in the range of 40% (Sloan et al. 2015).

Long-term treatment of methylmalonic aciduria includes a restricted protein diet to minimize production of propionyl-CoA from branched-chain amino acids, and IM cobalamin in patients that are clinically cobalamin-responsive. Both cyanocobalamin and hydroxocobalamin have been used successfully; treatment with adenosylcobalamin does not appear to have any advantage over these forms (Batshaw et al. 1984). Carnitine is used to prevent carnitine deficiency, and antibiotics may be given to decrease production of propionic acid by gut flora. Despite these treatments, MMA levels remain markedly elevated, and episodes of decompensation continue to occur. In addition to renal transplantation in patients with end-stage renal failure, liver transplantation (combined liver and kidney transplantation in patients with end-stage renal failure) has been advocated to increase metabolic control and prevent metabolic decompensation. Successful transplantation has been associated with decreased, but not normal, levels of MMA and improved metabolic control (Niemi et al. 2015; Sloan et al. 2015). However, neurological dysfunction and renal deterioration may continue despite the improved biochemical parameters.

On the basis of results of newborn screening, the estimated frequency of isolated methylmalonic aciduria has ranged from 1:22,000 to 1:115,000 births in different populations. The *mut* disorder is the most common form of isolated methylmalonic aciduria, with several hundred patients identified. Over 240 *MUT* mutations have been identified, including specific mutations common in North American Hispanic patients, patients of African descent, and Japanese patients (Worgan et al. 2006). Over 120 patients with the *cblA* disorder, and over 60 patients with the *cblB* disorder, have been reported.

4. Isolated Homocystinuria

Methylcobalamin is generated during the catalytic cycle of methionine synthase, which catalyzes remethylation of homocysteine to form methionine. Remethylation represents one of two pathways for eliminating potentially harmful homocysteine: transsulfuration, with conversion of homocysteine to cystathionine as the first step and leading to generation of cysteine, is the other. Mutations affecting either pathway result in accumulation of homocysteine in blood and urine. Disorders affecting remethylation result in decreased or low-normal levels of methionine in addition to elevated homocysteine, while those affecting transsulfuration (classic homocystinuria due to mutations affecting the *CBS* gene encoding cystathionine β-synthase) result in elevated methionine levels.

Homocysteine remethylation requires the presence of two proteins. Methionine synthase (5-methyltetrahydrofolate:homocysteine methyltransferase), encoded by the *MTR* gene on chromosome 1q43, catalyzes transfer of a methyl group from 5-methyltetrahydrofolate to enzyme-bound fully-reduced cob(I)alamin to form methylcobalamin, followed by transfer of the methyl group from methylcobalamin to homocysteine to form methionine and regenerate cob(I)alamin. During this catalytic cycle, cob(I)alamin may be spontaneously oxidized to the cob(II)alamin form; when this happens, methionine synthase reductase, encoded by the *MTRR* gene on chromosome 5p15.31, catalyzes reductive methylation of the cobalamin prosthetic group using S-adenosylmethionine as methyl donor to regenerate methylcobalamin. Isolated cobalamin-responsive homocystinuria is caused by mutations affecting one or the other of these proteins: the *cblG* disorder (MIM 250940) is caused by mutations affecting the *MTR* gene (Gulati et al. 1996; Leclerc et al. 1986), while the *cblE* disorder (MIM 236270) is caused by mutations affecting the *MTRR* gene (Leclerc et al. 1998).

Clinical presentation in the *cblE* and *cblG* disorders is virtually identical (Huemer et al. 2014a); differentiation requires either complementation analysis or sequencing of the *MTR* and *MTRR* genes. Patients usually present in the first year of life, but presentation later in childhood or adulthood has been reported. Megaloblastic anemia is almost universally present; neutropenia, thrombocytopenia or pancytopenia may also be observed. The presence of megaloblastic anemia can differentiate the *cblE* and *cblG* disorders from severe methylenetetrahydrofolate reductase deficiency, which also results in impaired homocysteine remethylation but is not characterized by megaloblastosis. Neurological findings are usually prominent, with developmental delay, cerebral atrophy, microcephaly, hypotonia, seizures and EEG abnormalities, hydrocephalus, ataxia and psychiatric symptoms all described; however, patients with no neurological findings have been reported (Vilaseca et al. 2003). Hypotonia may be a more frequent finding in the *cblG* disorder compared to the *cblE* disorder (Huemer et al. 2014a). Atypical hemolytic uremic syndrome and glomerulopathy have been described in some patients (Paul et al. 2013; Huemer et al. 2014a).

Serum total homocysteine is elevated in patients with the *cblE* and *cblG* disorders; values in the range of 30–250 μM have been reported (reference < 14 μM). These levels are higher than those typical of dietary cobalamin deficiency but lower than those typical of cystathionine β-synthase deficiency. Serum methionine is decreased or in the low-normal range. Serum and urine MMA and serum cobalamin levels are within the reference range. In cultured patient fibroblasts, synthesis of methylcobalamin from exogenous cyanocobalamin is decreased and methionine synthase function is reduced. Methionine synthase specific activity in extracts of patient cells is decreased under all assay conditions in the *cblG* disorder. In *cblE* fibroblast extracts, methionine synthase specific activity is within the reference range when the assay is performed under standard conditions, which include saturating levels of reducing agent, but is decreased compared to control extracts when the concentration of reducing agent is reduced (Watkins and Rosenblatt 1988; Watkins and Rosenblatt 1989).

Treatment of these disorders involves supplementation with cobalamin. Both cyanocobalamin and hydroxocobalamin, given orally or IM, have been effective. There has been no consensus on appropriate dosage. Oral betaine is used to reduce homocysteine level; betaine is the substrate for the liver enzyme betaine:homocysteine methyltransferase, which catalyzes cobalamin-independent homocysteine remethylation. Folate (folic acid or folinic acid) and methionine have also been given to some patients. With treatment, total homocysteine is reduced but remains above the reference range. Most patients continue to experience neurological deficits despite treatment. Longer delay between onset of clinical symptoms and starting treatment is associated with poorer outcome, emphasizing the importance of early treatment (Huemer et al. 2014). A patient diagnosed prenatally and treated with hydroxocobalamin during pregnancy and after birth has been less severely affected than his older brother (Rosenblatt et al. 1985).

Approximately 35 patients with the *cblE* disorder have been reported, and 28 different *MTRR* mutations have been described. The most common of these is a deep intronic sequence change, c.903+469T>C, which results in incorporation of 140 bp of intronic sequence into mRNA (Homolova et al. 2010). A c.1361C>T (p.S454L) mutation may be associated with a mild variant lacking neurological involvement (Vilaseca et al. 2003). Forty-seven *cblG* patients have been described, with 39 different *MTR* mutations identified. The most common is a c.3518C>T (p.P1173L) mutations which that represents approximately 27% of reported mutant alleles (Watkins et al. 2002).

5. Combined Methylmalonic Aciduria and Homocystinuria

Patients with defects affecting synthesis of both cobalamin coenzyme derivatives have elevated levels of both MMA and homocysteine in blood and urine. The frequency of combined methylmalonic aciduria and homocystinuria, determined by newborn screening, ranges from 1:59,000 to 1:71,000. The *cblC* disorder is by far the most frequent cause of combined methylmalonic aciduria and homocystinuria.

5.1 Transcobalamin receptor deficiency (MIM 613646)

Sequence variants in the *CD320* gene on chromosome 19p13.2, which encodes the transcobalamin receptor, have been identified and have been shown to reduce endocytosis of the cobalamin-transcobalamin complex (Quadros et al. 2010). Affected individuals typically have some degree of methylmalonic aciduria and homocystinuria. However, levels are not as high as seen in patients with other inborn errors of cobalamin metabolism, and no consistent clinical picture has been described. One individual homozygous for a c.262_264delGAG mutation was reported to have bilateral retinal occlusions, possibly the result of elevated total serum homocysteine (Karth et al. 2012). On the other hand, the same mutation has been described as a polymorphism in the Irish population, and asymptomatic homozygous individuals have been detected (Pangilinan et al. 2010). Currently it is not clear that any clinical disorder is consistently associated with *CD320* mutations.

5.2 cblF (MIM 277380) and cblJ (MIM 614857)

Transport of endocytosed cobalamin across the lysosomal membrane into the cytoplasm is impaired in patients with the *cblF* and *cblJ* disorders. Fibroblasts from patients with either disorder are characterized by accumulation of unmetabolized cobalamin in lysosomes with decreased synthesis of both adenosylcobalamin and methylcobalamin and decreased function of both cobalamin-dependent enzymes (Rosenblatt et al. 1985; Coelho et al. 2012). Patients typically have elevated levels of both MMA and homocysteine in blood and urine. Serum cobalamin levels are frequently low, possibly reflecting involvement of lysosomes in transit of absorbed cobalamin across enterocytes.

Patients with the *cblF* disorder have had variable clinical presentations. Findings recorded in more than one patient have included megaloblastic anemia, feeding difficulties, failure to thrive, developmental delay, persistent stomatitis, congenital heart defects and mild facial dysmorphology. Fifteen patients have been reported. Mutations in the *LMBRD1* gene on chromosome 6q13 (Rutsch et al. 2009) have been identified in all cases. A common c.1056delG (p.L352fsX18) *LMBRD1* mutation has been identified in patients from several ethnic groups and represents approximately two-thirds of mutant alleles in patients with the disorder.

cblF patients have been treated with IM or oral hydroxocobalamin or cyanocobalamin. This has usually resulted in normalization of biochemical and hematological parameters. Betaine, carnitine, folic acid and protein restriction have been utilized in some cases. Most patients have had good neurologic development on treatment. One patient treated with monthly injections of cyanocobalamin for several years was found at 14 years of age to have elevated serum MMA and total homocysteine, but no apparent hematologic, neurologic or developmental problems (Oladipo et al. 2011).

The two initial patients reported with the *cblJ* disorder had presentation early in life. One, detected by newborn screening, had feeding difficulties, hypotonia, lethargy and bone marrow suppression, while the second had feeding difficulties, macrocytic anemia and congenital heart defects (Coelho et al. 2012). Two Taiwanese patients were subsequently reported with later presentation (4 and 6 years of age). Both patients had skin hyperpigmentation and prematurely gray hair. One patient complained of occasional dizziness and headaches; the second presented with no symptoms beyond the skin and hair abnormalities. In both cases macrocytic anemia, hyperhomocysteinemia, and methylmalonic aciduria were present although in the second case these were looked for only after the causative gene had been identified by whole exome sequencing (Kim et al. 2012; Takeichi et al. 2015). Mutations in the *ABCD4* gene on chromosome 14q14.3 were identified in all reported *cblJ* patients by whole exome sequencing (Coelho et al. 2012; Kim et al. 2012; Takeichi et al. 2015). The Taiwanese patients were both homozygous for the same c.423C>G (p.Asn141Lys) mutation.

5.3 cblC (MIM 277400)

The *cblC* disorder is the most common inborn error of cobalamin metabolism, with at least 550 patients identified. It is the result of mutations in the *MMACHC* gene on chromosome 1p34.1 (Lerner-Ellis et al. 2006), which encodes a protein that binds cobalamin after its release from the lysosome and catalyzes removal of the upper axial ligand if one is present. Cells from *cblC* patients take up cobalamin but do not accumulate the vitamin within the cell, possibly because cobalamin that does not become associated with methylmalonyl-CoA mutase or methionine synthase cannot be retained. Synthesis of both cobalamin coenzyme derivatives is impaired, with reduced function of both methylmalonyl-CoA mutase and methionine synthase.

Patients with the *cblC* disorder have increased levels of both MMA and homocysteine in blood and urine, in the presence of serum vitamin B12 levels within the reference range. Serum MMA levels in the range of 100 µM (reference < 0.27 µM) are typical, with urine MMA 500–10,000 mg/g creatinine (reference < 3 mg/g creatinine). Plasma total homocysteine values in excess of 100 µM have been reported (reference < 14 µM) (Carrillo-Carrasco et al. 2012). Clinical presentation is quite variable. Most patients present within the first year of life with megaloblastic anemia (and in some cases neutropenia, thrombocytopenia or pancytopenia), failure to thrive, feeding difficulties, developmental delay, and neurological findings that can include EEG abnormalities and seizures, hypotonia, hydrocephalus, cerebral (or less frequently, cerebellar) atrophy, microcephaly and white matter disease (Rosenblatt et al. 1997; Carrillo-Carrasco et al. 2012; Carrillo-Carrasco and Venditti 2012; Fischer et al. 2014). Ophthalmologic changes are frequently associated with the *cblC* disorder, particularly early-onset cases, including a progressive, infantile onset "bull's eye" maculopathy (with a hypopigmented perimacular zone surrounded with a hyperpigmented ring) and a salt-and-pepper pigmentary retinopathy (Schimel and Mets 2006; Weisfeld-Adams et al. 2015). A review of published cases found evidence of maculopathy in 60% of early-onset *cblC* patients; 70% had nystagmus (Weisfeld-Adams et al. 2015). Optic atrophy (26%) and strabismus (21%) were seen less frequently.

There has been evidence of effects during the prenatal period in some patients, including intrauterine growth retardation (Robb et al. 1984), congenital heart defects (Profitlich et al. 2009) and mild facial dysmorphology (Cerone et al. 1999).

Later onset forms of the disorder have been difficult to diagnose due to non-specificity of findings at presentation (Huemer et al. 2014b). Presentation can occur during childhood, adolescence or adulthood (Shinnar and Singer 1984; Mitchell et al. 1986; Powers et al. 2001; Thauvin-Robinet et al. 2008). Asymptomatic individuals have been identified following diagnosis of the disorder in a sibling (Gold et al. 1996) or after identification of low carnitine levels on newborn screening of an unaffected infant (Lin et al. 2009). Neurological findings, including cognitive decline, seizures, ataxia, neuropathy, myelopathy, encephalopathy, or cerebral atrophy are often present in late-onset patients, but several have had no neurological findings (Huemer et al. 2014b). Several patients have presented with psychiatric problems. Atypical hemolytic uremic syndrome (renal thrombotic microangiopathy), often accompanied by pulmonary arterial hypertension, has been identified in a number of patients, most of which did not have neurological findings (Kömhoff et al. 2013). Macrocytosis with or without anemia has been identified in fewer than half of late-onset cases (Huemer et al. 2014b).

Treatment of the *cblC* disorder typically includes parenteral hydroxocobalamin and oral betaine; folinic acid and carnitine are also frequently given although their effectiveness is not established (Carrillo-Carrasco et al. 2012). Hydroxocobalamin has been shown to be more effective than cyanocobalamin in some *cblC* patients (Andersson and Shapira 1998). Treatment usually results in correction of biochemical and hematological symptoms, although extremely high dosages of hydroxocobalamin may be necessary for complete normalization of biochemical parameters (Carrillo-Carrasco et al. 2009); response of neurological symptoms is typically incomplete, and most early-onset patients are left with some degree of neurological dysfunction. Ocular manifestations of the disorder do not appear to respond to therapy (Weisfeld-Adams et al. 2015). Prognosis is better for patients with late-onset disease than those with onset during the first year of life, and many patients with early-onset disease have not survived despite treatment (Rosenblatt et al. 1997; Fischer et al. 2014). It is critical that treatment be started as early in life as possible; the effects of early identification of affected infants by newborn screening using tandem mass spectroscopy on disease severity and patient survival are not yet known. Two patients that were diagnosed and treated prenatally have had generally good outcomes compared to their affected siblings, although one developed nystagmus, hyperpigmented retinopathy and hypotonia (Huemer et al. 2005; Zhang et al. 2008).

Identification of large numbers of *cblC* patients has allowed recognition of genotype-phenotype correlations. Homozygosity for the c.271dupA or c.331C>T *MMACHC* mutation was associated with early-onset disease, while homozygosity for the c.394C>T mutation was associated with later-onset disease (Morel et al. 2006). Heterozygosity for mutations affecting the guanidine residue at position 276 of the gene has been associated with combined atypical hemolytic uremic syndrome and pulmonary arterial hypertension in a Dutch population (Kömhoff et al. 2013). The c.271dupA mutation occurs frequently in populations of European origin, representing nearly half of identified causal *MMACHC* alleles in such populations. In Chinese populations, this allele is rare and the c.609G>A represents approximately half of identified causal alleles (Liu et al. 2010).

The California newborn screening program has identified a higher incidence of the *cblC* disorder than had been apparent when affected individuals were identified when they became clinically symptomatic (Cusmano-Ozog et al. 2007). This suggests that a significant number of patients identified on newborn screening may have a late-onset form of the disease or will remain clinically asymptomatic.

5.4 cblD (MIM 277410)

The *cblD* disorder was originally identified in 1970 in two brothers with combined methylmalonic aciduria and homocystinuria and relatively mild neurological problems. The phenotype of fibroblasts from the brothers was indistinguishable from that of cells of *cblC* patients, but complementation analysis established that a different locus was affected (Willard et al. 1978). No additional patients with the disorder were recognized until 2004, when it was demonstrated that *cblD* patients could have quite different presentations: some patients had combined methylmalonic aciduria and homocystinuria, but others had isolated methylmalonic aciduria or homocystinuria. Complementation analysis demonstrated that all these patients had mutations affecting the same locus (Suormala et al. 2004). The underlying gene, *MMADHC* on chromosome 2q23.2, was subsequently identified (Coelho et al. 2008). Analysis of mutations identified in *cblD* patients with different presentations identified a relationship between type and location of mutations in the *MMADHC* gene and phenotype, leading to a hypothesis explaining genotype-phenotype correlation. Nonsense mutations in the 5' region of the gene result in early termination of translation and reinitiation of at one of two in-phase alternate start sites (codons 62 or 116) and production of a message lacking a mitochondrial translocation sequence but retaining cytoplasmic function, allowing synthesis of methylcobalamin; patients with this type of mutation present with isolated methylmalonic aciduria. Patients with missense mutations downstream of the alternative start site at codon 116 produce a protein that contains the mitochondrial translocation sequence but lacks cytoplasmic function; these patients present with isolated homocystinuria. Patients with nonsense mutations downstream of codon 116 produce no functional MMADHC protein and present with combined methylmalonic aciduria and homocystinuria (Coelho et al. 2008).

Nineteen patients with the *cblD* disorder have been described, including 5 with combined methylmalonic aciduria and homocystinuria, 7 with isolated homocystinuria and 7 with isolated methylmalonic aciduria (Atkinson et al. 2014). A patient with isolated methylmalonic aciduria originally described as a novel complementation class designated *cblH* (Watkins et al. 2000) was subsequently shown to have the *cblD* disorder (Coelho et al. 2008). Fifteen different MMADHC mutations have been identified, all present in only one or two families.

5.5 cblX (MIM 309541)

Among patients that received a diagnosis of the *cblC* disorder based on complementation analysis, a small number have not had any *MMACHC* mutations on sequence analysis. Whole exome sequencing of genomic DNA from one male patient in this group identified a hemizygous mutation in the *HCFC1* gene on chromosome Xq28 (Yu et al. 2013). Subsequently, sequence analysis identified causal *HCFC1* mutations in 14 additional male patients, all affecting residues in the Kelch domain of the HCFC1 protein (Yu et al. 2013; Gérard et al. 2015). *HCFC1* encodes a transcription co-regulator that forms part of a complex previously implicated in control of progress through the cell cycle; its role in regulation of cobalamin metabolism remains uncharacterized.

Clinically, patients with the *cblX* disorder tend to have a less severe metabolic defect than that observed in *cblC* patients, but have a more severe neurological phenotype, with severe developmental delay, intractable seizures and EEG abnormalities, choreoathetosis and microcephaly. Onset of symptoms frequently occurs *in utero*, with intrauterine growth retardation, nuchal lucencies on ultrasound, and congenital microcephaly reported (Yu et al. 2013; Gérard et al. 2015). Since the *cblX* disorder is inherited in an X-liked recessive manner, genetic counseling is different from that for the other inborn errors of cobalamin metabolism, which are all inherited in an autosomal recessive manner.

Keywords: Inborn errors, Methylmalonic acid, Homocysteine, Methionine, Somatic cell genetics, Megaloblastic anemia

Abbreviations

ABCD4	:	product of the *ABCD4* gene
AdoCbl	:	adenosylcobalamin
ATR	:	cobalamin adenosyltransferase (product of the *MMAB* gene)
Cbl	:	cobalamin
cbl⁺	:	cob(I)alamin
cbl⁺⁺	:	cob(II)alamin
HCFC1	:	product of the *HCFC1* gene, host factor 1
LMBD1	:	product of the *LMBRD1* gene

MCM	:	methylmalonyl-CoA mutase (product of the *MUT* gene)
MeCbl	:	methylcobalamin
MMAA	:	product of the *MMAA* gene
MMACHC	:	product of the *MMACHC* gene
MMADHC	:	product of the *MMADHC* gene
MTR	:	methionine synthase (product of the *MTR* gene)
MTRR	:	methionine synthase reductase (product of the *MTRR* gene)
TC	:	transcobalamin
TCblR	:	transcobalamin receptor

References

Alfares A, Dempsey Nunez L, Al-Thihli K, Mitchell J, Melançon S, Anastasio N, Ha KCH, Majewski J, Rosenblatt DS and Braverman N. 2011. Combined malonic and methylmalonic aciduria: exome sequencing reveals mutations in the ACSF3 gene in patients with a non-classic phenotype. J Med Genet. 48: 602–605.

Andersson HC and Shapira E. 1998. Biochemical and clinical response to hydroxocobalamin versus cyanocobalamin treatment in patients with methylmalonic acidemia and homocystinuria (*cblC*). J Pediatr. 132: 121–124.

Atkinson C, Miousse IR, Watkins D, Rosenblatt DS and Raiman JAJ. 2014. Clinical, biochemical, and molecular presentation in a patient with the cblD-homocystinuria inborn error of cobalamin metabolism. JIMD Rep. 17: 77–81.

Baker EH, Sloan JL, Hauser NS, Gropman AL, Adams DR, Toro C, Manoli I and Venditti CP. 2015. MRI characteristics of globus pallidus infarcts in isolated methylmalonic acidemia. Am J Neuroradiol. 36: 194–201.

Batshaw ML, Thomas GH, Cohen SR, Matalon R and Mahoney MJ. 1984. Treatment of the *cbl B* form of methylmalonic acidaemia with adenosylcobalamin. J Inher Metab Dis. 7: 65–68.

Carrillo-Carrasco N, Chandler RJ and Venditti CP. 2012. Combined methylmalonic acidemia and homocystinuria, cblC type. I. Clinical presentation, diagnosis and management. J Inher Metab Dis. 35: 91–102.

Carrillo-Carrasco N, Sloan J, Valle D, Hamosh A and Venditti CP. 2009. Hydroxocobalamin dose escalation improves metabolic control in cblC. J Inher Metab Dis. 32: 728–731.

Carrillo-Carrasco N and Venditti CP. 2012. Combined methylmalonic acidemia and homocystinuria, cblC type. II. Complications, pathophysiology, and outcomes. J Inher Metab Dis. 35: 103–114.

Cerone R, Schiaffino MC, Caruso U, Lupino S and Gatti R. 1999. Minor facial anomalies in combined methylmalonic aciduria and homocystinuria due to a defect in cobalamin metabolism. J Inher Metab Dis. 22: 247–250.

Cheng KH, Liu MY, Kao CH, Chen YJ, Hsiao KJ, Liu TT, Lin HY, Huang CH, Chiang CC, Ho HJ, Lin SP, Lee NC, Hwu WL, Lin JL, Hung PY and Niu DM. 2010. Newborn screening for methylmalonic acid by tandem mass spectroscopy: 7 years' experience from two centers in Taiwan. J Chin Med Assoc. 73: 314–318.

Coelho D, Kim JC, Miousse IR, Fung S, Du Moulin M, Buers I, Suormala T, Burda P, Frapolli M, Stucki M, Nürnberg P, Thiele H, Robenek H, Höhne W, Longo N, Pasquali M, Mengel E, Watkins D, Shoubridge EA, Majewski J, Rosenblatt DS, Fowler B, Rutsch F and Baumgartner MR. 2012. Mutations in *ABCD4* cause a new inborn error of vitamin B12 metabolism. Nature Genet. 44: 1152–1155.

Coelho D, Suormala T, Stucki M, Lerner-Ellis JP, Rosenblatt DS, Newbold RF, Baumgartner MR and Fowler B. 2008. Gene identification for the cblD defect of vitamin B12 metabolism. N Engl J Med. 358: 1454–1464.

Cosson MA, Benoist JF, Touati G, Déchaux M, Royer N, Grandin L, Jais JP, Boddaert N, Barbier V, Desguerre I, Campeau PM, Rabier D, Valayannopoulos V, Niaudet P and de Lonlay P. 2009. Long-term outcome in methylmalonic aciduria: a series of 30 French patients. Mol Genet Metab. 97: 172–178.

Coulombe JT, Shih VE and Levy HL. 1981. Massachusetts metabolic disorders screening program. II. Methylmalonic aciduria. Pediatrics. 67: 26–31.

Cusmano-Ozog K, Lorey F, Levine S, Martin M, Nicholas E, Packman S, Rosenblatt DS, Cederbaum S, Cowan TM and Enns GM. 2007. Cobalamin C disease identified by expanded newborn screening: the California experience. Mol Genet Metab. 90: 240.

Dobson CM, Gradinger A, Longo N, Wu X, Leclerc D, Lerner-Ellis J, Lemieux M, Belair C, Watkins D, Rosenblatt DS and Gravel RA. 2006. Homozygous nonsense mutation in the *MCEE* gene and siRNA suppression of methylmalonyl-CoA epimerase expression: a novel cause of mild methylmalonic aciduria. Mol Genet Metab. 88: 327–333.

Fischer S, Huemer M, Baumgartner M, Deodato F, Ballhausen D, Boneh A, Burlina AB, Cerone R, Garcia P, Gökcay G, Grünewald S, Häberle J, Jaeken J, Ketteridge D, Lindner M, Mandel H, Martinelli D, Martins EG, Schwab KO, Gruenert SC, Schwahn B, Sztriha L, Tomaske M, Trefz F, Vilarinho L, Rosenblatt DS, Fowler B and Dionisi-Vici C. 2014. Clinical presentation and outcome in a series of 88 patients with the cblC defect. J Inher Metab Dis. 37: 831–840.

Fowler B, Leonard JV and Baumgartner MR. 2008. Causes and diagnostic approaches to methylmalonic acidurias. J Inher Metab Dis. 31: 350–360.

Gérard M, Morin G, Bourillon A, Colson C, Mathieu S, Rabier D, Billette de Villemeur T, Ogier de Baulny H and Benoist JF. 2015. Multiple congenital anomalies in two boys with mutations in *HCFC1* and cobalamin disorder. Eur J Med Genet. 58: 148–153.

Gold R, Bogdahn U, Kappos L, Toyka KV, Baumgartner ER, Fowler B and Wendel U. 1996. Hereditary defect of cobalamin metabolism (homocystinuria and methylmalonic aciduria) of juvenile onset. J Neurol Neurosurg Psych. 60: 107–108.

Gravel RA, Mahoney MJ, Ruddle FH and Rosenberg LE. 1975. Genetic complementation in heterokaryons of human fibroblasts defective in cobalamin metabolism. Proc Natl Acad Sci USA. 72: 3181–3185.

Gu W, Koh W, Blumenfeld YJ, El-Sayed YY, Hudgens L, Hintz SR and Quake SR. 2014. Noninvasive prenatal diagnosis in a fetus at risk for methylmalonic acidemia. Genet Med. 16: 564–567.

Gulati S, Baker P, Li YN, Fowler B, Kruger W, Brody LC and Banerjee R. 1996. Defects in human methionine synthase in cblG patients. Hum Molec Genet. 5: 1859–1865.

Homolova K, Zavadakova P, Doktor TK, Schroeder LD, Kozich V and Andresen BS. 2010. The deep intronic c.903+469T>C mutation in the *MTRR* gene creates an SF2/ASF binding exonic splicing enhancer, which leads to pseudoexon activation and causes the cblE type of homocystinuria. Hum Mut. 31: 437–444.

Hörster F, Baumgartner MR, Viardot C, Suormala T, Burgard P, Fowler B, Hoffmann GF, Garbade SF, Kölker S and Baumgartner ER. 2007: Long-term outcome in methylmalonic acidurias is influenced by the underlying defect (mut⁰, mut⁻, cblA, cblB). Pediatr Res. 62: 225–230.

Huemer M, Bürer C, Jesina P, Kozich V, Landolt MA, Suormala T, Fowler B, Augoustides-Savopoulou P, Blair E, Brennerova K, Broomfield A, De Meirleir L, Gökcay G, Hennermann J, Jardine P, Koch J, Lorenzl S, Lotz-Havla AS, Noss J, Parini R, Peters H, Plecko B, Ramos FJ, Schlune A, Tsiakas K, Zerjav Tansek M and Baumgartner MR. 2014a. Clinical onset and course, response to treatment and outcome in 24 patients with the cblE or cblG remethylation defect complemented by genetic and *in vitro* enzyme study data. J Inher Metab Dis. doi: 10.1007/s10545-014-9803-7.

Huemer M, Scholl-Bürgi S, Hadaya K, Kern I, Beer R, Sepp K, Fowler B, Baumgartner MR and Karali D. 2014b. Three new cases of late-onset cblC defect and review of the literature illustrating when to consider inborn errors of metabolism beyond infancy. Orphanet J Rare Dis. 9: 161.

Huemer M, Simma B, Fowler B, Suormala T, Bodamer OA and Sass JO. 2005. Prenatal and postnatal treatment in cobalamin C defect. J Pediatr. 147: 469–472.

Karth P, Singh R, Kim J and Costakos D. 2012. Bilateral central retinal artery occlusions in an infant with hyperhomocysteinemia. J AAPOS. 16: 398–400.

Kim JC, Lee NC, Hwu PW, Chien YH, Fahiminiya S, Majewski J, Watkins D and Rosenblatt DS. 2012. Late onset of symptoms in an atypical patient with the *cblJ* inborn error of vitamin B12 metabolism: diagnosis and novel mutation revealed by exome sequencing. Mol Genet Metab. 107: 664–668.

Kömhoff M, Roofthooft MT, Westra D, Teertstra TK, Losito A, van de Car NCAJ and Berger RMF. 2013. Combined pulmonary arterial hypertension and renal thrombotic microangiopathy in cobalamin C deficiency. Pediatrics. 132: e540–e544.

Ktena YP, Paul SM, Hauser NS, Sloan JL, Gropman A, Manoli I and Venditti CP. 2015. Delineating the spectrum of impairments, disabilities, and rehabilitation needs in methylmalonic acidemia (MMA). Am J Med Genet. A doi: 10.1002/ajmg.a.37127.

Leclerc D, Campeau E, Goyette P, Adjalla CE, Christensen B, Ross M, Eydoux P, Rosenblatt DS, Rozen R and Gravel RA. 1996. Human methionine synthase: cDNA cloning and identification of mutations in patients of the *cblG* complementation group of folate/cobalamin disorders. Hum Molec Genet. 5: 1867–1874.

Leclerc D, Wilson A, Dumas R, Gafuik C, Song D, Watkins D, Heng HHQ, Rommens JM, Scherer SW, Rosenblatt DS and Gravel RA. 1998. Cloning and mapping of a cDNA for methionine synthase reductase, a flavoprotein defective in patients with homocystinuria. Proc Natl Acad Sci USA. 95: 3059–3064.

Lemieux B, Auray-Blais C, Giguère R, Shapcott D and Scriver CR. 1988. Newborn urine screening experience with over one million infants in the Quebec Network of Genetic Medicine. J Inher Metab Dis. 11: 45–55.

Lerner-Ellis JP, Tirone JC, Pawelek PD, Doré C, Atkinson JL, Watkins D, Morel CF, Fujiwara TM, Moras E, Hosack AR, Dunbar GV, Antonicka H, Forgetta V, Dobson CM, Leclerc D, Gravel RA, Shoubridge EA, Coulton JW, Lepage P, Rommens JM, Morgan K and Rosenblatt DS. 2006. Identification of the gene responsible for methylmalonic aciduria and homocystinuria, cblC type. Nature Genet. 38: 93–100.

Lin HJ, Neidich JA, Salazar D, Thomas-Johnson E, Ferreira BF, Kwong AM, Lin AM, Jonas AJ, Levine S, Lorey F and Rosenblatt DS. 2009. Asymptomatic maternal combined homocystinuria and methylmalonic aciduria (*cblC*) detected through low carnitine levels on newborn screening. J Pediatr. 155: 924–927.

Liu MY, Yang YL, Chang YC, Chiang SH, Lin SP, Han LS, Qi Y, Hsiao KJ and Liu TT. 2010. Mutation spectrum of MMACHC in Chinese patients with combined methylmalonic aciduria and homocystinuria. J Hum Genet. 55: 621–626.

Matsui SM, Mahoney MJ and Rosenberg LE. 1983. The natural history of the inherited methylmalonic acidemias. N Engl J Med. 308: 857–861.

Mitchell GA, Watkins D, Melançon SB, Rosenblatt DS, Geoffroy G, Orquin J, Homsy MB and Dallaire L. 1986. Clinical heterogeneity in cobalamin C variant of combined homocystinuria and methylmalonic aciduria. J Pediatr. 108: 410–415.

Montgomery JA, Mamer OA and Scriver CR. 1983. Metabolism of methylmalonic acid in rats. Is methylmalonyl-coenzyme a racemase deficiency symptomatic in man? J Clin Invest. 72: 1937–1947.

Morath MA, Hörster F and Sauer SW. 2013. Renal dysfunction in methylmalonic aciduias: review for the pediatric nephrologist. Pediatr Nephrol. 28: 227–235.

Morel CF, Lerner-Ellis JP and Rosenblatt DS. 2006. Combined methylmalonic aciduria and homocystinuria (*cblC*): phenotype-genotype correlations and ethnic-specific observations. Mol Genet Metab. 88: 315–321.

Morel CF, Watkins D, Scott P, Rinaldo P and Rosenblatt DS. 2005. Prenatal diagnosis for methylmalonic acidemia and inborn errors of vitamin B12 metabolism and transport. Mol Genet Metab. 86: 160–171.

Niemi AK, Kim IK, Krueger CE, Cowan TM, Baugh N, Farrell R, Bonham CA, Concepcian W, Esquivel CO and Enns GM. 2015. Treatment of methylmalonic acidemia by liver or combined liver-kidney transplantation. J Pediatr. doi: 10.1016/j.peds.2015.01.051.

Nizon M, Ottolenghi C, Valayannopoulos V, Arnoux JB, Barbier V, Habarou F, Desguerre I, Boddaert N, Bonnefont JP, Acqaviva C, Benoist JF, Rabier D, Touati G and de Lonlay P. 2013. Long-term neurological outcome of a cohort of 80 patients with classical organic acidurias. Orphanet J Rare Dis. 8: 148.

Oladipo O, Rosenblatt DS, Watkins D, Miousse IR, Sprietsma L, Dietzen DJ and Shinawi M. 2011. Cobalamin F disease detected by newborn screening and follow-up on a 14-year-old patient. Pediatrics. 128: e1636–e1640.

O'Shea CJ, Sloan JL, Wiggs EA, Pao M, Gropman A, Baker EH, Manoli I, Venditti CP and Snow J. 2012. Neurocognitive phenotype of isolated methylmalonic acidemia. Pediatrics. 129: e1541–e1551.

Ostergaard E. 2008. Disorders caused by deficiency of succinate-CoA ligase. J Inher Metab Dis. 31: 226–229.

Pangilinan F, Mitchell A, VanderMeer J, Molloy AM, Troendle J, Conley M, Kirke PN, Sutton M, Seqeira JM, Quadros EV, Mills JM and Brody LC. 2010. Transcobalamin II receptor polymorphisms are associated with increased risk for neural tube defects. J Med Genet. 47: 677–685.

Powers JM, Rosenblatt DS, Schmidt RE, Cross AH, Black JT, Moser AB, Moser HW and Morgan DJ. 2001. Neurological and neuropathologic heterogeneity in two brothers with cobalamin C deficiency. Ann Neurol. 49: 396–400.

Profitlich LE, Kirmse B, Wasserstein MP, Diaz GA and Srivastava S. 2009. High prevalence of structural heart disease in children with cblC-type methylmalonic aciduria and homocystinuria. Mol Genet Metab. 98: 344–348.

Quadros EV, Lai SC, Nakayama Y, Sequeira JM, Hannibal L, Wang S, Jacobsen DW, Fedosov SN, Wright E, Gallagher RC, Anastasio N, Watkins D and Rosenblatt DS. 2010. Positive newborn screen for methylmalonic aciduria identifies the first mutation in TCblR/CD320, the gene for cellular uptake of transcobalamin-bound vitamin B12. Hum Mut. 31: 924–929.

Robb RM, Dowton SB, Fulton AB and Levy HL. 1984. Retinal degeneration in vitamin B12 disorder associated with methylmalonic aciduria and sulfur amino acid abnormalities. Am J Ophthalmol. 97: 691–696.

Rosenblatt DS, Aspler AL, Shevell MI, Pletcher BA, Fenton WA and Seashore MR. 1997. Clinical heterogeneity and prognosis in combined methylmalonic aciduria and homocystinuria (*cblC*). J Inher Metab Dis. 20: 528–538.

Rosenblatt DS, Cooper BA, Pottier A, Lue-Shing H, Matiaszuk N and Grauer K. 1984. Altered vitamin B12 metabolism in fibroblasts from a patient with megaloblastic anemia and homocystinuria due to a new defect in methionine biosynthesis. J Clin Invest. 74: 2149–2156.

Rosenblatt DS, Cooper BA, Schmutz SM, Zaleski WA and Casey RE. 1985. Prenatal vitamin B12 therapy of a fetus with methylcobalamin deficiency (cobalamin E disease). Lancet. 325: 1127–1129.

Rosenblatt DS, Hosack A, Matiaszuk NV, Cooper BA and Laframboise R. 1985. Defect in vitamin B12 release from lysosomes: newly described inborn error of vitamin B12 metabolism. Science. 228: 1319–1321.

Rosenblatt DS and Watkins D. 2010. Prenatal diagnosis of miscellaneous biochemical disorders. pp. 614–627. In: A Milunsky and JM Milunsky (eds.). Genetic Disorders and the Fetus. Diagnosis, Prevention and Treatment. Wiley-Blackwell, London.

Rutsch F, Gailus S, Miousse IR, Suormala T, Sagné C, Reza Toliat M, Nürnber G, Wittkampf T, Buers I, Sharifi A, Stucki M, Becker C, Baumgartner M, Robenek H, Marquardt T, Höhne W, Gasnier B, Rosenblatt DS, Fowler B and Nürnberg P. 2009. Identification of a putative lysosomal cobalamin exporter mutated in the cblF inborn error of vitamin B12 metabolism. Nature Genet. 41: 234–239.

Schimel AM and Mets MB. 2006. The natural history of retinal degeneration in association with cobalamin C (cblC) disease. Ophthalmic Genet. 27: 9–14.

Shinnar S and Singer HS. 1984. Cobalamin C mutation (methylmalonic aciduria and homocystinuria) in adolescence. A treatable cause of dementia and myelopathy. N Engl J Med. 311: 451–454.

Sloan JL, Johnston JJ, Manoli I, Chandler RJ, Krause C, Carrillo-Carrasco N, Chandrasekaran SD, Sysol JR, O'Brien K, Hauser NS, Sapp JC, Dorward HM, Huizing M, NIH Intramural Sequencing Center Group, Barshop, BA, Berry SA, James PM, Champaigne NL, de Lonlay P,

Valayannopoulos V, Geschwind MD, Gavrilov DK, Nyhan WL, Biesecker LG and Venditti CP. 2011. Exome sequencing identifies ACSF3 as a cause of combined malonic and methylmalonic aciduria. Nature Genet. 43: 883–886.

Sloan JL, Manoli I and Veditti CP. 2015. Liver or combined liver-kidney transplantation for patients with isolated methylmalonic acidemia: who and when? J Pediatr. 166: 1455–1461.

Suormala T, Baumgartner MR, Coelho D, Zavadakova P, Kozich V, Koch HG, Berghauser M, Wraith JE, Burlina A, Sewell A, Herwig J and Fowler B. 2004. The cblD defect causes either isolated or combined deficiency of methylcobalamin and adenosylcobalamin synthesis. J Biol Chem. 279: 42742–42749.

Takeichi T, Hsu CK, Yang HS, Chen HY, Wong TW, Tsai WL, Chao SC, Lee JY, Akiyama M, Simpson MA and McGrath JA. 2015. Progressive hyperpigmentation in a Taiwanese child due to an inborn error of vitamin B12 metabolism (cblJ). Br J Dermatol. 172: 1111–1115.

Thauvin-Robinet C, Roze E, Couvreur G, Horellou MH, Sedel F, Grabli D, Bruneteau G, Toneti C, Masurel-Paulet A, Perennou D, Moreau T, Giroud M, Ogier de Baulny H, Giraudier S and Faivre L. 2008. The adolescent and adult form of cobalamin C disease: clinical and molecular spectrum. J Neurol Neurosurg Psych. 79: 725–728.

Tortorelli S, Turgeon CT, Lim JS, Baumgart S, Day-Salvatore DL, Abdenur J, Bernstein JA, Lorey F, Lichter-Konecki U, Oglesbee D, Raymond K, Matern D, Schimmenti L, Rinaldo P and Gavrilov DK. 2010. Two-tier approach to the newborn screening of methylenetetrahydrofolate reductase deficiency and other remethylation disorders with tandem mass spectrometry. J Pediatr. 157: 271–275.

Turgeon CT, Magera MJ, Cuthbert CD, Loken PR, Gavrilov DK, Tortorelli S, Raymond KM, Oglesbee D, Rinaldo P and Matern D. 2010. Determination of total homocysteine, methylmalonic acid, and 2-methylcitric acid in dried blood spots by tandem mass spectroscopy. Clin Chem. 56: 1686–1695.

Vilaseca MA, Vilarinho L, Zavadakova P, Vela E, Cleto E, Pineda M, Coimbra E, Suormala T, Fowler B and Kozich V. 2003. CblE type of homocystinuria: mild clinical phenotype in two patients homozygous for a novel mutation in the MTRR gene. J Inher Metab Dis. 26: 361–369.

Watkins D, Matiaszuk N and Rosenblatt DS. 2000. Complementation studies in the *cblA* class of inborn error of cobalamin metabolism: evidence for interallelic complementation and for a new complementation class (*cblH*). J Med Genet. 37: 510–513.

Watkins D and Rosenblatt DS. 1986. Failure of lysosomal release of vitamin B12: a new complementation group causing methylmalonic aciduria. Am J Hum Genet. 39: 404–408.

Watkins D and Rosenblatt DS. 1988. Genetic heterogeneity among patients with methylcobalamin deficiency. Definition of two complementation groups, cblE and cblG. J Clin Invest. 81: 1690–1694.

Watkins D and Rosenblatt DS. 1989. Functional methionine synthase deficiency (cblE and cblG): clinical and biochemical heterogeneity. Am J Med Genet. 34: 427–434.

Watkins D, Ru M, Hwang HY, Kim CD, Murray A, Philip NS, Kim W, Legakis H, Wai T, Hilton JF, Ge B, Doré C, Hosack A, Wilson A, Gravel RA, Shane B, Hudson TJ and Rosenblatt DS. 2002. Hyperhomocysteinemia due to methionine synthase deficiency, cblG: structure of the MTR gene, genotype diversity, and recognition of a common mutation, P1173L. Am J Hum Genet. 71: 143–153.

Weisfeld-Adams JD, McCourt EA, Diaz GA and Oliver SC. 2015. Ocular disease in the cobalamin C defect: a review of the literature and a suggested framework for clinical surveillance. Mol Genet Metab. doi: 10.1016/j.ymgme.2015.01.012.

Wilcken B, Wiley V, Hammond J and Carpenter K. 2003. Screening of newborns for inborn errors of metabolism by tandem mass spectrometry. N Engl J Med. 348: 2304–2312.

Willard HF, Mellman IS and Rosenberg LE. 1978. Genetic complementation among inherited deficiencies of methylmalonyl-CoA mutase activity: evidence for a new class of human cobalamin mutant. Am J Hum Genet. 30: 1–13.

Wong D, Tortorelli S, Bishop L, Sellars EA, Schimmenti LA, Gallant N, Prada CE, Hopkins RJ, Leslie ND, Berry SA, Rosenblatt DS, Fair AL, Matern D, Raymond K, Oglesbee D, Rinaldo P and

Gavrilov D. 2015. Outcomes of four patients with homocyateine remethuylation disorders detected by newborn screening. Genet Med. doi: 10.1038/gim.2015.45.

Worgan LC, Niles K, Tirone JC, Hofmann A, Verner A, Sammak A, Kucic T, Lepage P and Rosenblatt DS. 2006. Spectrum of mutations in *mut* methylmalonic acidemia and identification of a common Hispanic mutation and haplotype. Hum Mutat. 27: 31–43.

Yu HC, Sloan JL, Scharer G, Brebner A, Quintana A, Achilly NP, Manoli I, Coughlin CR, Geiger EA, Schneck U, Watkins D, Suormala T, Van Hove JLK, Fowler B, Baumgartner MR, Rosenblatt DS, Venditti CP and Shaikh TM. 2013. An X-linked cobalamin disorder caused by mutations in transcriptional coregulator *HCFC1*. Am J Hum Genet. 93: 506–514.

Zhang Y, Yang Y, Hasegawa Y, Yamaguchi S, Shi C, Song J, Sayami S, Liu P, Yan R, Dong J and Qin J. 2008. Prenatal diagnosis of methylmalonic aciduria by analysis of organic acids and total homocysteine in amniotic fluid. Chin Med J. 121: 216–219.

5

Conditions and Diseases that Cause Vitamin B12 Deficiency

Form Metabolism to Diseases

Emmanuel Andrès

1. Introduction

Although vitamin B12 was isolated almost 60 years ago, its metabolism remains incompletely defined. In practice, cobalamin metabolism is complex and requires many processes and steps, any one of which, if not present, may lead to vitamin B12 deficiency (Nicolas and Gueant 1994; Markle 1996). Table 1 describes through a synthetic view the different stages of vitamin B12 metabolism used in clinical practice and the corresponding causes of cobalamin deficiency (Andrès et al. 2004; Stabler 2013).

The present review summarizes the current knowledge on vitamin B12 metabolism and metabolic pathways in a clinical perspective, with a focus on the etiologies of cobalamin deficiency, especially in adult.

2. Vitamin B12 Ingestion and Related Disorders

2.1 Vitamin B12 sources and dietary recommendations

Vitamin B12 is produced exclusively by microbial synthesis in the digestive tract of animals. Therefore, animal products are the main sources of vitamin B12 in the human diet, in particular organ meats (liver, kidney) (Markle 1996;

Department of Internal Medicine, Diabetes and Metabolic Diseases, Hôpitaux Universitaires de Strasbourg, and Faculty of Medicine, Université de Strasbourg, Strasbourg, France.
Email: emmanuel.andres@chru-strasbourg.fr

Table 1. Stages of vitamin B12 metabolism and corresponding causes of cobalamin deficiency (Dali-Youcef and Andrès 2009).

Stages and actors in cobalamin metabolism	Causes of cobalamin deficiency
Intake solely through food	- Strict vegetarianism or poor diet, critically ill patients, elderly in institutions or in psychiatric hospitals
Digestion brings into play: - Haptocorrin - Gastric secretions (hydrochloric acid and pepsin) - Intrinsic factor - Pancreatic and biliary secretions - Enterohepatic cycle	- Gastrectomies - Pernicious anemia - Food-cobalamin malabsorption
Absorption brings into play: - Intrinsic factor - Cubilin, amionless - Calcium and energy	- Ileal resections and malabsorption - Pernicious anemia - Food-cobalamin malabsorption
Transport by transcobalaminII	- Congenital deficiency in transcobalamin II
Intracellular metabolism based on various intracellular enzymes	- Congenital deficiency in various intracellular enzymes (see Chapter 4)

Andrès et al. 2004; Stabler 2013). Other good sources are fish, eggs and dairy products. In foods, hydroxo-, methyl- and 5′-deoxyadenosyl-cobalamins are the main cobalamins present.

A typical Western diet contributes 3–30 μg of cobalamin per day. The *Food and Nutrition Board of the US Institute of Medicine* recommends dietary allowance of 2.4 μg per day for adults and 2.6 to 2.8 μg per day during pregnancy (Medicine 1998). The RDA did not distinguish between adults and elderly people though it is questionable whether an intake of 2.4 μg per day can maintain cobalamin status in elderly people with malabsorption.

2.2 Malnutrition and vegetarianism

Vitamin B12 deficiency caused by limited intake of vitamin dietary sources (which requires any animal product intake) is rare, even exceptional, in general population. Dietary causes of cobalamin deficiency are common in elderly people who are already malnourished, such as elderly patients in psychiatric hospitals (Matthews 1995) or those living in institutions who may consume inadequate amounts of vitamin B12-containing foods. This is also the case of certain groups of patients, especially patients who ingest a strict vegetarian diet (Pawlak et al. 2014). However, an inadequate intake is not the only explanation for the common cobalamin deficiency in elderly population. Food-cobalamin malabsorption is a significant participating factor in elderly people.

Studies focusing on elderly people, particularly those who are in institutions or who are sick and malnourished have suggested a vitamin B12 deficiency prevalence of 30–40% (Matthews 1995; van Asselt et al. 2000). The *Framingham* study demonstrated a prevalence of 12% among elderly people living in the community (Lindenbaum et al. 1994). Using stringent definition, we found that vitamin B12 deficiency had a prevalence of 5% in a group of patients followed or hospitalized in a tertiary reference hospital (Andrès et al. 2004).

In developing countries, in sub-Sahara (with marasmus kwashiorkor) or in several Latino countries, malnutrition may be one of or be the leading cause of cobalamin deficiency.

In practice, the diagnosis is based on patient interview and a comprehensive dietary survey (at least 1 week).

3. Food-cobalamin Digestion

3.1 Physiology of cobalamin absorption

Dietary vitamin B12, which is bound to proteins in food, is released in the acidic environment of the stomach where it is rapidly complexed to the binding protein and transporter haptocorrin (HC), also referred to as the R-binder or transcobalamin I (Figure) (Nicolas and Gueant 1994; Dali-Youcef and Andrès 2009). About 80% of circulating vitamin B12 are bound to HC and serum cobalamin levels show positive correlation to serum HC concentrations.

Although some unexplained low serum cobalamin concentrations were reported to be caused by mild to severe HC deficiencies, these abnormalities were not accompanied by clinical manifestations of cobalamin deficiency (Carmel 1983; Carmel 2003).

Cobalamin continues its route in the gastrointestinal track and dissociates from HC under the action of pancreatic proteases, followed by its association in the intestine with the intrinsic factor (also known as the S-binder) which is essential for ileal absorption of cobalamin (Figure) (Nicolas and Gueant 1994; Dali-Youcef and Andrès 2009). The intestinal absorption of cobalamin into the enterocytes takes place in the terminal ileum via intrinsic factor receptor cubilin. The amount of acid secretion in the gastrointestinal tract plays a critical role in binding of cobalamin to its transporting proteins.

Indeed, homozygous nonsense and missense mutations in the gene encoding the gastric intrinsic factor GIF were reported to cause hereditary juvenile cobalamin deficiency (Tanner et al. 2005). This metabolism step is also implicated in the physiopathology of the so called pernicious anemia (see the next section) (Nicolas and Gueant 1994).

3.1.1 Food-cobalamin malabsorption

Food-cobalamin malabsorption (FCM) is a syndrome characterized by the inability of the body to release cobalamin from food or intestinal transport proteins ("maldigestion"), particularly in the presence of hypochlorhydria

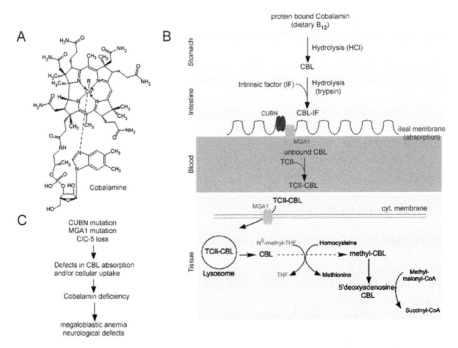

**Figure. (A): Structure of vitamin B12. (B): Various steps of vitamin B12 (cobalamin [cbl])
metabolism.** Dietary vitamin B12 is absorbed in food bound to proteins and undergo acidic
gastric digestion. The released cbl is attached to the R-binder haptocorrin and transported to the
intestine following pancreatic proteases processing. The unbound cbl is then associated in the
gut with the gastric-produced intrinsic factor (IF). This association is necessary for intestinal cbl
absorption through a complex of endocytic receptors and proteins, including the endocytic receptor
cubilin (CUBN) and the apical membrane protein amnionless (AMN), which is stabilized by two
other proteins, namely the receptor megalin/LRP-2 and its binding protein receptor associated
protein (RAP) which also binds to cubilin. After it absorption free cobalamin reaches the systemic
circulation where it associates with transcobalamin II (TCII) and subsequently the cbl-TCII complex
is uptaken in cells through it binding to megalin/LRP-2 and TCII receptor (TCII-R). Intracellularly,
the complex is dissociated following lysosomal digestion. Part of cbl serves as a cofactor for the
methionine synthase (MS) mediated transformation of homocysteine into methionine and for
methyl-tetrahydrofolate reductase-mediated formation of tetrahydrofolate (THF) a precursor of
purine and pyrimidine necessary for nucleic acid synthesis. The other fraction of cbl reaches the
mitochondria where it forms adenosyl-cbl, a cofactor for the methyl-malonyl mutase-mediated
catabolism of methyl-malonyl coA. **(C):** Mutations in genes encoding the intrinsic factor, cubilin,
amnionless or transcobalamin II or its receptor provoke defects in cbl absorption and/or cellular
uptake which translates into functional cbl deficiency and its clinical manifestations.

(Andrès et al. 2005; Carmel 1995). Ralph Carmel first characterized FCM in
cases of subtle cobalamin deficiencies (Carmel 1995). In our experience, this
syndrome accounted for 60 to 70% of cases of mild to severe vitamin B12
deficiency in elderly patients (Andrès et al. 2000). The principal characteristics
of this syndrome are listed in Table 2.

Table 2. Food-cobalamin malabsorption syndrome (Andrès et al. 2005).

Criteria for food-cobalamin malabsorption	Associated conditions or agents
- Low serum cobalamin (vitamin B12) levels - Normal results of Schilling test using free cyanocobalamin labeled with cobalt-58 or abnormal results of derived Schilling test ‡ - No anti-intrinsic factor antibodies - No dietary cobalamin deficiency	- Gastric disease: atrophic gastritis, type A atrophic gastritis, gastric disease associated with *Helicobacter pylori* infection, partial gastrectomy, gastric by-pass, vagotomy - Pancreatic insufficiency: alcohol abuse - Gastric or intestinal bacterial overgrowth: achlorhydria, tropical sprue, Ogilvie's syndrome, HIV - Drugs: antacids (H2-receptor antagonists and proton pump inhibitors) or biguanides (metformin) - Alcohol abuse - Sjögren's syndrome, systemic sclerosis - Ageing or idiopathic

‡ Derived Schilling tests use food-bound cobalamin (e.g., egg yolk, chicken and fish proteins).

FCM is caused primarily by atrophic gastritis (Andrès et al. 2005). Other factors that contribute to FCM are: chronic infection with *Helicobacter pylori* and intestinal microbial proliferation, situations in which cobalamin deficiency can be corrected by antibiotic treatment (Andrès et al. 2005; Kaptan et al. 2000); long-term ingestion of antiacids such as H2-receptor antagonists and proton-pump inhibitors, particularly among patients with Zollinger-Ellison syndrome (Jung et al. 2015), and long term intake of biguanides (metformin) (Bauman et al. 2000; Andrès et al. 2002). In addition, other FCM inducers include: chronic alcoholism, especially in malnourished patients; surgery or gastric reconstruction (e.g., bypass surgery for obesity); partial exocrine pancreatic failure; and rarely Sjögren's syndrome or systemic sclerosis (Andrès et al. 2005).

It is to note that in case of FCM, patients can absorb "unbound" cobalamin through intrinsic factor or passive diffusion mechanisms (Dali-Youcef and Andrès 2009; Andrès et al. 2005). Thus the recognition of the syndrome permits new developments of oral cobalamin therapy using free crystalized cobalamin that is readily absorbed (Andrès et al. 2008).

In practice, the diagnosis of FCM is to date based on the exclusion of the main others causes of vitamin B12 deficiency, in connection with the fact that the Schilling's test is no longer currently available (Stabler 2013). Health care providers need to be aware of FCM, since it is easy to treat and treatment can prevent serious late consequences of vitamin B12 deficiency.

3.2 Vitamin B12 absorption

3.2.1 Physiology

Absorption depends mainly on intrinsic factor (IF), which is secreted by the gastric mucosa. IF binds cobalamin forming a complex that is absorbed by the terminal ileum (Figure) (Nicolas and Gueant 1994; Dali-Youcef and Andrès 2009). This mechanism is responsible for at least 60% absorption on oral cobalamin (Nicolas and Gueant 1994). This complex is located at the apical side of brush-border membranes (BBMs) of polarized epithelia, such as the intestinal apical BBM.

It consists of the intrinsic factor-vitamin B12 receptor named cubilin, a 460 kDa peripheral membrane glycoprotein, encoded by the *CUBN* gene which was mapped to chromosomal region 10p12.33-p13 (Kozyraki et al. 1998), and the 48 kDa amnionless (AMN) protein encoded by the *AMN* gene, a gene essential for mouse gastrulation and localized on human chromosome 14 (Fyfe et al. 2004).

The human megalin/gp330/LRP-2 receptor, encoded by the *LRP-2* gene located on chromosome 2q24-q31 (Saito et al. 1994), is a giant endocytic receptor (600 kDa) of the low-density lipoprotein receptor (LDLR) family that was strongly suggested to play an important role in the stability of the cubilin/AMN complex (Dali-Youcef and Andrès 2009; Ahuja et al. 2008). It is noteworthy that ligands for megalin include apoE, lipoprotein lipase, lactoferrin, receptor-associated protein (RAP) among other proteins and that this interaction is Ca2+-dependent (Dali-Youcef and Andrès 2009; Moestrup et al. 1996). Cubilin and megalin are also expressed in the apical side of proximal tube and are considered responsible for cobalamin reuptake into the circulation. Importantly, the endoplasmic reticulum (ER) localized 39 kDa protein RAP, which binds to all members of the LDLR family but also in a region contiguous to the cobalamin-intrinsic factor binding region on the cubilin protein (Dali-Youcef and Andrès 2009; Birn et al. 1997), allows the processing of megalin where it binds to the newly synthesized megalin receptor in the ER and prevents the early binding of ligands and the aggregation of megalin receptors (Birn et al. 1997; Kristiansen et al. 1999; Christensen and Birn 2002).

3.2.2 Biermer's or Addison's disease

In adults, vitamin B12 deficiency is classically caused by Biermer's also named Addison's disease, formerly known as pernicious anemia (Rojas Hernandez and Oo 2015; Toh et al. 1997). This disorder is an auto-immune disease characterized by: the destruction of the gastric mucosa, especially fundal, associated with a primarily cell-mediated auto-immune process; and the presence of various antibodies, especially anti-intrinsic factor antibodies and gastric parietal anti-cell antibodies that target the H+/K+ ATPase α and β subunits (Rojas Hernandez and Oo 2015; Toh et al. 1997). Pernicious anemia has a genetic component (Rojas Hernandez and Oo 2015). Pernicious anemia is associated with other immunologic diseases such as Sjögren's syndrome, Hashimoto's disease, type 1 diabetes mellitus, and celiac disease (Rojas Hernandez and Oo 2015; Andrès et al. 2006).

In our experience, Biermer's disease accounted for 30 to 40% of cases of cobalamin deficiency in adults, and more than 60% in severe vitamin B12 deficiencies (Stabler 2013). In this later situation, hematological (e.g., macrocytic anemia) or psycho-neurological manifestations (e.g., medullar combined sclerosis) are commonly observed (Andrès et al. 2006).

In practice, the diagnosis of Biermer's or Addison's disease is based on the presence of intrinsic factor antibodies in serum (specificity > 98% and sensitivity around 50%) or biopsy-proven autoimmune atrophic gastritis (Table 3) (Nicolas and Gueant 1994; Rojas Hernandez and Oo 2015). The presence of *Helicobacter pylori* infection in gastric biopsies is an exclusion factor. It is important to note that of Biermer's disease require a long term gastric follow-up (upper-endoscopy with biopsies, every year in case of gastric lesions or every 2 to 5 years in the absence of detectable lesion), because this disorder favors the emergence of various cancers of the stomach (Nicolas and Gueant 1994; Stabler 2013).

Table 3. Biermer or Addison's disease (Nicolas and Gueant 1994; Rojas Hernandez and Oo 2015; Andrès et al. 2006).

Criteria for Biermer disease (pernicious anemia)	Associated conditions
- Low serum cobalamin (vitamin B12) levels - Abnormal results of Schilling test using free cyanocobalamin labeled with cobalt[58] ‡ - Presence of anti-intrinsic factor antibodies (sensibility of 50%, specificity > 98%) - Presence of an auto-immune gastritis (especially fundal), with absence of *Helicobacter pylori* in the status phase of the disease	- Auto-immune disorders as: Sjögren's syndrome, Hashimoto's disease, type 1 diabetes mellitus, and celiac disease - Predisposition of gastric cancer

‡ Schilling test is not used anymore in clinical practice.

3.2.3 Cobalamin malabsorption

Since the 1980s, the prevalence of vitamin B12 malabsorption declined, owing mainly to the decreasing frequency of gastrectomy (due to the provision of antacid drugs) and terminal small intestine surgical resection (Stabler 2013; Dali-Youcef and Andrès 2009). Several disorders commonly seen in gastroenterology practice might, however, be associated with cobalamin malabsorption. These disorders include exocrine pancreas' function deficiency following chronic pancreatitis (usually alcoholic), lymphomas or tuberculosis (of the intestine), celiac disease, Crohn's disease, Whipple's disease, and uncommon celiac disease (Stabler 2013).

In practice, the diagnosis is based on patient interview, a full clinical examination and digestive explorations in doubt.

3.2.4 Genetic disorders of cobalamin malabsorption

The endocytic receptor cubilin comprises a short N-terminal region followed by 8 epidermal growth factor (EGF) repeats and a large cluster of 27 CUB domains. Deletion mutant and immunoprecipitation experiments identified the CUB1-8 region as the binding domain for the vitamin B12-IF complex and the overlapping CUB13 and 14 domains as the binding region for the RAP protein (Dali-Youcef and Andrès 2009; Moestrup et al. 1996; Birn et al. 1997; Kristiansen et al. 1999; Christensen and Birn 2002).

Mutations in *CUBN* were reported to cause hereditary megaloblastic anemia 1 (MGA1), a rare autosomal recessive disorder affecting human subjects with neurological symptoms and juvenile megaloblastic anemia (Grasbeck 1960; Aminoff et al. 1999; Kristiansen et al. 2000). Two principal mutations were identified in Finnish patients (FM), a 3916C→T missense mutation named FM1 changing a highly conserved proline to leucine (P1297L) in CUB domain 8, suggesting that this proline is functionally crucial in cubilin, and one point mutation (FM2) in the intron interrupting CUB domain 6 responsible for in-frame insertions producing truncated cubilin. Interestingly normal size cubulin protein was identified in urine samples from homozygous FM1 patients, whereas a complete absence of the protein was reported in a patient homozygous for the FM2 mutation (Dali-Youcef and Andrès 2009; Aminoff et al. 1999).

Other mutations were also uncovered but were subsequently identified as polymorphisms after their detection in normal individuals in the general population. The cubilin P1297L mutation associated with hereditary MGA1 was reported to cause impaired recognition of the cobalamin-IF complex by cubilin (Kristiansen et al. 2000).

Moreover, mutation in *AMN* was reported in recessive hereditary MGA1 (Dali-Youcef and Andrès 2009; Kristiansen et al. 2000) and hence was demonstrated to be crucial for a functional cobalamin-IF receptor (Dali-Youcef and Andrès 2009). This study demonstrated that homozygous mutations affecting exons 1-4 of the human *AMN* gene translated into selective malabsorption of vitamin B12, a phenotype associated with hereditary MGA1. Another study reported *AMN* deletion mutants in dogs with selective intestinal malabsorption of cobalamin associated with urinary loss of low molecular weight protein reminiscent of the human Imerslund-Gräsbeck syndrome (IGS a.k.a. MGA1) (Grasbeck 1960; Aminoff et al. 1999). The authors showed that these mutations in the *AMN* gene abrogated AMN expression and blocked cubilin processing and targeting to the apical membrane (He et al. 2005). The essential AMN-cubilin interaction was recapitulated and validated in a heterologous cell-transfection model, hence explaining the molecular basis of intestinal cobalamin malabsorption syndrome (Dali-Youcef and Andrès 2009; Quadros 2010).

Of particular interest for the practitioner in this section of "malabsorption" is the observation that about 1 to 5% of free vitamin B12 (or *crystalline* cobalamin) is absorbed along the entire intestine by passive diffusion (Dali-Youcef and Andrès 2009). This absorption explains the mechanism underlying oral cobalamin treatment of vitamin B12 deficiencies (Andrès et al. 2010). Our working group has developed an effective oral treatment for FCM (Andrès et al. 2001) and pernicious anemia (Andrès et al. 2005) using crystalline cobalamin (cyanocobalamin). Oral cobalamin has been proposed as a way of avoiding the discomfort, inconvenience and cost of monthly injections.

3.3 Cobalamin transport to blood and tissues

3.3.1 Physiology

After vitamin B12 is absorbed at the BBM-blood barrier, it dissociates from the intrinsic factor and reaches the systemic circulation where it associates with transcobalamin II (TCII) (Figure) (Nicolas and Gueant 1994; Dali-Youcef and Andrès 2009; Quadros 2010). The kidney represents an essential organ were body vitamin B12 stores are maintained and studies demonstrated that the kidney regulates plasma vitamin B12 levels by maintaining a pool of unbound cobalamin that can be released in case of vitamin B12 deficiency (Dali-Youcef and Andrès 2009; Birn 2006). The tissular cobalamin-TCII complex uptake is achieved through megalin (LRP2)- and transcobalamin II receptor (TCII-R)-mediated endocytosis which plays a crucial role in cobalamin homeostasis (Andrès et al. 2006; Yammani 2003).

Following cobalamin-TCII cellular uptake, TCII undergoes lysosomal digestion, which allows cobalamin separation from TCII and its cytoplasmic transfer. It has been estimated that there is a delay ranging from 5 and 10 years between the onset of cobalamin deficiency and the appearance of clinical manifestations, due to large hepatic stores (> 1.5 mg) and the enterohepatic cycle ensuring re-absorption of the vitamin in the gastrointestinal tract (Nicolas and Gueant 1994; Dali-Youcef and Andrès 2009). Also the reabsorption of TCII-bound cobalamin in the proximal tubules limits the loss of B12 in urine.

The average vitamin B12 content is approximately 1.0 mg in healthy adults, with 20–30 µg found in the kidneys, heart, spleen and brain. Estimates of total vitamin B12 body content for adults range from 0.6 to 3.9 mg with mean values of 2–3 mg. The normal range of vitamin B12 plasma concentrations is 150–750 pg/ml, with peak levels achieved 8–12 hours after ingestion of a single dose of the vitamin.

Part of the cobalamin serves as a cofactor for methionine synthase-mediated homocysteine catabolism into methionine and methyltetrahydrofolate reductase (MTHFR)-mediated formation of the vitamin B9 biologically active form, tetrahydrofolate, which is then involved in the synthesis of purines and pyrimidines (Figure) (Dali-Youcef and Andrès 2009; Fowler 1998). The other part of vitamin B12 is transferred to the mitochondria where it is transformed into adenosyl-B12, an important cofactor in methylmalonyl-coenzyme A mutase-mediated formation of succinyl-CoA from methylmalonyl-CoA, the product of odd-chain fatty acid and some amino acid catabolism. Hence, cobalamin deficiency will cause homocysteine accumulation, increased methylmalonyl-CoA levels and decreased MTHFR activity. These changes are translated into several abnormalities including folate deficiency and subsequent inhibition of purines and pyrimidines formation essential for RNA and DNA synthesis (Dali-Youcef and Andrès 2009; Quadros 2010).

3.3.2 Genetic disorders of cobalamin transport

It is worth mentioning that TCII is responsible for the cellular uptake of B12 in most tissues and that TC deficiency is associated with severe megaloblastic anemia (Li et al. 1994; Teplitsky et al. 2003) and developmental disorders (see Chapter 4).

Impaired megalin function has not been associated with cobalamin deficiency so far; however inappropriate megalin signaling has been shown to cause deleterious effects as a consequence of cobalamin uptake inhibition in tissues. This was particularly the case where mutations in the human *LRP2* gene encoding megalin were recently described to cause Donnai-Barrow and facio-oculo-acoustico-renal syndromes. Patients affected with these rare autosomal recessive disorders display severe malformations with proteinuria (Kantarci et al. 2007).

The clinical manifestations of the metabolic abnormalities are hereditary megaloblastic anemia, neurological defects, malformations, increased cardiovascular thrombotic risk and renal disease, and methylmalonic acidemia (Dali-Youcef and Andrès 2009; Quadros 2010; Fowler 1998). Functional cobalamin deficiency can also be caused by defects in the intracellular processing of cobalamin such as abnormal lysosomal digestion of the TCII-cobalamin complex, and subsequent defective lysosomal release of cobalamin, and abnormalities in intracytoplasmic cobalamin metabolism with all the consequences on biochemical reactions in which cobalamin acts as an important cofactor (Dali-Youcef and Andrès 2009; Fowler 1998).

3.4 Treatment of cobalamin deficiency

3.4.1 Parenteral administration

The classic treatment for cobalamin deficiency is by parenteral administration —in most countries as intramuscular injections—in the form of cyanocobalamin and, more rarely, hydroxy or methyl cobalamin (Andrès et al. 2010). Hydroxocobalamin may have several advantages due to better tissue retention and storage.

However, the management concerning both the dose and schedule of administration varies considerably between countries (Andrès et al. 2010). In the USA and UK, doses of ranging from 100 to 1,000 μg per month (or every 2–3 months when hydroxocobalamin is given) are used for the duration of the patient's life. In France, treatment, involves the administration of 1,000 µg of cyanocobalamin per day for 1 week, followed by 1,000 µg per week for 1 month, followed by 1,000 µg per month, again, normally for the remainder of the patient's lifetime (Andrès et al. 2010).

3.4.2 New routes of administration

Since cobalamin is absorbed by intrinsic factor-independent passive diffusion, daily high-dose (pharmacological dose, of at least 1,000 µg per day) oral vitamin B12 (cyanocobalamin) can induce and maintain remission in patients with megaloblastic anemia (Lane and Rojas-Fernandez 2002). In cases of cobalamin deficiency other than those caused by nutritional deficiency, alternative routes of cobalamin administration have been proposed used with good effect such as via the oral and nasal passages (Quadros 2010; Slot et al. 1997).

These other routes of administration have been proposed as a way of avoiding the discomfort, inconvenience and cost of monthly injections (Andrès et al. 2010).

Three studies that fulfilled the criteria of evidence based-medicine supported the efficacy of oral cobalamin therapy (Kuzminski et al. 1998; Bolaman et al. 2003; Vidal-Aiaball et al. 2005). Two prospective randomized controlled studies comparing oral vitamin B12 *versus* intramuscular vitamin B12 treatment documented the efficacy of oral vitamin B12 as a curative treatment (Kuzminski et al. 1998; Bolaman et al. 2003). Kuzminski et al., in a prospective randomized trial including 38 patients, reported improvement of hematological parameters and vitamin B12 levels (mean value: 907 pg/mL), after 4 months of oral cyanocobalamin therapy using a much higher dose (i.e., 2,000 µg *per* day) (Kuzminski et al. 1998). Bolaman et al., in a prospective randomized trial of 60 patients, also reported significant improvement of hematological parameters and vitamin B12 levels (mean improvement: +140.9 pg/mL), after 3 months of daily 1,000 µg of oral cyanocobalamin therapy (Bolaman et al. 2003).

An evidence-based analysis by the Vitamin B12 *Cochrane Group* supports the efficacy of oral vitamin B12 as a curative treatment, with a daily dose between 1,000 and 2,000 µg vitamin B12 (Vidal-Alaball et al. 2005). In this analysis, serum vitamin B12 levels increased significantly in patients receiving oral vitamin B12 and both groups of patients (receiving oral and intramuscular treatment) showed an improvement in neurological symptoms. The *Cochrane Group* concludes that daily oral therapy *"may be as effective as intramuscular administration in obtaining short term haematological and neurological responses in vitamin B12 deficient patients"* (Vidal-Alaball et al. 2005).

Nevertheless to our knowledge, the effect of oral cobalamin treatment in patients presenting with severe neurological manifestations has not yet been adequately documented. Thus until this has been studied, parenteral cobalamin therapy is still to be recommended for such patients (Andrès et al. 2010).

Our working group *CARE B12* has developed an effective oral curative treatment in patients presenting with food-cobalamin malabsorption and pernicious anemia using crystalline cyanocobalamin (Andrès et al. 2010; Andrès et al. 2001). In the *princeps* study, we prospectively studied 10 patients with cobalamin deficiency and well-established food-cobalamin malabsorption who received 3,000 or 5,000 µg of oral crystalline cyanocobalamin once a week for at least 3 months (Andrès et al. 2001). After 3 months of treatment, all patients had increased hemoglobin levels (mean increase of 1.9 g/dL; 95% confidence interval: 0.9 to 3.9 g/dL; $p < 0.01$ compared with baseline), and decreased mean erythrocyte cell volume (mean decrease of 7.8 fL; 95% confidence interval: 0.9 to 16.5 fL; $p < 0.001$). However, 2 patients had only minor, if any, responses. Serum cobalamin levels were increased in all 8 patients in whom it was measured.

These results were also observed in a documented population of patients presenting with pernicious anemia (Andrès et al. 2005). We studied in an open study 10 patients with well-documented cobalamin deficiency related to Biermer's disease who daily received 1,000 µg of oral crystalline cyanocobalamin for at least 3 months. After 3 months of treatment, serum cobalamin levels were increased in all 9 patients in whom it was measured (mean increase of 117.4 pg/mL; $p < 0.0000003$ compared with baseline). Eight patients had increased hemoglobin levels (mean increase of 2.45 g/dL; $p < 0.01$). All 10 patients had decreased mean erythrocyte cell volume (mean decrease of 10.4 fL; $p < 0.003$). Three patients experienced clinical improvements.

Recently, several groups have also well-documented the long-term efficacy of oral cobalamin treatment, with a median follow up of 2.5 to 5 years (Andrès et al. 2010).

Table 4 lists our recommendations for the curative treatment of vitamin B12 deficiency (Andrès et al. 2010). We currently recommend a dose of 1,000 µg *per* day of oral cyanocobalamin, for life, in case of Biermer's disease. We recommend 1,000 µg *per* day of oral cyanocobalamin for 1 month, than 125 to 1,000 µg *per* day, in case of intake vitamin B12 deficiency or food-cobalamin malabsorption (Andrès et al. 2010). It is to keep in mind that the effect of oral cobalamin treatment in patients presenting with severe neurological manifestations or in case of genetic disorders in the metabolism of the cobalamin has not yet been adequately documented, both in our experience that in the literature.

The patients should be monitored long-term clinically and with biochemical markers, such as: serum vitamin B12 level with total homocysteine level (better, methyl malonic acid) or holotranscobalamin level (if available) (Andrès et al. 2010).

Table 4. Recommendations for vitamin B12 treatment (Lane and Rojas-Fernandez 2002).

	Pernicious anemia	Intake deficiency and food-cobalamin malabsorption
Parenteral administration (intramuscular)	Cyanocobalamin: – 1,000 µg *per* day for 1 week – than 1,000 µg *per* week for 1 month – than 1,000 µg *per* each month, for life (1,000 to 2,000 µg *per* day for at least 1 to 3 months in case of severe neurological manifestations)	Cyanocobalamin: – 1,000 µg *per* day for 1 week – than 1,000 µg *per* week for 1 month – than 1,000 µg *per* each 1 or 3 months, until the cobalamin deficiency cause is corrected (1,000 µg *per* day for at least 1 to 3 months in case of severe neurological manifestations)
Oral administration	Cyanocobalamin: 1,000 µg *per* day for life‡	Cyanocobalamin: – 1,000 µg *per* day for 1 month – than 125 to 1,000 µg *per* day, until the cobalamin deficiency cause is corrected‡

‡ The effect of oral cobalamin treatment in patients presenting with severe neurological manifestations and in case of genetic disorders of the cobalamin metabolism has not yet been adequately documented.

4. Conclusions

In this chapter, we presented the main etiologies of vitamin B12 deficiency in relation to different steps of the cobalamin transport and metabolism. However to date, many causes of cobalamin deficiency remained unknown. These causes include mutations in genes encoding important proteins of the cobalamin transport or metabolic pathway. Moreover, many clinically diagnosed vitamin B12 deficiency remain unexplained and molecular tools aimed at targeting genes involved in vitamin B12 absorption and cellular uptake signaling pathways will pave the way for new therapeutic approaches to efficiently treat functional cobalamin deficiency.

Competing interests: no conflict of interest.

Acknowledgments

We especially want to thank Dr. Nassim Dali-Youcef who contributed to this chapter through his expertise and assistance to our works on the management of vitamin B12 deficiency in adults.

Keywords: Absorption, food cobalamin, cobalamin transport, treatment, intrinsic factor, megalin, pernicious anemia, genetic defects, transcobalamin

Abbreviations

AM	:	amnionless
HC	:	haptocorrin
FCM	:	food-cobalamin malabsorption
IF	:	intrinsic factor
LDLR	:	low-density lipoprotein receptor
MGA1	:	megaloblastic anemia 1
MTHFR	:	methyltetrahydrofolate reductase
TCII	:	transcobalamin II
TCII-R	:	transcobalamin II receptor

References

Ahuja R, Yammani R, Bauer JA, Kalra S, Seetharam S and Seetharam B. 2008. Interactions of cubilin with megalin and the product of the amnionless gene (AMN): effect on its stability. Biochem J. 410(2): 301–8.

Aminoff M, Carter JE, Chadwick RB, Johnson C, Grasbeck R, Abdelaal MA et al. 1999. Mutations in CUBN, encoding the intrinsic factor-vitamin B12 receptor, cubilin, cause hereditary megaloblastic anaemia 1. Nat Genet. 21(3): 309–13.

Andrès E, Affenberger S, Vinzio S, Kurtz JE, Noel E, Kaltenbach G et al. 2005. Food-cobalamin malabsorption in elderly patients: clinical manifestations and treatment. Am J Med. 118(10): 1154–9.

Andrès E, Affenberger S, Zimmer J, Vinzio S, Grosu D, Pistol G et al. 2006. Current hematological findings in cobalamin deficiency. A study of 201 consecutive patients with documented cobalamin deficiency. Clin Lab Haematol. 28(1): 50–6.

Andrès E, Fothergill H and Mecili M. 2010. Efficacy of oral cobalamin (vitamin B12) therapy. Expert Opinion Pharmacotherapy. 11(3): 249–56.

Andrès E, Goichot B and Schlienger JL. 2000. Food-cobalamin malabsorption: a usual cause of vitamin B12 deficiency. Arch Intern Med. 160(11): 2061–2.

Andrès E, Henoun Loukili N, Noel E, Maloisel F, Vinzio S, Kaltenbach G et al. 2005. Oral cobalamin (daily dose of 1000 μg) therapy for the treatment of patients with pernicious anemia. An open label study of 10 patients. Curr Ther Research. 66(10): 13–22.

Andrès E, Kurtz JE, Perrin AE, Maloisel F, Demangeat C, Goichot B et al. 2001. Oral cobalamin therapy for the treatment of patients with food-cobalamin malabsorption. Am J Med. 111(9): 126–9.

Andrès E, Loukili NH, Noel E, Kaltenbach G, Abdelgheni MB, Perrin AE et al. 2004. Vitamin B12 (cobalamin) deficiency in elderly patients. CMAJ. 171(3): 251–4.

Andrès E, Noel E and Goichot B. 2002. Metformin-associated vitamin B12 deficiency. Arch Intern Med. 162(19): 2251–2.

Andrès E, Vogel T, Federici L, Zimmer J and Kaltenbach G. 2008. Update on oral cyanocobalamin (vitamin B12) treatment in elderly patients. Drugs Aging. (1); 25: 927–32.

Bauman WA, Shaw S, Jayatilleke E, Spungen AM and Herbert V. 2000. Increased intake of calcium reverses vitamin B12 malabsorption induced by metformin. Diabetes Care. 23(9): 1227–31.

Birn H, Verroust PJ, Nexo E, Hager H, Jacobsen C, Christensen EI et al. 1997. Characterization of an epithelial approximately 460-kDa protein that facilitates endocytosis of intrinsic factor-vitamin B12 and binds receptor-associated protein. J Biol Chem. 272(42): 26497–504.

Birn H. 2006. The kidney in vitamin B12 and folate homeostasis: characterization of receptors for tubular uptake of vitamins and carrier proteins. Am J Physiol Renal Physiol. 291(1): F22–36.

Bolaman Z, Kadikoylu G, Yukselen V, Yavasoglu I, Barutca S and Senturk T. 2003. Oral versus intramuscular cobalamin treatment in megaloblastic anemia: a single-center, prospective, randomized, open-label study. Clin Ther. 25: 3124–34.

Carmel R. 1983. R-binder deficiency. A clinically benign cause of cobalamin pseudodeficiency. JAMA. 250(14): 1886–90.

Carmel R. 1995. Malabsorption of food cobalamin. Baillieres Clin Haematol. 8(3): 639–55.

Carmel R. 2003. Mild transcobalamin I (haptocorrin) deficiency and low serum cobalamin concentrations. Clin Chem. 49(8): 1367–74.

Christensen EI and Birn H. 2002. Megalin and cubilin: multifunctional endocytic receptors. Nat Rev Mol Cell Biol. 3(4): 256–66.

Dali-Youcef N and Andrès E. 2009. An update on cobalamin deficiency in adults. QJM. (7); 102: 17–28.

Fowler B. 1998. Genetic defects of folate and cobalamin metabolism. Eur J Pediatr. 157 Suppl 2: S60–6.

Fyfe JC, Madsen M, Hojrup P, Christensen EI, Tanner SM, de la Chapelle A et al. 2004. The functional cobalamin (vitamin B12)-intrinsic factor receptor is a novel complex of cubilin and amnionless. Blood. 103(5): 1573–9.

Grasbeck R. 1960. Familiar selective vitamin B12 malabsorption with proteinuria. A pernicious anemia-like syndrome. Nord Med. 63(2): 322–3.

He Q, Madsen M, Kilkenney A, Gregory B, Christensen EI, Vorum H et al. 2005. Amnionless function is required for cubilin brush-border expression and intrinsic factor-cobalamin (vitamin B12) absorption *in vivo*. Blood. 106(4): 1447–53.

Imerslund O. 1960. Idiopathic chronic megaloblastic anemia in children. Acta Paediatr. 49(Suppl 119): 1–115.

Jung SB, Nagaraja V, Kapur A and Eslick GD. 2015. Association between vitamin B12 deficiency and long-term use of acid-lowering agents: a systematic review and meta-analysis. Intern Med J. 45(4): 409–16.

Kantarci S, Al-Gazali L, Hill RS, Donnai D, Black GC, Bieth E et al. 2007. Mutations in LRP2, which encodes the multiligand receptor megalin, cause Donnai-Barrow and facio-oculo-acoustico-renal syndromes. Nat Genet. 39(8): 957–9.

Kaptan K, Beyan C, Ural AU, Cetin T, Avcu F, Gulsen M et al. 2000. Helicobacter pylori—is it a novel causative agent in Vitamin B12 deficiency? Arch Intern Med. 160(9): 1349–53.

Kozyraki R, Kristiansen M, Silahtaroglu A, Hansen C, Jacobsen C, Tommerup N et al. 1998. The human intrinsic factor-vitamin B12 receptor, cubilin: molecular characterization and chromosomal mapping of the gene to 10p within the autosomal recessive megaloblastic anemia (MGA1) region. Blood. 91(10): 3593–600.

Kristiansen M, Aminoff M, Jacobsen C, de La Chapelle A, Krahe R, Verroust PJ et al. 2000. Cubilin P1297L mutation associated with hereditary megaloblastic anemia 1 causes impaired recognition of intrinsic factor-vitamin B(12) by cubilin. Blood. 96(2): 405–9.

Kristiansen M, Kozyraki R, Jacobsen C, Nexo E, Verroust PJ and Moestrup SK. 1999. Molecular dissection of the intrinsic factor-vitamin B12 receptor, cubilin, discloses regions important for membrane association and ligand binding. J Biol Chem. 274(29): 20540–4.

Kuzminski AM, Del Giacco EI, Allen RH, Stabler SP and Lindenbaum J. 1998. Effective treatment of cobalamin deficiency with oral cobalamin. Blood. 92: 1191–8.

Lane LA and Rojas-Fernandez C. 2002. Treatment of vitamin B12 deficiency anemia: oral versus parenteral therapy. Ann Pharmacother. 36: 1268–72.

Li N, Rosenblatt DS, Kamen BA, Seetharam S and Seetharam B. 1994. Identification of two mutant alleles of transcobalamin II in an affected family. Hum Mol Genet. 3(10): 1835–40.

Lindenbaum J, Rosenberg IH, Wilson PW, Stabler SP and Allen RH. 1994. Prevalence of cobalamin deficiency in the Framingham elderly population. Am J Clin Nutr. 60(1): 2–11.

Markle HV. 1996. Cobalamin. Crit Rev Clin Lab Sci. 33(4): 247–356.

Matthews JH. 1995. Cobalamin and folate deficiency in the elderly. Baillieres Clin Haematol. 8(3): 679–97.

Medicine I. 1998. Dietary reference intake of thiamin, riboflavin, niacin, vitamin B6, folate, vitamin B12, panthotenic acid, biotin and choline Food and Nutrition Board, Washington DC National Academies Press.

Moestrup SK, Birn H, Fischer PB, Petersen CM, Verroust PJ, Sim RB et al. 1996. Megalin-mediated endocytosis of transcobalamin-vitamin-B12 complexes suggests a role of the receptor in vitamin-B12 homeostasis. Proc Natl Acad Sci USA. 93(16): 8612–7.

Nicolas JP and Gueant JL. 1994. Absorption, distribution and excretion of vitamin B12. Ann Gastroenterol Hepatol (Paris). 30(6): 270–6, 81; discussion 81-2.

Pawlak R, Lester SE and Babatunde T. 2014. The prevalence of cobalamin deficiency among vegetarians assessed by serum vitamin B12: a review of literature. Eur J Clin Nutr. 68(5): 541–8.

Quadros EV. 2010. Advances in the understanding of cobalamin assimilation and metabolism. Br J Haematol. 148(2): 195–204.

Rojas Hernandez CM and Oo TH. 2015. Advances in mechanisms, diagnosis, and treatment of pernicious anemia. Discov Med. 19(104): 159–68.

Saito A, Pietromonaco S, Loo AK and Farquhar MG. 1994. Complete cloning and sequencing of rat gp330/"megalin", a distinctive member of the low density lipoprotein receptor gene family. Proc Natl Acad Sci USA. 91(21): 9725–9.

Slot WB, Merkus FW, Van Deventer SJ and Tytgat GN. 1997. Normalization of plasma vitamin B12 concentration by intranasal hydroxocobalamin in vitamin B12-deficient patients. Gastroenterology. 113: 430–3.

Stabler SP. 2013. Clinical practice. Vitamin B12 deficiency. N Engl J Med. 368(2): 149–60.

Tanner SM, Li Z, Perko JD, Oner C, Cetin M, Altay C et al. 2005. Hereditary juvenile cobalamin deficiency caused by mutations in the intrinsic factor gene. Proc Natl Acad Sci USA. 102(11): 4130–3.

Teplitsky V, Huminer D, Zoldan J, Pitlik S, Shohat M and Mittelman M. 2003. Hereditary partial transcobalamin II deficiency with neurologic, mental and hematologic abnormalities in children and adults. Isr Med Assoc J. 5(12): 868–72.

Toh BH, van Driel IR and Gleeson PA. 1997. Pernicious anemia. N Engl J Med. 337(20): 1441–8.

van Asselt DZ, Blom HJ, Zuiderent R, Wevers RA, Jakobs C, van den Broek WJ et al. 2000. Clinical significance of low cobalamin levels in older hospital patients. Neth J Med. 57(2): 41–9.

Vidal-AIaball J, Butler CC, Cannings-John R, Goringe A, Hood K, McCaddon A et al. 2005. Oral vitamin B12 versus intramuscular vitamin B12 for vitamin B12 deficiency. Cochrane Database Syst Rev. 20: CD004655.

Yammani RR, Seetharam S, Dahms NM and Seetharam B. 2003. Transcobalamin II receptor interacts with megalin in the renal apical brush border membrane. J Membr Biol. 193(1): 57–66.

6

Vitamin B12 Deficiency in Developing and Newly Industrialising Countries

Chittaranjan Yajnik,[1,*] *Urmila Deshmukh,*[2] *Prachi Katre*[1] and *Tejas Limaye*[1]

1. Introduction

Once considered a 'luxus' vitamin in humans (meaning ample or excess) (McLaren 1981), deficiency of vitamin B12 is increasingly recognised in different populations, and the developing countries are not an exception. Recent research in developing countries has shed the light on the effects of this vitamin on foetal growth and programming of non-communicable diseases. In the developed world, symptomatic B12 deficiency is predominantly due to malabsorption of the vitamin (pernicious anaemia or food B12 malabsorption in the elderly). In contrast, in the developing populations it is largely due to low intake of vitamin B12-rich animal origin foods and possibly to infectious gastrointestinal diseases (Allen 2008).

We will review the specific aspects of epidemiology, disease associations and public health significance of vitamin B12 deficiency in the developing countries. We will also highlight the importance of vitamin B12 nutrition across the lifecycle starting with genetic factors, and from pregnancy to infancy, childhood, adolescence, adulthood and elderly.

[1] Diabetes Unit, King Edward Memorial Hospital Research Centre, Pune, India.
[2] BioTRaK Research and Diagnostics Centre, Navi Mumbai, India.
* Corresponding author

2. Epidemiology of Vitamin B12 Deficiency

We referred to the World Bank classification of countries by income and development to define 'developing countries'. However, we appreciate that the differences in the nutritional status between countries are due to a variety of factors including economy, demography, ethnicity, religion, culture, dietary intake as well as body's requirements.

Data on vitamin B12 status is available only from a limited number of these countries and the quality of information is variable. Investigators have used different biomarkers, laboratory methods, and different cut-points for defining 'deficiency', thus limiting comparison across studies. Most studies have referred to small population or subpopulation samples and there is little nationally representative data (Figure 1). However, despite all limitations, the results on certain populations have been confirmed by several studies using different designs and methodologies, thus suggesting that vitamin B12 deficiency is a public health problem in many regions.

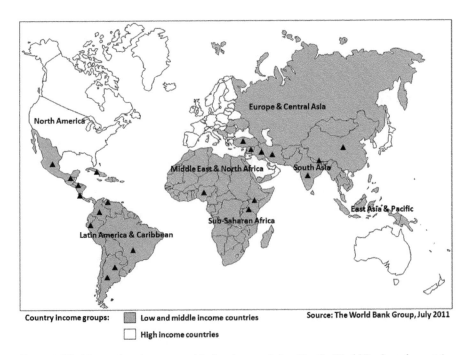

Figure 1. World map showing geographical regions as defined by the World Bank, and countries in those regions from where data about vitamin B12 status of the population is available (black triangles).

2.1 Asia and Pacific

Asian countries have a higher prevalence of vitamin B12 deficiency compared to other parts of the world. South Asians in the Indian sub-continent are particularly affected. High prevalence of abnormal vitamin B12 markers is seen in individuals resident in their home countries, as well as in those migrated to western countries. Both men and women are affected. Reports are available from north as well as south India and the adjoining country of Nepal (Ulak et al. 2014). The largest data is available in Indians (Refsum et al. 2001; Antony 2003; Yajnik et al. 2006; Rush et al. 2009). Many studies are clinic-based, while community-based studies are less common (Yajnik et al. 2006).

Deficiency in a substantial part of the populations has been reported in different studies across the lifecycle: in pregnant women (prevalence 30–70%) (Dwarkanath et al. 2013; Samuel et al. 2013; Krishnaveni et al. 2009; Yajnik et al. 2008; Pathak et al. 2007; Yusufji et al. 1973), preschoolers (~75%) (Taneja et al. 2007), adolescents (~50%) (Thomas et al. 2015; Gupta et al. 2015; Kapil and Sareen 2014) and in adults (~70%) (Yajnik et al. 2006). Countrywide studies are required to corroborate these findings. In Pune, vitamin B12 concentrations were lower in the urban middle class men compared to the slum dwellers, which could be related to differences in consumption of a non-vegetarian foods and to more common infections in the slum dwellers. Vegetarianism, sanctioned by religion and culture makes a significant contribution to B12 deficiency in Indians. The concentrations of plasma homocysteine and methylmalonic acid (MMA) (biochemical markers that increase in plasma of B12 deficient subjects) are substantially elevated in Indians compared with age-matched Europeans. In Indians, vitamin B12 deficiency is the main factor that explains hyperhomocysteinemia while folate deficiency is not common (Refsum et al. 2001).

Low vitamin B12 status has a particular health significance in pregnant women. A low vitamin B12 status has been reported in approximately third to two-thirds of pregnant women from India and Nepal (Yajnik et al. 2008; Stewart et al. 2011; Bondevik et al. 2001). B12 deficiency very likely exists from before conception and negatively affects health outcome of the newborns. Deficiency of vitamin B12 during pregnancy has been related to poor fetal growth [i.e., causes Intrauterine Growth Retardation (IUGR)], and birth defects [i.e., neural tube defects (NTD)]. Maternal B12 deficiency is the strongest predictor of low vitamin B12 status in the neonate which may have effects on several developmental domains (i.e., physical growth and neurodevelopment). Prolonged breast feeding is common in countries with food insecurity and is associated with low vitamin B12 status in the offspring, which reflects low maternal stores (Taneja et al. 2007; Pasricha et al. 2011; Lubree et al. 2012).

In East Asia, data is mostly available from China where vitamin B12 deficiency is prevalent in all age groups, though little data is available from pregnant women (Dang et al. 2014). Deficiency is more common in rural compared to the urban population probably due to lower income or dependence on local foods (Dang et al. 2014).

2.2 Central Asia and the Middle East

Vitamin B12 deficiency has been reported in pockets in these areas. In Sanliurfa province of Turkey, a high proportion (72%) of pregnant women and 41% of their babies (cord blood) were found to be vitamin B12 deficient; more than half of these were severely deficient (Koc et al. 2006). Clinical manifestations were common in infants (i.e., anemia, failure to thrive or neurological disorders). In a study from Tehran, a quarter of adult men and women (age 25–64 years) had low serum vitamin B12 levels (Fakhrzadeh et al. 2006). The prevalence of low serum B12 was two to three fold higher in the elderly compared with young subjects (Khodabandehloo 2015) and approximately one in five women in childbearing-age had low vitamin B12 levels (Abdollahi et al. 2008). High prevalence of vitamin B12 deficiency (12%), elevated MMA levels (49%), and hyperhomocysteinemia (61%) were reported in patients with coronary vascular diseases from Syria (Herrmann et al. 2003). Low intake of vitamin B12 from animal based-foods or a Mediterranean diet (primarily plant-based foods) could contribute to deficiency of this nutrient in these countries (Balci et al. 2014).

2.3 Sub-Saharan Africa

Many of the countries in this region belong to low or lower-middle income categories and have high prevalence of infection-related morbidities. Vitamin B12 deficiency has been reported only in some studies from this geographic region. In Uganda, young adults appear to have sufficient vitamin B12 status (Galukande et al. 2011), but approximately third of HIV infected pre-school children were B12 deficient (Ndeezi et al. 2011). High rates of vitamin B12 deficiency have been reported in school children from Kenya (70%) (Siekmann et al. 2003) and pregnant women from Nigeria (36%) (VanderJagt et al. 2011). Breast milk concentrations of vitamin B12 were low in rural Kenyan women (VanderJagt et al. 2011; Neumann et al. 2013; McLean et al. 2007), suggesting insufficient intake in the infants and high risk to develop deficiency at a very young age.

2.4 Latin America and Caribbean

Across studies in Latin America, ~40% of children and adults had deficient or marginal B12 status (Allen 2004), including a nationally representative sample of women and children that were studied in the 1999 Mexican National Nutrition Survey. Adolescent girls were one risk group for vitamin B12 deficiency (Brito et al. 2015). Recent national data from Mexico indicated that the prevalence of low B12 (< 148 pmol/L) was 8.5% in women (20–49 years) (Shamah et al. 2015) and 1.9% in preschoolers and 2.3% in school age children (Villalpando et al. 2015). However, regional differences in the risk of having low B12 were reported with the highest risk found for Mexico City (OR= 7.03, 95% CI 3.09–15.97) regions compared with the Northern region.

Despite the limitations of the data, there appears to be a consensus that low vitamin B12 status is an important public health problem in many countries in different parts of the world. This could have important implications for a variety of disorders across the life course. Investigative epidemiology may shed new light on the aetiology of these conditions, and interventions on a population level could play an important role in prevention of vitamin B12 deficiency and the associated disorders. Vitamin B12 deficiency continues to exist in many countries after folic acid fortification (Brito et al. 2015). Vitamin B12 is required for folate metabolism and the expected benefit of folic acid fortification programs in prevention of birth defects and other diseases will be limited in populations with common B12 deficiency.

3. Vitamin B12 Across the Life-course

Investigation of vitamin B12 status across the life-course allows an understanding of how the deficiency develops and is transferred to the next generation. This process is affected by several factors as shown in Figure 2.

An important realization is that, poor maternal vitamin status during pregnancy and lactation translates into a depleted storage in the child. After birth, continued low intake in the child because of low content in mother's milk, parent's household, cultural and economic factors play determinant roles in promoting a deficiency situation.

3.1 Pregnancy and lactation

Vitamin B12 deficiency is a major public health challenge in pregnant and lactating women from developing countries. The reasons are related to a low intake of animal-source foods, low frequency of supplements use during pregnancy and lactation, repeated pregnancies and short intervals between pregnancies. These factors are aggravated by socio-cultural factors such as early marriage and adolescent pregnancies.

The dietary intake of vitamin B12 during pregnancy is inadequate in many developing countries (Dwarkanath et al. 2013; Samuel et al. 2013; Yajnik et al. 2008; Balci et al. 2014; Halicioglu et al. 2012). Pregnant women achieved less than 50% of the Estimated Average Requirement (EAR) for vitamin B12 in some Asian and African countries (Torheim et al. 2010). Multiple nutrient deficiencies such as folate, riboflavin and choline are common. Deficiency of these nutrients is associated with hyperhomocysteinemia, pregnancy complications and adverse birth outcomes (Torheim et al. 2010). However,

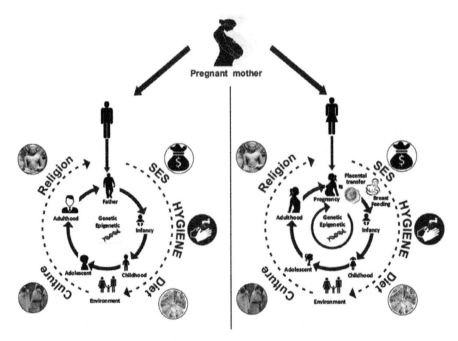

Figure 2. Mother exerts a triple influence on the vitamin B12 status of the offspring (male and female): (1) genetic, (2) direct transfer (transplacental and through breast milk) and (3) post-natal family environment [Socioeconomic status (SES), hygiene, diet, religion and culture]. Father exerts a double influence: (1) genetic and (2) family environment. A female child propagates the transplacental influence to the next generations).

fortification or supplementation programs in many countries have focused on folate, zinc and iron in pregnant women, while the importance of vitamin B12 is underestimated.

The fetus draws all its vitamin B12 requirements from the mother through active placental transfer. Circulating total vitamin B12 levels decrease during pregnancy and return to pre-pregnancy levels postpartum. These changes are partly explained by a decline in serum haptocorrin, while serum holotranscobalamin remains rather stable. Due to these marked physiological changes, concentrations of total serum vitamin B12 will depend on gestational age and using this marker to evaluate vitamin B12 status in pregnancy is not recommended. Estimates of vitamin B12 deficiency vary according to the country and criteria used to define deficient women, being low in Canada but high in many developing countries (Table 1).

Low vitamin B12 status was associated with higher body weight and higher plasma glucose concentrations (gestational diabetes) (Samuel et al. 2013; Krishnaveni et al. 2009) which is probably explained by unbalanced diet. Moreover, low maternal vitamin status during pregnancy in Indian women with gestational diabetes predicted higher prevalence of diabetes in the mother 5 years later (Krishnaveni et al. 2009).

Table 1. Prevalence of low serum vitamin B12 in pregnant women.

Country (Reference)	Definition of low serum vitamin B12*	Prevalence
Brazil (Barnabe et al. 2015)	< 148 pmol/L	8%
Colombia (Ramírez-Vélez et al. 2016)	< 148 pmol/L	19%
Nigeria (VanderJagt et al. 2011)	Holo-Transcobalamin < 40 pmol/L	36%
Argentina (Brito et al. 2015)	< 148 pmol/L	49%
Nepal (Bondevik et al. 2001)	< 150 pmol/L	49%
India (Duggan et al. 2014)	< 150 pmol/L	51%
India (Katre et al. 2010)	< 150 pmol/L	73%
India (Krishnaveni et al. 2009)	< 150 pmol/L	43%
India (Yajnik et al. 2008)	< 150 pmol/L	71%
India (Pathak et al. 2007)	< 148 pmol/L	74%
India (Yusufji et al. 1973)	< 104 pmol/L	52%
Parts of Turkey (Koc et al. 2006)	< 118 pmol/L	72%

* All were converted to pmol/L (1 ng/ml = 0.74 pmol/L).

Low amounts of vitamin B12 are transferred to the child via breast milk if the maternal intake of the vitamin is not sufficient. Low breast milk concentrations of vitamin B12 have been reported in women from rural Kenya (mean of 155 pmol/L at 6 months post-partum) (Neumann et al. 2013), Guatemala (≤ 50 pmol/L at 12 months postpartum) (Deegen et al. 2012), India (85 to 100 pmol/L) (Duggan et al. 2014; Jathar et al. 1970; Deodhar and Ramakrishna 1959), compared with 900 pmol/L in well-nourished American women (Thomas et al. 1980). However, these studies cannot be compared directly since vitamin B12 concentrations in breast milk show marked changes during lactation (Greibe et al. 2013). Moreover, measurement of vitamin B12 content in human milk is technically difficult because of the high content of haptocorrin that disturbs the assays if not removed prior to B12 assay (Lildballe et al. 2009).

In Pune Maternal Nutrition Study (Rao et al. 2001), a prospective study of maternal nutrition, fetal growth and risk of future non-communicable disease, maternal micronutrient nutrition showed a strong association with offspring birth size. Two thirds of the mothers had low vitamin B12 status; while folate deficiency was rare. Maternal homocysteine concentrations were negatively associated with B12 and folate levels. High homocysteine concentrations predicted intrauterine growth restriction. The association between maternal vitamin B12 deficiency and IUGR has been confirmed by other studies from South India (Muthayya et al. 2006).

Taken together, maternal vitamin B12 status starting from before conception and throughout pregnancy and lactation contributes to neonatal stores (Finkelstein et al. 2015; Murphy et al. 2007) and low status is associated with adverse health outcome in the mother and the child. Only few intervention studies on pregnant or lactating women are available, and long term follow up is essential to study the effect of vitamin B12 replacement on child health.

3.2 Childhood and adolescence

Studies on children from Pune and Delhi, in India have shown that maternal vitamin B12 status in pregnancy predicted offspring concentrations of vitamin B12 and homocysteine (Lubree et al. 2012). Prolonged breast feeding and the delay in introducing complementary animal based foods are also associated with low vitamin status in toddlers (Taneja et al. 2007; Ingrid Kristin et al. 2015). Low intake of animal food after weaning is another factor contributing to vitamin B12 deficiency in pre-school and school age children. The effect of maternal vitamin B12 status may persist up to adolescence (Author's unpublished data).

In rural Indian infants, the daily vitamin B12 intake from non-breast milk food source is substantially low and one in two consumes less than 75% of the recommended daily intake (0.2 µg/d) (Pasricha et al. 2011). It has been shown that the total daily intake of B12 and the frequency of consumption of meat, fish and micronutrient-enriched beverages are positively associated with plasma B12 levels in 10 years-old South Indian children (Christian et al. 2015). Genetic risk score (that included MTHFRC677T, MTHFR A1298C and FUT2 AG polymorphisms), poverty, low intake of animal-derived foods, geo-helminth infection, low vitamin A and folate status were important factors associated with low vitamin B12 status of Amazonian children aged < 10 y, with 4.5% children having B12 deficiency (Cobayashi et al. 2015).

Vitamin B12 blood markers show age-dependent changes even in well-nourished populations. A cross-sectional study of 700 children aged 4 d to 19 year consuming typical Western diet showed that vitamin B12 was low (median = 350 pmol/L) during the first 6 months, it decreased markedly after birth (median = 314 pmol/L) to reach a nadir (median = 217 pmol/L) between 6 wk and 6 months (Monsen et al. 2003). Thereafter, it increased to a maximum concentration at ~3–7 years, and the median concentration gradually decreased thereafter toward the concentrations observed in adults (Monsen et al. 2003). Several case reports demonstrate the importance of maintaining adequate vitamin B12 status during periods of rapid growth and development (Allen 2008; Dror and Allen 2011; Goraya et al. 2015). Vitamin B12 deficiency in infancy can cause failure to thrive, developmental delay, regression, progressive neurologic disorders, or hematologic symptoms such as anemia and macrocytosis. The symptoms may be evident as early as 3–4 months of age, but are often nonspecific and difficult to detect. The neurological symptoms may persist after treatment of vitamin deficiency. In children who start their life with depleted stores, a drop in vitamin B12 status in the first 6 month after birth or low intake in the first 2 years of life are expected to have serious effects on growth and development. Moreover, the substantial prevalence of vitamin B12 deficiency in adolescents from developing countries (Asia, Latin America and Caribbeans, Africa, etc.) could have profound effects on physical growth and reproductive function.

3.3 Adults and elderly

Studies in Pune, India have reported a high prevalence of vitamin B12 deficiency and hyperhomocysteinemia in adults and elderly people (Refsum et al. 2001; Yajnik et al. 2006). Vegetarians had 4.4 times higher risk for developing vitamin B12 deficiency and 3 times higher risk of hyperhomocysteinemia than non-vegetarians. The prevalence estimates of vitamin B12 deficiency in non-pregnant Guatemalan women (15–49 years) ranged between 12 to 26% (Rosenthal et al. 2015). Argentina and Colombia reported 12% and 37% prevalence, respectively (Brito et al. 2015). In Northwest China 46% of the women had B12 deficiency that was more frequent in rural than in the urban parts (Dang et al. 2014) probably due to consumption of locally produced foods that are low in vitamin B12.

Vitamin B12 deficiency in elderly is attributed to food-cobalamin malabsorption (Fernández-Bañares et al. 2009). The prevalence varies from 9% in Latin America and Caribbean region (Brito et al. 2015) to 70% in China (Zhang et al. 2014). On the other hand it was less frequent in South Indian elderly individuals who consumed B12 containing supplements (Shobha et al. 2011). High homocysteine concentrations in the elderly are associated with dementia or cognitive impairment without dementia, and the risk is lower in subjects with higher plasma vitamin B12 (Haan et al. 2007).

Gastric atrophy affects 40% of subjects older than 80 years and is the main factor that can cause food-cobalamin malabsorption. Other factors that contribute to cobalamin malabsorption in elderly people include: intestinal microbial proliferation; long term ingestion of biguanides (metformin) and antacids, including H2-receptor antagonists and proton pump inhibitors, chronic alcoholism, gastric reconstruction, and partial pancreatic exocrine failure (Dali-Yocef and Andreas 2009). Regular vitamin supplementation may protect against low status (Shobha et al. 2011) and its effect on neurological and other outcomes needs to be studied.

4. Risk Factors for Vitamin B12 Deficiency in Developing Populations

Table 2 summarizes risk factors associated with vitamin B12 deficiency in developing countries. The first and major cause of deficiency is an insufficient intake of the vitamin from the diet. Vitamin B12 dietary intake and status markers have shown a direct association in studies from developed world (Pawlak et al. 2013; Tucker et al. 2000). Consumption of animal products provides an intake between 3 and 22 µg B12 per day compared with 0 to 0.25 µg per day that can be achieved by strict vegetarians (Allen 2008).

Table 2. Common risk factors for vitamin B12 deficiency in developing populations.

Risk factor	Possible reasons of vitamin B12 deficiency
(A) Apparently healthy, free-living population	
1. Low intake of animal-source foods	Ethical, religious, cultural, or economic factors
2. Low intake and increased demand during pregnancy and lactation	Nausea and vomiting at early pregnancy, active transfer to fetus
3. Low stores at birth and increased demands during infancy and childhood	Poor maternal nutrition, low human milk B12, prolonged breastfeeding and late introduction of complementary foods or insufficient intake of animal products after weaning, less dependence on formula milk
4. Intestinal infections	*Helicobacter pylori*, bacterial overgrowth, helminthes
5. Drug interactions	Uncontrolled use of antibiotics, proton pump inhibitors, and other drugs that affect the gastrointestinal pH, and metformin
6. Lifestyle	Smoking and excessive alcohol consumption
(B) Sick or critically ill population	
1. Malabsorption	Tropical sprue, celiac disease
2. Systemic infections	HIV, tuberculosis
3. Elderly and critically ill people	Low intake, food B12 malabsorption (gastric atrophy), medications, and co-morbidities like depression that cause eating disorders

Recent genome-wide association studies have shown that several common genetic polymorphisms are associated with low vitamin B12 concentrations. Examples of these variants are alleles (rs492602 G, rs602662 and rs601338) of fucosyltransferase 2 (FUT2) enzyme, TT genotype of the methylenetetrahydrofolate reductase (MTHFR C677T) variant and the CC genotype of the MTHFR A1298C variant (Hazra et al. 2008, 2009).

Among individuals adhering to different types of vegetarian diet, the reported vitamin B12 deficiency is 45% in infants, up to 33% in children and adolescents, 17–39% in pregnant women, and up to 85% in adults and elderly, with higher deficiency in those adhering to vegan diet than in those consuming lacto-vegetarian, lacto-ovo-vegetarian or macrobiotic diets (Pawlak 2013). Dairy (predominantly milk) and meat are the most important contributors to vitamin B12 status followed by fish and shellfish (Brouwer-Brolsma et al. 2015). In countries with common vitamin B12 deficiency, regional, traditional foods could be used for enrichment with vitamin B12 to improve vitamin B12 status. For example, soybean-fermented foods and some seaweeds have been suggested as fortification vehicles in Korea (Kwak et al. 2010).

Tropical sprue, gastrointestinal infestations and infections (including *Helicobacter pylori*) are common in developing countries and are associated with low vitamin B12 concentrations. In adult patients who underwent upper gastrointestinal endoscopy, those who tested positive for *Helicobacter pylori* infection were more likely to have low vitamin B12 levels, anti-parietal cell anti-bodies and anti-intrinsic factor antibodies (Ayesh et al. 2013), suggesting that low vitamin B12 was due to malabsorption disorders.

5. Public Health Perspective of Vitamin B12 Deficiency and Evidence of Its Prevention

The World Health Organization has identified infants, preschool children, and pregnant and lactating women as the most vulnerable groups to develop micronutrient deficiencies (McLean et al. 2008). Maternal vitamin B12 deficiency is now considered an important contributor to folate-resistant neural tube defects (NTD). A multicentre case-control study from India highlighted a role of maternal B12 deficiency in the etiology of NTD (Godbole et al. 2011). A recent meta-analysis of 32 studies involving 1,890 mothers with NTD-affected pregnancies and 3,995 controls revealed that NTD-affected mothers had significantly higher plasma homocysteine and lower levels of plasma folate, red cell folate and vitamin B12 compared with the control mothers (Tang et al. 2015).

A study from South India reported that intake of vitamin B12 was inadequate in pregnant women throughout the pregnancy and low maternal vitamin B12 and folate intakes in the first trimester were independently associated with a higher risk of small-for-gestational age (SGA) in the offspring (Dwarkanath et al. 2013). In women who took high supplemental folic acid (> 1000 µg/d) in the second trimester and also were in the lowest tertile of vitamin B12:folate ratio had higher risk of SGA compared to those in the highest tertile. High maternal circulating homocysteine concentrations, largely due to low vitamin B12 status predicted smaller birth weight and IUGR in another Indian study (Yajnik et al. 2014). Therefore, it appears that disturbances in maternal 1-carbon metabolism caused by vitamin B12 deficiency and vitamin B12-folate imbalance associated with dietary and supplemental folate intake adversely affect fetal growth and development. The policy of folic acid prophylaxis and ongoing supplementation with folic acid in late pregnancy in B12 deficiency populations could be harmful to the fetus. Majority of women in India approach the doctor after pregnancy is well established, and the peri-conceptional window of prophylaxis is missed. Moreover, the most frequently used tablet in India contains 5 mg folic acid which is 12 times higher than the prescribed dose for NTD prophylaxis (400 mcg), many obstetricians prescribe more than one tablet a day in the belief that it is harmless because it is water soluble. These issues need to be addressed in national programmes as well as in clinical practice.

In a longitudinal study in India maternal folate was associated with offspring motor development and maternal vitamin B12 and folate both were associated with offspring mental and social development quotients at 2 years of age (Bhate et al. 2012). In another Indian birth cohort, higher maternal vitamin B12 status in pregnancy favourably associated with offspring short-term memory and sustained attention at 9 years of age (Bhate et al. 2008). In North Indian children aged 12 to 18 months, each 2-fold increment in B12 concentration was associated with a increment of 1.3 in mental development index score and each 2-fold increment in homocysteine or MMA concentration was associated with a decrement of 2.0 in the same score (Strand et al. 2013). Supplementation with vitamin B12 in pre-school children led to better motor and mental skills in pre-school children.

The new paradigm of DOHaD suggests that maternal health and nutrition are major determinants of lifelong health of the offspring. These may be mediated by influence of nutrients on structural or functional development of cells and tissues. Many of these effects may be mediated by epigenetic mechanisms, and DNA methylation is an important component of these. Both animal and human studies have highlighted a role for dietary methyl donors (vitamin B12, folate and others) in these mechanisms.

In the Pune Maternal Nutrition Study, in a rural population, over 60% of women had a low concentration of vitamin B12 in pregnancy. Low maternal vitamin B12 concentrations in pregnancy predicted higher insulin resistance (HOMA-IR) in the offspring at 6 years of age, those born to mothers with lowest vitamin B12 concentrations and highest erythrocyte folate concentrations were the most insulin resistant (Yajnik et al. 2008). Higher maternal folate concentrations were associated with higher adiposity in the offspring. A study in Nepal confirmed the association between maternal vitamin B12 deficiency and higher insulin resistance in the offspring, but folate was not associated with insulin resistance (Stewart et al. 2011).

Elevated homocysteine concentrations are associated with increased CVD risk (Boushey et al. 1995), though causality is still controversial. Low vitamin B12 status and hyperhomocysteinemia are associated with adverse lipid parameters (Adaikalakoteswari et al. 2014), and with coronary artery disease (Kumar et al. 2009) in Indians.

B-complex supplementation trials in the western non-vegetarian populations have failed to reduce the CVD. It is to be noted that these populations have a relatively lower level of homocysteine concentrations, are usually B12 replete, and the trials are usually for secondary or tertiary prevention. There is an urgent need of large scale trials in developing countries where vitamin B12 deficiency is common and an effect of supplementation on disease outcomes can be easier to confirm or dispute. It is noteworthy that a folic acid supplementation trial in hypertensive patients in China significantly reduced the risk of a stroke, emphasizing the need for population specific interventions (Huo et al. 2015).

6. Improving Vitamin B12 Status by Supplementation or Fortification

The overall long term strategy for controlling vitamin B12 deficiency is to promote consumption of foods rich in vitamin B12. Vegetarians have a substantial difficulty in achieving adequate daily intakes, and supplementation or food fortification need to be considered.

Vitamin B12-containing plant-derived food sources include green and purple lavers (Nori) and blue-green algae/cyanobacteria (Spirulina). Nori is considered suitable for humans whereas Spirulina contains pseudo-vitamin B12 which is biologically inactive in humans (Watanabe et al. 2007). Nori is not widely available. Therefore, milk remains the only acceptable animal source of B12 for vegetarians who consume it. A study from Pune reported improvement in B12 status in young, healthy, B12 deficient vegetarians by promoting regular intake of non-fortified milk (600 ml/day) that was estimated to deliver approximately 1.9 µg vitamin B12 daily (Naik et al. 2013). A trial studying effectiveness of food based micronutrient rich snack (dried fruits, leafy vegetables and milk powder) in non-pregnant young women reported no change in B12 concentrations (Kehoe et al. 2015), and similar intervention from before conception and throughout pregnancy in healthy women had no overall effect on B12 status (Potdar et al. 2015). Also micronutrient mix powder that can be directly added to grain products or milk has been used in several countries in Asia to deliver necessary nutrients for pregnant women or children.

Many widely available and regularly consumed foods have been used as vehicles to deliver vitamin B12. For example, fortified wheat flour (Winkels et al. 2008), milk (Dhonukshe et al. 2005; Kuriyan et al. 2016) and mineral water (Tapola et al. 2004) are widely available.

There are only a few vitamin B12 supplementation trials in developing countries. In a 12 months intervention with oral vitamin B12 (2 or 10 µg/d) in a deficient rural population, mean plasma homocysteine concentration decreased by 6 µmol/L after 2 µg/d, and by 7 µmol/l after 10 µg/d (Deshmukh et al. 2010). These results demonstrate that substantial benefits can be obtained in deficient populations by physiological dose supplementation. A trial from Uganda reported improvement in vitamin B12 and folate status after using multiple micronutrient supplementation (twice the recommended dietary allowance) in HIV-infected children (Ndeezi et al. 2011). A randomized controlled from India reported improvement of gross motor and problem-solving skills in Indian children aged 6–30 months after supplementing low doses of vitamin B12 and folic acid for 6 months (Kvestad et al. 2015). The supplements contained 150 µg folic acid or 1.8 µg vitamin B12 for age group > 12 months and half of this amount for the younger age groups. Children who had lower vitamin B12 status at baseline, were more likely to show measurable improvement in growth and other clinical outcomes (Kvestad et al. 2015). However, the supplements did not reduce infections in the children (Taneja et al. 2013).

Certain strains of probiotics synthesize water-soluble vitamins (folate, riboflavin and vitamin B12) (LeBlanc et al. 2011). Probiotic preparations could be a cost-effective alternative for vitamin fortification programmes. There are only few studies on the efficacy of such products in improving B12 status. A study from Egypt demonstrated the effectiveness of 42 days ingestion of probiotic yoghurt in improving vitamin B12, folate and hemoglobin levels in children (Mohammad et al. 2006). Fortified ice-cream and yogurt are available but may be more expensive. More research is needed in this area to determine a cost-effective population-specific food vehicle that can be used to improve vitamin B12 in targeted groups (school children, pregnant women, and elderly) and study the long term effect of this strategy.

7. Summary and Conclusions

We have discussed the current evidence on vitamin B12 deficiency in many developing countries, its causes, and the public health significance of this phenomenon. We also reviewed strategies that can be used to improve vitamin B12 status in developing countries and available studies showing positive effects on health. This area is however confounded with many methodological difficulties and lack of evidence. We have tried to provide constructive recommendations for improvement of the level of evidence, the comparability of the studies and ways to show the significance of prevention programmes.

Acknowledgments

We are grateful to participants, collaborators and funding agencies who have supported our studies on vitamin B12.

Keywords: Vitamin B12, cobalamin, diet, developing populations, public health, Developmental Origins of Health and Disease (DOHaD)

Abbreviations

CVD	:	Cardio Vascular Disease
DOHaD	:	Developmental Origins of Health and Diseases
IUGR	:	Intrauterine Growth Retardation
MMA	:	Methylmalonic Acid
MTHFR	:	Methylenetetrahydrofolate Reductase
NTD	:	Neural Tube Defects
SGA	:	Small-for-Gestational Age

References

Abdollahi Z, Elmadfa I, Djazayeri A, Sadeghian S, Freisling H, Mazandarani FS and Mohamed K. 2008. Folate, vitamin B12 and homocysteine status in women of childbearing age: baseline data of folic acid wheat flour fortification in Iran. Ann Nutr Metab. 53: 143–150.

Adaikalakoteswari A, Jayashri R, Sukumar N, Venkataraman H, Pradeepa R, Gokulakrishnan K, Anjana RM, McTernan PG, Tripathi G, Patel V, Kumar S, Mohan V and Saravanan P. 2014. Vitamin B12 deficiency is associated with adverse lipid profile in Europeans and Indians with type 2 diabetes. Cardiovasc Diabetol. 13: 129.

Allen LH. 2004. Folate and vitamin B12 status in the Americas. Nutr Rev. 62: S29–S33.

Allen LH. 2008. Causes of vitamin B12 and folate deficiency. Food Nutr Bull. 29: S20–34; discussion S35–7.

Antony AC. 2003. Vegetarianism and vitamin B12 (cobalamin) deficiency. Am J Clin Nutr. 78: 3–6.

Ayesh MH, Jadalah K, Awadi E, Alawneh K and Khassawneh B. 2013. Association between vitamin B12 level and anti-parietal cells and anti-intrinsic factor antibodies among adult Jordanian patients with *Helicobacter pylori* infection. Braz J Infect Dis. 17: 629–632.

Balci YI, Ergin A, Karabulut A, Polat A, Doğan M and Küçüktaşcı K. 2014. Serum vitamin B12 and folate concentrations and the effect of the Mediterranean diet on vulnerable populations. Pediatr Hematol Oncol. 31: 62–67.

Barnabé A, Aléssio AC, Bittar LF, de MoraesMazetto B, Bicudo AM, de Paula EV, Höehr NF and Annichino-Bizzacchi JM. 2015. Folate, vitamin B12 and Homocysteine status in the post-folic acid fortification era in different subgroups of the Brazilian population attended to at a public health care center. Nutr J. 14: 19.

Bhate V, Deshpande S, Bhat D, Joshi N, Ladkat R, Watve S, Fall C, de Jager CA, Refsum H and Yajnik C. 2008. Vitamin B12 status of pregnant Indian women and cognitive function in their 9-year-old children. Food Nutr Bull. 29: 249–254.

Bhate VK, Joshi SM, Ladkat RS, Deshmukh US, Lubree HG, Katre PA, Bhat DS, Rush EC and Yajnik CS. 2012. Vitamin B12 and folate during pregnancy and offspring motor, mental and social development at 2 years of age. J Dev Orig Health Dis. 3: 123–130.

Bondevik GT, Schneede J, Refsum H, Lie RT, Ulstein M and Kvåle G. 2001. Homocysteine and methylmalonic acid levels in pregnant Nepali women. Should cobalamin supplementation be considered? Eur J Clin Nutr. 55: 856–64.

Boushey CJ, Beresford SA, Omenn GS and Motulsky AG. 1995. A quantitative assessment of plasma homocysteine as a risk factor for vascular disease. Probable benefits of increasing folic acid intakes. JAMA. 274: 1049–1057.

Brito A, Mujica-Coopman MF, López de Romaña D, Cori H and Allen LH. 2015. Folate and vitamin B12 status in Latin America and the caribbean: An Update. Food Nutr Bull. 36: S109–118.

Brouwer-Brolsma EM, Dhonukshe-Rutten RA, van Wijngaarden JP, Zwaluw NL, Velde Nv and de Groot LC. 2015. Dietary sources of vitamin B12 and their association with vitamin B12 status markers in healthy older adults in the B-PROOF study. Nutrients. 7: 7781–7797.

Cobayashi F, Tomita LY, Augusto RA, D'Almeida V and Cardoso MA. 2015. ACTION Study Team. Genetic and environmental factors associated with vitamin B12 status in Amazonian children. Public Health Nutr. 18: 2202–10.

Christian AM, Krishnaveni GV, Kehoe SH, Veena SR, Khanum R, Marley-Zagar E, Edwards P, Margetts BM and Fall CH. 2015. Contribution of food sources to the vitamin B12 status of South Indian children from a birth cohort recruited in the city of Mysore. Public Health Nutr. 18: 596–609.

Dali-Youcef N and Andres E. 2009. An update on cobalamin deficiency in adults. QJM. 102: 17–28.

Dang S, Yan H, Zeng L, Wang Q, Li Q, Xiao S and Fan X. 2014. The status of vitamin B12 and folate among Chinese women: a population-based cross-sectional study in northwest China. PLoS One. 9: e112586.

Deegan KL, Jones KM, Zuleta C, Ramirez-Zea M, Lildballe DL, Nexo E and Allen LH. 2012. Breast milk vitamin B12 concentrations in Guatemalan women are correlated with maternal but not infant vitamin B12 status at 12 months postpartum. J Nutr. 142: 112–116.

Deodhar AD and Ramakrishnan CV. 1959. Studies on human lactation. II. Effect of socio-economic status on the vitamin content of human milk. Indian J Med Res. 47: 352–355.

Deshmukh US, Joglekar CV, Lubree HG, Ramdas LV, Bhat DS, Naik SS, Hardikar PS, Raut DA, Konde TB, Wills AK, Jackson AA, Refsum H, Nanivadekar AS, Fall CH and Yajnik CS. 2010. Effect of physiological doses of oral vitamin B12 on plasma homocysteine—A randomized, placebo-controlled, double-blind trial in India. Eur J Clin Nutr. 64: 495–502.

Dhonukshe-Rutten RA, van Zutphen M, de Groot LC, Eussen SJ, Blom HJ and van Staveren WA. 2005. Effect of supplementation with cobalamin carried either by a milk product or a capsule in mildly cobalamin-deficient elderly Dutch persons. Am J Clin Nutr. 82: 568–574.

Dror DK and Allen LH. 2011. The importance of milk and other animal-source foods for children in low-income countries. Food Nutr Bull. 32: 227–243.

Duggan C, Srinivasan K, Thomas T, Samuel T, Rajendran R, Muthayya S, Finkelstein JL, Lukose A, Fawzi W, Allen LH, Bosch RJ and Kurpad AV. 2014. Vitamin B12 supplementation during pregnancy and early lactation increases maternal, breast milk, and infant measures of vitamin B12 status. J Nutr. 144: 758–64.

Dwarkanath P, Barzilay JR, Thomas T, Thomas A, Bhat S and Kurpad AV. 2013. High folate and low vitamin B12 intakes during pregnancy are associated with small-for-gestational age infants in South Indian women: a prospective observational cohort study. Am J Clin Nutr. 98: 1450–8.

Fakhrzadeh H, Ghotbi S, Pourebrahim R, Nouri M, Heshmat R, Bandarian F, Shafaee A and Larijani B. 2006. Total plasma homocysteine, folate, and vitamin B12 status in healthy Iranian adults: the Tehranhomocysteine survey (2003–2004)/a cross—sectional population based study. BMC Public Health. 6: 29.

Fernández-Bañares F1, Monzón H and Forné M. 2009. A short review of malabsorption and anemia. World J Gastroenterol. 15: 4644–4652.

Finkelstein JL, Layden AJ and Stover PJ. 2015. Vitamin B12 and perinatal health. Adv Nutr. 6: 552–563.

Galukande M, Jombwe J, Fualal J, Baingana R and Gakwaya A. 2011. Reference values for serum levels of folic acid and vitamin B12 in a young adult Ugandan population. Afr Health Sci. 11: 240–243.

Godbole K, Gayathri P, Ghule S, Sasirekha BV, Kanitkar-Damle A, Memane N, Suresh S, Sheth J, Chandak GR and Yajnik CS. 2011. Maternal one-carbon metabolism, MTHFR and TCN2 genotypes and neural tube defects in India. Birth Defects Res A Clin Mol Teratol. 91: 848–856.

Goraya JS, Kaur S and Mehra B. 2015. Neurology of nutritional vitamin B12 deficiency in infants: Case series from India and literature review. J Child Neurol. 30: 1831–1837.

Greibe E, Lildballe DL, Streym S, Vestergaard P, Rejnmark L, Mosekilde L and Nexo E. 2013. Cobalamin and haptocorrin in human milk and cobalamin-related variables in mother and child: a 9-mo longitudinal study. Am J Clin Nutr. 98: 389–395.

Gupta Bansal P, Singh Toteja G, Bhatia N, Kishore Vikram N, Siddhu A, Kumar Garg A and Kumar Roy A. 2015. Deficiencies of serum ferritin and vitamin B12, but not folate, are common in adolescent girls residing in a slum in Delhi. Int J Vitam Nutr Res. 85: 14–22.

Haan MN, Miller JW, Aiello AE, Whitmer RA, Jagust WJ, Mungas DM, Allen LH and Green R. 2007. Homocysteine, B vitamins, and the incidence of dementia and cognitive impairment: results from the Sacramento Area Latino Study on Aging. Am J Clin Nutr. 85: 511–7.

Halicioglu O, Sutcuoglu S, Koc F, Ozturk C, Albudak E, Colak A, Sahin E and AsikAkman S. 2012. Vitamin B12 and folate statuses are associated with diet in pregnant women, but not with anthropometric measurements in term newborns. J Matern Fetal Neonatal Med. 25: 1618–1621.

Hazra A, Kraft P, Lazarus R, Chen C, Chanock SJ, Jacques P, Selhub J and Hunter DJ. 2009. Genome-wide significant predictors of metabolites in the one-carbon metabolism pathway. Hum Mol Genet. 18: 4677–87.

Hazra A, Kraft P, Selhub J, Giovannucci EL, Thomas G, Hoover RN, Chanock SJ and Hunter DJ. 2008. Common variants of FUT2 are associated with plasma vitamin B12 levels. Nat Genet. 40: 1160–2.

Herrmann W, Obeid R and Jouma M. 2003. Hyperhomocysteinemia and vitamin B12 deficiency are more striking in Syrians than in Germans-causes and implications. Atherosclerosis. 166: 143–150.

Huo Y, Li J, Qin X, Huang Y, Wang X, Gottesman RF, Tang G, Wang B, Chen D, He M, Fu J, Cai Y, Shi X, Zhang Y, Cui Y, Sun N, Li X, Cheng X, Wang J, Yang X, Yang T, Xiao C, Zhao G, Dong

Q, Zhu D, Wang X, Ge J, Zhao L, Hu D, Liu L and Hou FF. 2015. CSPPT Investigators. Efficacy of folic acid therapy in primary prevention of stroke among adults with hypertension in China: the CSPPT randomized clinical trial. JAMA. 313: 1325–35.

Ingrid Kristin Torsvik, Per Magne Ueland, Trond Markestad, Øivind Midttun and Anne-Lise Bjørke Monsen. 2015. Motor development related to duration of exclusive breastfeeding, B vitamin status and B12 supplementation in infants with a birth weight between 2000–3000 g, results from a randomized intervention trial. BMC Pediatr. 15: 218.

Jathar VS, Kamath SA, Parikh MN, Rege DV and Satoskar RS. 1970. Maternal milk and serum vitamin B12, folic acid, and protein levels in Indian subjects. Arch Dis Child. 45: 236–241.

Kapil U and Sareen N. 2014. Prevalence of ferritin, folate and vitamin B12 deficiencies amongst children in 5–18 years of age in Delhi. Indian J Pediatr. 81: 312.

Kehoe SH, Chopra H, Sahariah SA, Bhat D, Munshi RP, Panchal F, Young S, Brown N, Tarwande D, Gandhi M, Margetts BM, Potdar RD and Fall CH. 2015. Effects of a food-based intervention on markers of micronutrient status among Indian women of low socio-economic status. Br J Nutr. 113: 813–21.

Khodabandehloo N, Vakili M, Hashemian Z and Zare Zardini H. 2015. Determining functional vitamin B12 deficiency in the elderly. Iran Red Crescent Med J. 17: e13138.

Koc A, Kocyigit A, Soran M, Demir N, Sevinc E, Erel O and Mil Z. 2006. High frequency of maternal vitamin B12 deficiency as an important cause of infantile vitamin B12 deficiency in Sanliurfa province of Turkey. Eur J Nutr. 45: 291–7.

Krishnaveni GV, Hill JC, Veena SR, Bhat DS, Wills AK, Karat CL, Yajnik CS and Fall CH. 2009. Low plasma vitamin B12 in pregnancy is associated with gestational 'diabesity' and later diabetes. Diabetologia. 52: 2350–2358.

Kumar J, Garg G, Sundaramoorthy E, Prasad PV, Karthikeyan G, Ramakrishnan L, Ghosh S and Sengupta S. 2009. Vitamin B12 deficiency is associated with coronary artery disease in an Indian population. Clin Chem Lab Med. 47: 334–338.

Kuriyan R, Thankachan P, Selvam S, Pauline M, Srinivasan K, Kamath-Jha S, Vinoy S, Misra S, Finnegan Y and Kurpad AV. 2016. The effects of regular consumption of a multiple micronutrient fortified milk beverage on the micronutrient status of school children and on their mental and physical performance. Clin Nutr. 35: 190–8.

Kvestad I, Taneja S, Kumar T, Hysing M, Refsum H, Yajnik CS, Bhandari N and Strand TA. 2015. Folate and Vitamin B12 Study Group. Vitamin B12 and Folic Acid Improve Gross Motor and Problem-Solving Skills in Young North Indian Children: A Randomized Placebo-Controlled Trial. PLoS One. 10: e0129915.

Kwak CS, Lee MS, Oh Se In and Park SC. 2010. Discovery of novel sources of vitamin B12 in traditional Korean foods from nutritional surveys of centenarians. Curr Gerontol Geriatr Res. 2010: 374897.

LeBlanc JG, Laiño JE, del Valle MJ, Vannini V, van Sinderen D, Taranto MP, de Valdez GF, de Giori GS and Sesma F. 2011. B-group vitamin production by lactic acid bacteria—current knowledge and potential applications. J Appl Microbiol. 111: 1297–1309.

Lildballe DL, Hardlei TF, Allen LH and Nexo E. 2009. High concentrations of haptocorrin interfere with routine measurement of cobalamins in human serum and milk. A problem and its solution. Clin Chem Lab Med. 47: 182–187.

Lubree HG, Katre PA, Joshi SM, Bhat DS, Deshmukh US, Memane NS, Otiv SR, Rush EC and Yajnik CS. 2012. Child's homocysteine concentration at 2 years is influenced by pregnancy vitamin B12 and folate status. J Dev Orig Health Dis. 3: 32–38.

McLaren DS. 1981. The luxus vitamins—A and B12. Am J Clin Nutr. 34: 1611–6.

McLean E, de Benoist B and Allen LH. 2008. Review of the magnitude of folate and vitamin B12 deficiencies worldwide. Food Nutr Bull. (2 Suppl): S38–51.

McLean ED, Allen LH, Neumann CG, Peerson JM, Siekmann JH, Murphy SP, Bwibo NO and Demment MW. 2007. Low plasma vitamin B12 in Kenyan school children is highly prevalent and improved by supplemental animal source foods. J Nutr. 137: 676–82.

Mohammad MA, Molloy A, Scott J and Hussein L. 2006. Plasma cobalamin and folate and their metabolic markers methylmalonic acid and total homocysteine among Egyptian children before and after nutritional supplementation with the probiotic bacteria Lactobacillus acidophilus in yoghurt matrix. Int J Food Sci Nutr. 57: 470–480.

Monsen AL, Refsum H, Markestad T and Ueland PM. 2003. Cobalamin status and its biochemical markers methylmalonic acid and homocysteine in different age groups from 4 days to 19 years. Clin Chem. 49: 2067–75.

Murphy MM, Molloy AM, Ueland PM, Fernandez-Ballart JD, Schneede J, Arija V and Scott JM. 2007. Longitudinal study of the effect of pregnancy on maternal and fetal cobalamin status in healthy women and their offspring. J Nutr. 137: 1863–7.

Muthayya S, Kurpad AV, Duggan CP, Bosch RJ, Dwarkanath P, Mhaskar A, Mhaskar R, Thomas A, Vaz M, Bhat S and Fawzi WW. 2006. Low maternal vitamin B12 status is associated with intrauterine growth retardation in urban South Indians. Eur J Clin Nutr. 60: 791–801.

Naik S, Bhide V, Babhulkar A, Mahalle N, Parab S, Thakre R and Kulkarni M. 2013. Daily milk intake improves vitamin B12 status in young vegetarian Indians: an intervention trial. Nutr J. 12: 136.

Ndeezi G, Tumwine JK, Ndugwa CM, Bolann BJ and Tylleskär T. 2011. Multiple micronutrient supplementation improves vitamin B12 and folate concentrations of HIV infected children in Uganda: a randomized controlled trial. Nutr J. 10: 56.

Neumann CG, Oace SM, Chaparro MP, Herman D, Drorbaugh N and Bwibo NO. 2013. Low vitamin B12 intake during pregnancy and lactation and low breastmilk vitamin 12 content in rural Kenyan women consuming predominantly maize diets. Food Nutr Bull. 34: 151–159.

Pasricha SR, Black J, Muthayya S, Shet A, Bhat V, Nagaraj S, Prashanth NS, Sudarshan H, Biggs BA and Shet AS. 2010. Determinants of anemia among young children in rural India. Pediatrics. 126: e140–9.

Pasricha SR, Shet AS, Black JF, Sudarshan H, Prashanth NS and Biggs BA. 2011. Vitamin B12, folate, iron, and vitamin A concentrations in rural Indian children are associated with continued breastfeeding, complementary diet, and maternal nutrition. Am J Clin Nutr. 94: 1358–70.

Pathak P, Kapil U, Yajnik CS, Kapoor SK, Dwivedi SN and Singh R. 2007. Iron, folate, and vitamin B12 stores among pregnant women in a rural area of Haryana State, India. Food Nutr Bull. 28: 435–438.

Pawlak R, Parrott SJ, Raj S, Cullum-Dugan D and Lucus D. 2013. How prevalent is vitamin B(12) deficiency among vegetarians? Nutr Rev. 71: 110–7.

Potdar RD, Sahariah SA, Gandhi M, Kehoe SH, Brown N, Sane H, Dayama M, Jha S, Lawande A, Coakley PJ, Marley-Zagar E, Chopra H, Shivshankaran D, Chheda-Gala P, Muley-Lotankar P, Subbulakshmi G, Wills AK, Cox VA, Taskar V, Barker DJ, Jackson AA, Margetts BM and Fall CH. 2014. Improving women's diet quality preconceptionally and during gestation: effects on birth weight and prevalence of low birth weight-a randomized controlled efficacy trial in India (Mumbai Maternal Nutrition Project). Am J Clin Nutr. 100: 1257–1268.

Ramírez-Vélez R, Correa-Bautista JE, Martínez-Torres J, Meneses-Echávez JF and Lobelo F. 2016. Vitamin B12 concentrations in pregnant Colombian women: analysis of nationwide data 2010. BMC Pregnancy Childbirth. 16: 26.

Rao S, Yajnik CS, Kanade A, Fall CH, Margetts BM, Jackson AA, Shier R, Joshi S, Rege S, Lubree H and Desai B. 2001. Intake of micronutrient-rich foods in rural Indian mothers is associated with the size of their babies at birth: Pune Maternal Nutrition Study. J Nutr. 131: 1217–1224.

Refsum H, Yajnik CS, Gadkari M, Schneede J, Vollset SE, Örning L, Guttormsen AB, Joglekar A, Sayyad MG, Ulvik A and Ueland PM. 2001. Hyperhomocysteinemia and elevated methylmalonic acid indicate a high prevalence of cobalamin deficiency in Asian Indians. Am J Clin Nutr. 74: 233–241.

Rosenthal J, Lopez-Pazos E, Dowling NF, Pfeiffer CM, Mulinare J, Vellozzi C, Zhang M, Lavoie DJ, Molina R, Ramirez N and Reeve ME. 2015. Folate and Vitamin B12 Deficiency Among Non-pregnant Women of Childbearing-Age in Guatemala 2009–2010: Prevalence and Identification of Vulnerable Populations. Matern Child Health J. 19: 2272–85.

Rush EC, Chhichhia P, Hinckson E and Nabiryo C. 2009. Dietary patterns and vitamin B(12) status of migrant Indian preadolescent girls. Eur J Clin Nutr. 63: 585–7.

Samuel TM, Duggan C, Thomas T, Bosch R, Rajendran R, Virtanen SM, Srinivasan K and Kurpad AV. 2013. Vitamin B(12) intake and status in early pregnancy among urban South Indian women. Ann Nutr Metab. 62: 113–22.

Shamah-Levy T, Villalpando S, Mejía-Rodríguez F, Cuevas-Nasu L, Gaona-Pineda EB, Rangel-Baltazar E and Zambrano-Mujica N. 2015. Prevalence of iron, folate, and vitamin B12 deficiencies in 20 to 49 years old women: Ensanut 2012. Salud Publica Mex. 57: 385–93.

Shobha V, Tarey SD, Singh RG, Shetty P, Unni US, Srinivasan K and Kurpad AV. 2011. Vitamin B12 deficiency & levels of metabolites in an apparently normal urban south Indian elderly population. Indian J Med Res. 134: 432–439.

Siekmann JH, Allen LH, Bwibo NO, Demment MW, Murphy SP and Neumann CG. 2003. Kenyan school children have multiple micronutrient deficiencies, but increased plasma vitamin B12 is the only detectable micronutrient response to meat or milk supplementation. J Nutr. 133(11 Suppl 2): 3972S–3980S.

Stewart CP, Christian P, Schulze KJ, Arguello M, LeClerq SC, Khatry SK and West KP Jr. 2011. Low maternal vitamin B12 status is associated with offspring insulin resistance regardless of antenatal micronutrient supplementation in rural Nepal. J Nutr. 141: 1912–7.

Strand TA, Taneja S, Ueland PM, Refsum H, Bahl R, Schneede J, Sommerfelt H and Bhandari N. 2013. Cobalamin and folate status predicts mental development scores in North Indian children 12–18 mo of age. Am J Clin Nutr. 97: 310–7.

Taneja S, Bhandari N, Strand TA, Sommerfelt H, Refsum H, Ueland PM, Schneede J, Bahl R and Bhan MK. 2007. Cobalamin and folate status in infants and young children in a low-to-middle income community in India. Am J Clin Nutr. 86: 1302–9.

Taneja S, Strand TA, Kumar T, Mahesh M, Mohan S, Manger MS, Refsum H, Yajnik CS and Bhandari N. 2013. Folic acid and vitamin B12 supplementation and common infections in 6-30-mo-old children in India: a randomized placebo-controlled trial. Am J Clin Nutr. 98: 731–7.

Tang KF, Li YL and Wang HY. 2015. Quantitative assessment of maternal biomarkers related to one-carbon metabolism and neural tube defects. Sci Rep. 5: 8510.

Tapola NS, Karvonen HM, Niskanen LK and Sarkkinen ES. 2004. Mineral water fortified with folic acid, vitamins B6, B12, D and calcium improves folate status and decreases plasma homocysteine concentration in men and women. Eur J Clin Nutr. 58: 376–385.

Thomas D, Chandra J, Sharma S, Jain A and Pemde HK. 2015. Determinants of nutritional Anemia in adolescents. Indian Pediatr. 52: 867–9.

Thomas MR, Sneed SM, Wei C, Nail PA and Wilson M. 1980. Sprinkle EE 3rd. The effects of vitamin C, vitamin B6, vitamin B12, folic acid, riboflavin, and thiamin on the breast milk and maternal status of well-nourished women at 6 months postpartum. Am J Clin Nutr. 33: 2151–6.

Torheim LE, Ferguson EL, Penrose K and Arimond M. 2010. Women in resource-poor settings are at risk of inadequate intakes of multiple micronutrients. J Nutr. 140: S2051–S2058.

Tucker KL, Rich S, Rosenberg I, Jacques P, Dallal G, Wilson PW and Selhub J. 2000. Plasma vitamin B12 concentrations relate to intake source in the Framingham Offspring study. Am J Clin Nutr. 71: 514–522.

Ulak M, Chandyo RK, Adhikari RK, Sharma PR, Sommerfelt H, Refsum H and Strand TA. 2014. Cobalamin and folate status in 6 to 35 months old children presenting with acute diarrhea in Bhaktapur, Nepal. PLoS One. 9: e90079.

VanderJagt DJ, Ujah IAO, Ikeh EI, Bryant J, Pam V, Hilgart A, Crossey MJ and Glew RH. 2011. Assessment of the vitamin B12 status of pregnant women in Nigeria using plasma holotranscobalamin. Obstet Gynecol. 2011: 365894.

Villalpando S, Cruz Vde L, Shamah-Levy T, Rebollar R and Contreras-Manzano A. 2015. Nutritional status of iron, vitamin B12, folate, retinol and anemia in children 1 to 11 years old: Results of the Ensanut 2012. Salud Publica Mex. 57: 372–84.

Watanabe F. 2007. Vitamin B12 sources and bioavailability. Exp Biol Med. 232: 1266–1274.

Winkels RM, Brouwer IA, Clarke R, Katan MB and Verhoef P. 2008. Bread cofortified with folic acid and vitamin B12 improves the folate and vitamin B12 status of healthy older people: a randomized controlled trial. Am J Clin Nutr. 88: 348–355.

Yajnik CS, Deshpande SS, Jackson AA, Refsum H, Rao S, Fisher DJ, Bhat DS, Naik SS, Coyaji KJ, Joglekar CV, Joshi N, Lubree HG, Deshpande VU, Rege SS and Fall CH. 2008. Vitamin B12 and folate concentrations during pregnancy and insulin resistance in the offspring: The Pune Maternal Nutrition Study. Diabetologia. 51: 29–38.

Yajnik CS, Deshpande SS, Lubree HG, Naik SS, Bhat DS, Uradey BS, Deshpande JA, Rege SS, Refsum H and Yudkin JS. 2006. Vitamin B12 Deficiency and Hyperhomocysteinemia in rural and Urban Indians. J Assoc Physicians India. 54: 775–82.

Yajnik CS, Chandak GR, Joglekar C, Katre P, Bhat DS, Singh SN, Janipalli CS, Refsum H, Krishnaveni G, Veena S, Osmond C and Fall CH. 2014. Maternal homocysteine in pregnancy and offspring birthweight: epidemiological associations and Mendelian randomization analysis. Int J Epidemiol. 43: 1487–97.

Yusufji D, Mathan VI and Baker SJ. 1973. Iron, folate, and vitamin B12 nutrition in pregnancy: a study of 1000 women from southern India. Bull World Health Organ. 48: 15–22.

Zhang W, Li Y, Wang TD, Meng HX, Min GW, Fang YL, Niu XY, Ma LS, Guo JH, Zhang J, Sun MZ and Li CX. 2014. Nutritional status of the elderly in rural North China: a cross-sectional study. J Nutr Health Aging. 18: 730–736.

7

Vitamin B12 in Neurology and Aging

Andrew McCaddon[1,]* and *Joshua W Miller*[2]

1. Introduction

The classical pathophysiological manifestations of vitamin B12 (B12) deficiency are myriad and profound, primarily affecting the hematopoietic system (i.e., megaloblastic anemia, hypersegmented neutrophils and pancytopenia) and the nervous system (i.e., demyelination and neurodegeneration). Neurological and psychiatric manifestations of B12 deficiency are particularly prevalent in patients with autoimmune pernicious anaemia (PA)—a common cause of severe deficiency due to malabsorption of the vitamin. Memory loss, poor concentration, peripheral neuropathy, gait ataxia, depression and autonomic dysfunction often occur in such individuals, as well as some less well recognized features such as mild nominal aphasia and vertigo.

The advent of diagnostic tests using metabolic markers of B12 status, including plasma homocysteine and methylmalonic acid assays, has also revealed a surprisingly high prevalence of a rather more subtle form of B12 deficiency, particularly within the older population. Although this has been termed 'sub-clinical' deficiency, due to the absence of haematological features of the deficiency, it is often associated with cognitive impairment and Alzheimer's disease (AD). It is also reported in association with other

---

[1] Honorary Senior Research Fellow, School of Medicine, Cardiff University, Gwenfro Units 6/7, Wrexham Technology Park, Wrexham LL17 7YP, Wales, United Kingdom.
Email: mccaddon@sky.com
[2] Professor and Chair, Dept. of Nutritional Sciences, School of Environmental and Biological Sciences, Rutgers – The State University of New Jersey, 65 Dudley Road, New Brunswick, NJ, 08901, USA.
Email: jmiller@aesop.rutgers.edu
* Corresponding author

age-related neurodegenerative disorders including vascular dementia, Parkinson's disease and multiple sclerosis.

In this chapter, we compare and contrast the neurological features of 'classical' and 'sub-clinical' B12 deficiency and explore possible reasons why they might differ. The discrepancy could be attributable, at least partly, to differing aetiology; classical features might be more closely related to outright clinical B12 *deficiency* arising from severe malabsorption (as occurs in PA) or very low dietary intake. In contrast in some individuals, particularly older adults, subclinical deficiency might reflect chronic B12 *depletion* owing to moderate malabsorption or suboptimal intake, but also increased demand for the vitamin in response to age and disease-related oxidative stress. The general concept of any deficiency arising from low intake *or* excess demand is well recognized for other nutrients. A simple example would be iron deficiency arising from poor dietary intake compared with iron deficiency arising as a result of the increased demands of pregnancy. This principle is, of course, applicable to all essential nutrients, but the concept of 'excess demand' in relation to vitamin B12 is generally little recognized in the current literature.

The chapter begins with a description of the 'classical' neurological features of acquired B12 deficiency, followed by an explanation of the historical evolution of 'subclinical' deficiency. The surprisingly high prevalence of subclinical deficiency in older patients is discussed, along with its association with neurodegenerative diseases. Last, we consider the neurological features of clinical and subclinical deficiency, and discuss how these two apparently disparate conditions might be reconciled.

2. Classical Neurology of Clinical Vitamin B12 Deficiency

There are many excellent earlier reviews of the classical neurological signs and symptoms of B12 deficiency (Evans et al. 1983; Green and Miller 2014; Healton et al. 1991; Hector and Burton 1988). Neurological features are usually chronic, progressive, and can exhibit both peripheral and central (i.e., spinal cord and cerebral) manifestations (Savage and Lindenbaum 1995).

Typical peripheral symptoms are of altered sensation, including a usually symmetrical paraesthesiae of extremities, and gait ataxia. Some patients also develop reduced manual dexterity, poor vision, orthostatic dizziness, loss of taste or smell, and urinary or faecal incontinence and impotence, although the latter symptoms are rare.

Clinical signs include loss of cutaneous sensation in a 'glove and stocking' distribution, and impaired vibration sense and proprioception. Romberg's sign, where the patient requires visual input in order to stand steadily, is frequently positive.

Various cerebral and psychiatric manifestations of B12 deficiency are also described in the early literature including memory impairment, personality change, and even frank psychosis. These are discussed more fully later, since technological advances in the laboratory diagnosis of B12 deficiency have significantly broadened the clinical appreciation of its scope.

Several neuropathologists describe demyelinating lesions in the spinal cord and brain in PA. In 1900, Russell, Batten and Collier published the first full clinicopathological description of 'subacute combined degeneration of the spinal cord' (Russell et al. 1900). However, this also affects the brain and peripheral nervous system, leading Weir and Scott to suggest that it should perhaps be termed 'cobalamin deficiency associated neuropathy' (Weir and Scott 1999). Predominantly, the lesions affect the cervical and upper thoracic spinal cord and cerebrum, which appears mildly atrophic (Weir and Scott 1999). White matter in the spinal cord appears grey because of demyelination; there may also be cerebral atrophy. The white matter is 'sponge-like', especially in the dorsolateral columns of the spinal cord. Vacuoles surrounded by myelin-laden macrophages occur. This appearance was purported to have been termed a *lachen felden* by German physicians who first described the condition in the mid-nineteenth century (Weir and Scott 1999), but the origin of this terminology, which literally translates to "laugh fields" in English, is obscure.

Adams and Kubik confirmed that cerebral lesions also occur in PA, and commented that the cerebral and spinal lesions are almost identical (Adams and Kubik 1944). Ferraro et al. performed autopsies on five patients, which accidentally revealed PA in the course of psychosis and/or dementia praecox (Ferraro et al. 1945), commenting that:

> "The nerve cells presented acute, severe, chronic, ischaemic, oedematous and fatty changes in varying degree. The small blood vessels appeared to be increased in number and presented a mild endarteritis in all cases. The distribution of the glia of the white matter was markedly irregular. Often the glia nuclei gathered around the blood vessels, so that the course of the latter was well outlined even when the vascular walls were at a different level, and could not be seen. Frequently between the conglomeration of the glia nuclei and the blood vessels a small area of white matter, which frequently underwent a process of demyelination, was interposed. These areas of demyelination at times had the tendency to coalesce."

In vivo magnetic resonance imaging in PA also reveals the involvement of white matter in the disease; hyperintense signal is observed in the posterior column of the spinal cord and the periventricular region of the brain (Scherer 2003).

3. The Relationship between Haematological and Neurological Features of B12 Deficiency

B12 deficiency is primarily associated with megaloblastic anaemia. This is essentially historic, arising as a result of the discovery of B12 as the 'extrinsic factor', which corrected an otherwise fatal anaemia. Throughout the 20th century it was believed that haematological evidence of B12 deficiency *preceded* neuropsychiatric abnormalities. However, in the last few decades it was suggested that these might actually represent two major, and sometimes entirely separate, clinical syndromes (Beck 1988).

In a retrospective study of 369 B12 deficient patients Healton et al. found an inverse correlation between the degree of anaemia and the extent of neurological involvement (Healton et al. 1991), similar to an earlier report by Lindenbaum et al. (Lindenbaum et al. 1988). Not only was anaemia more severe in patients lacking nervous system involvement, but in those affected neurologically the haematocrit correlated directly with severity of neurologic dysfunction. The presence of neurological features in B12-deficient patients was not related to the overall severity of deficiency of the vitamin, leading to the conclusion that, in most patients, either neurologic or haematologic dysfunction predominates. It is still unclear why some patients with B12 deficiency present with neurological disorders in the absence of haematological changes and why some patients develop a predominantly cerebral picture, and still others a spinal or peripheral nerve disorder. A possible explanation for this discrepancy is discussed later in this chapter.

Many of our preconceptions of B12 deficiency stem from the initial observations of the vitamin's relationship with haematological abnormalities; indeed the term PA is often, incorrectly, considered to be synonymous with 'vitamin B12 deficiency.' Because of this, the diagnosis of B12 deficiency traditionally proceeds in three steps: (1) the recognition of an anaemia, (2) the observation that it is macrocytic, and (3) the confirmation of an underlying B12 deficiency. Although a final diagnosis of B12 deficiency generally hinges on the finding of low blood concentrations of the vitamin, the decision to carry out this measurement in the first place is usually based on clinical evidence, and often on the presence of macrocytic anaemia. However, it has become increasingly clear that macrocytic anaemia cannot be used as the sole criterion for pursuing such a diagnosis (Carmel 1988; Carmel 1990). If suspicion of these deficiencies were based on the presence of macrocytic anaemia alone, a substantial proportion of B12 deficient individuals would escape detection. This point is elegantly expressed by Carmel who notes that, "The proscription that vitamin B12 deficiency should not be diagnosed unless megaloblastic changes are found is akin to requiring jaundice to diagnose liver disease" (Carmel 2000).

'Textbook' cases of megaloblastic anaemia now occur infrequently, partly due to earlier presentation and diagnosis. An important development in relation to early diagnosis was the concept of B12 and folate deficiencies as being *gradually progressive* (Herbert 1987). Thus, before becoming clinically deficient patients traverse earlier stages beginning with subclinical asymptomatic deficiency (Carmel 2000). This set the scene for the development and clinical use of more sensitive and specific tests to try to detect such early and subtle forms of deficiency. As in any nutrient deficiency anaemia a sequence of changes occurs and laboratory test abnormalities arise during the course of the deficiency. These changes begin to develop once a critical level of depletion of body stores of the vitamin is reached (Herbert 1987). Lowering of total serum levels of the vitamin may be preceded by biochemical effects, demonstrable by changes in the levels of metabolites in the blood and urine. Tissue effects, such as megaloblastic changes, macrocytosis, anaemia, and neurological damage occur later.

4. Development and Implications of Metabolite and Carrier Protein Assays

Vitamin B12, but not folate, is required for the conversion of methylmalonyl CoA to succinyl CoA. Consequently, serum and urine levels of methylmalonate (a usually minor side-reaction product of methylmalonyl CoA metabolism) rise in B12 deficiency. On the other hand, both folate and B12 are required for the methionine synthase reaction, which converts homocysteine to methionine. Therefore, plasma levels of homocysteine increase in both folate and B12 deficiencies. Technological advancements in the 1980s led to fast and reliable assays for these metabolites. These made it possible to identify subtle and atypical forms of B12 deficiency. The increasing use of such tests soon confirmed that the clinical features of B12 deficiency, including its neurological manifestations, were far broader than previously recognized, and the concept of sub-clinical deficiency evolved. This is defined as a state in which metabolic evidence of deficiency exists without macrocytosis and neutrophil hypersegmentation (Carmel 1990).

Because complications arising from B12 deficiency can occur in individuals with apparently normal hematological values and blood levels of the vitamin, there has also become a need to redefine what exactly is meant by 'normal' values. With respect to the current definitions of the lower or upper limits of the normal range, several reports indicate that there is considerable overlap between extremes of the normal range distribution and a proportion of patients with other objective evidence of vitamin deficiency. Regland et al. commented that, "...in the realities of biology, the dichotomy of a laboratory, based on clear cut-off values, does not exist" (Regland et al. 1990).

Recent research efforts have been directed toward redefining the assessment of B12 status beyond simple measurement of total serum B12 to include metabolite assays, as well as the percentage of the total serum B12 bound to the blood transport protein, transcobalamin (a complex referred to as 'holotranscobalamin') (Hvas and Nexo 2005; Miller et al. 2006). This might be especially useful in older adults and psychogeriatric populations. For example, Nilsson et al. found that B12 deficiency as defined by abnormal levels of methylmalonate was relatively common in such individuals, and that the use of B12 levels alone was insufficient to detect all patients with metabolic evidence of this deficiency (Nilsson et al. 1997). Recently suggested approaches to more accurately determine B12 status include the use of inflection point data for the metabolite assays (Vogiatzoglou et al. 2009) and the use of a mathematically derived 'combined B12 factor' that utilizes all four B12-related assays—total serum B12, holotranscobalamin, methylmalonate and homocysteine—to estimate B12 status in individuals (Fedosov 2010; Fedosov et al. 2015; Lildballe et al. 2011).

5. Vitamin B12 Deficiency in Older Adults

A high prevalence of low-normal serum B12 levels has long been recognized in healthy elderly people lacking any clinical signs and symptoms of such deficiency (Baik and Russell 1999; Pennypacker et al. 1992; Thompson et al. 1989; Yao et al. 1992). Most studies were cross-sectional, although one longitudinal study of community-dwelling Swedish elderly showed that the prevalence of low serum levels of B12, defined as < 130 pmol/L, increased from 4.6% to 7.2% as the subjects aged from 70 to 81 years (Nilsson-Ehle et al. 1991). The mean annual decline was 3.4 pmol/L for men and 3.2 pmol/L for women. Metabolite assays confirmed that older adults often also had biochemical evidence of deficiency (Allen et al. 1995; Lindenbaum et al. 1994). For example, Lindenbaum et al determined whether the increased prevalence of low serum levels of B12 in older adults represented 'true' deficiency by assaying serum concentrations of methylmalonate and homocysteine in 548 surviving members of the original Framingham Study cohort (Lindenbaum et al. 1994). Serum B12 concentrations < 258 pmol/L were significantly more common in older adults, occurring in 40.5% of such subjects compared with only 17.9% of younger control subjects. Similarly, in a Swedish study of 224 randomly selected subjects aged 70 years or older, half had abnormal methylmalonate and homocysteine levels suggesting a latent tissue deficiency of B12 or folate (Bjorkegren and Svardsudd 2001).

It is now considered that vitamin B12 deficiency may affect 10%–15% of people over the age of 60 years, but these older adults frequently lack the classical signs and symptoms of such a deficiency. Most studies show that the prevalence of pathologically low values for serum B12 increases with increasing age. Clarke et al. found that the prevalence of B12 deficiency in the United Kingdom, whether defined as low serum B12 or 'metabolically significant' B12 deficiency, increases with age from about 5% among people aged 65–74 years to 10% or more among people over 75 years old (Clarke et al. 2004). Using methylmalonate concentrations as an indicator of B12 deficiency a remarkably high prevalence of approximately 40% has even been reported in some elderly populations (Bates et al. 2003; McCracken et al. 2006; Tangney et al. 2009).

Over the years, varying and conflicting opinions have been expressed regarding likely causes of poor vitamin B12 status in older adults. Elsborg et al. suggested that the deficiency might arise simply as a result of an insufficient intake of B12 due to poor dietary habits of older adults (Elsborg et al. 1976). Howard et al. later showed that poor diet could generally not be implicated, and suggested that non-dietary causes should always be sought to explain this common metabolic insufficiency state in older adults (Howard et al. 1998). In contrast, a more recent report confirmed that, even in older adults, plasma B12 concentrations are associated with dietary intake of the vitamin, and that the food source itself is an important determinant (Tucker et al. 2000).

The most commonly suggested factor to account for low B12 status in older adults is the increasing prevalence with age of atrophic gastritis and its associated hypochlorhydria or achlorhydria (Nilsson-Ehle 1998). The prevalence of this can range from 20% to as much as 50% depending on how the diagnosis is made and on which definitions are used. In the Framingham Heart Study, the prevalence of atrophic gastritis among 60-69 year olds and those > 80 years was 24% and 37%, respectively; diagnosis of atrophic gastritis was made using serum pepsinogen I and II concentrations as measured by radioimmunoassay (Krasinski et al. 1986). In this study, the serum ratio of pepsinogen I to pepsinogen II decreased progressively with increasing severity of atrophic gastritis. Fasting blood was obtained from 359 free-living and institutionalized elderly people (age range 60–99 years). A pepsinogen I/pepsinogen II ratio less than 2.9, indicating atrophic gastritis, was noted in 113 (31.5%) subjects. The prevalence of atrophic gastritis increased significantly with age. A significant increase in the prevalence of low serum B12 levels (p < 0.005) was observed with stepwise increases in the severity of atrophic gastritis. The authors concluded that: (1) serum pepsinogen I and pepsinogen II levels can be used to determine the prevalence and severity of atrophic gastritis; (2) atrophic gastritis is common in older adults, and (3) that it is associated with B12 deficiency and anaemia.

Atrophic gastritis results in a low acid-pepsin secretion by the gastric mucosa, which in turn results in a reduced release of free vitamin B12 from food proteins. However, the ability to absorb crystalline (supplemental) vitamin B12 remains intact. The hypochlorhydria of atrophic gastritis might also result in bacterial overgrowth of the stomach and small intestine, and these bacteria may bind and metabolize B12 for their own use (Allen and Stabler 2008; Carkeet et al. 2006).

There are conflicting studies concerning the presence of protein-bound B12 malabsorption in older adults. One study suggested that B12 absorption does not decline with age in healthy elderly (McEvoy et al. 1982). Scarlett et al. used a modified protein-bound cobalamin absorption test to study dietary B12 absorption in healthy adults of different age groups and patients with isolated low serum B12 concentrations (Scarlett et al. 1992). Dietary B12 absorption was significantly reduced in healthy adults aged 55–75 years compared with young adults, with a further reduction in those older than 75 years. However, the diagnostic value of this protein-bound B12 absorption test in older adults was limited by the frequent finding of reduced absorption in healthy elderly people with normal serum B12 concentrations. Conversely, Joosten et al. found that protein-bound B12 absorption was abnormal in only 9 (26%) of 34 elderly patients with low serum B12 (Joosten et al. 1993). They concluded that tests of protein-bound B12 absorption offered little advantage over the Schilling test in diagnosing B12 malabsorption in older patients.

Van Asselt et al. found no significant difference in free or protein-bound B12 absorption between healthy middle-aged and older adults, and suggested that the high prevalence of low B12 levels in older people cannot be explained by the aging process *nor* the presence of mild to moderate atrophic gastritis (van Asselt et al. 1996). The authors also investigated the association between atrophic gastritis and mild B12 deficiency in older adults by measuring serum B12 and dietary B12 intake, the presence and severity of atrophic gastritis, the presence of *helicobacter pylori* infection, and serum methylmalonate concentrations in 105 individuals (van Asselt et al. 1998). 23% of apparently healthy older adults (74–80 year olds) were B12 deficient as diagnosed by raised methylmalonate and low to low-normal B12 concentrations. Only one person had insufficient dietary intake lower than the recommended daily intake. Although 31% had atrophic gastritis, only 25% of the cases of mild B12 deficiency could be ascribed to this. Thus, in the majority of cases of mild B12 deficiency no explanation was apparent and it was suggested that other mechanisms to explain the mild B12 deficiency found in older people should be sought.

It has also been suggested that certain drugs, commonly used by older adults, might contribute to B12 deficiency. These include proton pump inhibitors, H2 antagonists and metformin (Abraham 2012; Liu et al. 2014).

One other possible mechanism should be mentioned. There appears to be some alteration in the holotranscobalamin delivery system in older adults (Marcus et al. 1987; Metz et al. 1996), although these changes appear to be marginal and hence unlikely to be of clinical significance (Gimsing et al. 1989). Nevertheless, it is possible that genetic variants of transcobalamin and its recently identified cell membrane receptor might play a contributory role in the aetiology of age-related B12 deficiency (McCaddon 2013; McCaddon et al. 2013).

6. Vitamin B12 Deficiency and Cognition/Dementia

Mental disturbance associated with B12 deficiency has long been recognized. In Thomas Addison's first description of PA he noted that, "…the mind occasionally wanders" (Addison 1849). Dementia was also reported in a few studies in the 1950's (Droller and Dossett 1959; Holmes 1956). In 1959, Droller and Dossett investigated a series of confusional states and non-organic senile dementias in geriatric patients and compared their serum B12 levels with that of older adults with normal cognitive function. Using a microbiological based assay they found lower values of B12 in senile dementia compared to controls. They also found no reduction of the vitamin with aging, and no difference in bodyweight between the cases and controls, suggesting that malnutrition was an unlikely cause of these low values.

In all the instances in the 1950's cognitive impairment failed to respond to B12 replacement, and it was unclear whether the deficiency simply co-existed with the dementia or was of any aetiological significance. Nevertheless, B12 deficiency was subsequently included in many medical textbooks as a reversible cause of dementia, despite the lack of definitive evidence at that time.

Byrne conducted a thorough review covering the literature through 1987 (Byrne 1987). She concluded that B12 deficiency was an infrequent cause of reversible dementia, occurring in only 4/188 subjects (2.1%) in the twelve studies reported in sufficient detail to permit close data analysis. The most commonly reported impairment was 'organic psychosis' and she commented that there was little evidence that cortical dementia is due to B12 deficiency or PA. Indeed, she suggested that it is probable that cases diagnosed as senile dementia have B12 deficiency as a co-existing nutritional deficiency. However, as can be inferred from the previous section, this is perhaps a naïve view since malnutrition is an uncommon cause of B12 deficiency.

Inada et al. performed one of the few post-mortem tissue studies of B12 and dementia (Inada et al. 1982). They measured the B12 content of brains in twelve autopsy cases of older patients with dementia. They studied histopathological changes, especially in the temporal and frontal lobes, and found that a decrease in B12 and transcobalamin in these brains was associated with neuronal loss, myelin degeneration, atrophy, ventricular dilatation and vascular lesions. They also studied 12 autopsy cases of vascular dementia and found that B12 temporal lobe content was markedly decreased. They suggested that their findings represented a "metabolic derangement or impaired transport of vitamin B12".

In 1988, Lindenbaum et al. published a report of patients with low B12 levels and various neuropsychiatric disorders, but without anaemia (Lindenbaum et al. 1988). Out of 37 patients, 36 had homocysteine levels more than 3 standard deviations above normal. This was a landmark study that stimulated much further clinical research. For example, shortly afterwards Bell et al. conducted a retrospective chart review to examine the relationship between cognitive measures and serum folate and B12 status in 102 elderly psychiatric inpatients (Bell et al. 1990). Although their medical records indicated good overall nutritional status, correlation analyses showed that those with below median values for both folate and B12 had significantly worse scores on the Mini-Mental State Examination, a screening test of global cognitive function. They concluded that lower levels of folate and B12, even within the putative normal range, may interact to produce CNS metabolic abnormalities affecting cognition.

Few intervention studies were conducted prior to the new millennium. Chiu observed that this perhaps explains why there are few guidelines concerning the treatment of low serum B12 in patients with dementia (Chiu 1996). This leads to the paradoxical situation where physicians routinely screen for deficiency, but subsequently ignore its discovery. The four key intervention studies are those of Martin et al. (Martin et al. 1992), Carmel et al. (Carmel et al. 1995), Teunisse et al. (Teunisse et al. 1996) and Eastley et al. (Eastley et al. 2000).

Martin et al. investigated the effects of B12 supplementation on cognitive performance in a group of 18 elderly participants with low serum B12 and evidence of cognitive impairment (Martin et al. 1992). Participants received 1,000 micrograms of cyanocobalamin intramuscularly daily for one week, weekly for 1 month, and then monthly for 6 months. Post-supplementation scores from the Mattis Dementia Rating Scale for 11 of the 18 participants showed improvement. However, only those with mild impairment and with symptoms for less than a year improved. Importantly, those who had been symptomatic for less than 6 months responded best, suggesting that age-related cognitive losses in B12 deficiency might be reversible if supplementation is initiated sufficiently early.

Carmel and colleagues evaluated B12, neuropsychological and electrophysiological indices in 13 older adults with dementia and low serum B12 levels before and after B12 supplementation (Carmel et al. 1995). Improvements were found for homocysteine and haemoglobin levels, neuropathological symptoms, EEG abnormalities, and visual evoked and somatosensory abnormalities. However, no improvements on neuropsychological tests of cognitive performance were observed.

Teunisse et al. studied 170 consecutive referrals to a memory clinic and found low B12 in 26 (15%), with 25 of the 26 patients with low B12 fulfilling diagnostic criteria for 'possible' AD (Teunisse et al. 1996). They treated these with intramuscular B12 according to a standard regime and reassessed 6 months later. They found no improvement and supplementation did not slow the progression of dementia. They concluded that, contrary to widely accepted belief, subnormal serum B12 is not a quantitatively important cause of reversible dementia. They used the cognitive subscale of the Cambridge Mental Disorders of the Elderly Examination (CAMDEX) questionnaire (CAMCOG), an interview for deterioration in daily living activities in dementia (IDDD), and a behavioural scale. Improvement had to be greater than the 68% confidence intervals calculated for the tests, smaller changes being interpreted as random variation. They conceded that these instruments might not have been sensitive enough to detect subtle effects that might require larger double blind placebo controlled studies. Unlike Martin et al., they did not find a time limited window of opportunity for effective intervention, but these patients had been demented for considerably longer and may well have had irreversible damage.

Last, Eastley et al. identified 125 patients out of 1,432 attending a memory disorders clinic who had low serum B12 (Eastley et al. 2000). They assessed 66 patients with dementia and 22 with cognitive impairment before and after B12 supplementation. There was no change in the dementia group, but patients in the cognitively impaired group improved on measures of verbal fluency.

Larner and Rakshi summarised this literature by concluding that, "…a low vitamin B12 is not an uncommon finding in patients with dementia or cognitive decline, but cases of dementia reversible with vitamin B12 therapy are extremely rare" (Larner and Rakshi 2001). They suggested that, "…in most cases a low blood vitamin B12 level in a demented patient is likely a coexistent rather than a causal abnormality."

Before commenting further on this general conclusion, mention should be made of the association between B12 deficiency and the specific dementia of AD. In particular, it will be seen that the new metabolite and carrier protein (holotranscobalamin) assays have played an important role in exploring this particular relationship, and have led to a wider study of B12 status in relation to dementia, to cognition in general, and to other chronic neurodegenerative diseases.

7. Vitamin B12 Deficiency and Alzheimer's Disease

In 1983 Van Tiggelen first suggested that B12 deficiency might sometimes occur specifically in association with AD (Van Tiggelen 1983). Van Tiggelen demonstrated that in 24 patients with senile dementia of Alzheimer-type (SDAT) there was a high incidence of pathologically low levels of B12 in CSF despite normal serum levels. He suggested this might indicate abnormal function of the choroid plexus, and possibly of the blood-brain barrier, in such patients.

Shortly afterwards, Cole and Prchal measured serum B12 levels in 20 subjects aged 65 years and over with Alzheimer-type dementia, 20 age-matched subjects with non-Alzheimer type dementia, and 20 age-matched non-demented subjects (Cole and Prchal 1984). They found that serum B12 levels were significantly lower and B12 deficiency was significantly more frequent in subjects with Alzheimer-type dementia, and that B12 levels were independent of age, sex, and haematological abnormality in these patients.

Karnaze and Carmel analyzed serum B12 retrospectively in 17 patients with primary degenerative dementia and 11 with specific demonstrable causes of dementia (secondary dementia) (Karnaze and Carmel 1987). The prevalence of low B12 levels was significantly higher in primary dementia (29% vs. 0% in secondary dementia). Because typical haematological findings of this deficiency were often absent, they prospectively studied two other patients with primary dementia and low B12 levels. Neither had megaloblastic anaemia; one had normal intestinal B12 absorption (as indicated by a Schilling test), while B12 absorption in the other was borderline impaired. Despite this absence of expected findings, the deoxyuridine suppression test, a functional indicator of B12 status, gave unequivocal biochemical evidence of B12 deficiency in both cases. Their survey of 28 patients thus established that low serum B12 levels are a frequent finding in patients with primary dementia. Their findings in the two prospectively studied cases (as well as in some of the patients in the survey) indicated that these levels are associated, in at least some cases, with a subclinical deficiency state rather than with severe deficiency such as that caused by PA. They noted that such subclinical deficiency states could not be identified by classic haematological criteria or by the Schilling test.

Regland et al. also found lower serum B12 levels in 56 patients with senile dementia of the Alzheimer type (SDAT) compared to 54 patients with vascular dementia (Regland et al. 1988). The frequency of low serum B12 in their patients with SDAT (23%) was much higher than an unselected population of 75-year old individuals (5%) in a separate study performed in the same geographical area and using the same reference values as their own study.

The following year Renvall et al. reported on the dietary intake and biochemical estimates of thiamine, riboflavin, folate, B12, protein, and iron in 22 free-living SDAT patients versus 41 cognitively normal controls aged 60 years and over (Renvall et al. 1989). The two groups did not differ in intake of these vitamins, although the SDAT group had lower serum transketolase (thiamin), red cell folate, and serum B12.

Ikeda et al. found lower cerebrospinal fluid (CSF) levels of B12 in 12 patients with AD compared with 10 patients with vascular dementia (Ikeda et al. 1990), and concluded that measuring CSF B12 levels in dementia patients might be a useful diagnostic tool. Regland et al. later showed that low CSF B12 levels might be explained by inactive B12 analogues somehow interfering with the transport of the vitamin. A lower ratio of active B12 to inactive analogues was found in demented patients (Regland et al. 1992). Further indirect evidence of disturbed CSF B12 metabolism was demonstrated by Bottiglieri et al., who found surprisingly low levels of S-adenosylmethionine in AD patients compared with patients with other neurological disease (Bottiglieri et al. 1990).

Therefore, by the early 1990s, it appeared that there might be an association between B12 deficiency and the specific dementia of AD. However, the nature of this association was unclear. Notwithstanding the observed associations between AD and B12 deficiency, as indicated by serum and CSF B12 levels, it was argued that because both conditions occur commonly in the elderly it was unsurprising that many subjects had a combination of the two (Stabler et al. 1997). One important criticism of these early papers was whether or not B12 deficiency in these patients was associated with genuine AD neuropathology. AD is essentially a histopathological diagnosis, and it had been argued that patients with a clinical diagnosis of the disease and low serum B12 might actually represent a previously undetected subgroup of patients with dementia secondary to B12 deficiency. However, the finding of low B12 values in patients with genetically determined familial AD confirmed that the two conditions could genuinely coexist (Kennedy et al. 1993; McCaddon and Kelly 1994). Thus, an interesting parallel was emerging between these two distinct diseases (B12 deficiency and AD). Both were increasingly recognized as being slowly progressive in nature, with subtle and previously undetectable preclinical stages.

Although it was acknowledged that B12 deficiency occurs commonly in the elderly, there were conflicting views as to its aetiology, and it seemed that explanations other than malnutrition or malabsorption were required to account for this. In the majority of cases no explanation was apparent. The haematological and neuropsychiatric features of B12 deficiency were often dissociated, but the underlying reasons for the predominant dysfunction of one or the other system in individual patients remained unclear. No convincing pathogenic mechanisms or hypotheses had been presented to account for such differences.

In summary, it was apparent that poor B12 status might at least partially contribute to cognitive decline in some elderly persons. The metabolic implications of B12 (and folate) deficiency for the nervous system suggested ways in which such deficiency might lead to the observed neurotransmitter and structural changes of AD (McCaddon and Kelly 1992; Regland and Gottfries 1992; Rosenberg and Miller 1992).

Metabolic evidence for such deficiency should therefore be common in AD. This was confirmed by the discovery of elevated blood homocysteine levels not only in clinically diagnosed AD cases (Joosten et al. 1997; McCaddon et al. 1998), but also in histopathologically confirmed AD (Clarke et al. 1998). In fact, patients with subsequent histopathological confirmation of AD had significantly lower vitamin B12 levels than clinically diagnosed patients. Joosten et al. also observed high levels of methylmalonic acid in their group of 52 AD patients (Joosten et al. 1997).

These initial studies prompted a substantial interest in metabolic evidence of B12 (and folate) deficiency in relation to other forms of dementia and to cognition in general. For example, in the following year, elevated blood homocysteine levels were also reported in patients with vascular dementia, as well as in individuals with mild cognitive impairment (Lehmann et al. 1999). In cross-sectional studies, blood homocysteine levels correlated with well-validated measures of cognition (Budge et al. 2000; Leblhuber et al. 2000; Lehmann et al. 1999; McCaddon et al. 1998; Miller et al. 2003a; Prins et al. 2002), and also with white matter hyperintensities (Hogervorst et al. 2002) and brain atrophy (Den Heijer et al. 2003; Sachdev 2005).

These relationships generally held true in later prospective studies. For example, homocysteine was shown to be an independent risk factor for cognitive decline not only in healthy elderly (McCaddon et al. 2001), but also in patients with established AD (Oulhaj et al. 2010). Homocysteine is also associated with an increased risk of incident dementia, including AD (Haan et al. 2007; Ravaglia et al. 2005; Seshadri et al. 2002; Zylberstein et al. 2011).

Perhaps most compelling are the recent findings of Smith and colleagues regarding the effects of B vitamin supplements on age-related brain loss and cognitive decline (de Jager et al. 2012; Douaud et al. 2013; Smith et al. 2010). In older adults diagnosed with mild cognitive impairment, high dose B vitamin supplements (folic acid, B12, and B6) slowed global and regional brain atrophy and prevented cognitive decline over a 2-year treatment period compared with placebo. Importantly, the protective effects were primarily observed in those individuals with elevated plasma homocysteine levels, i.e., in those with metabolic evidence of B vitamin inadequacy. Moreover, post-hoc statistical analysis suggested that it was the B12 in the active supplement that was primarily responsible for the lowering of homocysteine in the study participants, suggesting that inadequate B12 status contributes to brain atrophy and cognitive decline observed in older individuals with mild cognitive impairment.

The realization that total B12 is an insensitive indicator of vitamin B12 deficiency also led to studies of holotranscobalamin in relation to dementia. Patients with AD were found to have lower plasma holotranscobalamin levels compared with non-demented elderly, despite having similar plasma total B12 values (Johnston and Thomas 1997; Refsum and Smith 2003). In addition, the risk of incident AD over a seven-year period in a population-based cohort of older individuals was inversely associated with holotranscobalamin (Hooshmand et al. 2010), and baseline holotranscobalamin was directly associated with global cognitive performance, executive function and psychomotor speed (Hooshmand et al. 2012). The ratio of holotranscobalamin to total serum B12 may also be important; in older Latinos with elevated depressive symptoms the fraction of total B12 bound to transcobalamin was directly correlated with cognitive function scores suggesting a global effect of B12 status on cognition and mood (Garrod et al. 2008).

A more complete discussion of the numerous studies of the relationship between metabolic B12 deficiency, cognitive decline and dementia is beyond the scope of this chapter; the interested reader is referred to other comprehensive review articles (McCaddon and Miller 2015; Selhub et al. 2010; Smith 2006; Stanger et al. 2009).

8. Exacerbation of Vitamin B12 Deficiency by Excess Folic Acid

An evolving area of investigation is focusing on the possibility that excess folic acid exacerbates vitamin B12 deficiency. It has been known for several decades that folic acid supplements can reverse megaloblastic anaemia caused by B12 deficiency. However, B12 deficient patients treated in this manner would still be susceptible to the neurological manifestations of the deficiency, and thus folic acid supplements were said to 'mask' the B12 deficiency. Moreover, as reviewed by Reynolds (Reynolds 2002), the effect of the folic acid on B12-deficiency anaemia was often incomplete and temporary, and there was evidence that the folic acid actually precipitated or exacerbated the neurological manifestations. This possibility was brought to the fore in 2007 by Selhub and colleagues who found in cross-sectional analyses of data from the U.S. National Health and Nutrition Examination Survey (NHANES) that the risk of both anaemia and cognitive deficits was greater in older adults with low B12 status and high circulating folate levels than in those with low B12 status and non-elevated folate levels (Morris et al. 2007). In a follow-up study, it was shown that the combination of low B12 and high folate was also associated with the highest levels of methylmalonate and homocysteine (Selhub et al. 2007), a finding that was also observed by Miller et al. in cohort of older Latinos (Miller et al. 2009). More recently, it was shown in the Framingham Heart Study cohort that the decline in mini-mental state examination scores over an 8-year follow-up period was faster in those with low B12 and elevated folate than those with low B12 and non-elevated folate (Morris et al. 2012). Taken together, these findings suggest an exacerbating effect of high folate on B12 deficiency. However, the mechanism by which this occurs is unknown.

9. Vitamin B12 Deficiency and other Neurodegenerative Diseases

The link between metabolic B12 deficiency and dementia also prompted studies of its possible association with other chronic neurological diseases. To date, the majority of these have focused on blood homocysteine and/or methylmalonate levels. As yet, there are no clinical studies of holotranscobalamin in relation to these disorders.

Epilepsy. In animal studies, administering high doses of homocysteine induces seizures (Kubova et al. 1995). Very high homocysteine levels occur in patients with inborn errors of metabolism such as homocystinuria. These are also associated with seizures, but it is not clear whether more moderate levels of subclinical B12 deficiency are epileptogenic. A confounding factor is that many anticonvulsants adversely influence folate status thereby leading to hyperhomocysteinaemia (Schwaninger et al. 1999).

Parkinson's disease (PD). High homocysteine levels are found in patients with Parkinson's disease (Allain et al. 1995; Kuhn et al. 1998). There are also reports of higher methylmalonate levels in such patients (Levin et al. 2010; Toth et al. 2010). A potential confounder in Parkinson's disease is that such patients are treated with L-dopa, metabolism of which can increase homocysteine levels (Blandini et al. 2001; Miller et al. 2003b; Muller et al. 1999; Muller et al. 2001). The magnitude of the increase caused by L-dopa is exacerbated by low levels of B12, folate, and vitamin B6 (Miller et al. 2003b). Regardless of its aetiology, hyperhomocysteinaemic Parkinsonian patients are more likely to be depressed, and they perform less well on neuropsychometric tasks compared with normohomocysteinaemic patients (O'Suilleabhain et al. 2004). Dietary supplementation with B12 (and folate) reduces homocysteine levels in the disease (Lamberti et al. 2005). This could have important treatment implications, given that such patients have an increased risk of developing cognitive impairment and dementia (Zoccolella et al. 2010).

Multiple sclerosis. A relationship between B12 deficiency and multiple sclerosis has been suspected since the early 1980's (Reynolds 1992). In many of these cases haematological signs of B12 deficiency are either absent or minimal, serum B12 levels are borderline or low-normal, and in most cases the underlying cause of deficiency is unclear. Multiple Sclerosis is regarded as being an inflammatory disorder of unknown aetiology. Chronic immune reactions or recurrent myelin repair processes might increase demand for B12 in the disease (Reynolds et al. 1992). Homocysteine levels are increased in most (Ramsaransing et al. 2006; Russo et al. 2008; Teunissen et al. 2005), but not all studies (Goodkin et al. 1994; Teunissen et al. 2008). Higher levels in patients with multiple sclerosis are associated with impaired cognitive performance and depression (Triantafyllou et al. 2008).

Progressive supranuclear palsy/amyotrophic lateral sclerosis. Blood homocysteine levels are also elevated in patients with progressive supranuclear palsy and amyotrophic lateral sclerosis (ALS) (Levin et al. 2010), more markedly so in ALS patients with a shorter time to diagnosis. Higher levels may be linked to a more rapid disease progression (Zoccolella et al. 2008).

10. Reconciling the Neurology of 'Classical' and 'Subclinical' B12 Deficiency

How can the associations between subclinical B12 deficiency and such diverse neurodegenerative processes as AD, PD and multiple sclerosis be reconciled with the 'classical' neurological picture of B12 deficiency exemplified by PA? One possibility is that these associations, most strongly represented by increased homocysteine levels, perhaps reflect the effects of long-term exposure to oxidative stress, effectively resulting in chronic B vitamin *depletion*.

Neurodegenerative disorders are slowly progressive with a very long asymptomatic pre-clinical phase. For example, in the case of AD, biochemical changes are detectable more than twenty years before clinical signs of the disease (Bateman et al. 2012). These chronic disorders are also all associated with neuro-inflammation, characterised by microglial activation and release of a diverse array of inflammatory mediators such as cytokines, as well as generation of free radicals and other reactive oxidative species (Lucas et al. 2006).

In the brain, homocysteine is largely metabolized by B12-dependent methionine synthase. This enzyme system is readily inactivated by oxidative stress (Banerjee and Matthews 1990). Its reactivation requires S-adenosylmethionine as a methyl donor, generating S-adenosylhomocysteine and ultimately homocysteine in the process (Elmore and Matthews 2007). The important 'net' result is that oxidative stress *inevitably* increases homocysteine levels (McCaddon and Hudson 2007a; McCaddon et al. 2002) (Figures 1A and 1B).

Impaired methionine synthase activity also leads to secondary folate depletion because de-methylation of folate by methionine synthase is required for its cellular retention (McGing et al. 1978). Folate may also be oxidatively degraded as a result of chronic immune activation (Fuchs et al. 2001).

The discrepancy between the 'classical' neurology of deficiency (arising from malabsorption or low dietary intake) and that related to the association of 'subclinical' deficiency with neurodegenerative disease might be explained

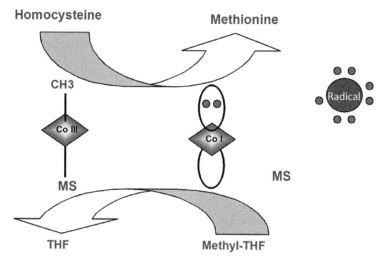

Figure 1A. In the methionine synthase (MS) reaction MS-bound methylcobalamin transfers its methyl group to homocysteine to generate methionine and a transient free cob(I)alamin intermediate. MS-bound methylcobalamin is regenerated when cob(I)alamin accepts a methyl group from methylfolate, generating free tetrahydrofolate (THF) in the process. Cob(I)alamin is vulnerable to oxidation by free radicals.

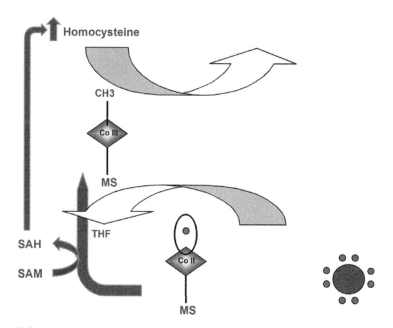

Figure 1B. Inactivation of MS occurs when free radicals oxidise cob(I)alamin to a cob(II)alamin species. Re-activation requires donation of a methyl group by S-adenosylmethionine (SAM), the universal methyl donor. The net effect is that homocysteine levels increase as a consequence of oxidative stress.

by this process. This could also account for the apparent dissociation of neurological and haematological features discussed earlier (McCaddon et al. 2004). In effect, such a model would account for a selective deficiency of B12 in nervous tissue, with relative sparing of haematopoiesis.

There is an alternative hypothesis. In animal models, B12 deficiency is associated with inflammation (Scalabrino et al. 2008). Because neurodegenerative conditions are to a great extent inflammatory diseases, it is possible that a suboptimal B12 status would exacerbate neurodegeneration by accelerating its inflammatory component (Miller 2002). This simply requires B12 deficiency to *co-exist* with neurodegenerative disease—a reasonable expectation based on prevalence estimates for both B12 deficiency and neurodegenerative disease in the elderly population.

At least in AD, harmful self-propagating cascades might also exist, whereby impaired homocysteine recycling due to inflammatory oxidative stress generates neurotoxic derivatives such as homocysteic acid, which promote further free radical formation (McCaddon and Hudson 2007a). Increased homocysteine, associated with hypomethylation of other substrates, can also contribute to neurofibrillary tangle (McCaddon and Hudson 2007b; Obeid et al. 2007) and amyloid plaque formation characteristic of the disease (Fuso and Scarpa 2011; Sontag et al. 2007) (Figure 2). Thus, B12 deficiency may not be a direct cause of neurodegenerative disease *per se*, but rather serves

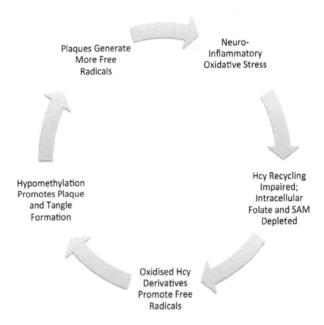

Plaques Generate
More Free
Radicals

Neuro-
Inflammatory
Oxidative Stress

Hcy Recycling
Impaired;
Intracellular
Folate and SAM
Depleted

Hypomethylation
Promotes Plaque
and Tangle
Formation

Oxidised Hcy
Derivatives
Promote Free
Radicals

Figure 2. Hypothesised self-propagating events in the interplay of neuroinflammatory oxidative stress and impaired homocysteine metabolism in Alzheimer's disease. This model integrates changes in Hcy and SAM levels with amyloid plaque and neurofibrillary tangle formation.

as a modifier or accelerant of the primary pathophysiological processes that underlie neurodegenerative disease.

11. The Patient's View

The Pernicious Anaemia Society (http://www.pernicious-anaemia-society. org/) is an active patient support group with over 4,000 members. In 2010 the Pernicious Anaemia Society produced a questionnaire concerning symptoms and treatment. Its members were invited to complete this, either online or via post (Hooper et al. 2014). Nearly 1,000 completed the questionnaire, revealing perceived difficulties and delays concerning the current diagnosis and treatment of B12 deficiency. The questionnaire also highlighted several interesting neuropsychiatric features of B12 deficiency, as seen from the patient's perspective.

A focus group studying the data synthesized the symptomatology of patients into descriptive phrases. For example, the tiredness and fatigue associated with B12 deficiency was described as a 'strange tiredness', quite unlike normal tiredness in that it is *far* more intense and often completely unrelieved after many hours of deep sleep. In many patients this fatigue seemed to persist even after haematological correction of deficiency by replacement therapy.

Confusion and poor concentration were described by the focus group as 'brain fog.' This is of course strikingly similar to the clinical, and meteorologically related clinical term—'clouding of consciousness.' Nominal aphasia was also frequently reported. This is concordant with recent reports of associations between methylmalonate concentration and verbal fluency (Lewis et al. 2005; McCracken et al. 2006).

PA patients frequently also experience memory impairment and absent-mindedness. Rather descriptively, the focus group termed this as the 'handbag in the fridge' syndrome. The peripheral neurological features of poor balance and coordination are described by patients as 'the shoulder bumps'; they often bump into walls, especially when descending stairs or showering. Some of the behavioural symptoms documented are also worth mentioning. These include mood swings, heightened emotions, irritability, frustration, impatience, a desire for solitude and an aversion to bright lights. These rarely feature in any textbook descriptions of the neuropsychiatric symptoms of B12 deficiency.

12. Summary

The concept of sub-clinical B12 deficiency evolved from advances in laboratory diagnostic techniques. Sub-clinical deficiency is highly prevalent in older adults. Although not associated with the usual clinical features of B12 deficiency, it is associated with an increased risk of cognitive decline, brain atrophy, and dementia. It is also observed in association with other neurodegenerative diseases.

Chronic neuroinflammatory oxidative stress might contribute to its pathogenesis, at least in some instances. Subclinical deficiency runs an unpredictable course, raising difficulties with its rational management. One suggestion is that if such patients have any subtle clinical signs or symptoms that are possibly related to the vitamin, they should be treated to achieve a desirable vitamin B12 blood level (Smith and Refsum 2012).

Last, many clinicians are aware of considerable individual variability in patient's responses to B12 replacement. Patients often continue to experience mild neurological symptoms such as poor memory, impaired concentration and fatigue even after adequate B12 replacement. The reason for this is currently unknown, but future research into the interplay between polymorphisms in transcobalamin and its cell membrane receptor might offer some explanation for this curious phenomenon (McCaddon 2013).

Keywords: neuropathy, dementia, Alzheimer disease, memory, cognitive decline, white matter, multiple sclerosis, epilepsy, Parkinson disease, neuroinflammation, hypomethylation, amyloid plaque

Abbreviations

AD	:	Alzheimer's disease
ALS	:	amyotrophic lateral sclerosis
CAMDEX	:	Cambridge Mental Disorders of the Elderly Examination
CSF	:	cerebrospinal fluid
MS	:	multiple sclerosis
NHANES	:	U.S. National Health and Nutrition Examination Survey
PA	:	pernicious anaemia
SAM	:	S-adenosylmethionine
SDAT	:	senile dementia of the Alzheimer type

References

Abraham NS. 2012. Proton pump inhibitors: potential adverse effects. Curr Opin Gastroenterol. 28(6): 615–620.

Adams RD and Kubik CS. 1944. Subacute degeneration of the brain in pernicious anaemia. N Engl J Med. 231: 1–9.

Addison T. 1849. Anemia: Disease of the suprarenal capsules. Lond Med Gaz. 43: 517.

Allain P, Le Bouil A, Cordillet E, Le Quay L, Bagheri H and Montastruc JL. 1995. Sulfate and cysteine levels in the plasma of patients with Parkinson's disease. Neurotoxicology. 16(3): 527–529.

Allen RH, Lindenbaum J and Stabler SP. 1995. High prevalence of cobalamin deficiency in the elderly. Trans Am Clin Climatol Assoc. 107: 37–45.

Allen RH and Stabler SP. 2008. Identification and quantitation of cobalamin and cobalamin analogues in human feces. Am J Clin Nutr. 87(5): 1324–1335.

Baik HW and Russell RM. 1999. Vitamin B12 deficiency in the elderly. Annu Rev Nutr. 19: 357–377.

Banerjee RV and Matthews RG. 1990. Cobalamin-dependent methionine synthase. FASEB J. 4(5): 1450–1459.

Bateman RJ, Xiong C, Benzinger TL, Fagan AM, Goate A, Fox NC, Marcus DS, Cairns NJ, Xie X, Blazey TM et al. 2012. Clinical and biomarker changes in dominantly inherited Alzheimer's Disease. N Engl J Med. 367(9): 795–804.

Bates CJ, Schneede J, Mishra G, Prentice A and Mansoor MA. 2003. Relationship between methylmalonic acid, homocysteine, vitamin B12 intake and status and socio-economic indices, in a subset of participants in the British National Diet and Nutrition Survey of people aged 65 y and over. Eur J Clin Nutr. 57(2): 349–357.

Beck WS. 1988. Cobalamin and the nervous system. N Engl J Med. 318(26): 1752–1754.

Bell IR, Edman JS, Marby DW, Satlin A, Dreier T, Liptzin B and Cole JO. 1990. Vitamin B12 and folate status in acute geropsychiatric inpatients: affective and cognitive characteristics of a vitamin nondeficient population. Biol Psychiatry. 27(2): 125–137.

Bjorkegren K and Svardsudd K. 2001. Serum cobalamin, folate, methylmalonic acid and total homocysteine as vitamin B12 and folate tissue deficiency markers amongst elderly Swedes—a population-based study. J Intern Med. 249(5): 423–432.

Blandini F, Fancellu R, Martignoni E, Mangiagalli A, Pacchetti C, Samuele A and Nappi G. 2001. Plasma homocysteine and l-dopa metabolism in patients with Parkinson disease. Clin Chem. 47(6): 1102–1104.

Bottiglieri T, Godfrey P, Flynn T, Carney MW, Toone BK and Reynolds EH. 1990. Cerebrospinal fluid S-adenosylmethionine in depression and dementia: effects of treatment with parenteral and oral S-adenosylmethionine. J Neurol Neurosurg Psychiatry. 53(12): 1096–1098.

Budge M, Johnston C, Hogervorst E, de Jager C, Milwain E, Iversen SD, Barnetson L, King E and Smith AD. 2000. Plasma total homocysteine and cognitive performance in a volunteer elderly population. Ann NY Acad Sci. 903: 407–410.

Byrne EJ. 1987. Reversible dementia. Int J Geriatr Psychiatry. 2: 73–81.

Carkeet C, Dueker SR, Lango J, Buchholz BA, Miller JW, Green R, Hammock BD, Roth JR and Anderson PJ. 2006. Human vitamin B12 absorption measurement by accelerator mass spectrometry using specifically labeled (14)C-cobalamin. Proc Natl Acad Sci U S A. 103(15): 5694–5699.

Carmel R. 1988. Pernicious anemia. The expected findings of very low serum cobalamin levels, anemia, and macrocytosis are often lacking. Arch Intern Med. 148(8): 1712–1714.

Carmel R. 1990. Subtle and atypical cobalamin deficiency states. Am J Hematol. 34(2): 108–114.

Carmel R. 2000. Current concepts in cobalamin deficiency. Annu Rev Med. 51: 357–375.

Carmel R, Gott PS, Waters CH, Cairo K, Green R, Bondareff W, DeGiorgio CM, Cummings JL, Jacobsen DW, Buckwalter G et al. 1995. The frequently low cobalamin levels in dementia usually signify treatable metabolic, neurologic and electrophysiologic abnormalities. Eur J Haematol. 54(4): 245–253.

Chiu HFK. 1996. Vitamin B12 deficiency and dementia. Int J Geriatr Psychiatry. 11: 851–858.

Clarke R, Grimley Evans J, Schneede J, Nexo E, Bates C, Fletcher A, Prentice A, Johnston C, Ueland PM, Refsum H et al. 2004. Vitamin B12 and folate deficiency in later life. Age Ageing. 33(1): 34–41.

Clarke R, Smith AD, Jobst KA, Refsum H, Sutton L and Ueland PM. 1998. Folate, vitamin B12, and serum total homocysteine levels in confirmed Alzheimer disease. Arch Neurol. 55(11): 1449–1455.

Cole MG and Prchal JF. 1984. Low serum vitamin B12 in Alzheimer-type dementia. Age Ageing. 13(2): 101–105.

de Jager C, Oulhaj A, Jacoby R, Refsum H and Smith AD. 2012. Cognitive and clinical outcomes of homocysteine lowering B vitamin treatment in mild cognitive impairment: a randomized controlled trial. Int J Geriatr Psychiatry. 27(6): 592–600.

Den Heijer T, Vermeer SE, Clarke R, Oudkerk M, Koudstaal PJ, Hofman A and Breteler MM. 2003. Homocysteine and brain atrophy on MRI of non-demented elderly. Brain. 126(Pt 1): 170–175.

Douaud G, Refsum H, de Jager CA, Jacoby R, Nichols TE, Smith SM and Smith AD. 2013. Preventing Alzheimer's disease-related gray matter atrophy by B-vitamin treatment. Proc Natl Acad Sci U S A. 110(23): 9523–9528.

Droller H and Dossett J. 1959. Vitamin B12 levels in senile dementia and confusional states. Geriatrics. 14(6): 367–373.

Eastley R, Wilcock GK and Bucks RS. 2000. Vitamin B12 deficiency in dementia and cognitive impairment: the effects of treatment on neuropsychological function. Int J Geriatr Psychiatry. 15(3): 226–233.

Elmore CL and Matthews RG. 2007. The many flavors of hyperhomocyst(e)inemia: insights from transgenic and inhibitor-based mouse models of disrupted one-carbon metabolism. Antioxid Redox Signal. 9(11): 1911–1921.

Elsborg L, Lund V and Bastrup-Madsen P. 1976. Serum vitamn B12 levels in the aged. Acta Med Scand. 200: 309–314.

Evans DL, Edelsohn GA and Golden RN. 1983. Organic psychosis without anemia or spinal cord symptoms in patients with vitamin B12 deficiency. Am J Psychiatry. 140(2): 218–221.

Fedosov SN. 2010. Metabolic signs of vitamin B(12) deficiency in humans: computational model and its implications for diagnostics. Metabolism. 59(8): 1124–1138.

Fedosov SN, Brito A, Miller JW, Green R and Allen LH. 2015. Combined indicator of vitamin B12 status: modification for missing biomarkers and folate status and recommendations for revised cut-points. Clinical Chemistry and Laboratory Medicine: CCLM/FESCC 53(8): 1215–1225.

Ferraro A, Arieti S and English WH. 1945. Cerebral changes in the course of pernicious anaemia and their relationship to psychiatric symptoms. J Neuropathol Exp Neurol. 4(3): 217–239.

Fuchs D, Jaeger M, Widner B, Wirleitner B, Artner-Dworzak E and Leblhuber F. 2001. Is hyperhomocysteinemia due to the oxidative depletion of folate rather than to insufficient dietary intake? Clin Chem Lab Med. 39(8): 691–694.

Fuso A and Scarpa S. 2011. One-carbon metabolism and Alzheimer's disease: is it all a methylation matter? Neurobiol Aging. 32(7): 1192–5.

Garrod MG, Green R, Allen LH, Mungas DM, Jagust WJ, Haan MN and Miller JW. 2008. Fraction of total plasma vitamin B12 bound to transcobalamin correlates with cognitive function in elderly Latinos with depressive symptoms. Clinical Chemistry. 54(7): 1210–1217.

Gimsing P, Melgaard B, Andersen K, Vilstrup H and Hippe E. 1989. Vitamin B12 and folate function in chronic alcoholic men with peripheral neuropathy and encephalopathy. J Nutr. 119(3): 416–424.

Goodkin DE, Jacobsen DW, Galvez N, Daughtry M, Secic M and Green R. 1994. Serum cobalamin deficiency is uncommon in multiple sclerosis. Arch Neurol. 51(11): 1110–1114.

Green R and Miller JW. 2014. Vitamin B12. pp. 447–489. In: J Zempleni, JW Suttie, JF Gregory III and P Stover (eds.). Handbook of Vitamins 5th ed. Boca Raton, FL: CRC Press.

Haan MN, Miller JW, Aiello AE, Whitmer RA, Jagust WJ, Mungas DM, Allen LH and Green R. 2007. Homocysteine, B vitamins, and the incidence of dementia and cognitive impairment: results from the Sacramento Area Latino Study on Aging. Am J Clin Nutr. 85(2): 511–517.

Healton EB, Savage DG, Brust JC, Garrett TJ and Lindenbaum J. 1991. Neurologic aspects of cobalamin deficiency. Medicine (Baltimore). 70(4): 229–245.

Hector M and Burton JR. 1988. What are the psychiatric manifestations of vitamin B12 deficiency? J Am Geriatr Soc. 36(12): 1105–1112.

Herbert V. 1987. The 1986 Herman award lecture. Nutrition science as a continually unfolding story: the folate and vitamin B12 paradigm. Am J Clin Nutr. 46(3): 387–402.

Hogervorst E, Ribeiro HM, Molyneux A, Budge M and Smith AD. 2002. Plasma homocysteine levels, cerebrovascular risk factors, and cerebral white matter changes (leukoaraiosis) in patients with Alzheimer disease. Arch Neurol. 59(5): 787–793.

Holmes J. 1956. Cerebral manifestations of vitamin B12 deficiency. BMJ. 2: 1394.

Hooper M, Hudson P, Porter F and McCaddon A. 2014. Patient journeys: diagnosis and treatment of pernicious anaemia. Br J Nurs. 23(7): 376–381.

Hooshmand B, Solomon A, Kareholt I, Leiviska J, Rusanen M, Ahtiluoto S, Winblad B, Laatikainen T, Soininen H and Kivipelto M. 2010. Homocysteine and holotranscobalamin and the risk of Alzheimer disease: a longitudinal study. Neurology. 75(16): 1408–1414.

Hooshmand B, Solomon A, Kareholt I, Rusanen M, Hanninen T, Leiviska J, Winblad B, Laatikainen T, Soininen H and Kivipelto M. 2012. Associations between serum homocysteine, holotranscobalamin, folate and cognition in the elderly: a longitudinal study. J Intern Med. 271(2): 204–212.

Howard JM, Azen C, Jacobsen DW, Green R and Carmel R. 1998. Dietary intake of cobalamin in elderly people who have abnormal serum cobalamin, methylmalonic acid and homocysteine levels. Eur J Clin Nutr. 52(8): 582–587.

Hvas AM and Nexo E. 2005. Holotranscobalamin—a first choice assay for diagnosing early vitamin B12 deficiency? J Intern Med. 257(3): 289–298.

Ikeda T, Furukawa Y, Mashimoto S, Takahashi K and Yamada M. 1990. Vitamin B12 levels in serum and cerebrospinal fluid of people with Alzheimer's disease. Acta Psychiatr Scand. 82(4): 327–329.

Inada M, Toyoshima M and Kameyama M. 1982. Cobalamin contents of the brains in some clinical and pathologic states. Int J Vitam Nutr Res. 52(4): 423–429.

Johnston CS and Thomas JA. 1997. Holotranscobalamin II levels in plasma are related to dementia in older people. J Am Geriatr Soc. 45(6): 779–780.

Joosten E, Lesaffre E, Riezler R, Ghekiere V, Dereymaeker L, Pelemans W and Dejaeger E. 1997. Is metabolic evidence for vitamin B12 and folate deficiency more frequent in elderly patients with Alzheimer's disease? J Gerontol A Biol Sci Med Sci. 52(2): M76–M79.

Joosten E, Pelemans W, Devos P, Lesaffre E, Goossens W, Criel A and Verhaeghe R. 1993. Cobalamin absorption and serum homocysteine and methylmalonic acid in elderly subjects with low serum cobalamin. Eur J Haematol. 51(1): 25–30.

Karnaze DS and Carmel R. 1987. Low serum cobalamin levels in primary degenerative dementia. Do some patients harbor atypical cobalamin deficiency states? Arch Intern Med. 147(3): 429–431.

Kennedy AM, Newman S, McCaddon A, Ball J, Roques P, Mullan M, Hardy J, Chartier-Harlin MC, Frackowiak RS, Warrington EK et al. 1993. Familial Alzheimer's disease. A pedigree with a mis-sense mutation in the amyloid precursor protein gene (amyloid precursor protein 717 valine-->glycine). Brain. 116(Pt 2): 309–324.

Krasinski SD, Russell RM, Samloff IM, Jacob RA, Dallal GE, McGandy RB and Hartz SC. 1986. Fundic atrophic gastritis in an elderly population. Effect on hemoglobin and several serum nutritional indicators. J Am Geriatr Soc. 34(11): 800–806.

Kubova H, Folbergrova J and Mares P. 1995. Seizures induced by homocysteine in rats during ontogenesis. Epilepsia. 36(8): 750–756.

Kuhn W, Roebroek R, Blom H, van Oppenraaij D and Muller T. 1998. Hyperhomocysteinaemia in Parkinson's disease. J Neurol. 245(12): 811–812.

Lamberti P, Zoccolella S, Armenise E, Lamberti SV, Fraddosio A, de Mari M, Iliceto G and Livrea P. 2005. Hyperhomocysteinemia in L-dopa treated Parkinson's disease patients: effect of cobalamin and folate administration. Eur J Neurol. 12(5): 365–368.

Larner AJ and Rakshi JS. 2001. Vitamin B12 deficiency and dementia. Eur J Neurol. 8(6): 730–731.

Leblhuber F, Walli J, Artner-Dworzak E, Vrecko K, Widner B, Reibnegger G and Fuchs D. 2000. Hyperhomocysteinemia in dementia. J Neural Transm. 107(12): 1469–1474.

Lehmann M, Gottfries CG and Regland B. 1999. Identification of cognitive impairment in the elderly: homocysteine is an early marker. Dement Geriatr Cogn Disord. 10(1): 12–20.

Levin J, Botzel K, Giese A, Vogeser M and Lorenzl S. 2010. Elevated levels of methylmalonate and homocysteine in Parkinson's disease, progressive supranuclear palsy and amyotrophic lateral sclerosis. Dement Geriatr Cogn Disord. 29(6): 553–559.

Lewis MS, Miller LS, Johnson MA, Dolce EB, Allen RH and Stabler SP. 2005. Elevated methylmalonic acid is related to cognitive impairment in older adults enrolled in an elderly nutrition program. J Nutr Elder. 24(3): 47–65.

Lildballe DL, Fedosov S, Sherliker P, Hin H, Clarke R and Nexo E. 2011. Association of cognitive impairment with combinations of vitamin B12-related parameters. Clin Chem. 57(10): 1436–1443.

Lindenbaum J, Healton EB, Savage DG, Brust JC, Garrett TJ, Podell ER, Marcell PD, Stabler SP and Allen RH. 1988. Neuropsychiatric disorders caused by cobalamin deficiency in the absence of anemia or macrocytosis. N Engl J Med. 318(26): 1720–1728.

Lindenbaum J, Rosenberg IH, Wilson PW, Stabler SP and Allen RH. 1994. Prevalence of cobalamin deficiency in the Framingham elderly population. Am J Clin Nutr. 60(1): 2–11.

Liu Q, Li S, Quan H and Li J. 2014. Vitamin B12 status in metformin treated patients: systematic review. PLoS One. 9(6): e100379.

Lucas SM, Rothwell NJ and Gibson RM. 2006. The role of inflammation in CNS injury and disease. Br J Pharmacol 147 Suppl. 1: S232–S240.

Marcus DL, Shadick N, Crantz J, Gray M, Hernandez F and Freedman ML. 1987. Low serum B12 levels in a hematologically normal elderly subpopulation. J Am Geriatr Soc. 35(7): 635–638.

Martin DC, Francis J, Protetch J and Huff FJ. 1992. Time dependency of cognitive recovery with cobalamin replacement: report of a pilot study. J Am Geriatr Soc. 40(2): 168–172.

McCaddon A. 2013. Vitamin B12 in neurology and ageing; Clinical and genetic aspects. Biochimie. 95(5): 1066–1076.

McCaddon A, Davies G, Hudson P, Tandy S and Cattell H. 1998. Total serum homocysteine in senile dementia of Alzheimer type. Int J Geriatr Psychiatry. 13(4): 235–239.

McCaddon A and Hudson P. 2007a. Alzheimer's disease, oxidative stress and B-vitamin depletion. Future Neurology. 2(5): 537–547.

McCaddon A, Hudson P, Abrahamsson L, Olofsson H and Regland B. 2001. Analogues, ageing and aberrant assimilation of vitamin B12 in Alzheimer's disease. Dement Geriatr Cogn Disord. 12(2): 133–137.

McCaddon A and Hudson PR. 2007b. Methylation and phosphorylation: a tangled relationship? Clin Chem. 53(6): 999–1000.

McCaddon A and Kelly CL. 1992. Alzheimer's disease: a 'cobalaminergic' hypothesis. Med Hypotheses. 37(3): 161–165.

McCaddon A and Kelly CL. 1994. Familial Alzheimer's disease and vitamin B12 deficiency. Age Ageing. 23(4): 334–337.

McCaddon A, McCracken C, Carr D, Hudson P, Moat S, Ellis R, Sequeira J and Quadros E. 2013. Transcobalamin receptor polymorphisms in the medical research council cognitive function and ageing study (MRC CFAS). J Inherit Metab Dis. 36(Suppl 1): S1–55.

McCaddon A and Miller JW. 2015. Assessing the association between homocysteine and cognition: reflections on Bradford Hill, meta-analyses, and causality. Nutr Rev. 73(10): 723–735.

McCaddon A, Regland B, Hudson P and Davies G. 2002. Functional vitamin B(12) deficiency and Alzheimer disease. Neurology. 58(9): 1395–1399.

McCaddon A, Tandy S, Hudson P, Gray R, Davies G, Hill D and Duguid J. 2004. Absence of macrocytic anaemia in Alzheimer's Disease. Clin Lab Haematol 26(4): 259–63.

McCracken C, Hudson P, Ellis R and McCaddon A. 2006. Methylmalonic acid and cognitive function in the medical research council cognitive function and ageing study. Am J Clin Nutr. 84(6): 1406–1411.

McEvoy AW, Fenwick JD, Boddy K and James OF. 1982. Vitamin B12 absorption from the gut does not decline with age in normal elderly humans. Age Ageing. 11(3): 180–183.

McGing P, Reed B, Weir DG and Scott JM. 1978. The effect of vitamin B12 inhibition *in vivo*: impaired folate polyglutamate biosynthesis indicating that 5-methyltetrahydropteroylglutamate is not its usual substrate. Biochem Biophys Res Commun. 82(2): 540–546.

Metz J, Bell AH, Flicker L, Bottiglieri T, Ibrahim J, Seal E, Schultz D, Savoia H and McGrath KM. 1996. The significance of subnormal serum vitamin B12 concentration in older people: a case control study. J Am Geriatr Soc. 44(11): 1355–1361.

Miller JW. 2002. Homocysteine, folate deficiency, and Parkinson's disease. Nutr Rev. 60(12): 410–413.

Miller JW, Garrod MG, Allen LH, Haan MN and Green R. 2009. Metabolic evidence of vitamin B12 deficiency, including high homocysteine and methylmalonic acid and low holotranscobalamin, is more pronounced in older adults with elevated plasma folate. Am J Clin Nutr. 90(6): 1586–1592.

Miller JW, Garrod MG, Rockwood AL, Kushnir MM, Allen LH, Haan MN and Green R. 2006. Measurement of total vitamin B12 and holotranscobalamin, singly and in combination, in screening for metabolic vitamin B12 deficiency. Clin Chem. 52(2): 278–285.

Miller JW, Green R, Ramos MI, Allen LH, Mungas DM, Jagust WJ and Haan MN. 2003a. Homocysteine and cognitive function in the Sacramento Area Latino Study on Aging. Am J Clin Nutr. 78(3): 441–447.

Miller JW, Selhub J, Nadeau MR, Thomas CA, Feldman RG and Wolf PA. 2003b. Effect of L-dopa on plasma homocysteine in PD patients: relationship to B-vitamin status. Neurology. 60(7): 1125–1129.

Morris MS, Jacques PF, Rosenberg IH and Selhub J. 2007. Folate and vitamin B12 status in relation to anemia, macrocytosis, and cognitive impairment in older Americans in the age of folic acid fortification. Am J Clin Nutr. 85(1): 193–200.

Morris MS, Selhub J and Jacques PF. 2012. Vitamin B12 and folate status in relation to decline in scores on the mini-mental state examination in the Framingham heart study. Journal of the American Geriatrics Society. 60(8): 1457–1464.

Muller T, Werne B, Fowler B and Kuhn W. 1999. Nigral endothelial dysfunction, homocysteine, and Parkinson's disease. Lancet. 354(9173): 126–127.

Muller T, Woitalla D, Hauptmann B, Fowler B and Kuhn W. 2001. Decrease of methionine and S-adenosylmethionine and increase of homocysteine in treated patients with Parkinson's disease. Neurosci Lett. 308(1): 54–56.

Nilsson K, Gustafson L, Faldt R, Anderson A, Vaara I, Nilsson R, Alm B and Hultberg B. 1997. Plasma methylmalonic acid in relation to serum cobalamin and plasma homocysteine in a psychogeriatric population and the effect of cobalamin treatment. Int J Geriatr Psychiatry. 12(1): 67–72.

Nilsson-Ehle H. 1998. Age-related changes in cobalamin (vitamin B12) handling. Implications for therapy. Drugs Aging. 12(4): 277–292.

Nilsson-Ehle H, Jagenburg R, Landahl S, Lindstedt S, Svanborg A and Westin J. 1991. Serum cobalamins in the elderly: a longitudinal study of a representative population sample from age 70 to 81. Eur J Haematol. 47(1): 10–16.

O'Suilleabhain PE, Sung V, Hernandez C, Lacritz L, Dewey RB, Jr., Bottiglieri T and Diaz-Arrastia R. 2004. Elevated plasma homocysteine level in patients with Parkinson disease: motor, affective, and cognitive associations. Arch Neurol. 61(6): 865–868.

Obeid R, Kasoha M, Knapp JP, Kostopoulos P, Becker G, Fassbender K and Herrmann W. 2007. Folate and methylation status in relation to phosphorylated tau protein(181P) and {beta}-amyloid(1-42) in cerebrospinal fluid. Clin Chem. 53(6): 1129–1136.

Oulhaj A, Refsum H, Beaumont H, Williams J, King E, Jacoby R and Smith AD. 2010. Homocysteine as a predictor of cognitive decline in Alzheimer's disease. Int J Geriatr Psychiatry. 25(1): 82–90.

Pennypacker LC, Allen RH, Kelly JP, Matthews LM, Grigsby J, Kaye K, Lindenbaum J and Stabler SP. 1992. High prevalence of cobalamin deficiency in elderly outpatients. J Am Geriatr Soc. 40(12): 1197–1204.

Prins ND, Den Heijer T, Hofman A, Koudstaal PJ, Jolles J, Clarke R and Breteler MM. 2002. Homocysteine and cognitive function in the elderly: the Rotterdam Scan Study. Neurology. 59(9): 1375–1380.

Ramsaransing GS, Fokkema MR, Teelken A, Arutjunyan AV, Koch M and De Keyser J. 2006. Plasma homocysteine levels in multiple sclerosis. J Neurol Neurosurg Psychiatry. 77(2): 189–192.

Ravaglia G, Forti P, Maioli F, Martelli M, Servadei L, Brunetti N, Porcellini E and Licastro F. 2005. Homocysteine and folate as risk factors for dementia and Alzheimer disease. Am J Clin Nutr. 82(3): 636–643.

Refsum H and Smith AD. 2003. Low vitamin B12 status in confirmed Alzheimer's disease as revealed by serum holotranscobalamin. J Neurol Neurosurg Psychiatry. 74(7): 959–961.

Regland B, Abrahamsson L, Blennow K, Gottfries CG and Wallin A. 1992. Vitamin B12 in CSF: reduced CSF/serum B12 ratio in demented men. Acta Neurol Scand. 85(4): 276–281.

Regland B, Abrahamsson L, Gottfries CG and Magnus E. 1990. Vitamin B12 analogues, homocysteine, methylmalonic acid, and transcobalamins in the study of vitamin B12 deficiency in primary degenerative dementia. Dementia. 1: 272–277.

Regland B and Gottfries CG. 1992. Slowed synthesis of DNA and methionine is a pathogenetic mechanism common to dementia in Down's syndrome, AIDS and Alzheimer's disease? Med Hypotheses. 38(1): 11–19.

Regland B, Gottfries CG, Oreland L and Svennerholm L. 1988. Low B12 levels related to high activity of platelet MAO in patients with dementia disorders. A retrospective study. Acta Psychiatr Scand. 78(4): 451–457.

Renvall MJ, Spindler AA, Ramsdell JW and Paskvan M. 1989. Nutritional status of free-living Alzheimer's patients. Am J Med Sci. 298(1): 20–27.

Reynolds EH. 1992. Multiple sclerosis and vitamin B12 metabolism. J Neuroimmunol. 40(2-3): 225–230.

Reynolds EH. 2002. Benefits and risks of folic acid to the nervous system. J Neurol Neurosurg Psychiatry. 72(5): 567–571.

Reynolds EH, Bottiglieri T, Laundy M, Crellin RF and Kirker SG. 1992. Vitamin B12 metabolism in multiple sclerosis. Arch Neurol. 49(6): 649–652.

Rosenberg IH and Miller J. 1992. Nutritional factors in physical and cognitive functions of elderly people. Am J Clin Nutr. 55: 1237s–1243s.

Russell JD, Batten FE and Collier J. 1900. Subacute combined degeneration of the cord. Brain. 23: 39–62.

Russo C, Morabito F, Luise F, Piromalli A, Battaglia L, Vinci A, Trapani L, V de MV, Morabito P, Condino F et al. 2008. Hyperhomocysteinemia is associated with cognitive impairment in multiple sclerosis. J Neurol. 255(1): 64–69.

Sachdev PS. 2005. Homocysteine and brain atrophy. Prog Neuropsychopharmacol Biol Psychiatry. 29(7): 1152–61.

Savage DG and Lindenbaum J. 1995. Neurological complications of acquired cobalamin deficiency: clinical aspects. Baillieres Clin Haematol. 8(3): 657–678.

Scalabrino G, Veber D and Mutti E. 2008. Experimental and clinical evidence of the role of cytokines and growth factors in the pathogenesis of acquired cobalamin-deficient leukoneuropathy. Brain Res Rev. 59(1): 42–54.

Scarlett JD, Read H and O'Dea K. 1992. Protein-bound cobalamin absorption declines in the elderly. Am J Hematol. 39(2): 79–83.

Scherer K. 2003. Images in clinical medicine. Neurologic manifestations of vitamin B12 deficiency. N Engl J Med. 348(22): 2208.

Schwaninger M, Ringleb P, Winter R, Kohl B, Fiehn W, Rieser PA and Walter-Sack I. 1999. Elevated plasma concentrations of homocysteine in antiepileptic drug treatment. Epilepsia. 40(3): 345–350.

Selhub J, Morris MS and Jacques PF. 2007. In vitamin B12 deficiency, higher serum folate is associated with increased total homocysteine and methylmalonic acid concentrations. Proc Natl Acad Sci USA. 104(50): 19995–20000.

Selhub J, Troen A and Rosenberg IH. 2010. B vitamins and the aging brain. Nutr Rev 68 Suppl. 2: S112–S118.

Seshadri S, Beiser A, Selhub J, Jacques PF, Rosenberg IH, D'Agostino RB, Wilson PW and Wolf PA. 2002. Plasma homocysteine as a risk factor for dementia and Alzheimer's disease. N Engl J Med. 346(7): 476–483.

Smith AD. 2006. Prevention of dementia: a role for B vitamins? Nutr Health. 18(3): 225–226.

Smith AD and Refsum H. 2012. Do we need to reconsider the desirable blood level of vitamin B12? J Intern Med. 271(2): 179–182.

Smith AD, Smith SM, de Jager C, Whitbread P, Johnston C, Agacinski G, Oulhaj A, Bradley KM, Jacoby R and Refsum H. 2010. Homocysteine-lowering by B-vitamins slows the rate of accelerated brain atrophy in mild cognitive impairment: a randomized controlled trial. PLoS One. 5(9): e12244.

Sontag E, Nunbhakdi-Craig V, Sontag JM, Diaz-Arrastia R, Ogris E, Dayal S, Lentz SR, Arning E and Bottiglieri T. 2007. Protein phosphatase 2A methyltransferase links homocysteine metabolism with tau and amyloid precursor protein regulation. J Neurosci. 27(11): 2751–2759.

Stabler SP, Lindenbaum J and Allen RH. 1997. Vitamin B12 deficiency in the elderly: current dilemmas. Am J Clin Nutr. 66(4): 741–749.

Stanger O, Fowler B, Piertzik K, Huemer M, Haschke-Becher E, Semmler A, Lorenzl S and Linnebank M. 2009. Homocysteine, folate and vitamin B12 in neuropsychiatric diseases: review and treatment recommendations. Expert Rev Neurother. 9(9): 1393–1412.

Tangney CC, Tang Y, Evans DA and Morris MC. 2009. Biochemical indicators of vitamin B12 and folate insufficiency and cognitive decline. Neurology. 72(4): 361–367.

Teunisse S, Bollen AE, van Gool WA and Walstra GJ. 1996. Dementia and subnormal levels of vitamin B12: effects of replacement therapy on dementia. J Neurol. 243(7): 522–529.

Teunissen CE, Killestein J, Kragt JJ, Polman CH, Dijkstra CD and Blom HJ. 2008. Serum homocysteine levels in relation to clinical progression in multiple sclerosis. J Neurol Neurosurg Psychiatry. 79(12): 1349–1353.

Teunissen CE, Van Boxtel MP, Jolles J, De Vente J, Vreeling F, Verhey F, Polman CH, Dijkstra CD and Blom HJ. 2005. Homocysteine in relation to cognitive performance in pathological and non-pathological conditions. Clin Chem Lab Med. 43(10): 1089–1095.

Thompson WG, Cassino C, Babitz L, Meola T, Berman R, Lipkin M, Jr. and Freedman M. 1989. Hypersegmented neutrophils and vitamin B12 deficiency. Hypersegmentation in B12 deficiency. Acta Haematol. 81(4): 186–191.

Toth C, Breithaupt K, Ge S, Duan Y, Terris JM, Thiessen A, Wiebe S, Zochodne DW and Suchowersky O. 2010. Levodopa, methylmalonic acid, and neuropathy in idiopathic Parkinson disease. Ann Neurol. 68(1): 28–36.

Triantafyllou N, Evangelopoulos ME, Kimiskidis VK, Kararizou E, Boufidou F, Fountoulakis KN, Siamouli M, Nikolaou C, Sfagos C, Vlaikidis N et al. 2008. Increased plasma homocysteine levels in patients with multiple sclerosis and depression. Ann Gen Psychiatry. 7: 17.

Tucker KL, Rich S, Rosenberg I, Jacques P, Dallal G, Wilson PW and Selhub J. 2000. Plasma vitamin B12 concentrations relate to intake source in the Framingham Offspring study. Am J Clin Nutr. 71(2): 514–522.

van Asselt DZ, De Groot LC, van Staveren WA, Blom HJ, Wevers RA, Biemond I and Hoefnagels WH. 1998. Role of cobalamin intake and atrophic gastritis in mild cobalamin deficiency in older Dutch subjects. Am J Clin Nutr. 68(2): 328–334.

van Asselt DZ, van den Broek WJ, Lamers CB, Corstens FH and Hoefnagels WH. 1996. Free and protein-bound cobalamin absorption in healthy middle-aged and older subjects. J Am Geriatr Soc. 44(8): 949–953.

Van Tiggelen CJM. 1983. Alzheimer's disease/alcohol dementia: association with zinc deficiency and cerebral vitamin B12 deficiency. J Orthomolecular Psychiatry. 13: 97–104.

Vogiatzoglou A, Oulhaj A, Smith AD, Nurk E, Drevon CA, Ueland PM, Vollset SE, Tell GS and Refsum H. 2009. Determinants of plasma methylmalonic acid in a large population: implications for assessment of vitamin B12 status. Clin Chem. 55(12): 2198–2206.

Weir DG and Scott JM. 1999. Brain function in the elderly: role of vitamin B12 and folate. Br Med Bull. 55(3): 669–682.

Yao Y, Yao SL, Yao SS, Yao G and Lou W. 1992. Prevalence of vitamin B12 deficiency among geriatric outpatients. J Fam Pract. 35(5): 524–528.

Zoccolella S, Lamberti SV, Iliceto G, Santamato A, Lamberti P and Logroscino G. 2010. Hyperhomocysteinemia in L-dopa treated patients with Parkinson's disease: potential implications in cognitive dysfunction and dementia? Curr Med Chem. 17(28): 3253–3261.

Zoccolella S, Simone IL, Lamberti P, Samarelli V, Tortelli R, Serlenga L and Logroscino G. 2008. Elevated plasma homocysteine levels in patients with amyotrophic lateral sclerosis. Neurology. 70(3): 222–225.

Zylberstein DE, Lissner L, Bjorkelund C, Mehlig K, Thelle DS, Gustafson D, Ostling S, Waern M, Guo X and Skoog I. 2011. Midlife homocysteine and late-life dementia in women. A prospective population study. Neurobiol Aging. 32(3): 380–386.

8

The Role of Cobalamin in the Central and Peripheral Nervous Systems
Mechanistic Insights
Elena Mutti

Acquired cobalamin (Cbl)-deficiency in human induces a neuropathy that is known as subacute combined degeneration or Cbl-deficient neuropathy. This neurological disease affects both the central-(CNS) and the peripheral nervous systems (PNS). The peripheral nerves and different columns of the spinal cord are particularly affected by the diseases. The histopathological hallmarks of Cbl-deficient neuropathy in the CNS are: (i) a diffuse but uneven vacuolation (the so-called "spongiform vacuolation") of the white matter; (ii) intramyelinic and interstitial edema of the white matter of the CNS (Agamanolis et al. 1978; Duffield et al. 1990; Scalabrino et al. 1990; Tredici et al. 1998); and (iii) reactive astrogliosis in both the white and grey matter (Agamanolis et al. 1978; Scalabrino et al. 1990; Tredici et al. 1998). The histopathological lesions affect especially the posterior and lateral columns of the spinal cord, but similar lesions have also been observed in brain white matter (i.e., leukoencephalopathy) (Chatterjee et al. 1996; Scalabrino 2001; Scherer 2003).

The histopathological and ultrastructural hallmarks of Cbl-deficient neuropathy in the PNS (peripheral neuropathy or polyneuropathy) are intramyelinic and interstitial edema, and gliosis (Kumar 2007; Scalabrino et al. 2008). Electrophysiological abnormalities have also been observed in the

Department of Biomedical Sciences for Health, University of Milan, Via Mangiagalli, 31, 20133 Milano, Italy.
Email: elena.mutti@libero.it

Table 1. Morphological abnormalities in the nervous system that have been associated with Cbl-deficiency.

CNS	References
Spongiform vacuolation of the white matters (especially the posterior and lateral columns of spinal cord)	Scalabrino et al. 1990, 1995
Intramyelinic and interstitial edema of the white matters (especially the posterior and lateral columns of spinal cord)	Agamanolis et al. 1978; Duffield et al. 1990; Scalabrino et al. 1990; Tredici et al. 1998
Reactive astrogliosis of the white and grey matters (especially the posterior and lateral columns of spinal cord)	Agamanolis et al. 1978; Scalabrino et al. 1990; Tredici et al. 1998
PNS	
Intramyelinic and interstitial edema	Kumar 2007; Scalabrino et al. 2008
Gliosis	Kumar 2007; Scalabrino et al. 2008
Electrophysiological abnormalities	Roos 1978; Fine et al. 1990; Saperstein and Barohn 2002

PNS (Roos 1978; Fine et al. 1990; Saperstein and Barohn 2002). Table 1 shows a summary of the most relevant morphological lesions that are associated with Cbl-deficient neuropathy in the nervous system.

The neurologic features of Cbl-deficient neuropathy typically include a spastic paraparesis or tetraparesis, extensor plantar response, and impaired perception of position and vibration. The involvement of the posterior and lateral columns of the spinal cord is responsible for the impairment of position sense, paraparesis and tetraparesis (Briani et al. 2013). Almost all patients have loss of vibratory sensation, often associated with diminished proprioception and cutaneous sensation and Romberg sign (Briani et al. 2013).

Besides neuropathy, Cbl deficiency has also been negatively related to cognitive function in healthy elderly subjects. Symptoms include slow mentation, memory impairment, and attention deficits (Briani et al. 2013). Several mechanism(s) have been discussed to explain the pathogenesis of Cbl-deficient neuropathy or cognitive dysfunction.

One of the key mechanisms is related to the two reactions mediated by Cbl as a cofactor. Cbl deficiency reduces the activity of the mitochondrial methylmalonyl Coenzyme A mutase and the cytosolic methionine synthase and thereby it causes accumulation of methylmalonic acid (MMA) and homocysteine, respectively. High concentrations of plasma homocysteine have been demonstrated in the brain of Cbl-deficient pig with symptoms of subacute combined degeneration (Cbl was inactivated by nitrous oxide) (Surtees 1993; Weir and Scott 1995). Moreover, hyperhomocysteinemia is associated with elevated S-adenosylhomocysteine (SAH) and low S-adenosylmethionine (SAM). SAM is an important methyl donor for numerous methyltransferases in the cell and SAH is a competitive inhibitor of these enzymes. Therefore, hyperhomocysteinemia is expected to cause hypomethylation in the nervous system and to have functional consequences.

The myelin sheath is a greatly extended and modified plasma membrane wrapped around the nerve axon in a spiral fashion. The myelin membranes originate from and are a part of the Schwann cells in the PNS and the oligodendroglial cells in the CNS. Myelin sheatis lipid-rich (approximately 80% lipid and 20% protein). 30% of the protein fraction constitutes myelin basic protein (Kim et al. 1997) and acts as electrical insulator thus facilitating conduction in axons (Morell and Quarles 1999). Abnormal changes in myelin structure and functions have been reported in Cbl deficiency and they were also related to alterations in methylation sites or the content of fatty acids in myelin protein or phospholipids contents, respectively.

Methylation of myelin basic protein plays a key role in maintaining myelin structure integrity and stability. Myelin basic protein is methylated at a single ariginine residue (arg-107) by a protein methylase (Protein arginine N-methyltransferase 1). The last step is SAM dependent (Bolander-Gouaille and Bottiglieri 2007). A decrease in SAM supply causes impairment in Arg methylation in myelin basic protein and thereby impairs myelin functions (Surtees 1993; Tefferi and Pruthi 1994; Weir and Scott 1995; Bottiglieri 1996).

In physiological conditions, the turn over of myelin lipids is rapid and myelin replacement is heavily dependent on fatty acid synthesis (Beck 1991). Normally, the *de novo* synthesis of fatty acids consists in the repetitive sequential addition of two-carbon units deriving from malonylCoA (three carbon-molecules). MMA that accumulates in a Cbl-deficiency is a branched four carbon-molecule that may compete with malonylCoA and thus it generates abnormal branched fatty acids or fatty acids with an odd number of carbon atoms (Beck 1991). These abnormal fatty acids could subsequently be incorporated into myelin and cause too fragile structure (Beck 1991; Scalabrino 2001; Naidich and Ho 2005). *In vivo* and *in vitro* studies have demonstrated the presence of abnormal branched fatty acids or fatty acids with an odd number of carbon atoms in Cbl-deficiency conditions. Nevertheless, it remains unproven whether the accumulation of such abnormal fatty acids in myelin directly accounts for myelopathy and polineuropathy of Cbl deficiency (Beck 1991; Scalabrino 2001; Naidich and Ho 2005), or whether this process is modifiable by administering Cbl that lowers MMA.

Elevated concentrations of homocysteine in Cbl-deficient patients can have several direct or indirect neurotoxic effects such as overstimulation of the N-methyl-D-aspartate receptors (Lipton et al. 1997; Diaz-Arrastia 2000). Activation of the N-methyl-D-aspartate receptors causes neuronal damage mediated by excessive Ca^{2+} influx and reactive oxygen species generation (Lipton et al. 1997).

Several lines of evidence suggested that increased plasma concentrations of MMA and homocysteine are not sufficient alone to cause the CNS lesions that are typically found in Cbl-deficiency. For example, the severity of the neuropathological damage in the white matter of spinal cord in a rat model of Cbl-deficiency did not correlate with the progressive accumulation of MMA and homocysteine. In humans with isolated hyperhomocysteinemia of any origin (e.g., due to defects in genes encoding for enzymes of homocysteine metabolism, without MMA elevation) or those with isolated methylmalonic acidemia not related to Cbl deficiency (due to an inherited methylmalonyl-CoA mutase deficiency) do not show the typical Cbl-deficient neuropathy (reviewed in Scalabrino 2005). This suggested that other mechanisms are involved.

Involvement of some cytokines and some neurotrophic growth factors in Cbl-deficient neuropathy have been explored because they are indispensable for proper development and maintenance of the nervous system, and their abnormalities are involved in several neuropathies.

Based on rat models, Scalabrino et al., argued that a derangement in the production of some cytokines and/or growth factors, due to Cbl fading, may be one of the central mechanisms in the pathogenesis of Cbl-deficient neuropathy (Scalabrino 2001, 2005). Cbl has been shown to down-regulate: (i) tumor necrosis factor (TNF)-α and nerve growth factor (NGF) levels and/or synthesis (Buccellato et al. 1999; Peracchi et al. 2001; Scalabrino et al. 2004; Scalabrino et al. 2006); (ii) the levels of the soluble(s) CD40:sCD40 ligand dyad (belonging to the TNF-α:TNF-α-receptor superfamily) (Veber et al. 2006). On the contrary, Cbl up-regulates: (i) epidermal growth factor (EGF) levels (Peracchi et al. 2001; Scalabrino et al. 1999; Scalabrino et al. 2004; Mutti et al. 2013); and (ii) interleukin (IL)-6 levels (Scalabrino et al. 2002).

Subsequently, it has been proved that the imbalance of cytokines and growth factor levels induced by Cbl deficiency in rat central nervous system is not marginal, but it is essential to the pathogenesis of the typical myelinolitic lesions in the white matter of spinal cord of acquired Cbl-deficient neuropathy.

In fact, in two rat models of Cbl-deficiency the typical myelinolitic lesions are cured by means of subcutaneous injections of the down-regulated proteins (EGF or IL-6) or by using monoclonal antibodies against the up-regulated proteins (TNF-α or NGF) (Buccellato et al. 1999; Scalabrino et al. 2000, 2006). In line with the proposed role for down regulation of EGF and up regulation of TNF-α or NGF in Cbl-deficient rat models, the typical myelinolitic lesions in the white matter of spinal cord of non-deficient rats have been reproduced by means of subcutaneous injections of TNF-α, NGF or anti-EGF antibodies (Buccellato et al. 1999; Scalabrino et al. 2000, 2006) showing that Cbl-deficiency causes neurological damages via cytokine and growth factors. Taken together, these rat models have shown that abnormally lower level of myelinotrophic cytokines or growth factors and higher level of neurotoxic cytokines or growth factors can be deleterious for myelin.

It has been hypothesized that the dysregulation in the synthesis and secretion of aforementioned cytokines and growth factors is induced by the ultrastructural abnormalities observed in glia cells (gliosis) from Cbl-deficient humans and animals, but if and how this comes about is still unclear (Scalabrino 2009).

Recently, another molecule attracted the attention of the researchers, the cellular prion protein (PrP^C). PrP^C is a sialoglycoprotein that appears to be involved in crucial functions in CNS and PNS, as it maintain the myelinated fibers, synaptic transduction, signal transduction, neuroprotection against some CNS injuries, and copper binding. PrP^C could be also involved in the pathogenesis of Cbl-deficient neuropathy. This suggestion is based on the fact that both Cbl-deficient neuropathy and diseases with prion abnormalities present the same typical lesions: myelin vacuolation and reactive astrocytosis (Scalabrino et al. 2011, 2012). Moreover, it has been demonstrated that Cbl contributes to maintain normal PrP^C levels in nervous system and abnormal levels of PrP^C are associated with myelin lesions and abnormal nerve conduction velocity (Scalabrino et al. 2011, 2012).

It is interesting to note that the decreased availability of the methyl donor SAM (consequence of reduced activity of methionine synthase) and the dysregulation of cytokines and/or growth factors, both observed in a Cbl-deficient status, are interrelated. In fact, epigenetic regulations, through DNA methylation, play a central role in linking between Cbl deficiency and the dysregulation of cytokine and/or growth factor. In line with a hypomethylation condition under Cbl-deficiency, SAM added to a murine monocyte cell line caused down regulation on the mRNA and protein level of members of the TNF-α family that are overexpressed under Cbl-deficiency conditions (Watson et al. 1999; Song et al. 2005).

Another potential mechanism that has been discussed in relation to Cbl deficiency neuropathy is the increase in Cbl analogues or the non-active Cbl corrinoides. These compounds share a similar chemical structure with Cbl and they are able to bind to one major Cbl transporter, haptocorrin. However, Cbl analogues can not bind Cbl-dependent enzymes and they are therefore unable to serve as cofactors for methylmalonyl Coenzyme A mutase and methionine synthase (Kolhouse et al. 1978). In line with this hypothesis, high concentrations of Cbl analogus have been reported in a group of patients with low Cbl levels and primarily and/or only neurologic symptoms as compared to patients with only and/or primarily hematologic abnormalities (Carmel et al. 1988). Likewise, patients affected by other neurological disorders like Alzheimer's disease had significantly higher ratio of Cbl analogue/corrinoid than control subjects (0.36 in patients with Alzheimer's disease vs 0.26 in the controls) (McCaddon et al. 2001). Therefore, increased analogues levels could derange the Cbl transport and metabolism carrying inert Cbl that is not available for the cell metabolism. Obviously, more studies are required to prove a potential role of this mechanism in neurological damage associated with Cbl deficiency.

Finally, Cbl shows an antioxidant and anti-inflammatory role (Manzanares and Hardy 2010) that could also be implicated in the pathogenesis of the Cbl-linked neuropathies. For example, thiolatoCbl derivatives have been found to protect against oxidative stress (induced by homocysteine or H_2O_2) in a cellular model (Birch et al. 2009). From the molecular point of view it has been found that Cbl reacts with superoxide at rates approaching those of superoxide dismutase itself, suggesting a probable mechanism by which Cbl modulates redox homeostasis and protects against oxidative stress (Suarez-Moreira et al. 2009). One potential mechanism explaining Cbl ameliorative effect on inflammation is its regulatory effect on all three nitric oxide synthases (NOS) (inducible (iNOS), endothelial (eNOS) and neuronal nitric oxide synthases (nNOS)) that promotes the more benign species and effects of nitric oxide (Wheatly 2006, 2007a, 2007b).

An *in vivo* study recently confirms this data demonstrating that hydroxoCbl produces in mice a complex, time- and organ-dependent, selective regulation of NOS/·nitric oxide during endotoxaemia (partially inhibits hepatic, but not lung, iNOS mRNA and promotes lung eNOS mRNA, but attenuates the hepatic rise in eNOS mRNA, whilst paradoxically promoting high iNOS/eNOS protein translation, but relatively moderate nitric oxide production), corollary regulation of downstream inflammatory mediators, and increased survival

Table 2. Principal potential mechanism(s) of the pathogenesis of Cbl-linked neuropathies.

Main mechanism	Mediators	Probable effect
Reduction of the Cbl biochemical functions	↑ homocysteine	Overstimulation N-methyl-D-aspartate receptors → neuronal damage
	↑MMA	Abnormal fatty acids → fragile myelin structure
	↓SAM	Decreased myelin methylation → fragile myelin structure
Derangement in the production of some cytokines and/or growth factors	↑TNF-α ↑NGF ↑sCD40:sCD40L	Myelinotoxicity
	↓ EGF ↓ IL-6	Decreased myelinotrophism
Derangement in the PrP^C level	PrP^C	Unknown
Increased inactive corrinoides that have no cofactor functions	↑corrinoid	Unknown
Reduction of antioxidant role	↑ superoxide	Oxidative stress
Reduction of anti-inflammatory role	NOS/NO	Chronic inflammation

Cbl: cobalamin; EGF: epidermal growth factor; IL-6: interleukin-6; L: ligand; MMA: methylmalonic acid; NGF: nerve growth factor; NO: nitric oxide; NOS: nitric oxide synthases; PrP^C: cellular prion protein; s: soluble; SAM: S-Adenosyl methionine; TNF-α: tumor necrosis factor-α; ↑ Increased level; ↓Decreased level.

(Sampaio et al. 2013). See Table 2 for a summary of the principal theories of the pathogenesis of Cbl-deficient neuropathy.

The role of oxidative stress in the pathogenesis of Alzheimer's disease has been repeatedly reported. Several studies have confirmed the association between low concentrations of plasma Cbl, elevated homocysteine and Alzheimer's disease (for a review Lopes da Silva et al. 2014). Cbl deficiency causes hyperhomocysteinemia that is itself associated with oxidative stress. Chronic cerebral oxidative stress may in turn cause depletion of B-vitamins thus explaining the association between Alzheimer's disease and low Cbl (McCaddon 2013). Though studies showing association do not prove causality, intervention studies have shown that a high-dose of Cbl plus folic acid and vitamins B_6 reduces not only homocysteine, but also cerebral atrophy by approximately 7 fold in those gray matter regions specifically vulnerable to the Alzheimer's process, including the medial temporal lobe (Douaud et al. 2013). A further intriguing study using cellular and *in-vivo* models of neurodegeneration suggested a decrease in release of Cbl from cell lysosomes after its uptake via the Cbl-receptor. Cbl utilization in the cell depends on its efficient release from the lysosomal compartment and subsequent delivery to the cytosol and mitochondria. Lysosomal acidification is defective in Alzheimer's disease and lysosomal proteolysis is disrupted in cellular model of Alzheimer diseases (presenilin 1 mutation). So Alzheimer's disease related lysosomal dysfunction may impair lysosomal Cbl transport with subsequent entrapment of Cbl in lysosome (Zhao et al. 2015).

Besides Cbl-deficient neuropathy and Alzheimer's disease, multiple sclerosis could also present Cbl abnormalities, although the exact pathogenetic role of Cbl in the diseases is unclear (Scalabrino 2005; McCaddon 2013). Concentrations of Cbl in body fluids from patients with multiple sclerosis revealed no conclusive results. While concentrations of Cbl were decreased in serum of patients with multiple sclerosis (Zhu et al. 2011), they were increased in cerebrospinal fluid in some subgroups of patients (patients with relapsing-remitting clinical courses) (Scalabrino et al. 2010), or even showed no alterations in serum Cbl level inother studies (Najafi et al. 2012). The current view is that multiple sclerosis that has several clinical phenotypes can show different associations with Cbl deficiency. Finally, increased homocysteine levels in patients with Parkinson's disease could be partly related to low cobalamin status or higher requirements, in particular those treated with L-dopa or Catechol O-methyltransferase inhibitors who have higher requirements for methyl groups from SAM (Qureshi et al. 2008; McCaddon 2013).

Summary and Conclusions

It is now certain the involvement of Cbl abnormalities in various neurological diseases and therefore the key role of Cbl in the proper functioning of the nervous system. This chapter summarizes the main mechanisms to explain the pathogenesis of Cbl-linked neuropathies and highlights a complicated Cbl pathway and multiple Cbl roles, often interrelated each other (i.e., coenzyme, regulatory of cytokines and growth factor levels, antioxidant and anti-inflammatory). In recent years, much has been discovered but to dispel last doubts, we argue that is necessary a new approach using new experimental models (i.e., mice treated with antivitamins or knockout mice), characterized by a selectively block of single Cbl metabolic actions and so able to define the exact role of each Cbl products.

Acknowledgments

E.M. would like to thank Prof. Giuseppe Scalabrino (University of Milan), who has introduced her in vitamin B12 field and involved her passionately in the vitamin B12 amazing and intricate pathogenetic role.

Keywords: Alzheimer's disease, antioxidant, anti-inflammatory, Cbl analogues, cytokine, growth factor, multiple sclerosis, nervous system, homocysteine, methylmalonic acid, Parkinson's disease

Abbreviations

Cbl	:	cobalamin
CNS	:	central nervous system
EGF	:	epidermal growth factor
IL-6	:	interleukin-6
MMA	:	methylmalonic acid
NGF	:	nerve growth factor
NOS	:	nitric oxide synthases
eNOS	:	endothelial NOS
iNOS	:	inducible NOS
nNOS	:	neuronal NOS
PNS	:	peripheral nervous system
PrP^C	:	cellular prion protein
s	:	soluble
SAH	:	S-Adenosylhomocysteine
SAM	:	S-Adenosylmethionine
TNF-α	:	tumor necrosis factor-α

References

Agamanolis DP, Victor M, Harris JW, Hines JD, Chester EM and Kark JA. 1978. An ultrastructural study of subacute combined degeneration of the spinal cord in vitamin B12-deficient rhesus monkeys. J Neuropathol Exp Neurol. 37(3): 273–299.
Beck WS. 1991. Neuropsychiatric consequences of cobalamin deficiency. Adv Int Med. 36: 33–56.

Birch CS, Brasch NE, McCaddon A and Williams JH. 2009. A novel role for vitamin B12: Cobalamins are intracellular antioxidants *in vitro*. Free Radical Biol Med. 47: 184–188.

Bolander-Gouaille C and Bottiglieri T. 2007. How can Homocysteine be Neurotoxic? *In*: Homocysteine. Related Vitamins and Neuropsychiatric Disorders. Springer-Verlag Paris.

Bottiglieri T. 1996. Folate, vitamin B12, and neuropsychiatric disorders. Nutr Rev. 54(12): 382–90.

Briani C, Dalla Torre C, Citton V, Manara R, Pompanin S, Binotto G and Adami F. 2013. Cobalamin deficiency: clinical picture and radiological findings. Nutrients. 5(11): 4521–39.

Buccellato FR, Miloso M, Braga M, Nicolini G, Morabito A, Pravettoni G, Tredici G and Scalabrino G. 1999. Myelinolytic lesions in spinal cord of cobalamin-deficient rats are TNF-alpha-mediated. FASEB J. 13(2): 297–304.

Carmel R, Karnaze DS and Weiner JM. 1988. Neurologic abnormalities in cobalamin deficiency are associated with higher cobalamin "analogue" values than are hematologic abnormalities. J Lab Clin Med. 111(1): 57–62.

Chatterjee A, Yapundich R, Palmer CA, Marson DC and Mitchell GW. 1996. Leukoencephalopathy associated with cobalamin deficiency. Neurology. 46(3): 832–834.

Diaz-Arrastia R. 2000. Homocysteine and neurologic disease. Arch Neurol. 57(10): 1422–7.

Douaud G, Refsum H, de Jager CA, Jacoby R, Nichols TE, Smith SM and Smith AD. 2013. Preventing Alzheimer's disease-related gray matter atrophy by B-vitamin treatment. Proc Natl Acad Sci U S A. 110(23): 9523–8.

Duffield MS, Phillips JI, Vieira-Makings E, Van der Westhuyzen J and Metz J. 1990. Demyelinisation in the spinal cord of vitamin B12 deficient fruit bats. Comp Biochem Physiol C. 96(2): 291–297.

Fine EJ, Soria E, Paroski MW, Petryk D and Thomasula L. 1990. The neurophysiological profile of vitamin B12 deficiency. Muscle Nerve. 13(2): 158–164.

Kim S, Lim IK, Park GH and Paik WK. 1997. Biological methylation of myelin basic protein: enzymology and biological significance. Int J Biochem Cell Biol. 29(5): 743–51.

Kolhouse JF, Kondo H, Allen NC, Podell E and Allen RH. 1978. Cobalamin analogues are present in human plasma and can mask cobalamin deficiency because current radioisotope dilution assays are not specific for true cobalamin. N Engl J Med. 12: 785–792.

Kumar N. 2007. Nutritional neuropathies. Neurol Clin. 25(1): 209–255.

Lipton SA, Kim WK, Choi YB, Kumar S, D' Emilia DM, Rayudu PV, Arnelle DR and Stamler JS. 1997. Neurotoxicity associated with dual actions of homocysteine at the N-methyl-D-aspartate receptor. Proc Natl Acad Sci U S A. 94(11): 5923–8.

Lopes da Silva S, Vellas B, Elemans S, Luchsinger J, Kamphuis P, Yaffe K, Sijben J, Groenendijk M and Stijnen T. 2014. Plasma nutrient status of patients with Alzheimer's disease: Systematic review and meta-analysis. Alzheimers Dement. 10(4): 485–502.

Manzanares W and Hardy G. 2010. Vitamin B12: the forgotten micronutrient for critical care. Curr Opin Clin Nutr Metab Care. 13(6): 662–8.

McCaddon A, Hudson P, Abrahamsson L, Olofsson H and Regland B. 2001. Analogues, ageing and aberrant assimilation of vitamin B12 in Alzheimer's disease. Dement Geriatr Cogn Disord. 12(2): 133–7.

McCaddon A. 2013. Vitamin B12 in neurology and ageing; clinical and genetic aspects. Biochimie. 95(5): 1066–76.

Morell P and Quarles RH. 1999. The Myelin Sheath. *In*: George J Siegel (ed.). Basic Neurochemistry: Molecular, Cellular and Medical. 6th edition. Lippincott-Raven, Philadelphia.

Mutti E, Lildballe DL, Kristensen L, Birn H and Nexo E. 2013. Vitamin B12 dependent changes in mouse spinal cord expression of vitamin B12 related proteins and the epidermal growth factor system. Brain Res. 1503: 1–6.

Naidich MJ and Ho SU. 2005. Case 87: Subacute combined degeneration. Radiology. 237(1): 101–5.

Najafi MR, Shaygannajad V, Mirpourian M and Gholamrezaei A. 2012. Vitamin B12 deficiency and multiple sclerosis; is there any association? Int J Prev Med. 3(4): 286–9.

Peracchi M, Bamonti Catena F, Pomati M, De Franceschi M and Scalabrino G. 2001. Human cobalamin deficiency: alterations in serum tumour necrosis factor-alpha and epidermal growth factor. Eur J Haematol. 67(2): 123–7.

Qureshi GA, Qureshi AA, Devrajani BR, Chippa MA and Syed SA. 2008. Is the deficiency of vitamin B12 related to oxidative stress and neurotoxicity in Parkinson's patients? CNS Neurol Disord Drug Targets. 7(1): 20–7.

Roos D. 1978. Neurological complications in patients with impaired vitamin B12 absorption following partial gastrectomy. Acta Neurol Scand. 59(suppl 69): 1–77.

Sampaio AL, Dalli J, Brancaleone V, D'Acquisto F, Perretti M and Wheatley C. 2013. Biphasic modulation of NOS expression, protein and nitrite products by hydroxocobalamin underlies its protective effect in endotoxemic shock: downstream regulation of COX-2, IL-1β, TNF-α, IL-6, and HMGB1 expression. Mediators Inflamm. 741804.

Saperstein DS and Barohn RJ. 2002. Peripheral neuropathy due to cobalamin deficiency. Curr Treat Options Neurol. 4: 197–201.

Scalabrino G. 2001. Subacute combined degeneration one century later. The neurotrophic action of cobalamin (vitamin B12) revisited. J Neuropathol Exp Neurol. 60(2): 109–120.

Scalabrino G. 2005. Cobalamin (vitamin B12) in subacute combined degeneration and beyond: traditional interpretations and novel theories. Exp Neurol. 192(2): 463–79.

Scalabrino G. 2009. The multi-faceted basis of vitamin B12 (cobalamin) neurotrophism in adult central nervous system: Lessons learned from its deficiency. Prog Neurobiol. 88(3): 203–20.

Scalabrino G, Monzio-Compagnoni B, Ferioli ME, Lorenzini EC, Chiodini E and Candiani R. 1990. Subacute combined degeneration and induction of ornithine decarboxylase in spinal cords of totally gastrectomized rats. Lab Invest. 62(3): 297–304.

Scalabrino G, Lorenzini EC, Monzio-Compagnoni B, Colombi RP, Chiodini E and Buccellato FR. 1995. Subacute combined degeneration in the spinal cords of totally gastrectomized rats. Ornithine decarboxylase induction, cobalamin status, and astroglial reaction. Lab Invest. 72(1): 114–23.

Scalabrino G, Nicolini G, Buccellato FR, Peracchi M, Tredici G, Manfridi A and Pravettoni G. 1999. Epidermal growth factor as a local mediator of the neurotrophic action of vitamin B12 (cobalamin) in the rat central nervous system. FASEB J. 13(14): 2083–90.

Scalabrino G, Tredici G, Buccellato FR and Manfridi A. 2000. Further evidence for the involvement of epidermal growth factor in the signaling pathway of vitamin B12 (cobalamin) in the rat central nervous system. J Neuropathol Exp Neurol. 59(9): 808–14.

Scalabrino G, Corsi MM, Veber D, Buccellato FR, Pravettoni G, Manfridi A and Magni P. 2002. Cobalamin (vitamin B12) positively regulates interleukin-6 levels in rat cerebrospinal fluid. J Neuroimmunol. 127(1-2): 37–43.

Scalabrino G, Carpo M, Bamonti F, Pizzinelli S, D'Avino C, Bresolin N, Meucci G, Martinelli V, Comi GC and Peracchi M. 2004. High tumor necrosis factor-alpha levels in cerebrospinal fluid of cobalamin-deficient patients. Ann Neurol. 56(6): 886–90.

Scalabrino G, Mutti E, Veber D, Aloe L, Corsi MM, Galbiati S and Tredici G. 2006. Increased spinal cord NGF levels in rats with cobalamin (vitamin B12) deficiency. Neurosci Lett. 396(2): 153–8.

Scalabrino G, Veber D and Mutti E. 2008. Experimental and clinical evidence of the role of cytokines and growth factors in the pathogenesis of acquired cobalamin-deficient leukoneuropathy. Brain Res Rev. 59: 42–54.

Scalabrino G, Galimberti D, Mutti E, Scalabrini D, Veber D, De Riz M, Bamonti F, Capello E, Mancardi GL and Scarpini E. 2010. Loss of epidermal growth factor regulation by cobalamin in multiple sclerosis. Brain Res. 1333: 64–71.

Scalabrino G, Mutti E, Veber D, Rodriguez Menendez V, Novembrino C, Calligaro A and Tredici G. 2011. The octapeptide repeat PrP^C region and cobalamin-deficient polyneuropathy of the rat. Muscle Nerve. 44(6): 957–67.

Scalabrino G, Veber D, Mutti E, Calligaro A, Milani S and Tredici G. 2012. Cobalamin (vitamin B12) regulation of PrP^C, PrP^C-mRNA and copper levels in rat central nervous system. Exp Neurol. 233(1): 380–90.

Scherer K. 2003. Images in clinical medicine. Neurologic manifestations of vitamin B12 deficiency. N Engl J Med. 348(22): 2208.

Song Z, Uriarte S, Sahoo R, Chen T, Barve S, Hill D and McClain C. 2005. S-adenosylmethionine (SAMe) modulates interleukin-10 and interleukin-6, but not TNF, production via the adenosine (A2) receptor. Biochim Biophys Acta. 1743(3): 205–13.

Suarez-Moreira E, Yun J, Birch CS, Williams JH, McCaddon A and Brasch NE. 2009. Vitamin B12 and redox homeostasis: cob(II)alamin reacts with superoxide at rates approaching superoxide dismutase (SOD). J Am Chem Soc. 131: 15078–15079.

Surtees R. 1993. Biochemical pathogenesis of subacute combined degeneration of the spinal cord and brain. J Inherit Metab Dis. 16(4): 762–70.

Tefferi A and Pruthi RK. 1994. The biochemical basis of cobalamin deficiency. Mayo Clin Proc. 69(2): 181–6.

Tredici G, Buccellato FR, Cavaletti G and Scalabrino G. 1998. Subacute combined degeneration in totally gastrectomized rats: an ultrastructural study. J Submicrosc Cytol Pathol. 30(1): 165–173.

Veber D, Mutti E, Galmozzi E, Cedrola S, Galbiati S, Morabito A, Tredici G, La Porta CA and Scalabrino G. 2006. Increased levels of the CD40:CD40 ligand dyad in the cerebrospinal fluid of rats with vitamin B12 (cobalamin)-deficient central neuropathy. J Neuroimmunol. 176(1-2): 24–33.

Watson WH, Zhao Y and Chawla RK. 1999. S-adenosylmethionine attenuates the lipopolysaccharide-induced expression of the gene for tumour necrosis factor alpha. Biochem J. 342(Pt 1): 21–5.

Weir DG and Scott JM. 1995. The biochemical basis of the neuropathy in cobalamin deficiency. Baillieres Clin Haematol. 8(3): 479–97.

Wheatley C. 2006. A scarlet pimpernel for the resolution of inflammation? The role of supra-therapeutic doses of cobalamin, in the treatment of systemic inflammatory response syndrome (SIRS), sepsis, severe sepsis, and septic or traumatic shock. Med Hypotheses. 67(1): 124–142.

Wheatley C. 2007a. The return of the Scarlet Pimpernel: cobalamin in inflammation II - cobalamins can both selectively promote all three nitric oxide synthases (NOS), particularly iNOS and eNOS, and, as needed, selectively inhibit iNOS and nNOS. J Nutr Environ Med. 16(3-4): 181–211.

Wheatley C. 2007b. Cobalamin in inflammation III—glutathionylcobalamin and thylcobalamin/adenosylcobalamin coenzymes: the sword in the stone? How cobalamin may directly regulate the nitric oxide synthases. J Nutr Environ Med. 16(3-4): 212–226.

Zhao H, Li H, Ruberu K and Garner B. 2015. Impaired lysosomal cobalamin transport in Alzheimer's disease. J Alzheimers Dis. 43(3): 1017–30.

Zhu Y, He ZY and Liu HN. 2011. Meta-analysis of the relationship between homocysteine, vitamin B12, folate, and multiple sclerosis. J Clin Neurosci. 18(7): 933–8.

9

Laboratory Markers and Diagnosis of Cobalamin Deficiency

Rima Obeid

1. Introduction

Cobalamin (Cbl, vitamin B12) deficiency causes megaloblastic anaemia and/or neurological disorders. Mild to moderate deficiency conditions are associated with several age-related diseases (i.e., neuropathy, dementia, brain atrophy, etc.). Though this association cannot be definitely regarded as causal, it can worsen the clinical picture. The clinical symptoms of cobalamin deficiency are considered late manifestations that can be irreversible. The availability of modern biomarkers and the cost-effective prevention via supplementation have dramatically reduced clinically manifested cobalamin deficiency. The necessity for prevention instead of treatment has shifted the focus to "subclinical vitamin B12 deficiency" that is expressed as biochemical disturbances in B12-dependent pathways in the absence of severe clinical symptoms.

Acquired cobalamin deficiency can be caused by a low intake, loss or an inability to absorb the vitamin (Figure 1). Low cobalamin status (i.e., biochemical abnormalities) is common in populations where the consumption of animal products is low. In Western countries, cobalamin deficiency is mostly caused by diseases in the gastrointestinal tract, such as disorders that affect the production or function of intrinsic factor or classical pernicious anaemia. Moreover, intestinal diseases that affect the terminal ileum and thereby the

Aarhus Institute of Advanced Studies, University of Aarhus, Høegh-Guldbergs Gade 6B, Building 1632, Dk-8000, Aarhus C, Denmark.
Email: rima.obeid@uks.eu

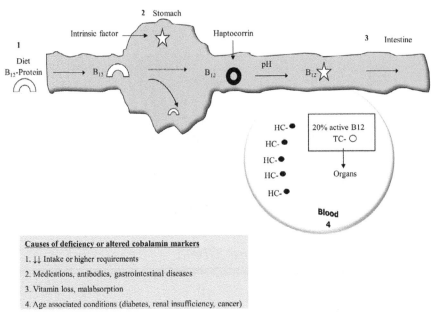

Figure 1. Vitamin B12 physiology and causes of deficiency. HC, haptocorrin; TC, transcobalamin.

absorption of cobalamin cause severe deficiency (see Chapter 5). There could be also age-related changes in cobalamin metabolism and handling. For example, elderly people may be unable to release cobalamin from food sources, even when they can absorb the vitamin from supplements or fortified foods. Therefore, the deficiency can affect individuals achieving dietary intakes above the Recommended Dietary Allowance (RDA) for vitamin B12 (2.4 µg/d for adults).

Modern cobalamin biomarkers are used to guide the routine diagnosis. They also have helped understanding how physiological factors or diseases affect cobalamin status (i.e., age, pregnancy, renal function, gastric resection).

This chapter will present current concepts concerning:

    I. targeted and non-targeted approaches to test for cobalamin deficiency;
    II. available markers, their biochemical origins, assays, and utility in order to make informed decision about cobalamin status;
    III. stepwise diagnosis strategies;
    IV. direct combination of several markers;
    V. unclassical or unexplained results;
    VI. other biosamples used for measuring cobalamin markers;
    VII. and the unresolved questions related to cobalamin markers and their utility.

2. History of Cobalamin Diagnosis and Modern Challenges

Cobalamin was identified many years after discovering its role in curing a deadly form of megaloblastic anaemia that was associated with severe neurological symptoms. Crude liver extract was discovered to contain an 'unknown liver factor' that cured anaemia (see Chapter 1). Thereafter, liver extracts had been used for many years to cure pernicious anaemia. However, the liver extract was sometimes less active depending on the way of processing and there was no measure to estimate the content of the liver factor in the extracts. In the early 1940s, the role of this 'unknown liver factor' in promoting the growth of certain microorganisms *in-vitro* was discovered and used to measure cobalamin concentrations in liver extracts and later in blood samples of deficient patients. Later, the microbiological assays have been developed to measure cobalamin concentrations in biological fluids from patients suspected to have deficiency.

Early tests that were used to diagnose cobalamin deficiency included: cobalamin levels in blood, examination of blood smears for the typical appearance of macrocytosis and polynucleated granulocytes, and examination of bone marrow for megaloblastic cells. The haematologic manifestations are not uniquely expressed in all patients, thus limiting their role as screening or diagnostic tools. The examination of bone marrow was more sensitive than the blood smear, but was too invasive to meet the requirements of modern diagnosis.

Recent discoveries related to cobalamin metabolism, physiology and transport have introduced candidate biomarkers to be used in patient's stratification in clinical settings. The modern laboratory biomarkers have been intensively studied in clinically manifested patients or in subclinical deficiency conditions. The diagnosis has been improved by introducing the metabolic markers and holotranscobalamin (holoTC). However, because all available markers have some limitations, their introduction also added new challenges to setting a diagnosis. There is a general agreement that none of the available markers is optimal for diagnosing cobalamin deficiency. Stepwise diagnostic algorithms or a direct combination of several markers have been suggested in recent years.

Cobalamin laboratory diagnostic tests have four main goals:

I. to identify individuals with cobalamin deficiency or subclinical deficiency;
II. clarify the cause of the deficiency;
III. ensure effectiveness and optimize treatment options (dose, route, frequency);
IV. and to monitor the status in people with long-term need for treatment.

3. Targeted and Non-targeted Testing of Cobalamin Status

3.1 In manifested clinical conditions (targeted-testing)

Several clinical conditions are likely to be caused by cobalamin deficiency or their progress can worsen if cobalamin deficiency co-exists (i.e., anaemia, cognitive dysfunction, polyneuropathy, cancer) (Table 1). Testing cobalamin markers in patients with anaemia is necessary even when megaloblastic cells are not detected or mean corpuscular volume (MCV) is normal. Iron and cobalamin deficiency may coincide. Therefore, normal or small size, but hypochromic red blood cells may dominate in this case. If cobalamin deficiency is not confirmed in the above mentioned conditions, there is no need for repeated testing, since the deficiency takes few years to develop.

Table 1. Candidate risk groups where testing for cobalamin status markers is highly recommended. Cobalamin deficiency can cause these conditions or worsen the prognosis.

Megaloblastic anaemia at any age
Unexplained anaemias even without megaloblastic changes
Unexplained neurological symptoms
Peripheral neuropathy with or without diabetes
Peripheral neuropathy after chemotherapy
Patients with cancer before starting the chemotherapy
Patients with mild cognitive dysfunction, Alzheimer disease, or dementia
Patients with depression
Anorexia nervosa and malnutrition
Recurrent pregnancy loss
Previous pregnancy with a neural tube defect outcome
Pregnancy complications in vegans and vegetarian women
Patients using metformin
Neurological manifestations (i.e., irritations, feeding difficulties, gross motor development, etc.) in infants of vegetarian, malnourished or deficient mothers

3.2 Health conditions that can cause cobalamin deficiency (targeted-testing)

Several chronic diseases that affect cobalamin absorption or cellular metabolism can cause severe deficiencies. Testing serum cobalamin biomarkers is recommended in patients affected with one of these conditions (Table 2). Early detection of abnormal cobalamin markers in selected target groups enables prevention of later clinical manifestations. If cobalamin deficiency is confirmed, monitoring the effect of oral supplementation is strongly recommended when patients have chronic diseases that affect cobalamin absorption. If B12 deficiency is not confirmed, repeated blood testing is recommended. Cobalamin deficiency can develop within few months in patients who cannot absorb the vitamin.

3.3 Non-targeted testing in the general population and asymptomatic subjects

The prevalence of abnormal cobalamin markers has been investigated in asymptomatic individuals of different ages, dietary habits, and countries

Table 2. Health conditions that cause cobalamin deficiency. Testing and repeated-testing of cobalamin status markers are recommended in the following conditions.

In patients with diseases likely to cause malabsorption of cobalamin

- Pernicious anaemia (intrinsic factor deficiency or antibodies against intrinsic factor).
- Anti-parietal cells antibodies (Lahner et al. 2009).
- Crohn's disease (Yakut et al. 2010).
- Gastric or ileum resection (Doscherholmen and Swaim 1973): causes intrinsic factor deficiency or inability to absorb Cbl in the intestine where intrinsic factor receptor is present.
- Unexplained cobalamin malabsorption: inability to release cobalamin from food proteins.
- Gastric atrophy.
- Hyperchlorhydia and achlorhydria: reduce Cbl absorption via intrinsic factor.
- Gastrointestinal manifestations such as irritable bowel syndrome, anorexia, and diarrhea.

In patients using certain drugs

- Proton pump inhibitors (i.e., omeprazole): reduce Cbl absorption that is pH-dependent (Hirschowitz et al. 2008; Lam et al. 2013).
- H2-blockers (i.e., cimetidine): reduce Cbl absorption by increasing the pH.
- Alcoholism: because of insufficient intake from foods and low Cbl absorption (Baker et al. 1998).
- Metformin: causes low serum levels of cobalamin for unknown reasons (Niafar et al. 2015).

Cbl, cobalamin

of origin (Bailey et al. 2013; Gonzalez-Gross et al. 2012; Yetley et al. 2011). Screening studies have identified subgroups of the population who are at risk for deficiency (i.e., low meat-eaters, vegetarians) and informed about potential health consequences. These studies have also guided policies or nutritional recommendations in sub-populations and specific age groups (Table 3).

Screening for cobalamin deficiency is not cost-effective or useful for health care professionals. For research, results of non-targeted screening should be interpreted with caution, since many studies used a single marker, did not account for confounders (i.e., renal dysfunction), and did not link the markers to any clinical outcome.

The explanation and impact of abnormal cobalamin markers differ across age groups, populations, and background diseases. For example, low cobalamin status was not related to the presence of intestinal infections in Guatemalan school children (Rogers et al. 2003b) and it did not explain differences in height, weight, hemoglobin, or hematocrit between the children (Rogers et al. 2003a). However, recent studies have shown that cobalamin supplementation improved weight gain, growth, and other developmental domains in Indian children who had low cobalamin at baseline (Kvestad et al. 2015; Strand et al. 2015). Abnormal cobalamin markers are not easily linked to measurable health outcomes. Long term follow-up or supplementation studies are scares, but are obviously needed, in particular from populations or subgroups with common abnormal markers.

Studies using non-targeted screening for abnormal cobalamin markers provided information about;

- the relationship between dietary intake and biomarkers (Bor et al. 2006; Siekmann et al. 2003) in order to guide nutritional recommendations,
- the effect of different lifestyles on cobalamin status markers (Herrmann et al. 2003),
- the effect of supplementation or fortification on cobalamin status (Duggan et al. 2014),
- the association between cobalamin status and specific health outcomes (i.e., growth in children, cognitive function in elderly, birth outcome in pregnant women),
- the effect of common diseases (i.e., cancer, *H. pylori*) on cobalamin status,
- differences in cobalamin status between populations or age groups (Eussen et al. 2012; Gonzalez-Gross et al. 2012).

Table 3. Main risk groups for cobalamin deficiency identified through non-targeted screening for abnormal cobalamin status markers. The aim of this approach is to guide health care providers, policy makers, and nutritional recommendations.

Risk group	Information gained from non-targeted screening	References
Strict vegetarians who are not using Cbl-supplements	Subjects with chronic low consumption of animal foods are at risk for Cbl deficiency and need to secure sufficient intake of Cbl through supplementation or fortification.	(Gilsing et al. 2010; Waldmann et al. 2004)
Newborns and infants from vegetarians or deficient mothers	Women with insufficient intake during pregnancy and lactation give birth to depleted children. Increasing Cbl intake should start during pregnancy and continue throughout lactation. Sufficient intake can be secured by consuming animal source diet, supplements or fortified food. Cbl is transferred to the child through the placenta or human milk and it prevents deficiency in the child.	see Chapters 10, 11
Individuals from low income countries	Cbl deficiency is common, it is caused insufficient intake from animal foods, it is related to chronic diseases (stunting, infections, etc.); the prevention can be achieved by dietary modifications or fortification.	see Chapter 6
Infants and children in general	Current dietary recommendations for infants are not evidence-based. Exclusively breastfed infants have a gap in cobalamin status between 4–6 months. Infant's cobalamin is likely to be low if lactation continued > 6 months.	see Chapter 11

Table 3. contd....

Table 3. contd.

Risk group	Information gained from non-targeted screening	References
Lactating and pregnant women in general	Dietary recommendations for pregnant and lactating women are not evidence-based.	see Chapter 10
Elderly in general	Dietary recommendations are currently not age-specific. Many elderly people have conditions or drugs that reduce Cbl absorption and increase the risk of deficiency.	see Chapter 5
Elderly with chronic morbidities	Elderly with dementia, depression, or malabsorption because of intestinal diseases or drugs may need long-term supplementation.	see Chapter 4

4. Laboratory Biomarkers of Cobalamin Deficiency

4.1 The biological origin of cobalamin status markers

Cobalamin in serum is bound to the glycoprotein carrier, haptocorrin (HC), and to a non-glycoprotein transporter called transcobalamin (TC). Haptocorrin binds approximately 70% of total serum cobalamins but the role of this protein in cobalamin transport and metabolism is not clear. Haptocorrin-bound-cobalamin is called holohaptocorrin (holoHC). Transcobalamin binds approximately 30% of serum cobalamin and this part is called holotranscobalamin (holoTC). In contrast to HC, the major part of TC in serum is unsaturated.

Total serum cobalamins represent the sum of HC- and TC-bound fractions. HoloTC (also called active-B12) represents the cobalamin fraction that is available for cellular uptake. HoloTC is taken up by the cell via transcobalamin receptor, also called CD320. A soluble form of CD320 (sCD320) has been recently detected in serum, urine and cerebrospinal fluid.

After cobalamin is delivered into the cell, it is used as a cofactor for 2 enzymes: methylcobalamin is a cofactor for the cytosolic enzyme, methionine synthase that converts homocysteine to methionine; adenosylcobalamin is a cofactor for the mitochondrial enzyme methylmalonyl-CoA mutase (Figure 2). Methylmalonic acid (MMA) is the end product of methylmalonyl-CoA that is otherwise converted into succinyl-CoA by the cobalamin-dependent mutase. Concentrations of homocysteine (tHcy) (i.e., the cytosolic reaction) and MMA (i.e., the product of the mitochondrial reaction) are elevated in blood of patients with cobalamin deficiency and are used as metabolic or functional markers of cobalamin status.

Cobalamin markers in blood

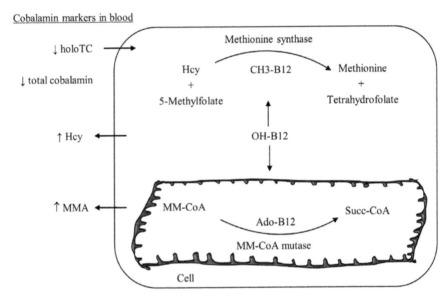

Figure 2. The biochemical bases of cobalamin markers. HoloTC delivers cobalamin into the cell. Homocysteine accumulates in the cell and plasma when methylcobalamin is low. Methylmalonic acid levels are elevated in blood when adenosylcobalamin is deficient. Ado-B12, adenosylcobalamin; CH3-B12, methylcobalamin; Hcy, homocysteine; HoloTC, holotranscobalamin; MMA, methylmalonic acid; MM-CoA, methylmalonyl-CoA; Succ-CoA, succinyl-CoA.

The cobalamin-binding proteins (total TC, total HC) have been suggested as potential markers of cobalamin deficiency or disease conditions (Gimsing and Nexo 1989). Immunological methods for direct measurement of HC (Morkbak et al. 2005) and TC (Nexo et al. 2000) have been developed in recent years, but their availability is limited, their clinical relevance is not clear, and therefore their utility is not well established. Currently, measuring the concentrations of HC and TC and their saturation levels in serum/plasma is not part of routine testing for cobalamin status, but they are sometimes measured in unexplained cases of too low or too high cobalamins.

4.2 Laboratory markers of cobalamin status

Exploring cobalamin status should start with collecting clinical and anamnestic information about the individual to identify potential causes of deficiency. Identifying causes related to low intake such as a strict vegan diet or malnutrition, gastrointestinal diseases, medications, or family history will help setting a diagnosis and optimizing prevention and treatment options.

When cobalamin deficiency is suspected, blood samples should be collected before starting the treatment. Otherwise, cobalamin markers will be changed and cannot provide reliable information about pre-treatment status.

Blood count is still used for preliminary screening for cobalamin deficiency since the deficiency is known to cause megaloblastic anaemia. This approach can be helpful when limited health resources are available and when measuring the biochemical markers is not possible. However, we need to recall that blood count is not sensitive or specific because many patients do not develop anaemia and anaemia can be caused by factors other than cobalamin deficiency.

Four primary cobalamin biomarkers will be discussed in the following session. These are: total cobalamins, holoTC (or active B12), MMA, and tHcy. These biomarkers are becoming more popular, but they have some limitations that hinder their utilization in certain clinical conditions. Moreover, there is no general agreement on reference ranges, combinations of the markers, and their utility in diagnosing vitamin deficiency in some critical cases, such as cancer, renal or liver diseases (Herrmann and Obeid 2013).

4.2.1 Total cobalamins

Analytical methods	Microbiological assays or methods using non-isotopic competitive protein binding assays are commercially available. Immunological methods have different detection techniques; electrochemiluminescence (i.e., Roche Cobas), or chemiluminescence detection (i.e., Architect i2000sr and ADVIA Centaur). The immunological methods measure generally higher concentrations than the microbiological assays (Arnaud et al. 1994; Ispir et al. 2015). The lower limit of detection for most assays is between 30–70 pmol/L. The automated methods show systematic differences, but they are still able to deliver informed decision (Ispir et al. 2015). The difference between methods is up to approximately 140 pmol/L (Ispir et al. 2015), thus limiting the possibility of combining data generated by using different analytical methods.
Biosamples	Serum or EDTA-plasma can be used; the two materials may show slight differences and they should not be used interchangeably. Other matrixes can be used after a prior extraction of cobalamin (i.e., human milk).
Preanalytical factors	Samples should be protected from light and kept at +4°C until analysis (< 1 week). Samples are stable for 1–2 months if stored at –20°C.

Cost effectiveness	Cbl tests are widely available, they have low running costs, and are available on several auto-analyzers. The cost-effectiveness is questionable given that measurement of total cobalamin cannot detect all deficient cases.
Reference intervals	Reference intervals vary between methods (Ispir et al. 2015) and there is no agreement on a population reference range. The following ranges have been suggested and are widely used: serum B12 levels < 148 pmol/L are considered low; 148–221 pmol/L are considered intermediate; and > 221 pmol/L are considered normal (Yetley et al. 2011). Elevated MMA was not an exclusion criterion in defining these cut-offs. Elevated MMA and lowered holoTC can occur in subjects with total cobalamins > 221 pmol/L.
Potential confounders	Sex differences in serum cobalamin have been reported (Fernandes-Costa et al. 1985), but sex-specific reference intervals were regarded as unnecessary. Levels of serum total cobalamin and its binding proteins change with age. However, age-specific reference intervals are not recommended (Gimsing and Nexo 1989).
Imprecision	Most automated methods show low imprecision of < 7% for between-day measurements (Ispir et al. 2015).
Method comparison	Total cobalamin differs between laboratories and analytical methods (Arnaud et al. 1994). The differences between methods range between 60 and 140 pmol/L.
Limitations	Serum cobalamin has a poor performance especially in the diagnosis critical range (low range). A low serum cobalamin can be found in non-deficient subjects (false positive) and normal cobalamin test results are commonly found in deficient subjects (false negative). The limitations of this marker can be reduced by combining it with a metabolic marker, such as MMA when cobalamin levels are in the low to intermediate range (< 250–300 pmol/L).

Interfering factors and changes not explained by cobalamin status	• Total cobalamin is elevated in blood of patients with myeloproliferative disorders, cancer, seriously ill patients, patients with advanced liver or renal diseases. Therefore, normal results of this marker cannot be considered as indicating sufficient cobalamin status in patients affected with these conditions. • Total cobalamin is very low in patients with haptocorrin deficiency without any functional meaning (Carmel 2003). • False low concentrations of serum cobalamin can be found in several conditions such as HIV infection and oral contraceptive use. • Immune-complexes that bind cobalamin in serum can give false results because of analytical interactions. • Total cobalamin is not a good marker for cobalamin status in pregnancy.

4.2.2 Holotranscobalamin (holoTC, or active B-12)

Analytical methods	Several assays for measuring holoTC in serum or plasma have been used. Earlier methods depended on ionic precipitation of TC followed by measurement of the cobalamin fraction that is trapped in the precipitate (Lindgren et al. 1999). Few methods used antibodies against TC instead of ionic separation (Lindemans et al. 1983), or measured the trapped vitamin B12 by an isotope dilution assay (RIA) (Loikas et al. 2003; Ulleland et al. 2002), or by a microbiological assay (Refsum et al. 2006). Other methods depended on removing apo-transcobalamin (unsaturated TC) with vitamin B12–coated beads followed by an enzyme-linked immunoassay measurement of holoTC in the supernatant (Nexo et al. 2002a). The holoTC-RIA and the holoTC–enzyme-linked immunosorbent assays give comparable results (Nexo et al. 2002a). The long term imprecisions for both assays are < 10%. The RIA assay results are also similar to those obtained with the use of a microbiological assay after precipitation

Analytical methods	(Refsum et al. 2006). The holoTC-RIA assay has been replaced by an assay that uses holoTC-specific monoclonal antibodies on the AxSYM platform (Abbott Labs, North Chicago, IL) (Orning et al. 2006). The results obtained by this method show a good agreement with those from the holoTC-RIA (Bamonti et al. 2010; Brady et al. 2008). More recently, the holoTC or active B-12 assay has become available on ADVIA Centaur® (Siemens).
Biosamples	Serum or EDTA-plasma can be used. Serum samples may show slightly lower holoTC than EDTA plasma (Nexo et al. 2000). Although this effect was not confirmed by all studies (Refsum et al. 2006), it is recommended that materials should not be used interchangeably in one study. For research purposes, holoTC can be measured in other biosamples (i.e., CSF, urine) by using the same common assays.
Preanalytical factors	HoloTC is stable in samples collected and stored for at least 16 months at −70°C (Loikas et al. 2003).
Cost effectiveness	HoloTC test is rather expensive. The test is available on ADVIA Centaur (Siemens), and AxSYM (Abbott) platforms. The running costs are expected to decline after that it recently became available on several platforms.
Reference intervals	There is no consensus on a reference interval. A commonly used lower cut-off value is: 35 pmol/L. Recent studies have defined a grey range for holoTC between 23–75 pmol/L. The grey range was based on combining best sensitivity and specificity of holoTC test to detect elevated MMA (as an outcome marker). When holoTC is used as a first line marker, and the result is below or within this intermediate range (23–75 pmol/L), a second marker such as MMA is strongly recommended in a following step (Herrmann and Obeid 2013).
Potential confounders	• Principally, lowered holoTC suggests a recent or an established depletion of cobalamin. Lowered holoTC alone does not necessary mean that the individual is deficient or requires treatment.

Potential confounders	• HoloTC concentrations are artificially elevated in renal insufficiency. Normal holoTC values in this case cannot be interpreted as suggesting normal cobalamin status.
Imprecision	The test has low day-to-day coefficient of variation (i.e., < 7% on AxSYM platform).
Limitations	Limitations related to test availability and costs are expected to be solved after that the active B12 assay has become available on different analytical platforms (i.e., ADIVA Centaur, AxSYM). Limitations related to confounding by some clinical conditions are related to cobalamin physiology, but can be partly solved by combining holoTC with MMA as explained later in this chapter.
Interfering factors and changes not explained by cobalamin status	• Serum holoTC shows a stepwise increase in individuals with renal insufficiency (low eGFR or high creatinine). • It shows elevation in aging, liver diseases, and cancer. • Using holoTC in pregnancy is recommended, because it shows no decline during pregnancy, as it is the case for total cobalamin.

4.2.3 Methylmalonic acid

Analytical methods	Different analytical methods are available with different sample preparation procedures (Stabler et al. 1986; Windelberg et al. 2005). Direct analysis by gas chromatography mass spectrometry (GCMS) after pre-separation on a solid phase and derivatization (silylation, cyclohexanol, chloroformate, and butanol) can improve the detection (Mineva et al. 2015). Liquid chromatography tandem mass spectrometry (LC-MS/MS) methods require smaller sample volume, shorter run times, and have higher sensitivity compared with the GCMS procedures (Pedersen et al. 2011). Recently, commercially available reagents for measuring MMA on LC-MS/MS systems became available and methods show excellent comparability with the GCMS method (Obeid et al. 2015).

Biosamples	Serum, EDTA-plasma, and urine can be used. Different specimen types and anticoagulant have been tested (serum, serum separator, K^{2+}EDTA plasma, Na-citrate plasma, and Na-heparin plasma. All anticoagulant systems show very similar results (Mineva et al. 2015).
Preanalytical factors	Blood samples should be centrifuged and separated from the cells and kept at +4°C until measurement. Samples can be stored at +4°C for few days or freezed for a longer time. MMA remains stable when samples are stored for up to 7 days at ambient temperature, or exposed to repeated freeze-thaw cycles (Mineva et al. 2015). MMA is stable for up to 14 years when stored at ≤ –70°C.
Cost effectiveness	The assay is rather expensive and still not widely available. The test has been adapted to LC-MS/MS systems which reduced the time and costs for sample preparation and analysis.
Reference intervals	There is no consensus on a cut-off limit for MMA. However, most studies have used an upper limit of approximately 300 nmol/L for serum/plasma MMA. A clinically relevant upper limit of 450 nmol/L has been suggested because elevated MMA concentrations can occur in the absence of clinical symptoms (Loikas et al. 2003). For MMA in urine a cut-off < 8 mmol/mol has been suggested (Gultepe et al. 2003). These values for serum or urine are not based on clinically manifested deficiency.
Imprecision	Quality control samples are commercially available. The assays on GCMS or LC-MS/MS systems show low imprecisions < 7% from day to day (Mineva et al. 2015).
Method comparison	Comparability of validated methods between labs is very good. Still, methods need standardisation.
Limitations	MMA test is not a suitable screening- or first line-marker. Measurement of MMA in biosamples is recommended as a second line marker in a stepwise algorithm that starts with holoTC or total cobalamin. Because MMA is influenced by renal dysfunction and bacterial production of propionic acid, it is not useful as a cobalamin marker in these conditions. However, this problem can be partly solved (see sessions on marker utility and unclassical results below).

Interfering factors and changes not explained by cobalamin status	• Renal dysfunction is the main cause of elevated MMA levels that are not necessarily explained by cobalamin deficiency. The size of MMA elevation is related to the degree of renal dysfunction, but not necessarily predicted by eGFR, creatinine or cystatin C. This is because cobalamin deficiency is also common in renal patients and can independently cause MMA elevation. • Intestinal bacterial overgrowth is known to cause elevation of MMA that is lowered after treatment with antibiotics, but not after vitamin B12.

4.2.4 Homocysteine

Analytical methods	There are several chromatographic and immunological methods, all of which are commercially available and applicable on popular analytical platforms (Ducros et al. 2002; Ubbink et al. 1991; Yu et al. 2000). However, the methods show variabilities and some of them were reported to have no discrimination power between patients and controls (Yu et al. 2000), or bad agreement in the upper range of tHcy (diagnosis critical range). The LC-MS/MS and GCMS methods show a good agreement and are considered the reference methods.
Biosamples	EDTA-plasma is recommended. Blood samples should be centrifuged and separated within 30 min of collection. Citrate plasma can be used for tHcy assay, but it may not be suitable if other markers are to be measured from the same tube (Hubner et al. 2007a; Tamura and Baggott 2008). Tubes with stabilizing agents have been introduced and became available in recent years.
Preanalytical factors	Blood sampling should take place under fasting conditions. Meals rich in methionine increase post-prandial tHcy within few hours. The blood should be placed on ice after collection, centrifuged and separated within 30 min of collection. Recently, tubes with stabilizing agents have been used for samples that need to be transported to the labs or in centres that do not have the possibility for immediate centrifugation (Hubner et al. 2007b). Red blood cells continue

Preanalytical factors	to release tHcy, causing an artificial increase if samples are placed at ambient temperature for > 1 hour. tHcy is stable upon thawing and refreezing. When the plasma is separated from the cells, tHcy remains stable for many years if stored at ≤ −20°C.
Cost effectiveness	The costs have been declined in recent years. The test is cost-effective for unselective screening for folate and cobalamin deficiency. However, when high levels are detected more specific tests are needed to explain the results (i.e., serum folate, holoTC).
Reference intervals	Generally recommended cut-off values are: < 12 µmol/L for adults; < 9 µmol/L for pregnant women, and < 8 µmol/L for children.
Imprecision	Most assays have low imprecision < 6% between days.
Method comparison	The immunological assays show differences, the chromatographic methods (GMCS, LC-MS/MS) are better comparable.
Limitations	• tHcy is not a specific marker for cobalamin status. It is affected by folate deficiency to a greater extent than by cobalamin deficiency. • tHcy is elevated in patients with renal dysfunction or those with elevated creatinine. • The concentrations of tHcy are affected by a recent methionine intake. Therefore, fasting levels are strongly recommended. • The necessity to centrifuge the samples within a short time limits the use of this marker in small, non-equipped health centres or large scale studies, when immediate sample processing is not possible.
Interfering factors and changes not explained by cobalamin status	• Renal dysfunction causes an artificial increase of tHcy levels in plasma. • Mutations in the folate cycle (i.e., MTHFR polymorphism) or the transsulfuration pathway (i.e., cystathionine beta synthase) can cause elevation of tHcy concentrations.

Interfering factors and changes not explained by cobalamin status	• Plasma tHcy is elevated also in folate and vitamin B6 deficiencies. • Plasma tHcy increases after acute coronary events and remains elevated for few weeks. • tHcy is elevated in patients with alcohol abuse.

Other cobalamin-related markers

Cells can internalize holoTC via a specific transcobalamin receptor (CD320). This protein has been recently discovered and a CD320 knockout mice model has been developed (Quadros 2010). A soluble form of transcobalamin receptor (sCD320) has been detected in serum and an immunological assay (ELISA) has been developed to measure its concentration. The molar concentrations of sCD320 in serum are approximately 2–3 fold higher than that of its ligand, holoTC (Abuyaman et al. 2013). Concentrations of sCD320 and that of holoTC show a weak positive correlation. sCD320 can probably bind its ligand, holoTC, with a low affinity. The biological importance of this potential low-affinity-binding needs further investigations. The concentrations of sCD320 in serum increase during pregnancy up to the 35 gestational weeks, before they start to decline. Studying changes of sCD320 concentrations in cobalamin deficiency or when cobalamin homeostasis is disturbed (i.e., renal insufficiency, cancer) is an interesting field for future research.

Other markers such as plasma concentrations of total transcobalamin and haptocorrin or their saturation levels can be measured in some unexplained cases, but these markers have limited importance for daily routine diagnosis.

Complementary tests

Few tests can be used to identify the cause of cobalamin deficiency if a low intake of animal foods is excluded (Table 4). However, some causes of deficiency may remain unexplained and could be related to 'food cobalamin malabsorption' or the acquired inability to release cobalamin from its protein-binding in foods.

4.3 Utility of the independent cobalamin markers and their combinations

No single test can differentiate between deficient and non-deficient cases with sufficient sensitivity and specificity (Bailey et al. 2011; Valente et al. 2011; Yetley et al. 2011). Diagnosing cobalamin deficiency in individuals with clinical manifestations (i.e., anemia or unexplained neurological symptoms) is straight forward. Available biomarkers can provide informed decision on cobalamin status in more advanced clinically manifested cases.

Table 4. Complementary tests used for detecting the cause of cobalamin deficiency or the optimal treatment route.

Test	Advantages	Limitations	References
Clarify the cause of deficiency			
Intrinsic factor antibodies	The specificity of the test is 100%	Low sensitivity; methods are not comparable	(Carmel 1996; Festen 1991; Snow 1999)
Parietal cell antibodies	Antibodies are found in early stages of pernicious anemia, but at later stages only 50% of the patients have positive results	Low specificity and sensitivity	(Bizzaro and Antico 2014; Lahner et al. 2009; Snow 1999)
Pepsinogen I, pepsinogen I /pepsinogen II ratio (atrophic gastritis)	Optimally used in combination with other markers*	Low specificity and sensitivity	(Agreus et al. 2012; Krasinski et al. 1986; Sipponen et al. 2003)
Gastrin (atrophic gastritis)	High sensitivity	Fasting sample needed; low specificity	(Hurwitz et al. 1997)
H. pylori	Is associated with cobalamin deficiency	A causal relationship is questionable	(Carmel et al. 2001a; Kaptan et al. 2000; Serin et al. 2002)
Clarify if patients can absorb oral doses of cobalamin			
Schilling test	Was once considered to be the gold standard since it tests the absorption of free Cbl and IF-bound Cbl	Depends on radioactive B12, is not used anymore	(Nickoloff 1988; Nilsson-Ehle et al. 1989)
CobaSorb: holoTC after a low dose of B12	Shows if subjects can absorb a low dose of crystalized B12 that is likely to be absorbed via IF-pathway. If holoTC increases, oral Cbl supplementation using physiological doses can be applied	The test does not answer the question of food Cbl malabsorption, results are reliable only at low baseline holoTC levels < 65 pmol/L	(Bor et al. 2005)
HoloTC after a single high dose of B12	Shows if subjects can absorb high doses of crystalized Cbl. If Cbl is absorbed, oral Cbl treatment can be applied using therapeutic doses	Does not inform about intrinsic factor mediated absorption, since at high doses Cbl will be absorbed by passive diffusion	Obeid et al. un-published data (Figure 12)

* Agreus et al. presented algorithm on using plasma levels of pepsinogen, gastrin and *H. pylori* to diagnose gastritis.

Early detection and prevention of cobalamin deficiency is recommended. However, diagnosis of a subtle deficiency is a debatable issue. The definition of "deficiency" in numerous recent publications is solely based on the presence of abnormal blood/or urine biomarkers. This approach is thought to enhance the prevention, but it may cause overtreatment, because there is no evidence that this metabolic condition will progress into clinically manifested disease if remained untreated.

The prevalence estimates for low cobalamin status vary between studies and depend on the primary marker used to define the deficiency, the cut-offs employed for each marker (Sobczynska-Malefora et al. 2014; Yetley et al. 2011), different types of samples such as urine or serum for MMA (Hill et al. 2013), and the degree of assays agreement for the same marker (Ispir et al. 2015). Therefore, available studies show large heterogeneity about the criteria used to define the deficiency (Herrmann and Obeid 2013; Sobczynska-Malefora et al. 2014). There are ongoing efforts to improve the diagnosis and reduce the heterogeneity by using algorithms (Herrmann and Obeid 2013) or directly combining the markers to develop a personalised cobalamin status-indicator or scores (Fedosov et al. 2015).

Combining blood cobalamin markers with health history, medications, and dietary habits is currently the best strategy for making a personalised decision. After setting the diagnosis, cobalamin deficiency should be treated. Current recommendations focus also on preventing the progress of the deficiency by administering low doses of the vitamin. Blood cobalamin biomarkers are useful for monitoring the success of the treatment.

Total cobalamin

Total cobalamin provides a rough estimation of cobalamin status. For appropriate interpretation of total serum cobalamin we need to recall that this marker represents a sum of 2 fractions of cobalamins; the one bound to haptocorrin and that is bound to transcobalamin. Total cobalamin levels < 300 pmol/L show poor performance in detecting subjects with elevated serum MMA (Joosten et al. 1993; Naurath et al. 1995) or those with low holoTC. Low holoTC levels can occur in subjects with normal to high normal serum cobalamin (Figure 3). In a study on vegetarians and omnivorous subjects, 'deficient subjects' were defined as those having an elevated MMA value > 271 nmol/L combined with a lowered holoTC < 35 pmol/L. A low serum total cobalamin showed a large overlap between deficient and non-deficient subjects (Figure 4). Furthermore, a low serum cobalamin level is a bad predictor of clinical symptoms in overt cobalamin deficiency. In the majority of cases, total cobalamin should be followed by a second test (i.e., MMA) to confirm or exclude a deficiency state.

High serum concentrations of cobalamin (> 600 pmol/L) are more common than low levels. Approximately 15% of all samples from non-supplemented individuals show concentrations of serum cobalamin above this suggestive

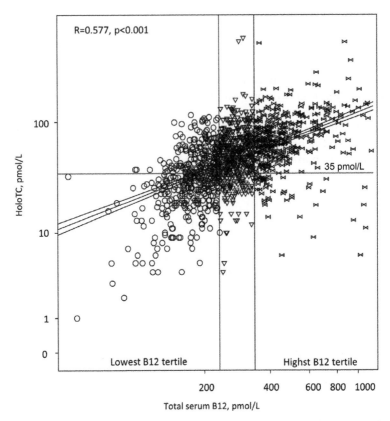

Figure 3. The correlation between concentrations of total cobalamin (divided into tertiles) and holoTC in serum. Though the correlation is significant, holoTC can be low when total B12 is in the upper normal range.

upper limit (Carmel et al. 2001c). The unspecific elevation of serum cobalamin is more common among patients with cancer, liver diseases, or renal diseases (see Chapter 13). In patients with diabetes, serum cobalamin and holoTC do not necessarily reflect changes in the metabolic markers, MMA and tHcy (Obeid et al. 2013). Similar findings were recently reported in patients with cancer (Vashi et al. 2016), or those with renal insufficiency (Obeid et al. 2005a). Therefore, elevated concentrations of serum cobalamin in some clinical conditions can hinder the diagnosis of a deficiency.

Holotranscobalamin

Compared with total serum cobalamin, concentrations of holoTC have higher sensitivity and specificity in detecting cases with elevated serum levels of MMA (Figure 5). The superiority of holoTC over total cobalamin has been confirmed by different studies (Nexo and Hoffmann-Lucke, 2011). However,

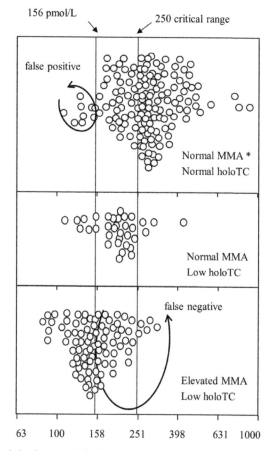

156 pmol/L

250 critical range

false positive

Normal MMA *
Normal holoTC

Normal MMA
Low holoTC

false negative

Elevated MMA
Low holoTC

Total cobalamin, pmol/L   63   100   158   251   398   631 1000

* Normal cobalamin status defined as MMA ≤ 271 nmol/L / holoTC > 35 pmol/L

Figure 4. Total serum cobalamin concentration is a poor predictor of a combination of elevated MMA and low holoTC (Normal cobalamin status defined as MMA ≤ 271 nmol/L/holoTC > 35 pmol/L). The figure includes data from omnivores and vegetarian subjects with normal renal function.

also holoTC has a limited value as a single marker since it shows marked changes following cobalamin depletion and repletion.

HoloTC has been used as a marker of recent changes in cobalamin intake and status. This function is mainly due to its increase in serum following minor changes in cobalamin intake (Bor et al. 2010; Lloyd-Wright et al. 2003). In vegetarians, serum concentrations of holoTC can reflect cobalamin depletion. The distribution of holoTC concentrations is shifted towards lower levels in the more strict dietary groups (Figure 6). Also, high concentrations of holoTC predict vitamin usage. All together, concentrations of holoTC can be used as a first line marker to screen for nutritional cobalamin deficiency or cobalamin depletion (Herrmann et al. 2003).

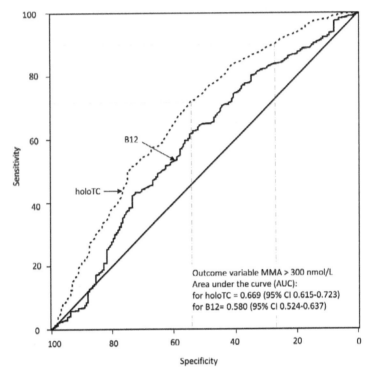

Figure 5. The Receiver Operating Characteristic (ROC) curves of total cobalamin and holoTC performance in detecting samples with elevated MMA > 300 nmol/L.

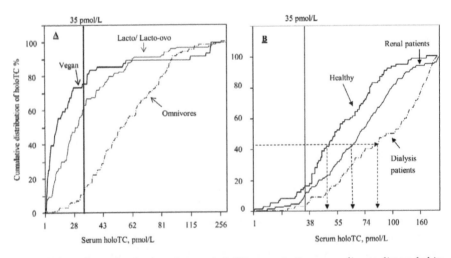

Figure 6. Cumulative distribution of serum holoTC concentrations according to dietary habits (A) and to renal function (B). A shift in holoTC distribution towards lower levels can be seen in vegans followed by lacto-ovo-vegetarians compared with the curve for omnivorous. In contrast, renal failure is associated with higher holoTC levels.

The cut-off values for holoTC of 35–40 pmol/L have been employed by several studies (Herrmann and Obeid 2009; Herrmann and Obeid 2013; Refsum et al. 2006). However, more recent studies have defined a grey range for holoTC between 25–50 pmol/L (Sobczynska-Malefora et al. 2014) or 23–75 pmol/L (Herrmann and Obeid 2013). If holoTC test is used for screening and if the concentrations are found to be below or within the grey range (< 50 or < 75 pmol/L), testing of serum MMA in a second step is highly recommended to explore cobalamin status.

Therefore, holoTC is a sensitive and specific marker for cobalamin status in dietary causes of deficiency. However, this marker can be lowered in a significant number of samples, without any metabolic sign that indicate deficiency (i.e., holoTC indicates cobalamin negative balance in this case). HoloTC is elevated in patients with renal insufficiency, but in this case it is not considered a good marker for cobalamin status. A direct or a stepwise combination of holoTC and MMA is currently the best available option.

Methylmalonic acid

Concentrations of MMA are elevated in serum and urine of patients with cobalamin deficiency (Rajan et al. 2002; Stabler et al. 1986). Serum MMA is a sensitive marker for clinically manifested cobalamin deficiency (Savage et al. 1994; Stabler et al. 1986). It is also the best available marker for monitoring the improvement of cobalamin-dependent metabolism after the start of treatment (Mansoor et al. 2013).

Reduced eGFRs or elevated serum creatinine levels are associated with higher serum concentrations of MMA, independent on cobalamin status (Hyndman et al. 2003; Obeid et al. 2005a). Therefore, a main limitation of using MMA as a maker for cobalamin status is its artificial increase in patients with renal dysfunction. However, we recommend that cobalamin treatment should be initiated when serum MMA levels are markedly elevated in patients with renal failure (MMA > 500 nmol/L). A decline of MMA level upon treatment will be evident within 1–2 weeks. A significant reduction of serum MMA suggests a pre-treatment deficiency (delta MMA ≥ 200 nmol/L).

MMA is excreted in urine and its urinary concentrations are elevated in patients with cobalamin deficiency (Gultepe et al. 2003; Hill et al. 2013). Earlier studies reported a smaller influence of renal function on urinary MMA (uMMA) after adjusting for urinary creatinine (Norman and Morrison 1993). In contrast to serum MMA, the use of uMMA as a functional marker for cobalamin is not well established. Moreover, determinants of uMMA levels and their relation to blood cobalamin markers (total cobalamin, holoTC, serum MMA), and their changes after treatment with cobalamin are not well investigated. The use of spot urine samples for MMA measurement has limitations, since this marker shows variations during the day and increases after food consumption (Rasmussen 1989), whereas serum MMA levels remain stable. Concentrations of uMMA in random urine show no correlation with serum MMA (Marcell et al. 1985). Therefore, it was suggested that 18- to 24-h urinary collection would be necessary to validly estimate the urinary excretion of MMA. Kwok et al. have shown that overnight fasting uMMA correlate with serum MMA and suggested using this marker to screen for cobalamin deficiency (Kwok et al. 2004). Concentrations of uMMA correlate with those of serum holoTC and cobalamin in elderly people. However, uMMA is affected by renal function, protein intake, and sex (Flatley et al. 2012).

Measurement of plasma tHcy is often used to support the diagnosis of cobalamin deficiency or to differentiate between cobalamin and folate deficiencies. An elevated plasma tHcy is commonly encountered during screening for cardio-vascular risk factors. When tHcy levels are elevated, physicians should investigate the causes by measuring folate and cobalamin status markers or looking for confounders (i.e., renal dysfunction). tHcy concentrations are frequently elevated ($\geq$ 12.0 µmol/L) in subjects who carry certain polymorphisms in the folate cycle and in subjects with renal insufficiency. tHcy test is not a primary diagnosis test for cobalamin deficiency, although some recent reports have used this marker to evaluate cobalamin status (Brito et al. 2016; Fedosov et al. 2015).

Several physiological and pathological conditions may affect cobalamin biomarkers, though changes of these markers are not primarily related to cobalamin status Table 5.

Table 5. Conditions that affect cobalamin markers and can influence the interpretation.

Factors	Expected effects	Reference
Age	↑ MMA and tHcy in elderly subjects. Both MMA and tHcy show less response to treatment in the elderly compared with young subjects.	(Obeid et al. 2004)
Sex	↓ holoTC in young women (< 45 years) compared with men.	(Refsum et al. 2006)
Smoking	↑ MMA in smokers and low response of MMA to B12 treatment in smokers and ex-smokers.	(Hill et al. 2013)
Genetic variants	TCblR G220R (rs2336573) associated with serum cobalamin, and TCN2 S348F (rs9621049) with homocysteine	(Kurnat-Thoma et al. 2015)
Ethnic origin	↑ cobalamin in Africans	(Carmel et al. 2001b)
Pregnancy	↓ holoHC but holoTC remains stable	(Morkbak et al. 2007)
Renal dysfunction	↑ MMA, tHcy, and holoTC	(Herrmann and Obeid 2013)
Type 2 diabetes	↑ MMA and tHcy	(Obeid et al. 2013)
Cancer	↑ serum total Cbl	
Liver diseases	↑ serum total Cbl	
Medications (i.e., metformin)	↓ holoTC and total vitamin B12, but no tHcy or MMA elevation	(Obeid et al. 2013)

Cbl, cobalamin

Stepwise Diagnosis—Algorithms

Recently, algorithms have been developed to improve the detection of a subtle cobalamin deficiency. The algorithms imply using serum cobalamin or holoTC as a first-line marker that is then followed by an MMA measurement, if the initial screening did not settle the diagnosis (Figure 7). This approach is considered cost-effective, since the second measurement is performed in a subgroup of the suspected samples and only when the first test does not answer the question.

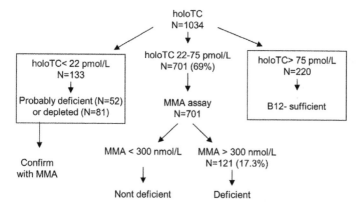

Total deficient samples identified using holoTC as a first line and MMA as a second line markers = 254

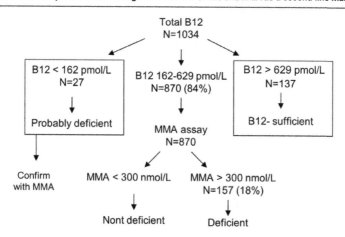

Total deficient samples identified using B12 as a first line marker and MMA as a second line marker = 184

Figure 7. Diagnostic algorithm that depends on screening using one marker (either holoTC or total B12) and a second marker (MMA) in subjects with low levels or those in the gray ranges. The numbers of cases identified through the second marker are indicated. Data are from (Herrmann and Obeid 2013).

Compared with using total cobalamin as a single marker, testing MMA in a second step will identify more cases that are classified as deficient. In contrast, when holoTC and MMA measurements are combined, fewer samples are expected to have abnormal results of both, compared with when using only holoTC test. Sobczynska-Malefora et al. have shown that out of 4175 test results for holoTC, 24% of the samples (n = 1019) were in the intermediate range of holoTC (25–50 pmol/L) (Sobczynska-Malefora et al. 2014) and those were further tested for MMA. An elevated MMA level (> 280 nmol/L, or > 360 nmol/L for elderly > 65 years) was identified in 244 samples (6% of the total population 4175) (Sobczynska-Malefora et al. 2014).

A direct combination of cobalamin biomarkers

A combination of at least 2 laboratory markers can improve the diagnosis of cobalamin deficiency. In contrast to a stepwise measurement of cobalamin markers, a direct combination of the markers has been suggested in recent years. Models based on a combination of total cobalamin, holoTC, MMA and tHcy have been developed aiming at stratification of subjects according to the so called combined cobalamin indicator or (cB12) (Fedosov et al. 2015). A new variable is created out of the 4 markers and this can further be adjusted for age and folate (i.e., known to influence tHcy concentrations). This cB12 index was considered as a 'metabolic fingerprint' for each individual. Generally, a cB12 index of ≥ 0 encode an adequate cobalamin status, and values < 0 encode one of three possibilities (low cobalamin, possibly deficient, or probably cobalamin deficient). Because of the high costs of the 4cB12 indicator, two and three-component indicators (2cB12 and 3cB12) were also tested when one or two of the 4 markers are missing. Four possibilities to describe cobalamin status were defined based on this approach: elevated cobalamin, adequate, low, possibly deficient, and probably deficient (Fedosov et al. 2015).

Cobalamin indicators have been recently used to verify the clinical response to cobalamin treatment in a study on elderly people (Brito et al. 2016). However, due to the combination of the markers, the costs for this approach are considerably higher than the stepwise or the consequent testing. MMA and holoTC tests are expensive and not widely available in routine labs, and homocysteine is less specific for cobalamin deficiency than MMA thus limiting the relevance of this approach for daily diagnosis. The advantages of cB12 compared with the single markers or the stepwise strategy and its potential clinical use are not clear yet.

5. Unexpected Results of Cobalamin Markers and their Interpretation

5.1 Elevated cobalamin markers in renal insufficiency

When testing for cobalamin deficiency, we theoretically expect that total cobalamin and holoTC concentrations will be low and concentrations of MMA and tHcy will be elevated. This figure may be different in patients with renal dysfunction where total cobalamin and holoTC can be high or in the upper normal ranges (> 300 pmol/L, > 50 pmol/L, respectively), while tHcy and MMA are also considerably elevated (> 15.0 μmol/L, > 300 nmol/L, respectively). Figure 8 shows that subjects with elevated creatinine show simultaneous elevation of serum holoTC, total cobalamin, and MMA compared with subjects who have lower creatinine levels. Serum concentrations of holoTC show positive correlation with serum creatinine, particularly in the high range of serum creatinine. Elevated serum concentrations of cobalamin or holoTC have been reported in kidney disorders (Areekul et al. 1995; Carmel et al. 2001c), but the reason for this increase is not well studied. Carmel et al. proposed impaired

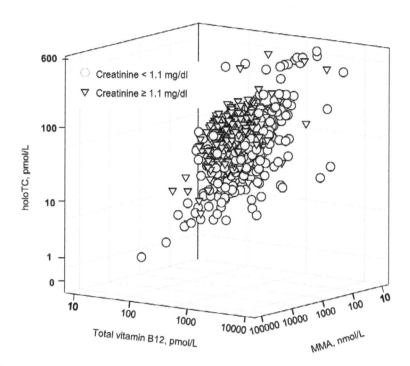

Figure 8. The relationship between serum concentrations of holoTC, total cobalamin and MMA according to 2 ranges of serum creatinine. The concentrations of all three cobalamin markers increase in plasma of patients with elevated creatinine.

cellular uptake of cobalamin by CD320 that is abundant in the kidney (Carmel et al. 2001c).

When total cobalamin or holoTC are measured to investigate cobalamin deficiency in patients with renal insufficiency, normal or high levels of the markers can give the wrong impression that cobalamin status is normal. However, serum MMA concentrations are markedly elevated and they can be lowered after treatment with pharmacological doses of cobalamin (Obeid et al. 2005a). Concentrations of tHcy in plasma also decline as a response to cobalamin and folate supplementation (Elian and Hoffer 2002; Moelby et al. 2000). In patients on hemodialysis, median MMA declined from 1077 nmol/L at baseline to 769 nmol/L after 4 weeks of treatment with i.v. 0.7 mg cyanocobalamin/3 per wk, but MMA was not normalised (Obeid et al. 2005a). The post-supplementation high MMA levels have been described in earlier studies on renal patients and concentrations up to 800 nmol/L were explained by kidney dysfunction (Moelby et al. 2000). Interestingly, the lowest concentrations of MMA achieved under B12-treatment were maintained in the 5 months following withdrawal of cyanocobalamin injections (Obeid et al. 2005a). Therefore, despite normal holoTC and total cobalamin at baseline, the biochemical marker, MMA, dramatically declined as a result of vitamin B12 treatment. Cobalamin deficiency does not account for the persistent increase of MMA post-supplementation (Figure 9).

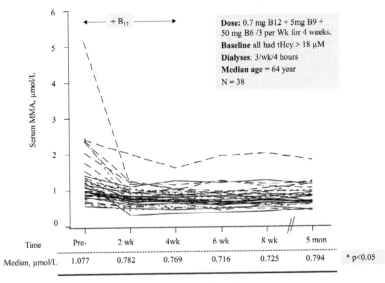

Time	Pre-	2 wk	4wk	6 wk	8 wk	5 mon	
Median, µmol/L	1.077	0.782	0.769	0.716	0.725	0.794	* p<0.05

Figure 9. Individual changes of serum MMA in 38 patients on haemodialysis who all had tHcy > 18.0 µmol/L at baseline. Administration of cobalamin for 4 weeks (0.7 mg/3 per week) lowered serum MMA from a median of 1077 nmol/L at baseline to 769 nmol/L after 4 weeks. This new median level was maintained after stopping the vitamin treatment for the follow up time of 5 months.

5.2 Cobalamin markers in elderly people

Diagnosis of cobalamin deficiency in elderly people have a significant impact on health, since several age-associated conditions and drugs can cause deficiency that is associated with hyperhomocysteinemia and higher risk of age-related diseases.

Age and the age-associated decline in renal function affect cobalamin markers. Concentrations of MMA and tHcy show a marked elevation in elderly people often without lowered levels of serum holoTC or total cobalamin (Figure 10) (Obeid et al. 2004). MMA and tHcy elevation in serum is only partly related to higher creatinine, since the differences with age remained significant after adjustment for serum creatinine. Concentrations of serum MMA show the expected negative association with serum holoTC and total cobalamin in elderly subjects (Herrmann et al. 2005). Serum MMA levels decline as a response to oral cobalamin supplementation in elderly people. However, it has been shown that high doses of oral cobalamin (i.e., 1 mg/d) are required in elderly subjects to normalise serum MMA or tHcy (Rajan et al. 2002).

We recommend combining a screening test (total cobalamin or optimally holoTC) with MMA test to verify functional cobalamin deficiency. In case of

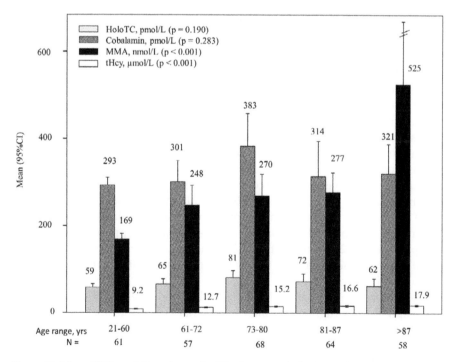

Figure 10. Mean (95% confidence intervals, CI) of serum cobalamin markers according to age groups. Cobalamin and holoTC are not significantly different between age groups, but MMA and tHcy are higher at older age. Higher concentrations of MMA and tHcy in older subjects remained significant after adjustment for serum creatinine (p < 0.001, univariate analysis of variance test).

contradicted findings (i.e., normal holoTC and elevated MMA), a marked lowering of MMA after treatment, suggests that a pre-treatment deficiency is very likely. Causes of cobalamin deficiency in elderly subjects need careful assessment of dietary and medical information. Food cobalamin malabsorption due to difficulties in releasing cobalamin from dietary proteins is common in the elderly, while free cobalamin can be absorbed from supplements or fortified foods.

5.3 Elevated total cobalamin in patients with cancer

Cobalamin deficiency can worsen the clinical condition in patients with cancer who receive chemotherapy. The anti-cancer drugs, antifolates, have many side effects such as anaemia, peripheral neuropathy, and severe liver and gastrointestinal toxicity. Though folinic acid is used for prevention of toxicity, cobalamin supplementation is not used routinely. Cobalamin deficiency can be causally involved in the side effects of chemotherapy such as peripheral-neuropathy (Schloss et al. 2015), depression, and anaemia. Diagnosis of cobalamin deficiency in patients with cancer is important for preventing further complications.

Serum cobalamin concentrations are high rather than low in patients with cancer (Carmel and Eisenberg 1977; Carmel and Hollander 1978; Mendelsohn et al. 1958). Patients with serum cobalamin levels > 300 pmol/L had poor overall survival and time to progression of colorectal cancer than patients with levels below this value (Bystrom et al. 2009). These results may theoretically suggest that cobalamin deficiency is not common and supplementation should not be recommended for patients with cancer. Elevated serum concentrations of cobalamin in patients with cancer is often wrongly interpreted as indicating normal status, without further assessment of MMA or tHcy. High serum cobalamin could be a result of elevated haptocorrin and transcobalamin or a release of cobalamin from the damaged cells in more advanced stages of cancer.

A recent study has tested the concentrations of total cobalamin, MMA, and tHcy in a group of 316 cancer patients without any prior selection according to the tumor type or other clinical conditions (Vashi et al. 2016). A lowered concentration of total cobalamin (defined as levels < 220 pmol/L) was found in approximately 2% of the patients. The mean total serum cobalamin in the whole patient group was 431 pmol/L, suggesting that the majority of the patients had a normal to high status. However, 11–17% of the patients had elevated concentrations of MMA or tHcy. Elevated levels of MMA and tHcy were explained by higher age, men sex, higher creatinine, and the presence of malnutrition (Vashi et al. 2016). Elevated MMA levels were also related to previous cancer treatments compared with newly diagnosed cancers. The highest MMA levels were found in patients with colorectal, prostate, or pancreas cancer (Vashi et al. 2016).

Serum cobalamin markers were measured in a group of patients with cancer that were treated with antifolates (Niyikiza et al. 2002). Higher concentrations of tHcy were related to more severe hematological toxicity, diarrhea, and neutropenia after antifolates. Patients who had high levels of tHcy, MMA, or a combination of both before starting the antifolate treatment, had higher prevalence of selected severe toxicities (i.e., infections, thrombocytopenia, neutropenia, diarrhea) (Niyikiza et al. 2002). Recent studies failed however, to reduce the toxicity after chemotherapeutics by pre-treatment supplementation of folate and cobalamin (Okuma et al. 2015). The current available studies lack the power to show this outcome. The efficacy of a pre-treatment with cobalamin and folate maybe affected by other factors such as the tumor type, age of the patient, a long-term deficiency, depression, ability to absorb, or existing drugs. Obviously, more studies are needed in this field.

Therefore, serum cobalamin is not an appropriate marker for evaluation of cobalamin status in patients with cancer. High values can be seen as a result of cancer and can be related to more advanced cases. In general, cobalamin metabolic markers show unexpected diversion from that of total cobalamin in patients with cancer. Cobalamin treatment in patients with cancer in order to prevent co-morbidities is an unresolved debate that obviously needs more randomized controlled studies.

5.4 Cobalamin markers in newborn and infants

Cobalamin deficiency is common in newborns of mothers that have had cobalamin deficiency or low intake during pregnancy and lactation. Clinical symptoms become manifested especially in breastfed infants at the age of 4–6 months (see Chapter 11). Infants present usually with anaemia and/ or neurological symptoms. Diagnosis of cobalamin deficiency in infants is important for prevention of irreversible or long-term damages.

Concentrations of total cobalamin and MMA in newborn infants show dependence on infant's age, feeding patterns, and maternal vitamin status. Concentrations of cobalamin and MMA are high in cord blood, but they decline with age up to adulthood. The highest concentrations of MMA are found around the age of 4–6 months. This period is also associated with a simultaneous decline in holoTC and cobalamin and an elevation in plasma tHcy, suggesting a gap in cobalamin status at this age. Low cobalamin status does not completely explain elevated MMA concentrations in infants. Propionate that are produced by intestinal bacteria and absorbed into the blood can cause elevation of serum MMA that is not related to cobalamin (Thompson et al. 1990).

Evaluation of cobalamin status in infants should consider a combination of several markers (total cobalamin, holoTC, MMA and tHcy). The commonly used reference ranges for cobalamin markers in adults cannot be extrapolated to infants. Testing tHcy and differential diagnosis of folate deficiency are important in some cases. The presence of relevant neurological or hematological symptoms, or severely elevated tHcy/ and or MMA that occur shortly after birth suggests that the deficiency could be related to congenital defects of cobalamin metabolism or transport. Causes of inherited conditions need further genetic and laboratory tests of blood, urine, and biopsies (Table 6) (see Chapter 4).

Table 6. Laboratory markers used for differential diagnosis of cobalamin genetic disorders.

Disorder	Gene	Biochemical findings	Other markers
Transcobalamin deficiency[1]	TCN2	↓ or undetectable holoTC	total B12 can be low or low normal (80% haptocorrin)
Transcobalamin receptor defects	CD320	↑ MMA, ↑ tHcy	normal holoTC, normal B12
cblA	MMAA	↑ MMA	normal holoTC, normal B12, normal tHcy
cblB	MMAB	↑ MMA	normal holoTC, normal B12, normal tHcy
cblC	MMACHC	↑ MMA, ↑ tHcy	normal holoTC, normal B12
cblD	MMADHC	↑ MMA or ↑ tHcy or both ↑ MMA, ↑ tHcy	normal holoTC, normal B12
cblE	MTRR	↑ tHcy	normal holoTC, normal B12, normal MMA
cblF	LMBRD1	↑ MMA, ↑ tHcy	normal holoTC, normal B12
cblG	MTR	↑ MMA, ↑ tHcy	normal holoTC, normal B12
cblJ	ABCD4	↑ MMA, ↑ tHcy	normal holoTC, normal B12
cblX	HCFC1	↑ MMA, ↑ tHcy	normal holoTC, normal B12
Methylmalonyl-CoA Mutase Deficiency	MUT	↑ MMA	normal holoTC, normal B12, normal tHcy

[1] In some cases holoTC can be present but not functional.

5.5 Cobalamin markers in pregnant women

Cobalamin deficiency during early pregnancy has been related to birth defects, early pregnancy loss, and pregnancy complications. Depleted women and those with low intake cannot transfer sufficient amount of the vitamin to their children. Screening for cobalamin deficiency is recommended in women with a history of birth defects, pregnancy complications, and those on a strict vegetarian diet, but not receiving supplements.

Concentrations of serum cobalamin decline during pregnancy mainly because of the decline in haptocorrin (Morkbak et al. 2007). Pregnant women upon delivery may have 50% lower total serum cobalamin compared to few weeks postpartum. In contrast to holohaptocorrin, holoTC remains stable during pregnancy. For this reason, holoTC is regarded as a better marker than total cobalamin for screening or monitoring cobalamin status during pregnancy (Morkbak et al. 2007; Murphy et al. 2007). A low holoTC test result should be followed by testing MMA in serum or urine. If MMA is elevated, supplementation of cobalamin to pregnant women can be recommended. In women who have an intact absorption system, physiological doses of cobalamin are sufficient to maintain maternal and child status. Pharmacological doses (> 500 μg/d) maybe required when serum MMA is above 500 nmol/L or when pernicious anemia is diagnosed during pregnancy. The decline in haptocorrin during pregnancy has no known clinical significance.

6. Monitoring the Effect of Cobalamin Treatment

6.1 Cobalamin markers: response to cobalamin intake or supplementation

Minor changes in cobalamin intake induce different responses in serum concentrations of cobalamin markers (Figure 11). When cobalamin supply becomes insufficient (depletion) or cobalamin is provided as supplements (repletion), the changes of the markers follow a time-dependent course.

A dietary intake of approximately 7 μg/d is necessary to maintain cobalamin status markers in young adults (Bor et al. 2010). Very small doses of cobalamin from supplements (approximately 3 μg/d) were associated with better cobalamin markers in vegans and lacto-ovo-vegetarians compared with non-supplemented individuals (Herrmann et al. 2003). Similarly, physiological

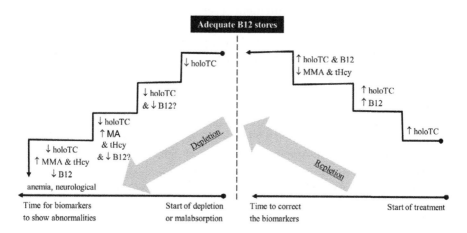

Figure 11. Theoretical time-progress of changes in cobalamin biomarkers under depletion and repletion conditions. Depending on the cause, the deficiency may take several years to develop after the start of depletion. In malabsorption disorders, the depletion develops to clinical symptoms within a relatively shorter time. The repletion causes a correction of serum holoTC as a first response. The increase of serum total cobalamin and the lowering of MMA and tHcy depend on the dose, frequency and route of administration.

oral doses of cobalamin (3–10 μg/d) are associated with corrections of all cobalamin biomarkers (Winkels et al. 2008). Therefore, very low doses can be used to prevent cobalamin deficiency in subjects who can absorb the vitamin.

Pharmacological doses of cobalamin provided as injections lower plasma concentrations of MMA and tHcy in deficient subjects within few days (Mansoor et al. 2013). Cyanocobalamin (i.m. 1 mg/week) was provided for 3 weeks to a group of 14 patients with severely elevated serum MMA levels. Concentrations of MMA and tHcy were strongly lowered (not normalised) few days after the first injection. After 21 days (total of 3 injections), concentrations of MMA were almost normalised (Mansoor et al. 2013). Also urinary MMA (uMMA) concentrations show a response after oral cyanocobalamin supplementation (Hill et al. 2013). Elderly people (n = 100 with serum cobalamin < 250 pmol/L and elevated uMMA > 1.5 μmol/mmol creatinine) were supplemented with oral cyanocobalamin (10 μg, 100 μg, or 500 μg) for 56 days. Plasma MMA and to a less extent, uMMA showed a significant response to supplementation (Hill et al. 2013).

Oral pharmacological doses of cobalamin (> 0.5 mg) are commonly used to treat or prevent deficiency. The absorption takes place partly via simple diffusion and can correct abnormal cobalamin status markers (Garcia et al. 2002; Sharabi et al. 2003; van Walraven et al. 2001). Oral supplementation of 1–2 mg cobalamin show effectiveness in correcting cobalamin markers in many, but not all patients. Therefore, there are no strict recommendations regarding the dose and route of administration of cobalamin supplements for patients with cobalamin deficiency. Physicians depend on trial-and-errors to find out the best treatment mode. Measurement of serum MMA levels is recommended for monitoring the outcome of the treatment.

There is a dose response relationship between cobalamin intake or supplemental dose and the increase of serum total cobalamin or the reduction of MMA and tHcy (Rajan et al. 2002). However, a metaanalysis has recently shown that the association between cobalamin doses and the changes in serum cobalamin markers is not linear over the range of intake or supplemental doses (Dullemeijer et al. 2013). For example, doubling cobalamin dietary intake was associated with approximately 11–13% higher serum cobalamin concentrations (Dullemeijer et al. 2013). Age was an important determinant for the response of serum cobalamin to supplements (Dullemeijer et al. 2013).

The effect of oral treatment with cobalamin on plasma cobalamin-binding proteins has been studied in non-deficient subjects over 84 days. Supplementation of vitamin B12 (400 μg/d) caused a fast and strong response in plasma holoTC, TC-saturation, and total TC. The highest levels of holoTC and TC-saturation were reached only 3 days after the start of oral supplements (means increase, +54% for holoTC and +82% for TC-saturation, respectively), whereas total-TC decreased by 16% to reach its lowest level after 3 days (Nexo et al. 2002b). HoloHC and total cobalamin increased by +20% and +28%, respectively, after 3 days but their levels continued to increase until the end of the study (84 days) (Nexo et al. 2002b). Changes of cobalamin binders and their saturation levels are currently used only for research questions but not routinely monitored in patients with deficiency.

The strong increase of serum holoTC and TC-saturation within 3 days suggested that these markers can be used to show recent cobalamin repletion, while holoHC and total cobalamin were discussed to reflect accumulation of the vitamin (Nexo et al. 2002b).

6.2 Using serum holotranscobalamin increase to test cobalamin absorption

As an alternative to the traditional Schlilling test, the change of serum holoTC following oral cobalamin has been used as a surrogate absorption test called the 'CobaSorb' test (Bor et al. 2005; Hardlei et al. 2010). Adults with normal cobalamin status were given three oral doses at 6-hours intervals (each dose 9 µg cobalamin) and changes of cobalamin markers were followed in serum samples collected over 3 days (von Castel-Roberts et al. 2007). The low dose of (3 x 9 µg) was chosen because it is believed to be actively transported via intrinsic factor. The increase of holoTC reached a maximum (+ 49% relative to baseline) at 24 hours while the changes in total cobalamin were less prominent (+ 15% from baseline) within 24 hours (von Castel-Roberts et al. 2007). The same test was applied to study cobalamin absorption in Indian subjects with a low cobalamin status (mean holoTC concentration at baseline ~ 11 pmol/L). Oral cyanocobalamin [3 x 10 µg or 3 x 2 µg doses] was administered within 6-hours intervals (Bhat et al. 2009). Serum holoTC levels increased in the 10 µg group from (mean ± SD) 9 ± 7 pmol/L to 54 ± 26 pmol/L and in the 2 µg group from 11 ± 9 pmol/L to 36 ± 19 pmol/L suggesting that this test can be used to verify cobalamin absorption in people with very low cobalamin status (Bhat et al. 2009). In contrast, the test appears to have limitations in subjects with holoTC > 60 pmol/L, because high holoTC shows limited response to cobalamin intake. Therefore, a negative test result should not be interpreted as 'cobalamin malabsorption' in subjects with high baseline holoTC.

Our group used the same principle of measuring the change of serum holoTC after cobalamin supplementation, with exception of using a single oral dose of 200 µg cyanocobalamin and measuring holoTC before and 6–8 hours after cyanocobalamin (Figure 12). HoloTC concentrations reached highest levels between 6–8 hours after the 200 µg dose. The increase of holoTC after this cobalamin loading dose can be interpreted as being able to benefit from oral supplements. This simple test is particularly helpful in patients who need to start oral treatment or those who shift from injections to oral treatment to ensure that they can absorb. This test does not make any assumption about the mechanisms of cobalamin absorption after 200 µg cyanocobalamin.

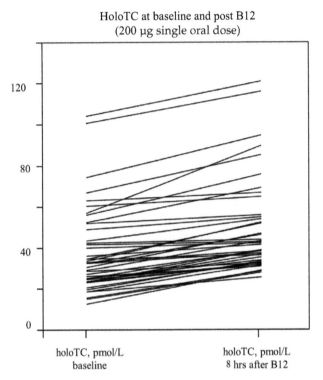

HoloTC at baseline and post B12
(200 µg single oral dose)

Figure 12. Baseline serum holoTC concentrations and 8 hours post-oral cobalamin single dose of 200 µg. HoloTC increased in all, except in 2 subjects. All participants were apparently healthy and did not report malabsorption disorders.

7. Differential Diagnosis

Nutritional deficiencies

Folate and cobalamin deficiencies can cause macrocytic anaemia (Aslinia et al. 2006). Patients with anaemia should be screened for nutritional deficiencies, among other possible causes. However, macrocytosis (i.e., elevated mean corpuscular volume) can be absent in many cases of folate and cobalamin deficiencies. This limits the sensitivity and specificity of blood count in detecting or differentiating between causes of anemia. Measurement of serum MMA is used for differential diagnosis of folate and cobalamin deficiencies, since the concentrations of this marker are hardly affected by folate deficiency. In contrast, tHcy concentrations are elevated in cobalamin and folate deficiencies but this marker is less affected by cobalamin deficiency than by folate deficiency, thus limiting its use as a marker for differential diagnosis of these two conditions.

Iron status markers (i.e., ferritin, transferrin, and transferrin saturation) are used for differential diagnosis of anaemia caused by iron deficiency. Serum and red blood cell folate are useful indexes of short and long term folate status, respectively. Serum folate concentrations can be elevated in subjects with cobalamin deficiency and show a decline after cobalamin treatment (Dierkes et al. 1999). This phenomenon is likely to be caused by a disturbance in the flow of the folate cycle (i.e., folate trap) when methylcobalamin as a cofactor for methionine synthase is deficient.

In population studies, differential diagnosis of nutritional deficiencies is more challenging since combined deficiencies of amino acids and micronutrients can be involved and metabolic adaptation to nutrient shortage is well known in populations with sessional variations of nutrients intake (Dominguez-Salas et al. 2013).

Genetic defects in cobalamin and folate pathways

Classical forms of cobalamin and folate genetic disorders are detected after birth or during early childhood. There are also late-onset forms that are expressed in adults. The manifestations in these genetic disorders are severe and involve both anemia and neurological signs. Moreover, the elevation of tHcy and/or MMA is severe compared with that seen in nutritional deficiencies. Early diagnosis is important since many defects are treatable by regular administration of the vitamin.

Several laboratory markers are used for differential diagnosis of genetic defects in cobalamin pathway. The combined or isolated elevation of tHcy and MMA gives important clues to the defect (Table 6). They can differentiate between cobalamin defects that affect methylation of tHcy to methionine (i.e., tHcy is elevated in this case) and those affecting the mitochondrial pathway (MMA is elevated). Elevated tHcy and MMA are seen in all defects that affect delivering cobalamin to the cell, such as defects in intrinsic factor or its receptor and defects in transcobalamin or its receptor. Isolated elevation of MMA is usually seen in disorders that affect intracellular delivery of cobalamin into the mitochondria or the cobalamin–dependent mitochondrial enzyme, methylmalonyl-CoA mutase. Several steps in cobalamin pathway have been discovered recently, but many transport or regulatory steps are still unknown.

Genetic defects in folate metabolism can cause similar clinical picture. The differential diagnosis of defects that affect cobalamin and folate pathways includes measurement of tHcy and MMA (MMA is not elevated in folate defects) in addition to other markers in the methionine and folate cycles. The biomarkers are generally used for orientation. In a following step, genetic testing for specific mutations is performed to identify the defect. Moreover, *in-vitro* experiments using patient's fibroblasts are still used in many labs in order to identify the nature of the defect by studying cobalamin uptake or its conversion to cofactors (Gulati et al. 1996; Stucki et al. 2012).

Biomarkers of cobalamin (mainly MMA and tHcy) are used to monitor the treatment effect in patients with genetic defects.

8. Cobalamin Markers in Other Biological Samples

Cerebrospinal fluid

Cobalamin is actively transported into the brain, but the exact process is poorly understood. Cobalamin is stored in the brain, but the total brain content of cobalamin is lower than that in the liver and the kidney (Birn et al. 2003). In contrast to serum cobalamin, most of cobalamin in cerebrospinal fluid (CSF) is bound to transcobalamin (Lazar and Carmel 1981). Concentrations of holoTC in CSF are approximately 20%–30% of that in serum. The median holoTC in CSF was 16 pmol/L vs. 70 pmol/L in serum of the same subjects (Obeid et al. 2007b). In CSF, the ratio of holoTC to albumin is higher than that in blood (Obeid et al. 2007b), suggesting that the amount of holoTC in CSF is over-represented compared to the blood.

Serum cobalamin concentration is the main determinant of CSF holoTC (Obeid et al. 2007b). CSF-holoTC shows a strong correlation with serum total cobalamin (correlation coefficient = 0.69, p < 0.001). The correlation between blood and CSF cobalamin contents was not found in some old studies (KIDD et al. 1963). However, old methods were probably not able to detect very low levels of cobalamin in CSF. Concentrations of holoTC in CSF can be influenced by polymorphisms in transcobalamin gene (Zetterberg et al. 2003). Also diseases are known to affect CSF cobalamin content. For example, CSF concentrations of cobalamin were significantly lower in patients with Alzheimer disease compared with age-matched controls (Ikeda et al. 1990). In contrast, serum cobalamin did not differ between the groups (Ikeda et al. 1990).

We observed no difference in CSF-holoTC between patients with and without dementia and no relationship between holoTC and P-tau 181 or amyloid beta 42 (i.e., both are markers of dementia) (Obeid et al. 2007a). CSF-holoTC concentrations did not decline with age or show significant differences between control patients and those with peripheral neuropathy, multiple sclerosis, or stroke. CSF-folate but not holoTC was the main determinant of CSF-tHcy (Obeid et al. 2007b). Therefore, CSF holoTC is unlikely to be a reliable marker for brain cobalamin-dependent reactions or a risk marker for diseases supported by low cobalamin status.

Similarly, concentrations of soluble transcobalamin receptor (sCD320) are detectable in CSF, and the levels are lower than in plasma (median 14 pmol/L vs 73 pmol/L) (Abuyaman and Nexo 2015). Although sCD320 levels in CSF showed correlation with markers of dementia, the study design (rest samples from the lab, pre-analytical conditions) has major limitations and no conclusions on causal associations can be made.

Differences between studies of CSF holoTC and its association with physiological or diseases conditions could be related to bias in selection of the control group or inclusion and exclusion criteria of the studies. Also differences in pre-analytical conditions (i.e., CSF contaminated with blood) could affect the results.

CSF-cobalamin markers and binders have currently no roles in diagnosis of deficiency conditions. Studies investigating their concentrations in CSF aim at understanding cobalamin brain physiology and identifying potential involvement in neurological disorders.

Human milk

Studies testing cobalamin content in human milk have provided valuable information on determinants of milk cobalamin as a sole source of this nutrient in breastfed infants. Studies on milk samples from deficient and non-deficient women have participated in shaping the nutritional recommendations for this nutrient in infants and children. Cobalamin intake and supplementation recommendations during pregnancy and lactation should be guided by studies on milk samples in relation to maternal serum cobalamin markers.

Human milk is not a typical sample matrix in a clinical laboratory, but rather a typical matrix for food analytics. Milk contains a high amount of apo-haptocorrin that has been reported to disturb most available cobalamin assay methods and can produce unreliably high or low values (Lildballe et al. 2009). Haptocorrin content increases during the course of lactation. There are also differences in milk composition and nutritional value according to the time of sample collection during one lactation session (foremilk, hindmilk). Results from different studies are difficult to compare due to methodological differences and also differences in nutritional statuses of the women from different countries.

Early methods have used different extraction procedures to remove proteins from the milk before measuring cobalamin content using regular methods. Jathar et al. released cobalamin from milk samples by proteolytic degradation prior to measurement by a microbiological assay (Jathar et al. 1970). The extraction was performed using papain and sodium cyanide and incubation under toluene at 37°C overnight (Jathar et al. 1970). The study reported very low concentrations of cobalamin in serum of Indian women and no correlation with milk cobalamin. Milk cobalamin was lower in this study (Jathar et al. 1970) than levels reported in earlier studies (Collins et al. 1951). Several modifications of the extraction method were used such as a hot extraction of the sample with ethanol or digestions using papain prior to the assay (Adjalla et al. 1994; Specker et al. 1990). Analysis of human milk using a competitive protein binding assay was able to distinguish milk samples from B12-deficient and well-nourished lactating women (McPhee et al. 1988). However, comparison between different studies is still not possible.

Recently, milk cobalamin assay was improved by using cobinamide-sepharose to remove apo-haptocorrin from the fat-reduced whey fraction before applying commercially available cobalamin assays such as methods on Centaur, Architect and Cobas analysers (Lildballe et al. 2009). HC concentrations below 10 nmol/L do not interfere with cobalamin assay. Milk samples analysed without prior extraction of apo-haptocorrin by using different commercially available immunoassay systems coupled with chemiluminescence detection resulted in too high (Centaur analyzer, Siemens) or too low (Architect i2000 analyzer, Abbott; Cobas 6000 E immunoassay system, Roche Diagnostics) concentrations of B12 (Lildballe et al. 2009).

When apo-haptocorrin cannot be removed in a preliminary step, the methods of choice for direct measurement of human milk cobalamin is the competitive protein binding assay coupled with chemiluminescence detection (Deegan et al. 2012; Hampel et al. 2014; Israel-Ballard et al. 2008).

In summary, human milk analytic has improved over the last years. Milk cobalamin assessment is helpful to understand cobalamin status and requirements in infants and mothers and derive cobalamin intake recommendations. However, there are still many sources of heterogeneity. There is an urgent need for standardisations of the time of sample collection after birth and during one lactation session (foremilk, hindmilk); the pre-extraction methods; and the final assay method of cobalamin. The analytical methods for human milk are not used for routine testing and the importance of routine testing of milk has not been established yet.

Cord blood

Concentrations of cobalamin and its binding proteins were studied in samples collected from umbilical venous blood after birth and were compared with maternal concentrations. The results provided evidence about the strong positive association between maternal and child vitamin status and changes in cobalamin homeostasis throughout the pregnancy or after birth. Studies performed on cord blood did not aim at diagnosing a deficiency condition.

Concentrations of total cobalamin in cord blood are higher than in the mother (Molloy et al. 2002; Obeid et al. 2005b). This is explained by high concentrations of total HC, holoHC, and holoTC while total TC is lower in cord blood than in the mother (Obeid et al. 2006). The main predictors of holoTC in cord blood were gestational age and maternal cobalamin concentrations.

High concentrations of cobalamin and holoTC in cord blood do not agree with the high levels of tHcy or MMA in cord blood (Obeid et al. 2006). We found that cord blood concentrations of MMA were higher than those of the mothers, but were unrelated to the concentrations of both holoTC and holoHC in the child (Obeid et al. 2005b; Obeid et al. 2006). Therefore, MMA concentrations in cord blood cannot be considered as a marker for cobalamin status since they can be influenced by other unknown factors.

Urine

Concentrations of MMA are measured in urine samples using several GCMS or LC-MS/MS methods. This test is used as part of cobalamin diagnosis in many labs as discussed above. Total cobalamin and holoTC have been also measured in human urine, but their use in routine diagnosis is not established.

The kidney plays a central role as a storage organ for cobalamin. Cobalamin is filtered through the kidney and reabsorbed to enter the blood again (Birn 2006; Moestrup et al. 1996). Kidney diseases have a strong effect on cobalamin biomarkers in blood and urine.

Studying cobalamin concentrations in urine may shed the light on factors affecting cobalamin kidney homeostasis and causes of elevated serum cobalamin and holoTC concentrations in patients with renal dysfunction. Only few studies have tested urine cobalamin concentrations.

We have measured urine concentrations of holoTC (uholoTC) in patients with diabetes and apparently healthy controls. In general, uholoTC levels are very low and show high levels of between-subjects variations (Table 7). The mean (SD) concentrations were not different between the groups [6.1 (14.1) pmol/l in diabetics vs. 6.7 (10.4) pmol/l in the controls] (Obeid et al. unpublished data). Concentrations of uholoTC showed a positive correlation with urinary albumin in patients with diabetes ($r = 0.535$, $p < 0.001$), and RBC-B12 ($r = 0.512$, $p < 0.001$), and a negative correlation with serum MMA ($r = -0.331$, $p = 0.003$), but no correlation with serum holoTC or total cobalamin. These correlations were not significant in the control subjects. Therefore, markers that show higher intracellular cobalamin or cobalamin stores (higher with RBC-B12 and lower MMA) were associated with higher uholoTC, suggesting that uholoTC is a way to excrete excess cobalamin in subjects with sufficient stores. Several questions concerning uholoTC remain open; such as potential circadian changes, day-to-day variations, influence of dietary intake or supplementation, and influence of renal diseases and drugs (diuretics). The concentrations of uholoTC are in the lowest range of most available analytical assays which may be associated with higher analytical imprecision.

The soluble form of transcobalamin receptor (sCD320) has been measured in serum (pmol/L) and urine (pmol/mmol creatinine) of young women

Table 7. Serum and urinary cobalamin markers in control subjects and patients with diabetes.

	Controls	Diabetes	p
Age, years	65 ± 8	66 ± 9	0.263
uHoloTC, pmol/L	6.7 ± 10.4	6.1 ± 14.1	0.518
urine albumin, mg/l	7.0 ± 9.2	29.6 ± 86.1	0.353
uHoloTC/ualbumin ratio	1.11 ± 1.65	0.50 ± 0.87	0.651
Serum holoTC, pmol/L	63 ± 37	76 ± 55	0.080
Serum MMA, nmol/L	239 ± 157	307 ± 274	0.018
Serum total B12, pmol/L	296 ± 117	291 ± 102	0.835
RBC-B12, pmol/L	260 ± 50	233 ± 42	0.001
Serum creatinine, mg/dl	0.85 ± 0.18	0.93 ± 0.29	0.169

Results are shown as mean ± SD. P values are according to Mann-Whitney test. Controls (n = 50); patients (n = 80). uholoTC and serum holoTC were measured by using AxSYM active B12 assay. The method for red blood cell (RBC)-B12 extraction and assay in addition to the original study were published before (Obeid et al. 2013).

throughout pregnancy (Abuyaman et al. 2013). Concentrations of sCD320 increased in serum and in urine up to the 35 gestational week, thereafter they declined. Urinary sCD320 (pmol/mmol creatinine) correlated with serum sCD320 (Spearman's correlation = 0.647, P < 0.001). The absolute molar concentrations of urinary sCD320 were higher than that in serum and correlated with urinary creatinine (Abuyaman et al. 2013). sCD320 shows some affinity to bind holoTC, but the clinical significance of measuring serum or urinary sCD320 is currently not established.

9. Unresolved Issues Surrounding Cobalamin Biomarkers

From the current knowledge and the challenges that cobalamin diagnosis optimally needs to meet, the following questions represent an extraction of unresolved issues for research to focus on in the following years:

- What are the biological variations and indexes of individuality for cobalamin markers in health and disease conditions?
- Most subjects with elevated MMA or lowered holoTC or B12 will be asymptomatic; but how stable are these markers over the time, if subjects are not treated with cobalamin?
- Which marker or a combination of markers can better predict specific clinical outcomes or recovery of clinical signs in deficient patients?
- Which marker should be tested in patients with cancer?
- Recommendations are needed on diagnosis, treatment and prevention of cobalamin deficiency in patients with cancer.
- What are the mechanisms of elevated holoTC in renal diseases and cancer?
- Which factors can affect uMMA and what is their influence on using this marker in routine diagnosis?
- Are there any advantages of using uMMA compared with serum MMA in specific disease groups or age groups (i.e., infants)?
- Should haptocorrin and transcobalamin be part of diagnosing cobalamin deficiency or its causes? (when?)
- What are the mechanisms of the changes of cobalamin markers and binders during pregnancy?
- What are the causes of the changes in the binders and saturation levels during early life?
- Is there any diagnostic value for measuring sCD320 in biological samples (serum, urine, CSF)?
- What are the biological roles of haptocorrin and transcobalamin in human milk? Why they change during lactation?
- Standardisations of the pre-analytical and analytical conditions are needed before combining or comparing data on Cbl status markers from different labs.

10. Summary and Conclusions

- In recent years, diagnosis of cobalamin deficiency aimed at detecting subjects with 'biochemical abnormalities' or subtle cobalamin deficiency.
- Targeted testing of cobalamin markers is recommended in subjects with diseases that cause a deficiency or diseases that might be worsened if cobalamin deficiency co-exists.
- Studies measuring cobalamin markers in 'at risk populations' have provided valuable information on the relation between intake and markers, prevalence of the deficiency, and response to intervention. However, screening asymptomatic individuals is not recommended for health care providers.
- The availability of methods to measure cobalamin fractions and its metabolic markers has improved the diagnosis. Currently, stepwise measurement of cobalamin markers, holoTC and MMA, is the best recommended approach to make an informed decision on cobalamin status when cobalamin deficiency is suspected.
- The general improvement of the diagnosis is facing new challenges related to unclassical or contradicted results of the biomarkers in some clinical conditions such as renal patients, liver diseases or cancer.
- Elevated serum concentrations of cobalamin in these conditions disagree with elevated MMA. There are many open questions related to the necessity to treat 'elevated MMA'. However, subtle cobalamin deficiency 'despite normal serum cobalamin' can be involved in co-morbidities of some diseases, such as patients with diabetes, cancer or renal insufficiency.

Keywords: deficiency, diagnosis, biomarkers, algorithm, blood samples, assay, dose, treatment, subclinical, functional markers, holotranscobalamin, total cobalamin, methylmalonic acid, homocysteine

Abbreviations

CSF	:	cerebrospinal fluid
CV	:	coefficient of variation
GCMS	:	gas chromatography mass spectrometry
HC	:	haptocorrin
holoHC	:	holohaptocorrin
HoloTC	:	holotranscobalamin
MCV	:	mean corpuscular volume
MMA	:	methylmalonic acid
LC-MS/MS	:	liquid chromatography tandem mass spectrometry
RDA	:	Recommended Dietary Allowance
ROC	:	Receiver Operating Characteristic
TC	:	transcobalamin
tHcy	:	total homocysteine

References

Abuyaman O, Andreasen BH, Kronborg C, Vittinghus E and Nexo E. 2013. The soluble receptor for vitamin B12 uptake (sCD320) increases during pregnancy and occurs in higher concentration in urine than in serum. PLoS One. 8: e73110.

Abuyaman O and Nexo E. 2015. The soluble transcobalamin receptor (sCD320) is present in cerebrospinal fluid and correlates to dementia-related biomarkers tau proteins and amyloid-beta. Scand J Clin Lab Invest. 75: 514–518.

Adjalla c, Lambert D, Benhayoun S, Berthelsen JG, Nicolas JP, Gueant JL and Nexo E. 1994. Forms of cobalamin and vitamin B12 analogs in maternal plasma, milk, and cord plasma. In: p 406–410.

Agreus L, Kuipers EJ, Kupcinskas L, Malfertheiner P, Di MF, Leja M, Mahachai V, Yaron N, van OM, Perez PG, Rugge M, Ronkainen J, Salaspuro M, Sipponen P, Sugano K and Sung J. 2012. Rationale in diagnosis and screening of atrophic gastritis with stomach-specific plasma biomarkers. Scand J Gastroenterol. 47: 136–147.

Areekul S, Churdchu K, Cheeramakara C, Wilairatana P and Charoenlarp P. 1995. Serum transcobalamin II levels in patients with acute and chronic renal failure. J Med Assoc Thai. 78: 191–196.

Arnaud J, Cotisson A, Meffre G, Bourgeay-Causse M, Augert C, Favier A, Vuillez JP and Ville G. 1994. Comparison of three commercial kits and a microbiological assay for the determination of vitamin B12 in serum. Scand J Clin Lab Invest. 54: 235–240.

Aslinia F, Mazza JJ and Yale SH. 2006. Megaloblastic anemia and other causes of macrocytosis. Clin Med Res. 4: 236–241.

Bailey RL, Carmel R, Green R, Pfeiffer CM, Cogswell ME, Osterloh JD, Sempos CT and Yetley EA. 2011. Monitoring of vitamin B12 nutritional status in the United States by using plasma methylmalonic acid and serum vitamin B12. Am J Clin Nutr. 94: 552–561.

Bailey RL, Durazo-Arvizu RA, Carmel R, Green R, Pfeiffer CM, Sempos CT, Carriquiry A and Yetley EA. 2013. Modeling a methylmalonic acid-derived change point for serum vitamin B12 for adults in NHANES. Am J Clin Nutr. 98: 460–467.

Baker H, Leevy CB, DeAngelis B, Frank O and Baker ER. 1998. Cobalamin (vitamin B12) and holotranscobalamin changes in plasma and liver tissue in alcoholics with liver disease. J Am Coll Nutr. 17: 235–238.

Bamonti F, Moscato GA, Novembrino C, Gregori D, Novi C, De GR, Galli C, Uva V, Lonati S and Maiavacca R. 2010. Determination of serum holotranscobalamin concentrations with the AxSYM active B(12) assay: cut-off point evaluation in the clinical laboratory. Clin Chem Lab Med. 48: 249–253.

Bhat DS, Thuse NV, Lubree HG, Joglekar CV, Naik SS, Ramdas LV, Johnston C, Refsum H, Fall CH and Yajnik CS. 2009. Increases in plasma holotranscobalamin can be used to assess vitamin B12 absorption in individuals with low plasma vitamin B12. J Nutr. 139: 2119–2123.

Birn H. 2006. The kidney in vitamin B12 and folate homeostasis: characterization of receptors for tubular uptake of vitamins and carrier proteins. Am J Physiol Renal Physiol. 291: F22–F36.

Birn H, Nexo E, Christensen EI and Nielsen R. 2003. Diversity in rat tissue accumulation of vitamin B12 supports a distinct role for the kidney in vitamin B12 homeostasis. Nephrol Dial Transplant. 18: 1095–1100.

Bizzaro N and Antico A. 2014. Diagnosis and classification of pernicious anemia. Autoimmun Rev. 13: 565–568.

Bor MV, Cetin M, Aytac S, Altay C and Nexo E. 2005. Nonradioactive vitamin B12 absorption test evaluated in controls and in patients with inherited malabsorption of vitamin B12. Clin Chem. 51: 2151–2155.

Bor MV, Lydeking-Olsen E, Moller J and Nexo E. 2006. A daily intake of approximately 6 microg vitamin B12 appears to saturate all the vitamin B12-related variables in Danish postmenopausal women. Am J Clin Nutr. 83: 52–58.

Bor MV, von Castel-Roberts KM, Kauwell GP, Stabler SP, Allen RH, Maneval DR, Bailey LB and Nexo E. 2010. Daily intake of 4 to 7 microg dietary vitamin B12 is associated with steady concentrations of vitamin B12-related biomarkers in a healthy young population. Am J Clin Nutr. 91: 571–577.

Brady J, Wilson L, McGregor L, Valente E and Orning L. 2008. Active B12: a rapid, automated assay for holotranscobalamin on the Abbott AxSYM analyzer. Clin Chem. 54: 567–573.

Brito A, Verdugo R, Hertrampf E, Miller JW, Green R, Fedosov SN, Shahab-Ferdows S, Sanchez H, Albala C, Castillo JL, Matamala JM, Uauy R and Allen LH. 2016. Vitamin B12 treatment of asymptomatic, deficient, elderly Chileans improves conductivity in myelinated peripheral nerves, but high serum folate impairs vitamin B12 status response assessed by the combined indicator of vitamin B12 status. Am J Clin Nutr. 103: 250–257.

Bystrom P, Bjorkegren K, Larsson A, Johansson L and Berglund A. 2009. Serum vitamin B12 and folate status among patients with chemotherapy treatment for advanced colorectal cancer. Ups J Med Sci. 114: 160–164.

Carmel R. 1996. Prevalence of undiagnosed pernicious anemia in the elderly. Arch Intern Med. 156: 1097–1100.

Carmel R. 2003. Mild transcobalamin I (haptocorrin) deficiency and low serum cobalamin concentrations. Clin Chem. 49: 1367–1374.

Carmel R, Aurangzeb I and Qian D. 2001a. Associations of food-cobalamin malabsorption with ethnic origin, age, Helicobacter pylori infection, and serum markers of gastritis. Am J Gastroenterol. 96: 63–70.

Carmel R, Brar S and Frouhar Z. 2001b. Plasma total transcobalamin I. Ethnic/racial patterns and comparison with lactoferrin. Am J Clin Pathol. 116: 576–580.

Carmel R and Eisenberg L. 1977. Serum vitamin B12 and transcobalamin abnormalities in patients with cancer. Cancer. 40: 1348–1353.

Carmel R and Hollander D. 1978. Extreme elevation of transcobalamin II levels in multiple myeloma and other disorders. Blood. 51: 1057–1063.

Carmel R, Vasireddy H, Aurangzeb I and George K. 2001c. High serum cobalamin levels in the clinical setting—clinical associations and holo-transcobalamin changes. Clin Lab Haematol. 23: 365–371.

Collins RA, Harper AE, Scheiber M and Elvehjem C. 1951. The folic acid and vitamin B12 content of the milk of various species. J Nutr. 43: 313–321.

Deegan KL, Jones KM, Zuleta C, Ramirez-Zea M, Lildballe DL, Nexo E and Allen LH. 2012. Breast milk vitamin B12 concentrations in guatemalan women are correlated with maternal but not infant vitamin B12 status at 12 months postpartum. J Nutr. 142: 112–116.

Dierkes J, Domrose U, Ambrosch A, Schneede J, Guttormsen AB, Neumann KH and Luley C. 1999. Supplementation with vitamin B12 decreases homocysteine and methylmalonic acid but also serum folate in patients with end-stage renal disease. Metabolism. 48: 631–635.

Dominguez-Salas P, Moore SE, Cole D, da Costa KA, Cox SE, Dyer RA, Fulford AJ, Innis SM, Waterland RA, Zeisel SH, Prentice AM and Hennig BJ. 2013. DNA methylation potential: dietary intake and blood concentrations of one-carbon metabolites and cofactors in rural African women. Am J Clin Nutr. 97: 1217–1227.

Doscherholmen A and Swaim WR. 1973. Impaired assimilation of egg Co 57 vitamin B12 in patients with hypochlorhydria and achlorhydria and after gastric resection. Gastroenterology. 64: 913–919.

Ducros V, Demuth K, Sauvant MP, Quillard M, Causse E, Candito M, Read MH, Drai J, Garcia I and Gerhardt MF. 2002. Methods for homocysteine analysis and biological relevance of the results. J Chromatogr B Analyt Technol Biomed Life Sci. 781: 207–226.

Duggan C, Srinivasan K, Thomas T, Samuel T, Rajendran R, Muthayya S, Finkelstein JL, Lukose A, Fawzi W, Allen LH, Bosch RJ and Kurpad AV. 2014. Vitamin B12 supplementation during pregnancy and early lactation increases maternal, breast milk, and infant measures of vitamin B12 status. J Nutr. 144: 758–764.

Dullemeijer C, Souverein OW, Doets EL, van der Voet H, van Wijngaarden JP, de Boer WJ, Plada M, Dhonukshe-Rutten RA, In 't Veld PH, Cavelaars AE, de Groot LC, van 't Veer P. 2013. Systematic review with dose-response meta-analyses between vitamin B12 intake and European Micronutrient Recommendations Aligned's prioritized biomarkers of vitamin B12 including randomized controlled trials and observational studies in adults and elderly persons. Am J Clin Nutr. 97: 390–402.

Elian KM and Hoffer LJ. 2002. Hydroxocobalamin reduces hyperhomocysteinemia in end-stage renal disease. Metabolism. 51: 881–886.

Eussen SJ, Nilsen RM, Midttun O, Hustad S, Ijssennagger N, Meyer K, Fredriksen A, Ulvik A, Ueland PM, Brennan P, Johansson M, Bueno-de-Mesquita B, Vineis P, Chuang SC, Boutron-Ruault MC, Dossus L, Perquier F, Overvad K, Teucher B, Grote VA, Trichopoulou A, Adarakis G, Plada M, Sieri S, Tumino R, de Magistris MS, Ros MM, Peeters PH, Redondo ML, Zamora-Ros R, Chirlaque MD, Ardanaz E, Sonestedt E, Ericson U, Schneede J, Van GB, Wark PA, Gallo V, Norat T, Riboli E and Vollset SE. 2012. North-south gradients in plasma concentrations of B-vitamins and other components of one-carbon metabolism in Western Europe: results from the European Prospective Investigation into Cancer and Nutrition (EPIC) Study. Br J Nutr. 1–12.

Fedosov SN, Brito A, Miller JW, Green R and Allen LH. 2015. Combined indicator of vitamin B12 status: modification for missing biomarkers and folate status and recommendations for revised cut-points. Clin Chem Lab Med. 53: 1215–1225.

Fernandes-Costa F, van TS and Metz J. 1985. A sex difference in serum cobalamin and transcobalamin levels. Am J Clin Nutr. 41: 784–786.

Festen HP. 1991. Intrinsic factor secretion and cobalamin absorption. Physiology and pathophysiology in the gastrointestinal tract. Scand J Gastroenterol Suppl. 188: 1–7.

Flatley JE, Garner CM, Al-Turki M, Manning NJ, Olpin SE, Barker ME and Powers HJ. 2012. Determinants of urinary methylmalonic acid concentration in an elderly population in the United Kingdom. Am J Clin Nutr. 95: 686–693.

Garcia A, Paris-Pombo A, Evans L, Day A and Freedman M. 2002. Is low-dose oral cobalamin enough to normalize cobalamin function in older people? J Am Geriatr Soc. 50: 1401–1404.

Gilsing AM, Crowe FL, Lloyd-Wright Z, Sanders TA, Appleby PN, Allen NE and Key TJ. 2010. Serum concentrations of vitamin B12 and folate in British male omnivores, vegetarians and vegans: results from a cross-sectional analysis of the EPIC-Oxford cohort study. Eur J Clin Nutr. 64: 933–939.

Gimsing P and Nexo E. 1989. Cobalamin-binding capacity of haptocorrin and transcobalamin: age-correlated reference intervals and values from patients. Clin Chem. 35: 1447–1451.

Gonzalez-Gross M, Benser J, Breidenassel C, Albers U, Huybrechts I, Valtuena J, Spinneker A, Segoviano M, Widhalm K, Molnar D, Moreno LA, Stehle P and Pietrzik K. 2012. Gender and age influence blood folate, vitamin B12, vitamin B6, and homocysteine levels in European adolescents: the Helena Study. Nutr Res. 32: 817–826.

Gulati S, Baker P, Li YN, Fowler B, Kruger W, Brody LC and Banerjee R. 1996. Defects in human methionine synthase in cblG patients. Hum Mol Genet. 5: 1859–1865.

Gultepe M, Ozcan O, Avsar K, Cetin M, Ozdemir AS and Gok M. 2003. Urine methylmalonic acid measurements for the assessment of cobalamin deficiency related to neuropsychiatric disorders. Clin Biochem. 36: 275–282.

Hampel D, Shahab-Ferdows S, Domek JM, Siddiqua T, Raqib R and Allen LH. 2014. Competitive chemiluminescent enzyme immunoassay for vitamin B12 analysis in human milk. Food Chem. 153: 60–65.

Hardlei TF, Morkbak AL, Bor MV, Bailey LB, Hvas AM and Nexo E. 2010. Assessment of vitamin B(12) absorption based on the accumulation of orally administered cyanocobalamin on transcobalamin. Clin Chem. 56: 432–436.

Herrmann W and Obeid R. 2009. Holotranscobalamin—an early marker for laboratory diagnosis of vitamin B12 deficiency. European Haematology. 2: 2–6.

Herrmann W and Obeid R. 2013. Utility and limitations of biochemical markers of vitamin B12 deficiency. Eur J Clin Invest. 43: 231–237.

Herrmann W, Obeid R, Schorr H and Geisel J. 2005. The usefulness of holotranscobalamin in predicting vitamin B12 status in different clinical settings. Curr Drug Metab. 6: 47–53.

Herrmann W, Schorr H, Obeid R and Geisel J. 2003. Vitamin B12 status, particularly holotranscobalamin II and methylmalonic acid concentrations, and hyperhomocysteinemia in vegetarians. Am J Clin Nutr. 78: 131–136.

Hill MH, Flatley JE, Barker ME, Garner CM, Manning NJ, Olpin SE, Moat SJ, Russell J and Powers HJ. 2013. A vitamin B12 supplement of 500 mug/d for eight weeks does not normalize urinary methylmalonic acid or other biomarkers of vitamin B12 status in elderly people with moderately poor vitamin B12 status. J Nutr. 143: 142–147.

Hirschowitz BI, Worthington J and Mohnen J. 2008. Vitamin B12 deficiency in hypersecretors during long-term acid suppression with proton pump inhibitors. Aliment Pharmacol Ther. 27: 1110–1121.

Hubner U, Schorr H, Eckert R, Geisel J and Herrmann W. 2007a. Stability of plasma homocysteine, S-adenosylmethionine, and S-adenosylhomocysteine in EDTA, acidic citrate, and Primavette collection tubes. Clin Chem. 53: 2217–2218.

Hubner U, Schorr H, Eckert R, Geisel J and Herrmann W. 2007b. Stability of plasma homocysteine, S-adenosylmethionine, and S-adenosylhomocysteine in EDTA, acidic citrate, and Primavette collection tubes. Clin Chem. 53: 2217–2218.

Hurwitz A, Brady DA, Schaal SE, Samloff IM, Dedon J and Ruhl CE. 1997. Gastric acidity in older adults 1. JAMA. 278: 659–662.

Hyndman ME, Manns BJ, Snyder FF, Bridge PJ, Scott-Douglas NW, Fung E and Parsons HG. 2003. Vitamin B12 decreases, but does not normalize, homocysteine and methylmalonic acid in end-stage renal disease: a link with glycine metabolism and possible explanation of hyperhomocysteinemia in end-stage renal disease. Metabolism. 52: 168–172.

Ikeda T, Furukawa Y, Mashimoto S, Takahashi K and Yamada M. 1990. Vitamin B12 levels in serum and cerebrospinal fluid of people with Alzheimer's disease. Acta Psychiatr Scand. 82: 327–329.

Ispir E, Serdar MA, Ozgurtas T, Gulbahar O, Akin KO, Yesildal F and Kurt I. 2015. Comparison of four automated serum vitamin B12 assays. Clin Chem Lab Med. 53: 1205–1213.

Israel-Ballard KA, Abrams BF, Coutsoudis A, Sibeko LN, Cheryk LA and Chantry CJ. 2008. Vitamin content of breast milk from HIV-1-infected mothers before and after flash-heat treatment. J Acquir Immune Defic Syndr. 48: 444–449.

Jathar VS, Kamath SA, Parikh MN, Rege DV and Satoskar RS. 1970. Maternal milk and serum vitamin B12, folic acid, and protein levels in Indian subjects. Arch Dis Child. 45: 236–241.

Joosten E, van den Berg A, Riezler R, Naurath HJ, Lindenbaum J, Stabler SP and Allen RH. 1993. Metabolic evidence that deficiencies of vitamin B12 (cobalamin), folate, and vitamin B-6 occur commonly in elderly people. Am J Clin Nutr. 58: 468–476.

Kaptan K, Beyan C, Ural AU, Cetin T, Avcu F, Gulsen M, Finci R and Yalcin A. 2000. Helicobacter pylori—is it a novel causative agent in Vitamin B12 deficiency? Arch Intern Med. 160: 1349–1353.

kidd HM, Gould CE and Thomas JW. 1963. Free and total vitamin B12 in cerebrospinal fluid. Can Med Assoc J. 88: 876–881.

Krasinski SD, Russell RM, Samloff IM, Jacob RA, Dallal GE, McGandy RB and Hartz SC. 1986. Fundic atrophic gastritis in an elderly population. Effect on hemoglobin and several serum nutritional indicators. J Am Geriatr Soc. 34: 800–806.

Kurnat-Thoma EL, Pangilinan F, Matteini AM, Wong B, Pepper GA, Stabler SP, Guralnik JM and Brody LC. 2015. Association of transcobalamin II (TCN2) and transcobalamin II-receptor (TCblR) genetic variations with cobalamin deficiency parameters in elderly women. Biol Res Nurs. 17: 444–454.

Kvestad I, Taneja S, Kumar T, Hysing M, Refsum H, Yajnik CS, Bhandari N and Strand TA. 2015. Vitamin B12 and folic acid improve gross motor and problem-solving skills in young North Indian children: A randomized placebo-controlled trial. PLoS One. 10: e0129915.

Kwok T, Cheng G, Lai WK, Poon P, Woo J and Pang CP. 2004. Use of fasting urinary methylmalonic acid to screen for metabolic vitamin B12 deficiency in older persons. Nutrition. 20: 764–768.

Lahner E, Norman GL, Severi C, Encabo S, Shums Z, Vannella L, Fave GD and Annibale B. 2009. Reassessment of intrinsic factor and parietal cell autoantibodies in atrophic gastritis with respect to cobalamin deficiency. Am J Gastroenterol. 104: 2071–2079.

Lam JR, Schneider JL, Zhao W and Corley DA. 2013. Proton pump inhibitor and histamine 2 receptor antagonist use and vitamin B12 deficiency. JAMA. 310: 2435–2442.

Lazar GS and Carmel R. 1981. Cobalamin binding and uptake *in vitro* in the human central nervous system. J Lab Clin Med. 97: 123–133.

Lildballe DL, Hardlei TF, Allen LH and Nexo E. 2009. High concentrations of haptocorrin interfere with routine measurement of cobalamins in human serum and milk. A problem and its solution. Clin Chem Lab Med. 47: 182–187.

Lindemans J, Schoester M and van Kapel J. 1983. Application of a simple immunoadsorption assay for the measurement of saturated and unsaturated transcobalamin II and R-binders. Clin Chim Acta. 132: 53–61.

Lindgren A, Kilander A, Bagge E and Nexo E. 1999. Holotranscobalamin—a sensitive marker of cobalamin malabsorption. Eur J Clin Invest. 29: 321–329.

Lloyd-Wright Z, Hvas AM, Moller J, Sanders TA and Nexo E. 2003. Holotranscobalamin as an indicator of dietary vitamin B12 deficiency. Clin Chem. 49: 2076–2078.

Loikas S, Lopponen M, Suominen P, Moller J, Irjala K, Isoaho R, Kivela SL, Koskinen P and Pelliniemi TT. 2003. RIA for serum holo-transcobalamin: method evaluation in the clinical laboratory and reference interval. Clin Chem. 49: 455–462.

Mansoor MA, Stea TH, Schneede J and Reine A. 2013. Early biochemical and hematological response to intramuscular cyanocobalamin therapy in vitamin B(12)-deficient patients. Ann Nutr Metab. 62: 347–353.

Marcell PD, Stabler SP, Podell ER and Allen RH. 1985. Quantitation of methylmalonic acid and other dicarboxylic acids in normal serum and urine using capillary gas chromatography-mass spectrometry. Anal Biochem. 150: 58–66.

McPhee AJ, Davidson GP, Leahy M and Beare T. 1988. Vitamin B12 deficiency in a breast fed infant. Arch Dis Child. 63: 921–923.

Mendelsohn RS, Watkin DM, Horbett AP and Fahey JL. 1958. Identification of the vitamin B12-binding protein in the serum of normals and of patients with chronic myelocytic leukemia. Blood. 13: 740–747.

Mineva EM, Zhang M, Rabinowitz DJ, Phinney KW and Pfeiffer CM. 2015. An LC-MS/MS method for serum methylmalonic acid suitable for monitoring vitamin B12 status in population surveys. Anal Bioanal Chem. 407: 2955–2964.

Moelby L, Rasmussen K, Ring T and Nielsen G. 2000. Relationship between methylmalonic acid and cobalamin in uremia. Kidney Int. 57: 265–273.

Moestrup SK, Birn H, Fischer PB, Petersen CM, Verroust PJ, Sim RB, Christensen EI and Nexo E. 1996. Megalin-mediated endocytosis of transcobalamin-vitamin-B12 complexes suggests a role of the receptor in vitamin-B12 homeostasis. Proc Natl Acad Sci U S A. 93: 8612–8617.

Molloy AM, Mills JL, McPartlin J, Kirke PN, Scott JM and Daly S. 2002. Maternal and fetal plasma homocysteine concentrations at birth: the influence of folate, vitamin B12, and the 5,10-methylenetetrahydrofolate reductase 677C-->T variant. Am J Obstet Gynecol. 186: 499–503.

Morkbak AL, Hvas AM, Milman N and Nexo E. 2007. Holotranscobalamin remains unchanged during pregnancy. Longitudinal changes of cobalamins and their binding proteins during pregnancy and postpartum. Haematologica. 92: 1711–1712.

Morkbak AL, Pedersen JF and Nexo E. 2005. Glycosylation independent measurement of the cobalamin binding protein haptocorrin. Clin Chim Acta. 356: 184–190.

Murphy MM, Molloy AM, Ueland PM, Fernandez-Ballart JD, Schneede J, Arija V and Scott JM. 2007. Longitudinal study of the effect of pregnancy on maternal and fetal cobalamin status in healthy women and their offspring. J Nutr. 137: 1863–1867.

Naurath HJ, Joosten E, Riezler R, Stabler SP, Allen RH and Lindenbaum J. 1995. Effects of vitamin B12, folate, and vitamin B6 supplements in elderly people with normal serum vitamin concentrations. Lancet. 346: 85–89.

Nexo E, Christensen AL, Hvas AM, Petersen TE and Fedosov SN. 2002a. Quantification of holo-transcobalamin, a marker of vitamin B12 deficiency. Clin Chem. 48: 561–562.

Nexo E, Christensen AL, Petersen TE and Fedosov SN. 2000. Measurement of transcobalamin by ELISA. Clin Chem. 46: 1643–1649.

Nexo E and Hoffmann-Lucke E. 2011. Holotranscobalamin, a marker of vitamin B12 status: analytical aspects and clinical utility. Am J Clin Nutr. 94: 359S–365S.

Nexo E, Hvas AM, Bleie O, Refsum H, Fedosov SN, Vollset SE, Schneede J, Nordrehaug JE, Ueland PM and Nygard OK. 2002b. Holo-transcobalamin is an early marker of changes in cobalamin homeostasis. A randomized placebo-controlled study. Clin Chem. 48: 1768–1771.

Niafar M, Hai F, Porhomayon J and Nader ND. 2015. The role of metformin on vitamin B12 deficiency: a meta-analysis review. Intern Emerg Med. 10: 93–102.

Nickoloff E. 1988. Schilling test: physiologic basis for and use as a diagnostic test. Crit Rev Clin Lab Sci. 26: 263–276.

Nilsson-Ehle H, Landahl S, Lindstedt G, Netterblad L, Stockbruegger R, Westin J and Ahren C. 1989. Low serum cobalamin levels in a population study of 70- and 75-year-old subjects. Gastrointestinal causes and hematological effects. Dig Dis Sci. 34: 716–723.

Niyikiza C, Baker SD, Seitz DE, Walling JM, Nelson K, Rusthoven JJ, Stabler SP, Paoletti P, Calvert AH and Allen RH. 2002. Homocysteine and methylmalonic acid: markers to predict and avoid toxicity from pemetrexed therapy. Mol Cancer Ther. 1: 545–552.

Norman EJ and Morrison JA. 1993. Screening elderly populations for cobalamin (vitamin B12) deficiency using the urinary methylmalonic acid assay by gas chromatography mass spectrometry [see comments]. Am J Med. 94: 589–594.

Obeid R, Geisel J and Herrmann W. 2015. Comparison of two methods for measuring methylmalonic acid as a marker for vitamin B12 deficiency. In: Diagnosis: Walter de Gruyter. pp. 67–72.

Obeid R, Jung J, Falk J, Herrmann W, Geisel J, Friesenhahn-Ochs B, Lammert F, Fassbender K and Kostopoulos P. 2013. Serum vitamin B12 not reflecting vitamin B12 status in patients with type 2 diabetes. Biochimie. 95: 1056–1061.

Obeid R, Kasoha M, Knapp JP, Kostopoulos P, Becker G, Fassbender K and Herrmann W. 2007a. Folate and methylation status in relation to phosphorylated tau protein(181P) and beta-amyloid(1-42) in cerebrospinal fluid. Clin Chem. 53: 1129–1136.

Obeid R, Kostopoulos P, Knapp JP, Kasoha M, Becker G, Fassbender K and Herrmann W. 2007b. Biomarkers of folate and vitamin B12 are related in blood and cerebrospinal fluid. Clin Chem. 53: 326–333.

Obeid R, Kuhlmann MK, Kohler H and Herrmann W. 2005a. Response of homocysteine, cystathionine, and methylmalonic acid to vitamin treatment in dialysis patients. Clin Chem. 51: 196–201.

Obeid R, Morkbak AL, Munz W, Nexo E and Herrmann W. 2006. The cobalamin-binding proteins transcobalamin and haptocorrin in maternal and cord blood sera at birth. Clin Chem. 52: 263–269.

Obeid R, Munz W, Jager M, Schmidt W and Herrmann W. 2005b. Biochemical indexes of the B vitamins in cord serum are predicted by maternal B vitamin status. Am J Clin Nutr. 82: 133–139.

Obeid R, Schorr H, Eckert R and Herrmann W. 2004. Vitamin B12 status in the elderly as judged by available biochemical markers. Clin Chem. 50: 238–241.

Okuma Y, Hosomi Y, Watanabe K, Takahashi S, Okamura T and Hishima T. 2015. Gemcitabine in patients previously treated with platinum-containing chemotherapy for refractory thymic carcinoma: radiographic assessment using the RECIST criteria and the ITMIG recommendations. Int J Clin Oncol. 21: 531–538.

Orning L, Rian A, Campbell A, Brady J, Fedosov SN, Bramlage B, Thompson K and Quadros EV. 2006. Characterization of a monoclonal antibody with specificity for holo-transcobalamin. Nutr Metab (Lond). 3: 3.

Pedersen TL, Keyes WR, Shahab-Ferdows S, Allen LH and Newman JW. 2011. Methylmalonic acid quantification in low serum volumes by UPLC-MS/MS. J Chromatogr B Analyt Technol Biomed Life Sci. 879: 1502–1506.

Quadros EV. 2010. Advances in the understanding of cobalamin assimilation and metabolism. Br J Haematol. 148: 195–204.

Rajan S, Wallace JI, Brodkin KI, Beresford SA, Allen RH and Stabler SP. 2002. Response of elevated methylmalonic acid to three dose levels of oral cobalamin in older adults. J Am Geriatr Soc. 50: 1789–1795.

Rasmussen K. 1989. Studies on methylmalonic acid in humans. I. Concentrations in serum and urinary excretion in normal subjects after feeding and during fasting, and after loading with protein, fat, sugar, isoleucine, and valine. Clin Chem. 35: 2271–2276.

Refsum H, Johnston C, Guttormsen AB and Nexo E. 2006. Holotranscobalamin and total transcobalamin in human plasma: determination, determinants, and reference values in healthy adults. Clin Chem. 52: 129–137.

Rogers LM, Boy E, Miller JW, Green R, Rodriguez M, Chew F and Allen LH. 2003a. Predictors of cobalamin deficiency in Guatemalan school children: diet, Helicobacter pylori, or bacterial overgrowth? J Pediatr Gastroenterol Nutr. 36: 27–36.

Rogers LM, Boy E, Miller JW, Green R, Sabel JC and Allen LH. 2003b. High prevalence of cobalamin deficiency in Guatemalan school children: associations with low plasma holotranscobalamin II and elevated serum methylmalonic acid and plasma homocysteine concentrations. Am J Clin Nutr. 77: 433–440.

Savage DG, Lindenbaum J, Stabler SP and Allen RH. 1994. Sensitivity of serum methylmalonic acid and total homocysteine determinations for diagnosing cobalamin and folate deficiencies. Am J Med. 96: 239–246.

Schloss JM, Colosimo M, Airey C and Vitetta L. 2015. Chemotherapy-induced peripheral neuropathy (CIPN) and vitamin B12 deficiency. Support Care Cancer. 23: 1843–1850.

Serin E, Gumurdulu Y, Ozer B, Kayaselcuk F, Yilmaz U and Kocak R. 2002. Impact of Helicobacter pylori on the development of vitamin B12 deficiency in the absence of gastric atrophy. Helicobacter. 7: 337–341.

Sharabi A, Cohen E, Sulkes J and Garty M. 2003. Replacement therapy for vitamin B12 deficiency: comparison between the sublingual and oral route. Br J Clin Pharmacol. 56: 635–638.

Siekmann JH, Allen LH, Bwibo NO, Demment MW, Murphy SP and Neumann CG. 2003. Kenyan school children have multiple micronutrient deficiencies, but increased plasma vitamin B12 is the only detectable micronutrient response to meat or milk supplementation. J Nutr. 133: 3972S–3980S.

Sipponen P, Laxen F, Huotari K and Harkonen M. 2003. Prevalence of low vitamin B12 and high homocysteine in serum in an elderly male population: association with atrophic gastritis and Helicobacter pylori infection. Scand J Gastroenterol. 38: 1209–1216.

Snow CF. 1999. Laboratory diagnosis of vitamin B12 and folate deficiency: a guide for the primary care physician [see comments]. Arch Intern Med. 159: 1289–1298.

Sobczynska-Malefora A, Gorska R, Pelisser M, Ruwona P, Witchlow B and Harrington DJ. 2014. An audit of holotranscobalamin ("Active" B12) and methylmalonic acid assays for the assessment of vitamin B12 status: application in a mixed patient population. Clin Biochem. 47: 82–86.

Specker BL, Black A, Allen L and Morrow F. 1990. Vitamin B12: low milk concentrations are related to low serum concentrations in vegetarian women and to methylmalonic aciduria in their infants. Am J Clin Nutr. 52: 1073–1076.

Stabler SP, Marcell PD, Podell ER, Allen RH and Lindenbaum J. 1986. Assay of methylmalonic acid in the serum of patients with cobalamin deficiency using capillary gas chromatography-mass spectrometry. J Clin Invest. 77: 1606–1612.

Strand TA, Taneja S, Kumar T, Manger MS, Refsum H, Yajnik CS and Bhandari N. 2015. Vitamin B12, folic acid, and growth in 6- to 30-month-old children: a randomized controlled trial. Pediatrics. 135: e918–e926.

Stucki M, Coelho D, Suormala T, Burda P, Fowler B and Baumgartner MR. 2012. Molecular mechanisms leading to three different phenotypes in the cblD defect of intracellular cobalamin metabolism. Hum Mol Genet. 21: 1410–1418.

Tamura T and Baggott JE. 2008. *In vitro* formation of homocysteine in whole blood in the presence of anticoagulants. Clin Chem. 54: 1402–1403.

Thompson GN, Chalmers RA, Walter JH, Bresson JL, Lyonnet SL, Reed PJ, Saudubray JM, Leonard JV and Halliday D. 1990. The use of metronidazole in management of methylmalonic and propionic acidaemias. Eur J Pediatr. 149: 792–796.

Ubbink JB, Hayward Vermaak WJ and Bissbort S. 1991. Rapid high-performance liquid chromatographic assay for total homocysteine levels in human serum. J Chromatogr. 565: 441–446.

Ulleland M, Eilertsen I, Quadros EV, Rothenberg SP, Fedosov SN, Sundrehagen E and Orning L. 2002. Direct assay for cobalamin bound to transcobalamin (holo-transcobalamin) in serum. Clin Chem. 48: 526–532.

Valente E, Scott JM, Ueland PM, Cunningham C, Casey M and Molloy AM. 2011. Diagnostic accuracy of holotranscobalamin, methylmalonic acid, serum cobalamin, and other indicators of tissue vitamin B status in the elderly. Clin Chem. 57: 856–863.

van Walraven C, Austin P and Naylor CD. 2001. Vitamin B12 injections versus oral supplements. How much money could be saved by switching from injections to pills? Can Fam Physician. 47: 79–86.

Vashi P, Edwin P, Popiel B, Lammersfeld C and Gupta D. 2016. Methylmalonic acid and homocysteine as indicators of vitamin B12 deficiency in cancer. PLoS One. 11: e0147843.

von Castel-Roberts KM, Morkbak AL, Nexo E, Edgemon CA, Maneval DR, Shuster JJ, Valentine JF, Kauwell GP and Bailey LB. 2007. Holo-transcobalamin is an indicator of vitamin B12 absorption in healthy adults with adequate vitamin B12 status. Am J Clin Nutr. 85: 1057–1061.

Waldmann A, Koschizke JW, Leitzmann C and Hahn A. 2004. Homocysteine and cobalamin status in German vegans. Public Health Nutr. 7: 467–472.

Windelberg A, Arseth O, Kvalheim G and Ueland PM. 2005. Automated assay for the determination of methylmalonic acid, total homocysteine, and related amino acids in human serum or plasma by means of methylchloroformate derivatization and gas chromatography-mass spectrometry. Clin Chem. 51: 2103–2109.

Winkels RM, Brouwer IA, Clarke R, Katan MB and Verhoef P. 2008. Bread cofortified with folic acid and vitamin B-12 improves the folate and vitamin B-12 status of healthy older people: a randomized controlled trial. Am J Clin Nutr. 88: 348–355.

Yakut M, Ustun Y, Kabacam G and Soykan I. 2010. Serum vitamin B12 and folate status in patients with inflammatory bowel diseases. Eur J Intern Med. 21: 320–323.

Yetley EA, Pfeiffer CM, Phinney KW, Bailey RL, Blackmore S, Bock JL, Brody LC, Carmel R, Curtin LR, Durazo-Arvizu RA, Eckfeldt JH, Green R, Gregory JF, III, Hoofnagle AN, Jacobsen DW, Jacques PF, Lacher DA, Molloy AM, Massaro J, Mills JL, Nexo E, Rader JI, Selhub J, Sempos C, Shane B, Stabler S, Stover P, Tamura T, Tedstone A, Thorpe SJ, Coates PM, Johnson CL and Picciano MF. 2011. Biomarkers of vitamin B12 status in NHANES: a roundtable summary. Am J Clin Nutr. 94: 313S–321S.

Yu HH, Joubran R, Asmi M, Law T, Spencer A, Jouma M and Rifai N. 2000. Agreement among four homocysteine assays and results in patients with coronary atherosclerosis and controls. Clin Chem. 46: 258–264.

Zetterberg H, Nexo E, Regland B, Minthon L, Boson R, Palmer M, Rymo L and Blennow K. 2003. The transcobalamin (TC) codon 259 genetic polymorphism influences holo-TC concentration in cerebrospinal fluid from patients with Alzheimer disease. Clin Chem. 49: 1195–1198.

10

Cobalamin During Pregnancy and Lactation

Rima Obeid,[1,*] *Pol Solé-Navais*[2,3] and *Michelle M Murphy*[2,3]

1. Introduction

Maternal and foetal nutritional requirements are exceptionally high during pregnancy to meet the associated physiological and metabolic demands. As is the case for many nutrients, cobalamin stores in foetal tissues are established during pregnancy. Current cobalamin requirements during pregnancy are based on estimates of the total amount of cobalamin assumed to accumulate in the foetal liver from studies published between 1962 and 1975. Pregnancy cobalamin intake recommendations are based on studies that used maternal plasma or serum total cobalamin or urinary methylmalonic acid excretion in the child as markers of cobalamin insufficiency. Cobalamin deficiency is highly prevalent in countries like India due to the widely practised vegetarian diet. Recent studies using functional markers of cobalamin status such as plasma holotranscobalamin (holoTC) and methylmalonic acid (MMA) during pregnancy, lactation and early life have shown that suboptimal cobalamin intake and status also affects women with omnivorous diets in industrialised countries. Many studies have shown that low cobalamin status is associated with birth defects and developmental problems in the children. However, cobalamin supplementation during pregnancy and lactation has received relatively little attention compared to folic acid supplementation.

---
[1] Aarhus Institute of Advanced Studies, University of Aarhus, Høegh-Guldbergs Gade 6B, building 1632, Dk-8000, Aarhus C, Denmark.
  Email: rima.obeid@uks.eu
[2] Area of Preventive Medicine and Public Health, Department of Basic Medical Sciences, Faculty of Medicine and Health Sciences, Universitat Rovira i Virgili, IISPV, Spain.
[3] CIBER (CB06/03) ISCIII.
  Email: pol.sole@urv.cat, michelle.murphy@urv.cat
* Corresponding author

This chapter summarizes the current knowledge and gaps in knowledge regarding cobalamin status and requirements during pregnancy and lactation.

2. Cobalamin During Pregnancy and Lactation

2.1 Classical view of cobalamin requirements during pregnancy and lactation

The Food and Nutrition Board of the Institute of Medicine (IOM), Department of Medicine, USA has defined the Recommended Dietary Allowances (RDAs) for cobalamin for pregnant women based on the amount accumulated by the foetus throughout pregnancy, the assumption that cobalamin absorption (Hellegers et al. 1957; Robertson and Gallagher 1983) is increased during pregnancy and that only the newly absorbed cobalamin from the diet is transferred to the foetus (Institute of Medicine 1998).

The following interpretations were used to justify cobalamin requirements during pregnancy (Institute of Medicine 1998):

1. Cobalamin concentrations decline during pregnancy, but this was not thought to be a reflection of depletion of maternal reserves. However, the decline in serum cobalamin concentrations during the first trimester was not only due to haemodilution. Also, haptocorrin increases during the second and third trimesters, and transcobalamin II increases sharply in the third trimester to about 30% higher than in non-pregnant, non-lactating women (Fernandes-Costa and Metz 1982).
2. Cobalamin is actively transported from the mother to the foetus and accumulates in the placental and foetal tissues, implying that pregnant women have higher requirements than non-pregnant women. The recently absorbed vitamin from the maternal diet is transported to the foetus and is a more important cobalamin source than maternal liver stores. Studies on infants from strict vegetarian mothers have shown that cobalamin deficiency is manifested in the child by the age of 4–6 months (Specker et al. 1990) suggesting that cobalamin stores in the child can be depleted fast (few weeks up to 6 months).
3. It has been estimated that an average of 0.07 to 0.14 nmol/day (0.1 to 0.2 µg/day) of cobalamin accumulates in the foetus. The estimates were based on studies of liver cobalamin content in infants born to cobalamin replete women (Vaz et al. 1975) and the assumption that the liver contains half of the total body cobalamin content. It has also been assumed that the total amount of cobalamin that accumulates in the foetus during pregnancy is unlikely to cause depletion of maternal stores.

To account for a foetal cobalamin deposition of 0.1 to 0.2 µg/day throughout pregnancy and more efficient maternal absorption of the vitamin during pregnancy, 0.2 µg/day was added to the estimated average requirement (EAR) of non-pregnant women (2.0 µg/day) to cover the anticipated foetal needs. The RDA is defined as equal to the EAR plus 2 coefficients of variation (CV) to cover the requirements of 98% of pregnant women. Therefore, the RDA for cobalamin during pregnancy is 120% of the EAR or 2.6 µg/day (Institute of Medicine 1998).

The estimated intake requirements during lactation were based on the loss of cobalamin in milk or milk cobalamin content. The average amount of cobalamin excreted in maternal milk is approximately 0.33 µg/day during the first 6 months of lactation. This amount declines by approximately 25% after 6 months (0.25 µg/day). The EAR for lactating women was set by adding 0.33 µg/day of cobalamin to the EAR of 2.0 µg/day for non-pregnant women and the RDA for cobalamin during lactation is set to 2.8 µg/day (Institute of Medicine 1998).

2.2 Cobalamin homeostasis during normal pregnancy

Pregnancy associated haemodilution, enhanced renal function and hormonal changes lead to reduced concentrations in many blood biomarkers during pregnancy (Faupel-Badger et al. 2007). These established physiological factors should be considered when assessing changes in cobalamin biomarkers during pregnancy. Pregnancy associated changes in cobalamin binding proteins, active cobalamin and metabolic markers have been addressed in several studies.

The phenomenon of cobalamin decline during pregnancy was observed in very old studies dating back to the era of the discovery of the vitamin (Heinrich 1954). In a series of serum measurements performed on 25 pregnant women with low to intermediate cobalamin status (130–400 ng/mL), serum cobalamin fell sharply between months 7 and 9 without a parallel decrease in haemoglobin. Serum cobalamin concentration had increased again and returned to baseline by the end of the sixth week after delivery (Izak et al. 1957). Animal studies have shown that maternal cobalamin stores in the liver and kidney are depleted during pregnancy and that the vitamin accumulates in the foetus and the placenta (Brown et al. 1977). This might contribute to the pregnancy-associated decline in serum cobalamin in humans but the hypothesis has not yet been tested in women.

Up to 60% of the absorbed vitamin had accumulated in the foetus and placenta following a subcutaneous cobalamin injection in pregnant rat dams three days before delivery (Hellegers et al. 1957). A separate study showed that the accumulation of cobalamin in the foetoplacental unit accounted for more than 90% of the intestinal uptake of the vitamin (Brown et al. 1977). Cobalamin uptake by the foetoplacental unit, following a high oral dose of the vitamin at 18–21 days of pregnancy, did not differ between mice that had received a cobalamin injection during early pregnancy and those that had not. This suggests a high capacity of the placenta for extracting and binding newly injected cobalamin from the blood.

Some animal studies have suggested that intestinal absorption of cobalamin is increased during pregnancy which could be a physiological way to fulfil the high requirements. In a series of studies on mice, Brown et al. have shown that cobalamin absorption in the ileum is upregulated as pregnancy progresses (Brown et al. 1977). This was attributed to the upregulation of intrinsic factor mediated uptake (Brown et al. 1977; Robertson and Gallagher 1983) and not to increased passive diffusion, because higher doses of cyanocobalamin were needed to saturate the absorption mechanisms in pregnant compared to non-pregnant mice. The greater increase in total serum cobalamin concentrations following a single oral dose of 1 mg cyanocobalamin, in pregnant compared to non-pregnant women, was attributed to enhanced absorption of the vitamin (Hellegers et al. 1957). However, an oral dose of 250 µg of cobalamin resulted in no detectable changes in plasma total cobalamin between pregnant and non-pregnant women (Hellegers et al. 1957). An orally administered single dose of 0.5 mg cobalamin increased plasma cobalamin by 198 µg (146 nmol) compared to 128 µg (94 nmol) in pregnant compared to non-pregnant women, respectively and an oral dose of 1 mg by 263 µg (194 nmol) and 158 µg (117 nmol), respectively (Hellegers et al. 1957). However, the high doses of cobalamin used by Hellegers et al. (1, 0.5, and 0.25 mg) are likely to be partly absorbed via simple diffusion, independently of the intrinsic factor receptor that has an estimated capacity to bind approximately 10 µg cobalamin per meal. Moreover, the assays used for cobalamin determinations in old studies were not sensitive enough to detect small changes, and serum cobalamin fractions were not investigated. A recent study using the CobaSorb test (testing the increase in serum holoTC in response to oral cyanocobalamin) failed to demonstrate increased absorption of small physiological amounts of cyanocobalamin (3 x 9 µg for 2 days) in pregnant women between the first and third trimesters (Greibe et al. 2011). Therefore, there is no direct evidence, to date, for an increase in maternal cobalamin absorption in human pregnancy.

Animal studies suggested that despite the assumed increased absorption during pregnancy, maternal liver and kidney stores were depleted, while cobalamin accumulated in the foetoplacental tissues (Brown et al. 1977). Other studies have shown that only newly available cobalamin accumulates in the foetoplacental organs (Hellegers et al. 1957). Thus it is apparent that maternal intake during pregnancy is the determining factor of the amount available to the foetus. This intake must be sufficient to maintain maternal stores and cover foetal requirements. A human study showed that the drop in plasma cobalamin observed between the first and second trimesters in the placebo group did not occur in mothers that received oral cobalamin supplements of 50 µg/d from 10 gestational weeks to 6 weeks postpartum, in a double blind, randomised, placebo-controlled trial (Duggan et al. 2014). Despite falling in both groups between the 2nd and 3rd trimesters, plasma cobalamin remained higher in the supplemented group. No differences were seen in plasma MMA or total homocysteine (tHcy) concentrations, both of which are affected by cobalamin status, between the two groups. This suggests uptake of the supplementary cobalamin by the foetus rather than supporting maternal stores. Furthermore, plasma MMA and tHcy were lower in children born to supplemented mothers at 6 weeks postpartum. However, we still do not know how cobalamin is partitioned between maternal and foetal tissues particularly when mothers are cobalamin deficient.

Several longitudinal studies on uncomplicated pregnancies have shown that serum concentrations of total plasma cobalamin decline during pregnancy (Koebnick et al. 2002; Murphy et al. 2007), and return to pre-pregnancy levels after birth (Milman et al. 2006). Recent studies reported changes in cobalamin binders which may explain lower total serum cobalamin in late pregnancy. In line with this, following the initial reductions in plasma total cobalamin and holoTC between preconception and 8 weeks of pregnancy, holoTC remained relatively stable for the rest of pregnancy but approximately 35% lower than at preconception. In contrast, total cobalamin continued to decline during the same period (Murphy et al. 2007). On the other hand, transcobalamin (TC)-saturation decreased during pregnancy (Koebnick et al. 2002). Additional studies have shown that the concentrations of total TC increase but holoTC levels remain stable throughout pregnancy (Greibe et al. 2011; Morkbak et al. 2007). The decline in total serum cobalamin has been explained by a reduction in total haptocorrin (HC) and holohaptocorrin (holoHC) that constitutes the major fraction of serum total cobalamin (Greibe et al. 2011; Morkbak et al. 2007). Unchanged cobalamin absorption in the intestine following an oral cobalamin load, at different stages of pregnancy, failed to explain stable holoTC concentrations during pregnancy (Greibe et al. 2011). In contrast to plasma holoTC, holoHC represents the non-active cobalamin or the one with no known functions. HoloHC is considered to be a non-functional part despite it binds most of plasma cobalamin.

Collectively, the results show stable holoTC and a decline in holoHC during pregnancy. The relatively stable holoTC during pregnancy may be perceived as reflecting no change in metabolic cobalamin markers during pregnancy. The mechanisms involved are not known yet but it is possible that the decline in holoHC is due to the mobilisation of maternal stores to maintain a stable holoTC that ensures cobalamin supply to the developing foetus.

Blood MMA and tHcy concentrations increase in situations of cobalamin deficiency. A study of Danish pregnant women showed that a 28% reduction in total cobalamin (from median 225 to 161 pmol/L) between 18 to 39 gestational weeks was associated with an increase in plasma tHcy (median 6.4 to 7.7 µmol/L) and MMA (median 0.11 to 0.14 µmol/L) (Milman et al. 2006). Another study showed that while holoTC remained stable, a decline in total serum cobalamin from week 13 to 36 of pregnancy was associated with an increase in tHcy (median 3.8 to 5.0 µmol/L) and MMA (0.10 to 0.16 µmol/L) (Greibe et al. 2011). This confirmed the results from a previous longitudinal study from pre-conception throughout pregnancy that reported reductions in total cobalamin, holoTC and MMA between preconception and 20 gestational weeks. Subsequently, holoTC remained relatively unchanged for the remainder of pregnancy, but total cobalamin decreased and MMA increased (Murphy et al. 2007). However, the increase in MMA from pre-conception to week 32 was significantly lower in pregnant women that started pregnancy with plasma holoTC > 67 pmol/L (Murphy et al. 2007). Plasma volume expansion during pregnancy is unlikely to explain changes of cobalamin markers in different directions. Accordingly, correction of plasma MMA for plasma dilution during pregnancy (haematocrit change from preconception) did not alter the results (Murphy et al. 2007).

It appears that changes in both MMA and homocysteine reflect a decline in maternal tissue cobalamin supply as pregnancy progresses. In line with this, the total pregnancy increase in MMA was lower in mothers that started pregnancy with higher holoTC (Murphy et al. 2007). Higher cobalamin requirements and transport to the foetus during pregnancy may explain this phenomenon, as reflected by higher correlations between maternal total cobalamin and cord blood MMA as pregnancy progresses. Given that infants with TC deficiency are healthy at birth, an intriguing hypothesis would be that holoHC enhances cobalamin uptake by maternal tissues during pregnancy thus explaining holoHC reduction in maternal blood.

Changes in cobalamin binders appear to continue throughout late pregnancy. A study of German pregnant women just before delivery (> 37 gestational weeks) showed large inter-subject variations in cobalamin binders within the short time window of 37–40 gestational weeks (Obeid et al. 2006). Maternal TC-saturation decreased with increasing gestational age during this period which may be explained by a slight increase in maternal total TC combined with a decrease in holoTC (Obeid et al. 2006). Accelerated cobalamin transport (via holoTC) to the foetus in late pregnancy may enhance the synthesis of more TC to capture cobalamin. However, many gaps in knowledge remain and need to be addressed in future studies (Table 1).

Table 1. Cobalamin in pregnancy: gaps in knowledge.

• How is recent cobalamin intake partitioned between maternal and foetoplacental tissues in cobalamin deficient or depleted women?
• Should the recommendations for maternal cobalamin intake during pregnancy be higher for women with depleted stores? How should these requirements be defined?
• What factors explain holohaptocorrin lowering and stable holotranscobalamim, but increased methylmalonic acid and homocysteine in late pregnancy?
• What is the function of haptocorrin in general and in pregnancy in particular?
• Why are children with transcobalamin deficiency healthy after birth? Can holohaptocorrin transport cobalamin specifically during pregnancy?
• Is holohaptocorrin regulated by hormones?
• Are cobalamin transporters and receptors hormone regulated?

2.3 Interaction between cobalamin and other nutrients during pregnancy

Cobalamin and folate metabolism are intimately related. The folate cycle converges with the methionine cycle at the point of homocysteine remethylation to methionine by the cobalamin-dependent 5-methyltetrahydrofolate-homocysteine methyltransferase (MTR, EC 2.1.1.13). The folate cycle provides the methyl group that binds with cobalamin to form methylcobalamin. This methyl group is then used to convert homocysteine to methionine and in losing its methyl group, 5-methyltetrahydrofolate is converted to tetrahydrofolate. Purine and pyrimidine synthesis and epigenetic regulation in growing cells are directly affected by folate and methionine metabolism. Cobalamin deficiency during pregnancy will impair both the conversion of 5-methyltetrahydrofolate to tetrahydrofolate and of homocysteine to methionine. This condition can impair DNA-replication, impair cell growth and division thus causing anaemia or other serious health conditions in the mother and the child.

Pregnancy-related anaemia is common in many parts of the world where micronutrient deficiencies are common. Megaloblastic anaemia in pregnancy has been attributed to folate deficiency since the discovery of folate in 1948 and the famous studies on folate deficient Indian women who developed severe anaemia in pregnancy (Wills 1931). Although folate deficiency appears to be a more common cause of anaemia in pregnancy, mean serum cobalamin concentrations were found to be lower in pregnant women with megaloblastic anaemia compared to those without (Giles 1966). Moreover, co-supplementation with cobalamin improves the response to folic acid treatment in pregnant patients with megaloblastic anaemia (Berry 1955; Moore et al. 1955; Tasker 1955).

The results of some studies in adult men and women have led to debate on the possible adverse effects of high folate status in cobalamin deficient subjects. Plasma MMA and tHcy were higher in cobalamin deficient old adults with high folate status than in those with normal folate status (Selhub et al. 2007). Moreover, Morris and collaborators observed a higher prevalence of anaemia in cobalamin deficient old adults with high (plasma folate > 59 nmol/L) compared to normal folate status (Morris et al. 2007). Mandatory fortification of flour with folic acid has been in place in the USA since January 1998 (Food and Drug Administration and Department of health and human services, 1996) and numerous other countries have also implemented similar policies (Chen and Rivera 2004; Hertrampf and Cortes 2004). The delayed detection of cobalamin deficiency when folic acid intake is high (due to the absence of apparent megaloblastic anaemia) has been reported to be unlikely in the US (Qi et al. 2014). No increase in cobalamin deficiency in the absence of macrocytosis was observed between pre- and post- folic acid fortification in US older adults (4.2% vs 4.1%; adjusted prevalence ratios, 95% CI: 0.96, 0.65–1.43).

The effects of folic acid on metabolic and clinical markers of cobalamin status have not been specifically investigated in pregnant women exposed to high folic acid intakes (through fortification or targeted supplementation). Providing high doses of folic acid (1–5 mg) during early pregnancy is a common practice in some countries like India and Sri-Lanka (Gomes et al. 2010). Prenatal folic acid supplementation is recommended without considering underlying cobalamin status. Imbalance in folate and cobalamin status may occur when high doses of folic acid are recommended during early pregnancy or throughout pregnancy. Women with cobalamin deficiency are not able to utilize the large doses of folic acid unless they are also supplemented with cobalamin.

In an Indian study, women were given daily supplements of 5 mg folic acid during the first trimester and then 0.5 mg for the rest of pregnancy. Cobalamin intake was below 1.2 µg/d in 25% of the women in the first trimester and in 10–11% for the rest of pregnancy. The offspring of women that took > 1000 µg/d of folic acid and also had a low cobalamin to total folate intake ratio were at increased risk of being small for gestational age (Dwarkanath et al. 2013). A possible explanation could lie in the methylfolate trap hypothesis in which folate is trapped in the folate cycle when cobalamin status is low. As a result, purine and pyrimidine synthesis may be impaired. An imbalance in cobalamin/folate status in the mother could influence imprinting in the embryo with long-term health consequences.

In India, where cobalamin status is relatively low, high maternal red blood cell folate has been associated with high adiposity (Yajnik et al. 2008) and with increased risk of insulin resistance in the child (Krishnaveni et al. 2014; Yajnik et al. 2008). These intriguing observations were replicated in another Indian study in which the increased risk of gestational diabetes in cobalamin deficient women was further increased with increasing folate status (Krishnaveni et al. 2014). Gestational diabetes prevalence in cobalamin deficient women was 5.4, 10.5 and 10.9% from the lowest to the highest tertile of plasma folate (Krishnaveni et al. 2014). However, folate status did not modify the association between cobalamin status and gestational diabetes. It was suggested that adiposity drove the association between maternal cobalamin and gestational diabetes because BMI was higher in the women with cobalamin deficiency than those with normal cobalamin status and the associations were no longer significant in the models that adjusted for BMI. In a UK study of 995 pregnant women, those with cobalamin deficiency at 28 gestational weeks had higher BMI and insulin resistance compared to cobalamin replete mothers (Knight et al. 2015). In this case, folate status was negatively associated with BMI. Women with the lowest cobalamin (plasma cobalamin < 170 pmol/L) and folate (plasma folate < 10.3 nmol/L) status had the highest BMI and the converse was true for women with the highest cobalamin (> 238 pmol/L) and folate status (> 18.3 nmol/L).

The remethylation of homocysteine to methionine is mainly performed by MTR, in a reaction in which 5-methyltetrahydrofolate provides the methyl group that is bound to the cobalamin cofactor of the enzyme. Homocysteine can also be remethylated by betaine homocysteine methyltransferase (BHMT, EC 2.1.1.5), where betaine provides the methyl group required. Under normal circumstances, the ubiquitous MTR pathway prevails, but the BHMT pathway appears to be upregulated when folate status is low in late pregnancy in non-supplemented women (Fernandez-Roig et al. 2013). The folate and BHMT pathways show strong interaction depending on folate and cobalamin status.

The BHMT pathway compensates for mild folate or cobalamin deficiency by mobilizing methyl groups of choline and betaine. Cobalamin has also been linked with choline and betaine metabolism during pregnancy in Canadian women (Wu et al. 2013). Marginal cobalamin deficiency (< 220 pmol/L) was associated with a lower rise in plasma free choline between mid-gestation and labour in folate replete Canadian women. Moreover, women with marginal cobalamin deficiency had lower plasma choline, betaine and dimethylglycine concentrations at 36 gestational weeks than women with cobalamin > 220 pmol/L. In this case, the results did not suggest upregulation of the BHMT activity, but reduced endogenous choline synthesis; due to lower availability of methyl groups in the low cobalamin status group. Similar effects on the BHMT pathway might be expected for low folate or cobalamin statuses, given that in both circumstances, tHcy concentrations increase. However, cobalamin status was not very low in the Canadian study (Wu et al. 2013) and mandatory fortification of flour with folic acid in Canada has been in place since 1998, so folate status was adequate throughout pregnancy in this study. There is no mandatory fortification policy with folic acid in Spain, and plasma folate fell sharply when folic acid supplement use was stopped by approximately 50% of the women at the end of the 1st trimester in the Spanish study (Fernandez-Roig et al. 2013). Nutrients imbalance is also an important aspect in pregnant women with fluctuations in the intake because of seasonal availability of foods. A study on African non-pregnant women with varying intakes of folate, betaine, and choline over the year has found profound effect on DNA-methylation (Dominguez-Salas et al. 2013).

Taken together, nutrients imbalance is a critical issue in countries where cobalamin or folate deficiency is a public health concern, but also when folic acid is provided at higher doses to cobalamin deficiency pregnant women. This is mainly because folic acid supplementation is unlikely to be effective in prevention of birth defects in women with cobalamin deficiency.

2.4 Cobalamin, pregnancy complications and birth outcome

Unlike the clear recommendation to use folic acid-containing supplements before conception or during the first trimester, there are no clear recommendations for using cobalamin. Moreover, most prenatal supplements contain either no cobalamin or a low amount that is unlikely to correct cobalamin deficiency. Vegetarian women or women from developing countries with insufficient cobalamin status are at increased risk of adverse pregnancy outcome, but screening for cobalamin deficiency is not practised regularly. Low plasma cobalamin and elevated MMA were reported to be highly prevalent during the first trimester of pregnancy in different Indian studies (Duggan et al. 2014; Samuel et al. 2013). Another recent study reported low serum cobalamin (< 148 pmol/L) in 18% of Guatemalan women of fertile age and the prevalence was higher in those with low socioeconomic status (Rosenthal et al. 2015).

Early and recurrent abortion

Maternal cobalamin deficiency has been related to increased risk of early and recurrent miscarriage in several reports (Candito et al. 2003; Hubner et al. 2008). Increased plasma MMA was associated with increased probability of spontaneous miscarriage in 100 Brazilian women (Odds ratio = 3.80; 95% Confidence intervals: 1.36–10.62). Normal preconception cobalamin concentrations ($\geq$ 258 pmol/L) were associated with a 60% reduced risk of preterm birth (odds ratio: 0.4; 95% CI: 0.2–0.9) but did not affect pregnancy loss in Chinese women (Ronnenberg et al. 2002).

Anaemia

Anaemia has been known to be a common complication of pregnancy in tropical countries since the early 1930s (Berry 1955). Although cobalamin deficiency was not the main factor affecting the severe cases, cobalamin administration (i.e., 200–250 µg initially then 50 µg weekly) alleviated the symptoms in many cases (Berry 1955), suggesting that screening for cobalamin deficiency in anaemic women could offer an option to improve maternal and child health.

Birth weight and growth rate

Low pregnancy serum cobalamin has also been associated with low birth weight (Frery et al. 1992; Relton et al. 2005). However, in a Norwegian study, cord cobalamin decreased with increasing birth weight quartile. Cord cobalamin was 28% lower in babies in the highest than in those in the lowest birth weight quartile (Hay et al. 2010). In line with the observations for birth weight, cord cobalamin was also negatively associated with birth length and head circumference (Hay et al. 2010). In contrast, cord MMA was not associated with any of the newborn measurements suggesting that its high levels in the cord might not reflect cobalamin status in the foetus or may reflect the inability to store the vitamin because of the low growth rate.

Birth defects

Cobalamin deficiency has been associated with neural tube defects (NTD) in populations with a high prevalence of deficiency. NTDs are severe birth defects caused by failure of the neural tube to close during early foetal development. The prevalence of neural tube defects is 6.7 to 8.2 per thousand live births in some parts of India (Cherian et al. 2005) where cobalamin and other micronutrient deficiencies are common (Antony 2003) and may be causally related to NTDs. The role of folic acid in preventing neural tube defects is well established. However, the protective effect of folic acid supplementation may be limited when the mother has cobalamin deficiency. Alternative sources of methyl groups such as choline, methionine, and betaine can be expected to support one-carbon metabolism in cobalamin deficient women. However, in situations of vegetarianism or poverty those alternative sources will be limited as well.

Amniotic fluid cobalamin concentration was consistently lower in pregnancies with NTDs compared to controls in a wide range of reports, but the evidence for an effect of low maternal cobalamin was moderate (Ray 2004). An Irish study of 171 mothers with NTD-affected pregnancies, 107 women with a previous NTD birth but whose current pregnancy was unaffected and 901 controls reported that plasma cobalamin < 250 ng/L (185 pmol/L) increased the risk of NTD by 2.5- to 3 fold, after adjusting for folate (Molloy et al. 2009). The authors recommended that women should aim for plasma cobalamin concentrations > 300 ng/L (221 pmol/L) at preconception in order to reduce the incidence of NTDs (Molloy et al. 2009).

The World Health Organization (WHO) has recently recommended more research on the added value of cobalamin in the prevention of neural tube defects and the lowest optimal red blood cell folate to achieve the greatest reduction in NTDs was set at 1000 nmol/L (World Health Organization 2015). This cut-off was first proposed by Daly et al. in 1995 (Daly et al. 1995), and was supported afterwards by Crider et al. and Marchetta et al. (Crider et al. 2014; Marchetta et al. 2015). By using population data from two studies in non-fortified regions in China (n = 247831), Crider et al., determined the optimal red blood cell folate for NTD prevention (Crider et al. 2014). The WHO guidelines also addressed the importance of studying the role of cobalamin deficiency in birth defects especially in countries where the deficiency is endemic.

2.5 Critical view of cobalamin requirements during pregnancy and lactation

Some of the assumptions made when estimating cobalamin requirements during pregnancy and lactation were based on historical studies and the results have not been confirmed yet. In our view, current cobalamin intake recommendations may be inadequate, in particular for women who start pregnancy with low status.

First, it was assumed that recently consumed cobalamin is transported to the foetus. However, the minimum sufficient intake in pregnancy if the mother is cobalamin depleted or deficient is not known. Therefore, the problem is likely to worsen if a pregnant woman is deficient because of low intake or malabsorption. Second, how the vitamin is absorbed and distributed when pregnant women are deficient or depleted is not known either. The assumed enhanced absorption has not been proven, and there is no data on the efficacy of cobalamin re-uptake through the kidney or other elimination pathways. Third, the assumption that cobalamin absorption increases during pregnancy is based on studies in mice (Brown et al. 1977) or rats (Hellegers et al. 1957), but recent studies haven't provided further evidence of an increase in cobalamin absorption between weeks 13, 24, and 36 of gestation (Greibe et al. 2011). Therefore, if the absorption of the vitamin does not increase, it is questionable whether the estimated EARs and RDAs for pregnant women are adequate, particularly in the case of women with cobalamin deficiency. The recommended intake of cobalamin for pregnant and lactating women needs to be re-considered based on current knowledge.

Numerous clinical manifestations of cobalamin deficiency in infants of vegetarian women have been described. The symptoms appear a few months after birth, when infant demands exceed vitamin stores. In the majority of cases, mothers have been constantly consuming a vegetarian diet for many years before becoming pregnant. Therefore, a constantly low intake during pregnancy has to be partitioned between the mother and the child, and in most cases, the deficiency is manifested in the child, rather than in the mother. This suggested that the cobalamin intake of vegetarian women during pregnancy should be increased to at least the same level as in omnivorous women. Pregnant vegetarian women and low meat eaters have shown significantly lower plasma cobalamin and higher homocysteine at any gestational age, with plasma homocysteine strongly predicted by serum cobalamin in vegetarians (Koebnick et al. 2004). The intake of cobalamin in pregnant lacto-ovo-vegetarian women was 2.5 µg/d (just below the RDA for pregnant women, 2.6 µg/day) versus 5.3 µg/d in omnivorous women (Koebnick et al. 2004). Based on these results, vegetarian women should at least double their cobalamin intake to decrease the likelihood that their infants will have cobalamin deficiency shortly after birth.

Studies on lactating women have shown that their serum cobalamin was 13–42% lower than in non-pregnant women, suggesting depletion of cobalamin stores during lactation (Jathar et al. 1970) (Figure 1). The cobalamin content in breast milk of Indian women was not significantly different between vegetarian and non-vegetarian women (mean = 91 vs 103 pg/ml or 67 vs 76 pmol/L). However, Indian lactating women had low milk cobalamin concentrations (mean 100 pg/ml or 74 pmol/L) compared to studies from other countries such as the USA (mean = 410 pg/ml or 303 pmol/L) or France (mean = 300 pg/ml or 222 pmol/L) (Collins et al. 1951; Jathar et al. 1970).

It is not known how cobalamin is distributed between milk and maternal tissues when the mother is deficient. Firm conclusions on cobalamin requirements during pregnancy and lactation cannot be drawn yet given the gaps in knowledge in this field (Table 2). However, it is becoming evident that the current recommendations are too low to cover the requirements during pregnancy of many women and their infants up to weaning time, especially in deficient lactating women.

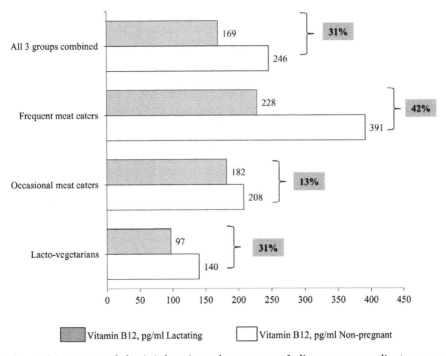

Vitamin B12, pg/ml Lactating   Vitamin B12, pg/ml Non-pregnant

Figure 1. Mean serum cobalamin in lactating and non-pregnant Indian women according to groups of dietary habits. In general, lactating women had 13–42% lower cobalamin levels compared to non-pregnant women. The difference appears to be larger in women with high or low compared to intermediate frequency of meat intake. Adapted from (Jathar et al. 1970). Conversion factor for cobalamin: 1 pg/mL = 0.7380 pmol/L.

Table 2. Gaps in knowledge regarding cobalamin requirements during pregnancy and lactation.

Open questions	Background
What are the EARs for pregnant and lactating women depending on whether they start pregnancy with adequate or inadequate stores?	The estimations need to consider measurements of functional status markers of the vitamin. Deficient women may not transfer the same amount of cobalamin into foetal tissues or milk as non-deficient women.
Is cobalamin absorption from food (protein bound) and from supplements (free) increased during pregnancy? Is there a rapid tissue clearance of the newly absorbed cobalamin?	The effects of a controlled cobalamin-rich diet on plasma holoTC and other metabolic markers should be tested at different stages of pregnancy.
Is cobalamin extracted more efficiently through the kidney in pregnant women who have low intake or deficiency?	The kidney has a major role in cobalamin homeostasis.
Should the calculations of the EAR consider maintenance of maternal cobalamin stores (i.e., holoTC) during pregnancy instead of being based on the amount thought to be transported to the foetus?	HoloTC declines in the 1st trimester and remains stable afterwards. The mechanisms and implications of maintenance of holoTC supply to the foetus throughout pregnancy are not well studied. The data on the amount of cobalamin transported to the foetus has not been confirmed.

3. Cobalamin Markers and Binding Proteins in Infants at Birth and Thereafter

Foetal cobalamin stores are established from early gestation throughout pregnancy through active transport of cobalamin from the mother. Animal studies have shown that cobalamin administered during pregnancy (orally or by injection) is actively transported and accumulates in the placenta and foetal organs (Giugliani et al. 1985; Graber et al. 1971; Hellegers et al. 1957; Ullberg et al. 1967). Little is known about other factors that influence cobalamin transport via the placenta or cobalamin accumulation in the child. There is currently no explanation for the facts that infants with congenital TC deficiency are asymptomatic at birth but symptoms such as vomiting, failure to thrive, lethargy and pallor develop few weeks afterwards (Kaikov et al. 1991) suggesting that the transport or exchange of functional cobalamin was normal up to birth. If not treated, neurological dysfunction with severe mental retardation may develop.

Cobalamin binding proteins and biochemical markers have been studied in cord blood and maternal blood at birth (Giugliani et al. 1985; Guerra-Shinohara et al. 2004; Obeid et al. 2006). A cohort study on healthy pregnant women with term deliveries and uncomplicated pregnancies has shown that total TC is lower in cord serum compared with that in the mother (Obeid et al. 2006). In contrast, the concentrations of total HC, holoHC and holoTC are higher in cord serum than in the mother's serum (Obeid et al. 2006). The concentration of cord serum holoTC did not differ between tertiles of gestational age at birth, whereas holoHC concentrations were higher in babies born later (despite all deliveries being at term) (Obeid et al. 2006). Maternal and cord serum holoTC are strongly and positively associated. However, the concentration of MMA in cord serum was generally high [median (range) = 373 (138–878) nmol/L] and varied relatively little (~18%) between the lowest and highest quintile of maternal holoTC (Obeid et al. 2006) (Figure 2). Cord serum MMA correlated inversely with holoTC ($r = -0.345$, $p = 0.001$) (Figure 3), but were not completely predicted by maternal holoTC or holoHC (Obeid et al. 2006).

Maternal serum concentration of cobalamin is a strong predictor of holoTC and TC saturation in cord serum (Obeid et al. 2006). Conversely, cord serum concentrations of total HC, holoHC and HC saturation were not significantly related to maternal cobalamin concentrations (Obeid et al. 2006) suggesting that HC and holoHC are not transferred from the mother to the baby. A late redistribution of cobalamin between TC and HC before birth has been suggested (Obeid et al. 2006) probably promoting cobalamin binding to HC rather than to TC in the foetus before birth.

In a study on cobalamin deficient women from Brazil, maternal plasma cobalamin and creatinine at delivery were significant predictors of cord plasma MMA but cord plasma cobalamin was not (Guerra-Shinohara et al. 2004). However, these factors did not account for all of the variability in MMA concentrations between subjects, suggesting that other factors may affect MMA

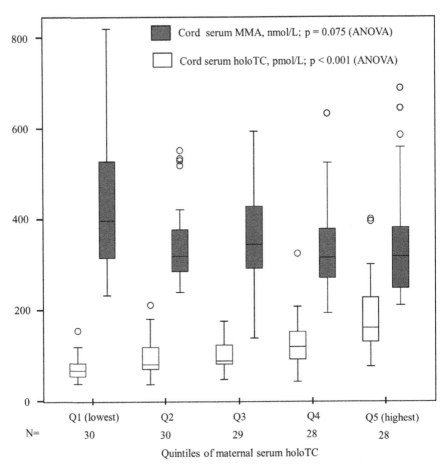

Figure 2. The relationship between cord and maternal serum holoTC and methylmalonic acid (MMA) at delivery. The concentrations of holoTC in cord serum increase with increasing maternal serum holoTC (p between maternal holoTC quintiles < 0.001). The relationship between cord serum MMA and maternal holoTC was less clear (p between maternal holoTC quintiles = 0.075). Maternal serum holoTC (range, pmol/L) = Q1 (19–33); Q2 (34–42); Q3 (43–52); Q4 (53–64); Q5 (65–244). Original study published in (Obeid et al. 2006).

concentrations. MMA is a mitochondrial metabolite that is linked to energy metabolism, gluconeogenesis and Krebs cycle. Therefore, MMA elevation in cord blood could be related to other connected metabolic pathways. After birth, the intake of branched chain amino acids that are present at high concentrations in human milk can increase MMA levels by increased production of propionate that is converted into MMA.

Studies on pregnant women with poor cobalamin status have reported that the offspring are already cobalamin deficient at birth. 72% of pregnant women had low serum cobalamin (< 118 pmol/L) in a study conducted

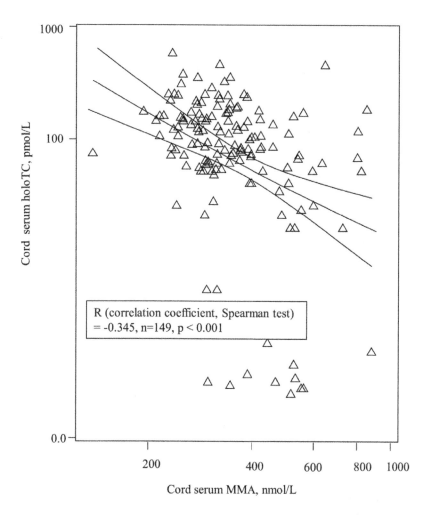

Figure 3. The relationship between cord serum MMA and holoTC in 149 pregnant German women just before delivery and their infants a birth. All deliveries were at term, with normal birth weight and without complications. The x and y axis are represented in the logarithmic scale. Original study was published in (Obeid et al. 2006).

in Southeast Turkey (Koc et al. 2006). Maternal and cord serum cobalamin (mean 96 and 153 pmol/L, respectively) were both very low (Koc et al. 2006). Serum cobalamin was < 118 pmol/L in 41% of the cords and < 90 pmol/L in 23% of them (Koc et al. 2006). A second study on Brazilian mothers with low socioeconomic status reported not only lower cord serum cobalamin, but also higher homocysteine and a lower methylation ratio [S-adenosylmethionine (SAM): S-adenosylhomocysteine (SAH)] with decreasing maternal cobalamin

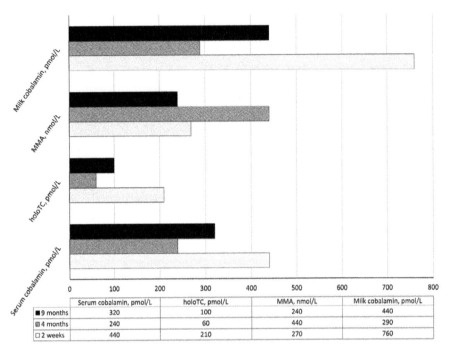

	Serum cobalamin, pmol/L	holoTC, pmol/L	MMA, nmol/L	Milk cobalamin, pmol/L
■ 9 months	320	100	240	440
▨ 4 months	240	60	440	290
☐ 2 weeks	440	210	270	760

Figure 4. Changes in maternal serum cobalamin markers and milk concentrations of total cobalamin from week 2 to month 9 of lactation. The figure is based on data from Greibe et al. (Greibe et al. 2013). Cobalamin was measured in milk after capturing HC on Sepharose and removing it (Greibe et al. 2013).

status at labour (Guerra-Shinohara et al. 2004). These results suggest possible adverse effects of low maternal cobalamin status on methylation pathways (DNA-methylation, phospholipids metabolism, etc.) in the offspring. Multiple linear regression analysis showed that maternal cobalamin was inversely associated with cord tHcy and cord MMA and positively correlated with the SAM:SAH ratio in the cord. Multiple logistic regression analysis showed that the odds of low cord SAM:SAH were increased by fivefold for maternal serum cobalamin < 102 pmol/L. Maternal plasma cobalamin during late pregnancy and at labour was also inversely associated with cord MMA in a Spanish pregnancy study where maternal cobalamin status at labour was higher than in the previous two studies. By 32 gestational weeks, plasma MMA was higher than at preconception and the total pregnancy increase in plasma MMA was higher when plasma holoTC at preconception was below the median (67 pmol/L) (Murphy et al. 2007). The observed changes in MMA concentrations were attributed to both the effects of pregnancy and of cobalamin status. Both maternal holoTC and cobalamin were inversely associated with MMA in the cord, in the same study (Murphy et al. 2007).

4. Determinants of Cobalamin Status During Lactation

4.1 Cobalamin intake and breast milk cobalamin content

As is the case for cobalamin transported to the foetal tissues, recently consumed cobalamin by the mother is available for transport through breast milk. In women with no apparent cobalamin deficiency, recent cobalamin intake or the amount of the vitamin that is absorbed appears to be more important than maternal stores as a determinant of milk cobalamin. Therefore, if maternal cobalamin intake is adequate, there is no concern that her stores will be depleted during lactation.

Lactating women with very low breast milk cobalamin are usually those on a strict vegetarian diet or women from developing countries who have a chronically low intake of cobalamin-containing foods. They are at increased risk of developing cobalamin deficiency during lactation (Black et al. 1994; Shapiro et al. 1965). Few longitudinal studies on deficient lactating women and milk cobalamin content or cobalamin status in their offspring are available to date.

Maternal use of supplements containing cobalamin is a strong positive determinant of breast milk cobalamin concentration. Human milk cobalamin concentration increased by approximately 150 fold following a single cobalamin injection of 50 µg, but the effect lasted only for 7 days (Srikantia and Reddy 1967). Median breast milk cobalamin concentration was approximately double in the treated versus placebo group in Indian pregnant women that were supplemented with 50 µg/d of cobalamin group from < 14 gestational weeks until 6 weeks postpartum. Their infants also had lower MMA and higher cobalamin compared to the placebo group. This effect was not maintained when the supplementation was stopped (Duggan et al. 2014). Feeding lactating women from New York with a standard diet (6 µg/d of cobalamin from diet plus 2.6 µg/d from a multivitamin supplement) for 10 weeks from 5 weeks postpartum did not increase breast milk cobalamin content (Bae et al. 2015). Maternal plasma concentration of holoTC did increase, tHcy decreased after the intervention, but cobalamin remained unchanged. These results remain unexplained. As cobalamin concentrations in breast milk vary during lactation (Greibe et al. 2013), the inclusion of a control group would have helped to determine whether no differences found with intervention, were due to the prevention of cobalamin decline. There were numerous differences in the study design and baseline characteristics of the participants in the New York study compared to the previous Indian study. These included the intervention regime (timing, duration and dose), and importantly, baseline cobalamin status of the participants, many of whom had low or deficient cobalamin status in the Indian study. The final post-supplementation milk cobalamin concentrations, though substantially higher than in the placebo group, in the treated women (median = 186 pmol/L) in the Indian study were still lower than the baseline concentrations (geometric mean = 318 pmol/L) in the New York study. Other studies (Casterline et al. 1997; Patel and Lovelady 1998) also reported a less impressive relationship between maternal serum and milk cobalamin than in Duggan et al.'s study that used high dose cobalamin supplements. Serum and milk cobalamin concentrations in Indian lactating vegetarian were almost 50% of those found in omnivorous lactating women (264 vs 531 pmol/L for serum and 277 vs 544 pmol/L for milk) and between dietary cobalamin intake and supplements, their total average intake of cobalamin was 4.6 μg/d (Patel and Lovelady 1998).

Mexican pregnant women with low to marginal cobalamin status were studied during pregnancy and again during lactation (Black et al. 1994). Low cobalamin was found in 15% of the women during pregnancy (plasma cobalamin < 74 pmol/L) and in 30% during lactation (plasma cobalamin < 103 pmol/L) (Black et al. 1994) suggesting that a significant portion of those women became depleted during lactation. Milk cobalamin concentrations were < 362 pmol/L in 31 of the 50 Mexican women (Black et al. 1994), which is the cut-off found by Specker et al. to be associated with elevated concentrations of MMA in the urine of infants from vegetarian women (Specker et al. 1988; Specker et al. 1990). The presence of anaemia was associated with lower concentrations of breast milk cobalamin in this study (mean 285 in anaemic vs. 418 nmol/L in non-anaemic women; p < 0.05) (Black et al. 1994). Breast milk cobalamin and maternal plasma cobalamin were moderately correlated (r = 0.48, p = 0.06). Plasma cobalamin concentrations during pregnancy correlated with that during lactation (r = 0.53, p < 0.05). Milk cobalamin content did not differ between women with high versus low plasma cobalamin (Black et al. 1994), suggesting that cobalamin is transferred to breast milk independently of maternal cobalamin status.

Methodological inaccuracy of measuring cobalamin content in human milk and dependency of milk concentrations on recent maternal intake makes comparison between different studies difficult, and caution should be taken when interpreting the results. Nevertheless, milk cobalamin content is the strongest determinant of cobalamin status in exclusively breastfed infants. A strong correlation between human milk, maternal serum and infant serum cobalamin has been reported in infants with manifested cobalamin deficiency (Srikantia and Reddy 1967). The risk of cobalamin deficiency in infants of cobalamin deficient mothers during the most important stage of neurodevelopment (first 24 months) warrants serious concern.

4.2 Cobalamin and its binding proteins in human milk from cobalamin deficient and sufficient women

The high content of unsaturated HC in milk complicates the analysis of cobalamin in milk. Furthermore, differences in analytical techniques, sample collection and timing of sample collection as well as maternal nutritional status between studies lead to highly variable results (Hampel et al. 2014). It is of critical importance to achieve human milk cobalamin concentrations that ensure sufficient supply of the vitamin to the infant. We also need data on the effect of maternal supplementation during pregnancy and/or lactation on status in the infant.

The major cobalamin binding protein in human milk is HC. Almost all cobalamin in human milk is known to be bound to HC (Sandberg et al. 1981), a protein also known to bind cobalamin analogues. However, human milk was reported to contain no cobalamin analogues, in contrast to human serum and cord blood (Adjalla et al. 1994). Human milk HC content increases during the course of lactation. When unbound HC (i.e., apo-HC) is higher than 10 pmol/L, the binding capacity of HC interferes with the assay methods (Lildballe et al. 2009). Cobalamin measurement in human milk has been improved when unsaturated HC was removed prior to cobalamin measurement (Greibe et al. 2013; Lildballe et al. 2009). A cleaning step prior to the assay involves removing apo-HC from the sample by absorption to cobinamid-coated Sepharose (Lildballe et al. 2009). Therefore, assays that do not eliminate unbound HC prior to cobalamin measurements will produce unreliable results on cobalamin levels.

Differences in vitamin composition of colostrum, transitional milk and mature milk and changes in milk composition (between foremilk and hindmilk) during a single feed have been investigated. In a longitudinal study on 9 lactating women, cobalamin content (measured using radioisotope dilution assay) were not different between fore- and hindmilk but declined from 0.35 to 0.25 nmol/L over 3 months of lactation whereas cobalamin binding capacity increased from 29.6 to 43.8 nmol/L over the same period (Trugo and Sardinha 1994). The increased cobalamin binding capacity is mainly due to the increase in milk HC during lactation. Another study measured milk cobalamin after removing HC from the samples in a longitudinal study of 25 lactating women (Greibe et al. 2013). Milk samples were tested at 2 weeks and at 4 and 9 months postpartum. At these three time points, both cobalamin and HC contents were higher in hindmilk than in foremilk (14%, 9%, and 9% more HC in hindmilk at 2 week, 4 and 9 months). Milk cobalamin concentrations changed substantially as lactation progressed from week 2 to 9 months (Figure 4) (Greibe et al. 2013). The lowest levels of milk cobalamin were observed at 4 months where it was associated with the lowest serum holoTC and highest MMA in breastfed children (Figure 4) (Greibe et al. 2013), suggesting that milk cobalamin at 4 months did not supply sufficient cobalamin to perform all of its metabolic function.

The definition of adequate cobalamin intake in lactating women urgently needs to be revisited. Studies are also needed on breast milk cobalamin content, how it is affected by maternal cobalamin intake and status and its effects on health outcomes in the child.

5. Summary and Conclusions

We have learned that the concentrations of cobalamin markers at birth (i.e., in cord blood) are predicted by maternal blood markers (strong positive association). After birth, a sufficient intake of cobalamin through lactation or formula milk can maintain adequate cobalamin status in the infant up to the age of 4–6 months. Exclusively breastfed infants show a dramatic decline in serum cobalamin and holoTC associated with an increase in MMA and homocysteine between 4 and 6 months of age. There appears to be a fall in infant cobalamin status that corresponds with a simultaneous decrease in milk cobalamin content. Biochemical markers of cobalamin status in the infant suggest that cobalamin intake at this age can be insufficient. Low cobalamin status in infants is strongly related to low maternal status during pregnancy and breastfeeding.

Breast milk is the only source of cobalamin for exclusively breast-fed infants. The content of cobalamin in human milk is predicted by recent maternal intake of the vitamin rather than by maternal vitamin stores. Recent cobalamin intake is thought to be readily available for transportation to the milk. Studies have shown that cobalamin depletion in strict vegetarian women and otherwise deficient women from developing countries can result in serious health consequences in exclusively breastfed infants. However, even asymptomatic lactating women on an omnivorous diet may have depleted vitamin stores and low intake or malabsorption resulting in the insufficient delivery of the vitamin to the infant.

Keywords: Pregnancy, lactation, placenta, intake, human milk, folate, neural tube defects, birth weight, maternal, recommendations, cobalamin markers

Abbreviations

AI	:	adequate intake
IOM	:	Institute of Medicine
RDA	:	Recommended Dietary Allowance
HoloTC	:	holotranscobalamin
HC	:	haptocorrin
TC	:	transcobalamin
holoHC	:	holohaptocorrin
CV	:	coefficient of variation
BHMT	:	betaine homocysteine methyltransferase
MMA	:	methylmalonic acid
MTR	:	5-methyltetrahydrofolate-homocysteine methyltransferase
NTDs	:	neural tube defects
SAM	:	S-adenosylmethionine
SAH	:	S-adenosylhomocysteine
tHcy	:	total homocysteine

References

Adjalla c, Lambert D, Benhayoun S, Berthelsen JG, Nicolas JP, Gueant JL and Nexo E. 1994. Forms of cobalamin and vitamin B12 analogs in maternal plasma, milk, and cord plasma. J Nutr Biochem. 8: 406–410.

Antony AC. 2003. Vegetarianism and vitamin B12 (cobalamin) deficiency. Am J Clin Nutr. 78: 3–6.

Bae S, West AA, Yan J, Jiang X, Perry CA, Malysheva O, Stabler SP, Allen RH and Caudill MA. 2015. Vitamin B12 status differs among pregnant, lactating, and control women with equivalent nutrient intakes. J Nutr. 145: 1507–1514.

Berry CG. 1955. Anaemia of pregnancy in Africans of Lagos. Br Med J. 2: 819–823.

Black AK, Allen LH, Pelto GH, de Mata MP and Chavez A. 1994. Iron, vitamin B12 and folate status in Mexico: associated factors in men and women and during pregnancy and lactation. J Nutr. 124: 1179–1188.

Brown J, Robertson J and Gallagher N. 1977. Humoral regulation of vitamin B12 absorption by pregnant mouse small intestine. Gastroenterology. 72: 881–888.

Candito M, Magnaldo S, Bayle J, Dor JF, Gillet Y, Bongain A and Van OE. 2003. Clinical B12 deficiency in one case of recurrent spontaneous pregnancy loss. Clin Chem Lab Med. 41: 1026–1027.

Casterline JE, Allen LH and Ruel MT. 1997. Vitamin B12 deficiency is very prevalent in lactating Guatemalan women and their infants at three months postpartum. J Nutr. 127: 1966–1972.

Chen LT and Rivera MA. 2004. The Costa Rican experience: reduction of neural tube defects following food fortification programs. Nutr Rev. 62: S40–S43.

Cherian A, Seena S, Bullock RK and Antony AC. 2005. Incidence of neural tube defects in the least-developed area of India: a population-based study. Lancet. 366: 930–931.

Collins RA, Harper AE, Scheiber M and Elvehjem C. 1951. The folic acid and vitamin B12 content of the milk of various species. J Nutr. 43: 313–321.

Crider KS, Devine O, Hao L, Dowling NF, Li S, Molloy AM, Li Z, Zhu J and Berry RJ. 2014. Population red blood cell folate concentrations for prevention of neural tube defects: bayesian model. BMJ. 349: g4554.

Daly LE, Kirke PN, Molloy A, Weir DG and Scott JM. 1995. Folate levels and neural tube defects. Implications for prevention. JAMA. 274: 1698–1702.

Dominguez-Salas P, Moore SE, Cole D, da Costa KA, Cox SE, Dyer RA, Fulford AJ, Innis SM, Waterland RA, Zeisel SH, Prentice AM and Hennig BJ. 2013. DNA methylation potential: dietary intake and blood concentrations of one-carbon metabolites and cofactors in rural African women. Am J Clin Nutr. 97: 1217–1227.

Duggan C, Srinivasan K, Thomas T, Samuel T, Rajendran R, Muthayya S, Finkelstein JL, Lukose A, Fawzi W, Allen LH, Bosch RJ and Kurpad AV. 2014. Vitamin B12 supplementation during pregnancy and early lactation increases maternal, breast milk, and infant measures of vitamin B12 status. J Nutr. 144: 758–764.

Dwarkanath P, Barzilay JR, Thomas T, Thomas A, Bhat S and Kurpad AV. 2013. High folate and low vitamin B12 intakes during pregnancy are associated with small-for-gestational age infants in South Indian women: a prospective observational cohort study. Am J Clin Nutr. 98: 1450–1458.

Faupel-Badger JM, Hsieh CC, Troisi R, Lagiou P and Potischman N. 2007. Plasma volume expansion in pregnancy: implications for biomarkers in population studies. Cancer Epidemiol Biomarkers Prev. 16: 1720–1723.

Fernandes-Costa F and Metz J. 1982. Levels of transcobalamins I, II, and III during pregnancy and in cord blood. Am J Clin Nutr. 35: 87–94.

Fernandez-Roig S, Cavalle-Busquets P, Fernandez-Ballart JD, Ballesteros M, Berrocal-Zaragoza MI, Salat-Batlle J, Ueland PM and Murphy MM. 2013. Low folate status enhances pregnancy changes in plasma betaine and dimethylglycine concentrations and the association between betaine and homocysteine. Am J Clin Nutr. 97: 1252–1259.

Food and Drug Administration, and Department of health and human services. 1996. Food standards: Amendment of standards of identity for enriched grain products to require addition of folic acid. Federal Register. 61: 8781–8797.

Frery N, Huel G, Leroy M, Moreau T, Savard R, Blot P and Lellouch J. 1992. Vitamin B12 among parturients and their newborns and its relationship with birthweight. Eur J Obstet Gynecol Reprod Biol. 45: 155–163.

Giles C. 1966. An account of 335 cases of megaloblastic anaemia of pregnancy and the puerperium. J Clin Pathol. 19: 1–11.

Giugliani ER, Jorge SM and Goncalves AL. 1985. Serum vitamin B12 levels in parturients, in the intervillous space of the placenta and in full-term newborns and their interrelationships with folate levels. Am J Clin Nutr. 41: 330–335.

Gomes TS, Lindner U, Tennekoon KH, Karandagoda W, Gortner L and Obeid R. 2010. Homocysteine in small-for-gestational age and appropriate-for-gestational age preterm neonates from mothers receiving folic acid supplementation. Clin Chem Lab Med. 48: 1157–1161.

Graber SE, Scheffel U, Hodkinson B and McIntyre PA. 1971. Placental transport of vitamin B12 in the pregnant rat. J Clin Invest. 50: 1000–1004.

Greibe E, Andreasen BH, Lildballe DL, Morkbak AL, Hvas AM and Nexo E. 2011. Uptake of cobalamin and markers of cobalamin status: a longitudinal study of healthy pregnant women. Clin Chem Lab Med. 49: 1877–1882.

Greibe E, Lildballe DL, Streym S, Vestergaard P, Rejnmark L, Mosekilde L and Nexo E. 2013. Cobalamin and haptocorrin in human milk and cobalamin-related variables in mother and child: a 9-mo longitudinal study. Am J Clin Nutr. 98: 389–395.

Guerra-Shinohara EM, Morita OE, Peres S, Pagliusi RA, Sampaio Neto LF, D'Almeida V, Irazusta SP, Allen RH and Stabler SP. 2004. Low ratio of S-adenosylmethionine to S-adenosylhomocysteine is associated with vitamin deficiency in Brazilian pregnant women and newborns. Am J Clin Nutr. 80: 1312–1321.

Hampel D, Shahab-Ferdows S, Domek JM, Siddiqua T, Raqib R and Allen LH. 2014. Competitive chemiluminescent enzyme immunoassay for vitamin B12 analysis in human milk. Food Chem. 153: 60–65.

Hay G, Clausen T, Whitelaw A, Trygg K, Johnston C, Henriksen T and Refsum H. 2010. Maternal folate and cobalamin status predicts vitamin status in newborns and 6-month-old infants. J Nutr. 140: 557–564.

Heinrich HC. 1954. Biochemical principles of diagnosis and therapy of vitamin B12 deficiency in man and in domestic animals. II. Studies on vitamin B12 metabolism in man during pregnancy and lactation. Klin Wochenschr. 32: 205–209.

Hellegers A, Okuda K, Nesbitt RE Jr, Smith DW and Chow BF. 1957. Vitamin B12 absorption in pregnancy and in the newborn. Am J Clin Nutr. 5: 327–331.

Hertrampf E and Cortes F. 2004. Folic acid fortification of wheat flour: Chile. Nutr Rev. 62: S44–S48.

Hubner U, Alwan A, Jouma M, Tabbaa M, Schorr H and Herrmann W. 2008. Low serum vitamin B12 is associated with recurrent pregnancy loss in Syrian women. Clin Chem Lab Med. 46: 1265–1269.

Institute of Medicine. 1998. Dietary Reference Intakes for Thiamin, Riboflavin, Niacin, Vitamin B6, Folate, Vitamin B12, Pantothenic Acid, Biotin, and Choline. In: USA: Washington, DC, National Academy Press. pp. 390–422.

Izak G, Rachmilewitz M, Stein Y, Berkovici B, Sadovsky A, Aronovitch Y and Grossowicz N. 1957. Vitamin B12 and iron deficiencies in anemia of pregnancy and puerperium. AMA Arch Intern Med. 99: 346–355.

Jathar VS, Kamath SA, Parikh MN, Rege DV and Satoskar RS. 1970. Maternal milk and serum vitamin B12, folic acid, and protein levels in Indian subjects. Arch Dis Child. 45: 236–241.

Kaikov Y, Wadsworth LD, Hall CA and Rogers PC. 1991. Transcobalamin II deficiency: case report and review of the literature. Eur J Pediatr. 150: 841–843.

Knight BA, Shields BM, Brook A, Hill A, Bhat DS, Hattersley AT and Yajnik CS. 2015. Lower Circulating B12 Is Associated with Higher Obesity and Insulin Resistance during Pregnancy in a Non-Diabetic White British Population. PLoS One. 10: e0135268.

Koc A, Kocyigit A, Soran M, Demir N, Sevinc E, Erel O and Mil Z. 2006. High frequency of maternal vitamin B12 deficiency as an important cause of infantile vitamin B12 deficiency in Sanliurfa province of Turkey. Eur J Nutr. 45: 291–297.

Koebnick C, Heins UA, Dagnelie PC, Wickramasinghe SN, Ratnayaka ID, Hothorn T, Pfahlberg AB, Hoffmann I, Lindemans J and Leitzmann C. 2002. Longitudinal concentrations of vitamin B(12) and vitamin B(12)-binding proteins during uncomplicated pregnancy. Clin Chem. 48: 928–933.

Koebnick C, Hoffmann I, Dagnelie PC, Heins UA, Wickramasinghe SN, Ratnayaka ID, Gruendel S, Lindemans J and Leitzmann C. 2004. Long-term ovo-lacto vegetarian diet impairs vitamin B12 status in pregnant women. J Nutr. 134: 3319–3326.

Krishnaveni GV, Veena SR, Karat SC, Yajnik CS and Fall CH. 2014. Association between maternal folate concentrations during pregnancy and insulin resistance in Indian children. Diabetologia. 57: 110–121.

Lildballe DL, Hardlei TF, Allen LH and Nexo E. 2009. High concentrations of haptocorrin interfere with routine measurement of cobalamins in human serum and milk. A problem and its solution. Clin Chem Lab Med. 47: 182–187.

Marchetta CM, Devine OJ, Crider KS, Tsang BL, Cordero AM, Qi YP, Guo J, Berry RJ, Rosenthal J, Mulinare J, Mersereau P and Hamner HC. 2015. Assessing the association between natural food folate intake and blood folate concentrations: a systematic review and Bayesian meta-analysis of trials and observational studies. Nutrients. 7: 2663–2686.

Milman N, Byg KE, Bergholt T, Eriksen L and Hvas AM. 2006. Cobalamin status during normal pregnancy and postpartum: a longitudinal study comprising 406 Danish women. Eur J Haematol. 76: 521–525.

Molloy AM, Kirke PN, Troendle JF, Burke H, Sutton M, Brody LC, Scott JM and Mills JL. 2009. Maternal vitamin B12 status and risk of neural tube defects in a population with high neural tube defect prevalence and no folic Acid fortification. Pediatrics. 123: 917–923.

Moore HC, Lillie EW and Gatenby PB. 1955. The response of megaloblastic anaemia of pregnancy to vitamin B12. Ir J Med Sci. 106–116.

Morkbak AL, Hvas AM, Milman N and Nexo E. 2007. Holotranscobalamin remains unchanged during pregnancy. Longitudinal changes of cobalamins and their binding proteins during pregnancy and postpartum. Haematologica. 92: 1711–1712.

Morris MS, Jacques PF, Rosenberg IH and Selhub J. 2007. Folate and vitamin B12 status in relation to anemia, macrocytosis, and cognitive impairment in older Americans in the age of folic acid fortification. Am J Clin Nutr. 85: 193–200.

Murphy MM, Molloy AM, Ueland PM, Fernandez-Ballart JD, Schneede J, Arija V and Scott JM. 2007. Longitudinal study of the effect of pregnancy on maternal and fetal cobalamin status in healthy women and their offspring. J Nutr. 137: 1863–1867.

Obeid R, Morkbak AL, Munz W, Nexo E and Herrmann W. 2006. The cobalamin-binding proteins transcobalamin and haptocorrin in maternal and cord blood sera at birth. Clin Chem. 52: 263–269.

Patel KD and Lovelady CA. 1998. Vitamin B12 status of east indian vegetarian lactating women living in the united states. Nutr Res. 18: 1839–1946.

Qi YP, Do AN, Hamner HC, Pfeiffer CM and Berry RJ. 2014. The prevalence of low serum vitamin B12 status in the absence of anemia or macrocytosis did not increase among older U.S. adults after mandatory folic acid fortification. J Nutr. 144: 170–176.

Ray JG. 2004. Folic acid food fortification in Canada. Nutr Rev. 62: S35–S39.

Relton CL, Pearce MS and Parker L. 2005. The influence of erythrocyte folate and serum vitamin B12 status on birth weight. Br J Nutr. 93: 593–599.

Robertson JA and Gallagher ND. 1983. Increased intestinal uptake of cobalamin in pregnancy does not require synthesis of new receptors. Biochim Biophys Acta. 757: 145–150.

Ronnenberg AG, Goldman MB, Chen D, Aitken IW, Willett WC, Selhub J and Xu X. 2002. Preconception homocysteine and B vitamin status and birth outcomes in Chinese women. Am J Clin Nutr. 76: 1385–1391.

Rosenthal J, Lopez-Pazos E, Dowling NF, Pfeiffer CM, Mulinare J, Vellozzi C, Zhang M, Lavoie DJ, Molina R, Ramirez N and Reeve ME. 2015. Folate and Vitamin B12 Deficiency Among Non-pregnant Women of Childbearing-Age in Guatemala 2009–2010: Prevalence and Identification of Vulnerable Populations. Matern Child Health J. 19: 2272–2285.

Samuel TM, Duggan C, Thomas T, Bosch R, Rajendran R, Virtanen SM, Srinivasan K and Kurpad AV. 2013. Vitamin B(12) intake and status in early pregnancy among urban South Indian women. Ann Nutr Metab. 62: 113–122.

Sandberg DP, Begley JA and Hall CA. 1981. The content, binding, and forms of vitamin B12 in milk. Am J Clin Nutr. 34: 1717–1724.

Selhub J, Morris MS and Jacques PF. 2007. In vitamin B12 deficiency, higher serum folate is associated with increased total homocysteine and methylmalonic acid concentrations. Proc Natl Acad Sci U S A. 104: 19995–20000.

Shapiro J, Alberts HW, Welch P and Metz J. 1965. Folate and vitamin B12 deficiency associated with lactation. Br J Haematol. 11: 498–504.

Specker BL, Black A, Allen L and Morrow F. 1990. Vitamin B12: low milk concentrations are related to low serum concentrations in vegetarian women and to methylmalonic aciduria in their infants. Am J Clin Nutr. 52: 1073–1076.

Specker BL, Miller D, Norman EJ, Greene H and Hayes KC. 1988. Increased urinary methylmalonic acid excretion in breast-fed infants of vegetarian mothers and identification of an acceptable dietary source of vitamin B12. Am J Clin Nutr. 47: 89–92.

Srikantia SG and Reddy V. 1967. Megaloblastic anaemia of infancy and vitamin B12. Br J Haematol. 13: 949–953.

Tasker PW. 1955. Correlation of serum-vitamin B12 levels and urinary folic acid in nutritional megaloblastic anaemia. Lancet. 269: 61–63.

Trugo NM and Sardinha F. 1994. Cobalamin and cobalamin binding capacity in human milk. pp. 23–33.

Ullberg S, Kristoffersson H, Flodh H and Hanngren A. 1967. Placental passage and fetal accumulation of labelled vitamin B12 in the mouse. Arch Int Pharmacodyn Ther. 167: 431–449.

Vaz PA, Torras V, Sandoval JF, Dillman E, Mateos CR and Cordova MS. 1975. Folic acid and vitamin B12 determination in fetal liver. Am J Clin Nutr. 28: 1085–1086.

Wills L. 1931. Treatment of "pernicious anaemia of pregnancy" and "tropical anaemia" with special reference to yeast extract as a curative agent. Nutrition. 7: 323–327.

World Health Organization. 2015. Guidelines: Optimal serum and red blood cell folate concentrations in women of reproductive age for prevention of neural tube defects. In: World Health Organization, editor.

Wu BT, Innis SM, Mulder KA, Dyer RA and King DJ. 2013. Low plasma vitamin B12 is associated with a lower pregnancy-associated rise in plasma free choline in Canadian pregnant women and lower postnatal growth rates in their male infants. Am J Clin Nutr. 98: 1209–1217.

Yajnik CS, Deshpande SS, Jackson AA, Refsum H, Rao S, Fisher DJ, Bhat DS, Naik SS, Coyaji KJ, Joglekar CV, Joshi N, Lubree HG, Deshpande VU, Rege SS and Fall CH. 2008. Vitamin B12 and folate concentrations during pregnancy and insulin resistance in the offspring: the Pune Maternal Nutrition Study. Diabetologia. 51: 29–38.

11

Vitamin B12 After Birth and During Early Life

Rima Obeid,[1,*] *Pol Solé-Navais*[2,3] *and Michelle M Murphy*[2,3]

1. Introduction

Cobalamin supplementation during pregnancy and lactation has received less attention than folic acid supplementation. However, low maternal cobalamin status has been related to birth defects and neurodevelopmental disorders in the child. Recent studies in infants have shown that cobalamin intake and status is a matter of concern. Cobalamin deficiency in pregnant or lactating women results in a lower supply of the vitamin to the foetus/infant. Children who start their lives with insufficient cobalamin stores are likely to develop deficiency symptoms during the first few weeks or months after birth. Classical risk groups are infants born to unsupplemented vegan or vegetarian women. However, cobalamin reserves may be depleted during pregnancy in many asymptomatic women on Western diets, thus leading to increased probability of depleted reserves in their newborn children. Deficiency in early life could be due to maternal factors such as poor dietary habits, low intake of animal food products or malabsorption disorders. Cobalamin deficiency in early life should be suspected in children with unexplained anaemia, growth retardation or neurodevelopmental disorders. Information regarding pre- and postnatal

---

[1] Aarhus Institute of Advanced Studies, University of Aarhus, Høegh-Guldbergs Gade 6B, building 1632, Dk-8000, Aarhus C, Denmark.
Email: rima.obeid@uks.eu
[2] Area of Preventive Medicine and Public Health, Department of Basic Medical Sciences, Faculty of Medicine and Health Sciences, Universitat Rovira i Virgili, IISPV, Spain.
[3] Centro de Investigación Biomédica en Red (CIBER; CB06/03), Instituto de Salud Carlos III (ISCIII).
Email: pol.sole@urv.cat, michelle.murphy@urv.cat
* Corresponding author

maternal health and dietary habits as well as infant feeding regime and patterns is essential for diagnosis. Cobalamin deficiency in children can be caused by prolonged lactation, late introduction or insufficient intake of complementary animal foods, gastrointestinal infections or parasites. In populations with low socioeconomic status, cobalamin deficiency in children concurs with other deficiencies (i.e., zinc, iron, folate, etc.) and the symptoms can be subtle or unspecific.

Deficiency in newborns can be prevented by low dose cobalamin supplementation (≤ 50 µg/d) throughout pregnancy and lactation. Changing maternal dietary habits to enhance cobalamin intake can also reduce deficiency risk in the child. Clinical manifestations of cobalamin deficiency in children should be treated with therapeutic doses of cobalamin (> 0.5 mg). A complete recovery is possible, but largely dependent on the severity of symptoms and the timing of intervention during the course of the disease.

This chapter summarises the current knowledge and gaps in knowledge of cobalamin requirements and status after birth and during early childhood.

2. Determinants of Cobalamin Status at Birth and Afterwards

Cobalamin is actively transported from the mother to the child during pregnancy (Giugliani et al. 1985; Graber et al. 1971; Hellegers et al. 1957; Ullberg et al. 1967). Maternal cobalamin status is the main positive predictor of newborn infant status (Figure 1). Maternal plasma total cobalamin as well as plasma holotranscobalamin (holoTC) in the last trimester of pregnancy are inversely associated with cord methylmalonic acid (MMA) and maternal MMA increases gradually as pregnancy progresses, possibly due to mobilisation of maternal cobalamin reserves (Murphy et al. 2007). It has been suggested that recently consumed cobalamin by the mother is transferred to the foetus. Therefore, sufficient cobalamin intake during pregnancy may prevent offspring deficiency in early life.

Very low concentrations of cobalamin are detected in cord blood when the mothers are deficient (Koc et al. 2006). Low cobalamin levels at birth are associated with elevated cord blood homocysteine and MMA and lower methylation index (Guerra-Shinohara et al. 2004). Therefore, methyl-dependent biological processes (DNA-methylation, etc.) may be disturbed even in the absence of clinical manifestations of cobalamin deficiency in the mother.

Low maternal status during pregnancy remains a significant predictor of low child cobalamin in pre-school age children. This is likely to be explained by low child cobalamin stores and continued insufficient dietary intake after weaning (Bhate et al. 2008).

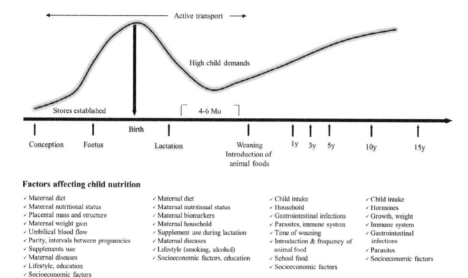

Factors affecting child nutrition

✓ Maternal diet	✓ Maternal diet	✓ Child intake	✓ Child intake
✓ Maternal nutritional status	✓ Maternal nutritional status	✓ Household	✓ Hormones
✓ Placental mass and structure	✓ Maternal biomarkers	✓ Gastrointestinal infections	✓ Growth, weight
✓ Maternal weight gain	✓ Maternal household	✓ Parasites, immune system	✓ Immune system
✓ Umbilical blood flow	✓ Supplement use during lactation	✓ Time of weaning	✓ Gastrointestinal
✓ Parity, intervals between pregnancies	✓ Maternal diseases	✓ Introduction & frequency of	infections
✓ Supplements use	✓ Lifestyle (smoking, alcohol)	animal food	✓ Parasites
✓ Maternal diseases	✓ Socioeconomic factors, education	✓ School food	✓ Socioeconomic factors
✓ Lifestyle, education		✓ Socioeconomic factors	
✓ Socioeconomic factors			

Figure 1. Building child cobalamin stores from pregnancy to adolescence. Stores are high at birth, but become depleted during the first 4–6 months in breastfed infants. Child cobalamin status after weaning depends on frequency and amount of animal foods. Maternal cobalamin status continues to be a main determinant of child status during early childhood because of low stores, and continued low intake in the same household.

3. Cobalamin Requirements in Infants

3.1 Classical definition of adequate intakes of cobalamin in infants

The Food and Nutrition Board of the Institute of Medicine (IOM), Department of Medicine, USA defined the cobalamin adequate intake (AI) for infants based on studies of human milk cobalamin concentrations from different mother—breastfed infant cohorts (Institute of Medicine 2000). A recent review highlights the need to review the AI given that it is based on few, relatively small and outdated mother-infant cohorts that measured milk cobalamin content using different sample collection protocols and assay techniques that had not been validated for analysing milk (Hampel and Allen 2016). Furthermore, milk cobalamin determinations are susceptible to interference from haptocorrin so the milk should be pre-treated to remove this (Lildballe et al. 2009). Nevertheless, for the purposes of providing recommendations for AI, average milk cobalamin concentrations were compared between mother—infant cohorts representing unsupplemented, well-nourished mothers (Trugo and Sardinha 1994), supplemented mothers of low socioeconomic status

(Donangelo et al. 1989) and unsupplemented strictly vegetarian mothers (Specker et al. 1990). It should be noted that the timing of collection of the samples and the breastfeeding regime itself (stage of exclusive or partial, and duration) varied even within the same cohorts. The association between maternal milk cobalamin concentration and urinary MMA in the offspring of vegan mothers was also considered (Specker et al. 1988a). The highest mean cobalamin concentration was observed in the supplemented cohort 0.91 µg/L (Donangelo et al. 1989) and the lowest in the vegetarian mothers (0.31 µg/L) (Specker et al. 1990). In unsupplemented vegan mothers, milk cobalamin concentration < 0.49 µg/L was associated with elevated urinary MMA in the offspring (Specker et al. 1988a). It is unclear where the chosen concentration of 0.42 µg/L at 2 months from Trugo & Sardinha's cohort, comes from because no data for this time point is provided in the cited reference. However, based on 0.42 µg/L as indicative of the observed concentration in healthy well-nourished mothers, and an average intake of 780 mL/d of human milk, the AI for cobalamin in infants aged 0 to 6 months was estimated to be 0.33 µg/d and rounded up to 0.4 µg/d. The AI for children from 6 months to up to 1 year was extrapolated from this estimate (Institute of Medicine 2000).

Subsequent studies and improvements in analytical techniques provide evidence that the current AI should be reconsidered, and these will be discussed later in the chapter. More recently, milk cobalamin content was compared between mothers with deficient (< 150 pmol/L), marginally deficient (150–220 pmol/L) and normal (< 220 pmol/L) plasma cobalamin status, from the same cohort that was relatively large compared to earlier studies, and by the same lab, using improved techniques to account for haptocorrin interference (Deegan et al. 2012). Values below the detection limit of the assay (≤ 50 pmol/L) were observed in 65% of the mothers and the authors questioned the validity of using techniques designed to perform serum cobalamin determinations to measure cobalamin in milk. Nevertheless, the study provides evidence that following 12 months of breastfeeding, milk cobalamin content was lower in deficient and marginally deficient mothers compared to mothers with normal cobalamin status. While maternal serum cobalamin correlated with milk cobalamin content, there was no correlation between milk cobalamin content and infant serum cobalamin.

3.2 Critical view of adequate intakes of cobalamin for infants

Maternal or child plasma cobalamin status markers were not considered when defining AI of cobalamin in infants. Despite referring to a study that investigated the association between milk cobalamin concentration and urinary MMA in the infant (Specker et al. 1988a), it is not clear how or whether this study was considered in the final recommendation. Specker and colleagues reported increased urinary MMA in exclusively breastfed infants of 13 vegetarian mothers. The average age of infants was 7 months (range 2–14 months) (Specker et al. 1988a; Specker et al. 1990). A wide overlap has been reported between urinary MMA from vegetarian and omnivorous infants. Additionally, urinary MMA showed large between-subject variation and was different from earlier studies (Specker et al. 1990). Dietary and physiological factors that affect urinary MMA have not been extensively studied. Large inter-individual variations in serum MMA have been reported in the first year of life (Monsen et al. 2003). Therefore, infant age at the time of cobalamin marker testing appears to be a critical factor.

Subsequent longitudinal studies on cobalamin markers after birth showed that cobalamin status decreases between the age of 4–6 months in exclusively breastfed infants, even in those born to non-deficient mothers. Urinary MMA was determined in 204 healthy newborn infants at birth and later at 2 and 6 months (Karademir et al. 2007). From birth to 2 months, urinary MMA increased from 12 to 26 mmol/mol creatinine and plasma homocysteine increased from 7.1 to 12.8 µmol/L in breastfed and formula fed infants, with a slight decrease until 6 months (20 mmol/mol creatinine and 8.1 µmol/L, respectively) (Karademir et al. 2007). The causes of elevated MMA concentrations are not well understood, but could be due to higher cobalamin requirements or physiological- or growth-related factors (Karademir et al. 2007). Moreover, elevated MMA may also be caused by increased propionate production by intestinal bacteria (Thompson et al. 1990), rather than by liver or kidney immaturity (Bjorke-Monsen et al. 2008).

The bioavailability of cobalamin from human milk is unknown, particularly when milk haptocorrin content increases a few months after starting breastfeeding. Methodological shortcomings, such as the effect of the high milk haptocorrin content on cobalamin determinations, will have affected the earlier studies on which AI for cobalamin in infants was based (Lildballe et al. 2009). The IOM opted not to consider the study on cobalamin milk content in low-income Brazilian women when defining the AI for infants. In that study, women had received prenatal cobalamin-containing supplements (Donangelo et al. 1989). The study reported an average breast milk cobalamin concentration of 0.91 µg/L at 1 month postpartum (Donangelo et al. 1989). Lildballe et al. measured cobalamin after removing unsaturated haptocorrin from breast milk samples collected 1–3 months postpartum from 24 healthy women with low socioeconomic status from the San Francisco Bay Area (Lildballe et al. 2009). Lildballe et al. reported cobalamin concentrations similar to those in the Brazilian study [mean (range) = 0.87 (0.16–3.7) nmol/L] corresponding to 1.2 µg/L (Lildballe et al. 2009). Thus, the mean milk cobalamin concentration of 1.2 µg/L observed after removing haptocorrin (Lildballe et al. 2009) is approximately 2.5 fold higher than the 0.42 µg/L defined by the IOM as AI for cobalamin in infants. The cobalamin levels in milk measured after haptocorrin removal (mean 0.87 nmol/L) are higher compared to earlier studies in healthy, non-vegetarian women [0.38 nmol/L (Specker et al. 1990); 0.6 nmol/L (Sakurai et al. 2005); and 0.71 nmol/L (Sandberg et al. 1981)]. A recent longitudinal study on Danish women estimated daily cobalamin intake from breast milk to be: 0.7 µg/d at 2 wk of age (from ~700 mL milk/d) and

0.3 µg/d at 4 mo of age (from ~800 mL milk/d) (Greibe et al. 2013a). In comparison, the recommended dietary allowances (RDA) for infants aged 0–6 mo was set at 0.4 µg by IOM. Therefore, it is likely that the average cobalamin concentration in breast milk used by the IOM is underestimated due to methodological and/or physiological factors.

Associations between plasma markers of cobalamin status in mothers and their breastfed infants and maternal milk cobalamin concentrations were only recently studied (Deegan et al. 2012; Duggan et al. 2014). Despite evidence suggesting that cobalamin can be delivered from maternal blood to breast milk (Bijur and Desai 1985), it is not known how efficient this process is during prolonged lactation and in depleted mothers (i.e., vegetarians). In a study in Indian women, milk cobalamin content was not significantly lower in vegetarians compared to low- or frequent meat-eaters. However, all of the women were cobalamin depleted which probably explains the lack of association between maternal plasma and milk cobalamin (Figure 2) (Bijur and Desai 1985). To what extent the transport of cobalamin to milk can maintain adequate child metabolic functioning if maternal stores are depleted or very low, remains unclear.

Establishing the AI for infants should consider maternal cobalamin intake and status markers, human milk cobalamin, and infant status markers across the age range of breastfeeding, to account for the dramatic decline in child cobalamin status between 4–6 months (Figures 3, and 4). At the critical 4–6 months age, clinical manifestations of cobalamin deficiency can become apparent.

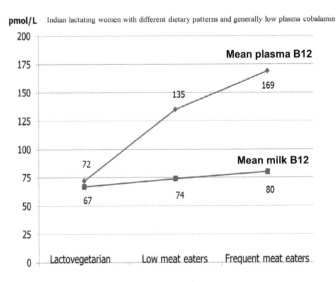

Figure 2. Cobalamin content in maternal milk according to women's diet. Milk cobalamin is not significantly affected by the strictness of animal food restriction or by plasma cobalamin. In this study from India, all participating women, even meat eaters had low mean plasma cobalamin (Bijur and Desai 1985).

In summary, a re-evaluation of AI for cobalamin in infants is necessary. The association between maternal cobalamin intake and breast milk cobalamin content as well as their effects on health outcomes in the children should be investigated. Table 1 summarises the gaps in knowledge regarding sufficient intake of cobalamin in infants and identifies open questions to be answered.

Table 1. Cobalamin intake requirements for infants: gaps in knowledge.

Open issues	What evidence is required?
Analytical methods to measure milk cobalamin	Analytical methods should be standardised for the purposes of valid cross-comparison of different populations.
Choice of cobalamin marker to base evidence for adequacy on pregnancy/lactation/infants?	A combination of several markers is recommended. Urinary MMA alone lacks specificity, shows inter-study variability, and how it is affected by biological factors unrelated to cobalamin status has not been thoroughly studied.
Age that infants should be tested at	Infant cobalamin status markers change during the first few months after birth and there is a physiological decline in cobalamin status between 4–6 months, coinciding with changes in human milk composition. This suggests that AI should be determined for subgroups according to age (0–4 months; 4–6 months; 6–12 months).
What are the short and long term consequences of marginally inadequate cobalamin intake in the first year of life?	Only a few studies from India are available (growth and weight gain); no large scale studies are available.

4. Cobalamin Status in Infants During Lactation and Prolonged Lactation

Human milk cobalamin is considered to be the only source of the vitamin in exclusively breastfed infants. The composition of cobalamin and its binding proteins change in milk over the course of lactation (Figure 3). However, the effect of these changes on cobalamin bioavailability in the child is unknown. The dramatic changes in cobalamin status markers between birth and early childhood depend on maternal supplement use, infant stores at birth, requirements, feeding patterns, and introduction of complementary foods from animal sources. Infants born to well-nourished women have high plasma cobalamin at birth (Hay et al. 2010; Obeid et al. 2006). In contrast, infants born to unsupplemented vegetarian mothers are depleted at birth (Specker et al. 1988a; Specker et al. 1990). Child urinary MMA is predicted by maternal diet and low maternal serum cobalamin (Specker et al. 1988a; Specker et al. 1990). Since recently consumed cobalamin is transferred to the child through milk, maternal supplementation could prevent deficiency in the child.

Dramatic changes in serum concentrations of cobalamin binders, fractions, and metabolic markers occur a few months after birth. A longitudinal follow up study of infants from birth to 2 years of age showed a decline in cobalamin status around the age of 4–6 months. Norwegian infants were investigated

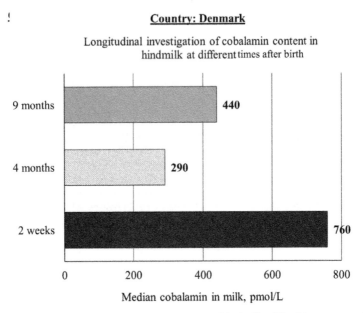

Figure 3. Longitudinal changes of cobalamin content in hindmilk of Danish women at different ages from week 2 to month 9 of lactation. Cobalamin was measured in milk after capturing haptocorrin on Sepharose and removing it (Greibe et al. 2013a).

immediately after birth (cord blood), and at the age of 6, 12, and 24 months according to their feeding patterns (Hay et al. 2008) (Figure 5). Independently of these, serum concentrations of holoTC and total cobalamin were highest in cord blood, declined to their lowest levels at 6 months and increased again by 24 months (Hay et al. 2008). Serum holoTC declines after birth to reach its lowest levels at approximately 4 to 6 months of age (Greibe et al. 2013b; Hay et al. 2010). This decrease is associated with a parallel increase in plasma MMA and tHcy, suggesting declining intracellular cobalamin status (Bjorke Monsen et al. 2001; Monsen et al. 2003). Age-related changes in cobalamin markers occur in asymptomatic infants and are known to be more pronounced in breastfed compared to formula fed infants (Figure 4). In addition, this low intracellular cobalamin status leads to the elevated plasma folate observed in this age group as a result of the "methyl folate" trap. The reduction of both serum tHcy and folate following cobalamin injections further supports this hypothesis (Bjorke-Monsen et al. 2008). Elevated serum tHcy and MMA in infants at 6 months confirm low cobalamin status (Greibe et al. 2013a; Hay et al. 2008). In children with sufficient folate status, serum cobalamin < 200 pmol/L is the main determinant of plasma homocysteine (van Beynum et al. 2005).

There is also no established "reference range" for cobalamin markers at or after birth. The association between cobalamin depletion and subtle signs of deficiency is not well studied. To date, no longitudinal studies have assessed the association between early postnatal cobalamin status and subsequent

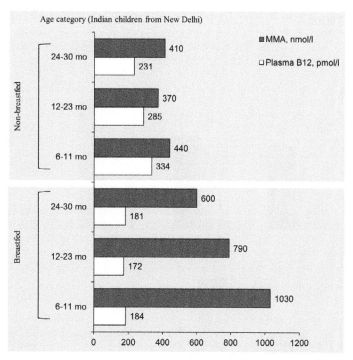

Figure 4. Plasma concentrations of cobalamin and methylmalonic acid (MMA) in Indian children are shown according to their age and breastfeeding patterns. MMA was significantly different between age groups in the breastfed category ($p < 0.05$); and plasma cobalamin and MMA were significantly different between age groups in the non-breastfed category ($p < 0.05$). The study included children from women residing in a low-to-middle income community in North India (Taneja et al. 2007).

neurological development in the child. It remains unknown whether the decline in cobalamin status following birth is associated with subtle clinical disorders.

The influence of feeding patterns on infant cobalamin status has been extensively studied in different populations. Cobalamin status is generally lower in breastfed compared to formula fed infants (Hay et al. 2008; Ulak et al. 2014) (Figures 4 and 5). Almost all available infant formulas are enriched with cobalamin, though the amounts of cobalamin added are not based on strong scientific evidence. Formula fed infants show an increase in plasma cobalamin and holoTC from birth to 12, and 24 months (Hay et al. 2008). In contrast to the strong decline in holoTC observed in breastfed infants at 6 months, non-breastfed infants maintained rather stable holoTC concentrations up to 24 months (Hay et al. 2008). Lactating Norwegian women that took cobalamin-containing supplements provided more cobalamin to their infants compared to nonusers of supplements (Hay et al. 2008) (Figure 5). The influence

Country: Norway

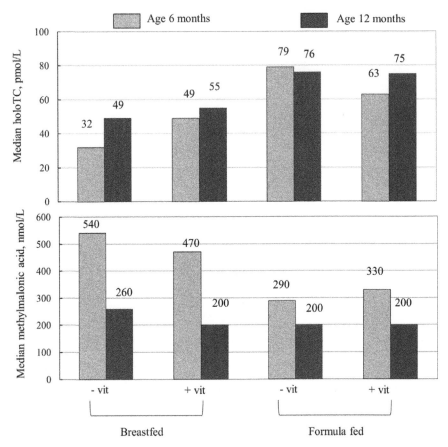

Figure 5. Longitudinal study on serum holoTC and methylmalonic acid levels at 6 and 12 months of age according to breastfeeding patterns and maternal supplement use. Lowest cobalamin status (highest MMA and lowest holoTC) was found in breastfed infants at the age of 6 months from non-supplemented women (-vit). Formula feeding was associated with higher cobalamin status and less differences between the age groups at 6 and 12 months. Country of origin: Norway, figure is modified from Hay et al. (Hay et al. 2008).

of maternal supplementation on child cobalamin markers is less evident in children receiving formula milk that already contains a substantial amount of cyanocobalamin (Figure 5) (Hay et al. 2008).

It is important to note that unsupplemented lactating women from Western countries (Greibe et al. 2013a) generally have higher serum and milk cobalamin status compared to those from low-income countries or low socioeconomic status. Cobalamin deficiency in newborn children from well-nourished women is not a major health concern. However, deficiency in newborns is a public health problem in countries with endemic deficiency because it can affect important developmental domains in the first 2 years of life.

In Nepal, where cobalamin deficiency is prevalent, breastfed children in the 6–11, 12–23, and 24–35 months age groups had lower plasma cobalamin and higher MMA than non-breastfed infants (Ulak et al. 2014). The greatest differences were observed in the youngest group where the breastfed infants were unlikely to receive complementary foods from animal sources (Figure 6).

In a study of Guatemalan lactating women and their infants aged 3 months, maternal serum cobalamin was related to infant urinary MMA (Casterline et al. 1997). The median cobalamin intake of the mothers was 1.47 µg/d and the infants showed reduced ponderal and linear growth rates (length-for-age and weight-for-age z-scores). Low serum cobalamin concentrations

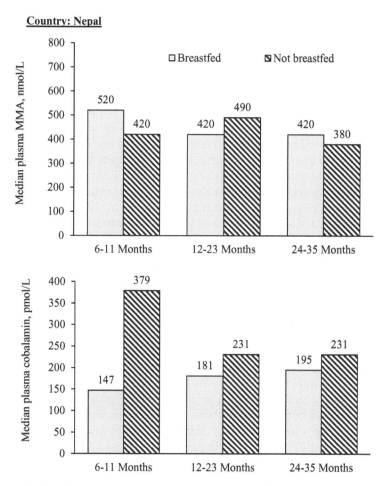

Figure 6. Median plasma cobalamin markers are shown in children from Nepal according to their age and feeding patterns. The study confirms that the lowest cobalamin status is found in breastfed infants of the youngest group (6–11 months). Extended breastfeeding in depleted women and late introduction of animal foods can cause deficiency at this age. The results are from (Ulak et al. 2014).

(< 148 pmol/L) were detected in 13% of the women and marginally low levels (148–221 pmol/L) in 33% (Casterline et al. 1997). Mean breast milk cobalamin concentration was 690 pmol/L, but inter-individual variation was high. Applying the cut-off for adequate milk cobalamin suggested by Specker et al. (≥ 362 pmol/L), 28 out of 64 of the women had insufficient milk cobalamin. Infants with elevated urinary MMA (> 23 µmol/mmol creatinine) received milk with lower cobalamin content than infants with urinary MMA below this concentration (median milk cobalamin = 411 ± 248 vs 705 ± 488 pmol/L respectively) (Casterline et al. 1997). There was a large overlap in milk cobalamin concentrations between mothers of infants with elevated and non-elevated urinary MMA, suggesting that the infants of many women with low milk cobalamin had normal urinary MMA.

Prolonged breastfeeding in cobalamin depleted infants further increases the risk of deficiency. In line with this, infants who were still breastfed at the age of 24 months had lower cobalamin and holoTC and higher homocysteine and MMA compared to those on a common diet (Hay et al. 2008).

In communities with micronutrient insufficiency, infants become strongly depleted the longer lactation lasted or with delayed introduction of/or insufficient complementary foods (Pasricha et al. 2011). Prolonged breastfeeding up to 24 months or beyond is common practice in countries with widespread food insecurity and particularly in mothers who are themselves depleted of several nutrients, including cobalamin. Poor life conditions and low food security in India is associated with cobalamin deficiency in the child due to prolonged breast feeding (Pasricha et al. 2011) and prolonged breast feeding is a strong predictor of low serum cobalamin concentrations in children from poor communities (Taneja et al. 2007). In a study of ~2400 New Delhi children aged 6–30 months, plasma cobalamin levels were 75% lower in breastfed (aged 6–11 months) when compared to non-breastfed (median = 185 vs 334 pmol/L) and low serum cobalamin (< 150 pmol/L) was very prevalent (36%) in breastfed compared to non-breastfed children (9%). Cobalamin concentrations even declined with age in non-breastfed children (median = 334 pmol/L in 6–11 months age group; 285 pmol/L in 12–23 months age group and 231 pmol/L in 24–30 months age group), possibly due to low intake of animal foods after weaning (Taneja et al. 2007).

Cobalamin deficiency is only one component of several nutritional deficiencies that occur in the context of poverty. A high prevalence of prolonged breastfeeding (86% of the children) was observed in Indian children (age 6–30 months, mean age = 16 months) from families of low to middle socioeconomic status (Kvestad et al. 2015). Stunting was observed in 40% of the children and 31% were underweight. Low serum cobalamin (< 200 pmol/L) and elevated homocysteine (> 10 µmol/L) were observed in 33% and 59% of the children, respectively. Pasricha et al. (2011), investigated the effect of prolonged breastfeeding on infant's cobalamin status in rural communities from Southern India (Pasricha et al. 2011). Anaemia was observed in 63% of the mothers (haemoglobin < 12 g/dl), 29% of the children (age 12–24 months) were stunted, 32% were underweight, and 20% were wasted (low weight-for-length). Inflammation and deficiency of other essential nutrients (iron, folate, vitamin A) were also common. Child's serum cobalamin was negatively associated with prolonged lactation and positively associated with regular meat intake. Cobalamin intake was lower in children still being breastfed at a mean age of 16.2 months compared to children who were fully weaned (mean age 18.5 months) (Pasricha et al. 2011).

Similar results were reported in a study of lactating Guatemalan Landino mother-infant pairs at 12 months postpartum (Deegan et al. 2012). Serum cobalamin was < 150 pmol/L in 35% of the women and marginally deficient (150–220 pmol/L) in a further 35%. The median cobalamin intake of the women was 2.6 µg/d, thus 50% of the women did not meet the Recommended Dietary Allowance (RDA) of cobalamin for lactating women. Milk cobalamin content was below the assay detection limit of 50 pmol/L in 65% of the women. In those with serum cobalamin > 220 pmol/L, milk cobalamin content was still very low (median = 69 pmol/L). Determinants of infant serum cobalamin at this age were limited to maternal serum cobalamin and intake of the vitamin from complementary foods. Milk cobalamin content was not a significant determinant of infant serum cobalamin at 12 months. In contrast, maternal cobalamin intake and serum cobalamin were significant determinants of milk cobalamin content at 12 months. Feeding 12 months old infants with cow's milk as a source of cobalamin was associated with a better cobalamin status than feeding with breast milk alone (Deegan et al. 2012).

Overall, maternal cobalamin status is the strongest determinant of cobalamin biomarkers in exclusively breastfed infants. A decline in infant cobalamin status is evident at 4–6 months, thus questioning the adequacy of cobalamin intake from breast milk beyond this age. Prolonged lactation does not appear to supply sufficient cobalamin to infants of non-supplemented mothers. Introduction and quality of complementary foods as well as maternal intake are strong predictors of cobalamin status in infants older than 6 months and are likely to be influenced by socioeconomic status and vegetarian diet.

5. Causes of Cobalamin Deficiency in Infants and Children

The RDAs for cobalamin for pregnant and lactating women is 2.6 µg/d and 2.8 µg/d, respectively compared with 2.4 µg/d for non-pregnant, non-lactating women (Institute of Medicine 1998). Non-inherited cases of cobalamin deficiency in newborns and infants are mainly explained by inadequate maternal cobalamin intake (Dagnelie et al. 1989; Reghu et al. 2005; Specker et al. 1988b; Specker et al. 1990). Restriction of cobalamin intake during pregnancy and lactation such as in vegetarians causes depleted cobalamin stores in the child (Miller et al. 1991; Specker et al. 1988b; Specker et al. 1990; Specker 1994). Cobalamin deficiency in infants of strict vegetarian women can occur at low-normal maternal serum cobalamin level (Graham et al. 1992). Six-year old Indian children born to women with very low serum cobalamin during pregnancy (< 77 pmol/L) had lower serum cobalamin (median = 216 vs 247 pmol/L, p < 0.01) than children born to women with serum cobalamin > 244 pmol/L (Bhate et al. 2008) (Figure 7).

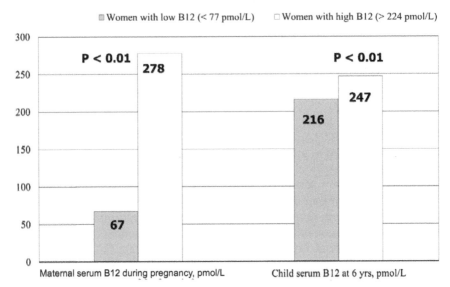

Figure 7. Low serum cobalamin in pregnant Indian women predicts child serum cobalamin at the age of 6 years. The figure is modified from (Bhate et al. 2008). The y-axis shows the average concentrations of serum cobalamin (pmol/L) in the women and their children.

Additionally, maternal cobalamin malabsorption is a risk factor for cobalamin deficiency in the child (Hoey et al. 1982; Johnson, Jr. and Roloff 1982; Wardinsky et al. 1995). Common causes of malabsorption are pernicious anaemia, achlorhydria, ileal damage, gastric bypass surgery, and infection with helicobacter pylori or intestinal parasites. Women treated with metformin before conception or during pregnancy are likely to have low serum cobalamin (Greibe et al. 2013b), but the effect of using this drug on child cobalamin status is currently unknown.

Besides the effect of maternal factors on cobalamin status markers in the child, there are well described age-related fluctuations of plasma total cobalamin, MMA, and tHcy in children and adolescents aged 4 days to 19 years (Monsen et al. 2003; Ueland and Monsen 2003). The strong association between maternal and child cobalamin status beyond the age of lactation is explained by socioeconomic status or feeding the infants on a vegetarian diet. Predictors of cobalamin status in pre-school age and school age children include cobalamin status at birth, feeding patterns (breast milk, formula), and introduction, quality and frequency of animal based complementary foods after weaning.

6. Cobalamin Deficiency in Children in Different Countries

Micronutrient deficiencies in children are widespread. Poverty, low consumption of animal foods and low education level are predictors of low serum cobalamin (Rosenthal et al. 2015). Avoiding animal products for ethical or religious reasons can cause cobalamin deficiency in children. Hidden hunger in Western societies can occur in immigrants who have suffered long-term economic problems or who maintain their dietary habits in their new countries of residence (Carmel et al. 2002; Chambers et al. 2000).

In many economically disadvantaged countries, spending money on a healthy diet has been traditionally regarded as a waste (Berry 1955). Marginally deficient or deficient plasma cobalamin has been reported in up to 40% of men and women across the age range in African (McLean et al. 2007; Siekmann et al. 2003), Southeast Asian (Bondevik et al. 2001; Chanarin et al. 1985; Gomber et al. 1998), Middle Eastern (Herrmann et al. 2005; Obeid et al. 2002), and Latin American (Black et al. 1994; Casterline et al. 1997) studies.

Low meat consumption is a strong predictor of low cobalamin concentrations (McLean et al. 2007). In a study of Kenyan children, the odds ratio (OR) and 95% confidence interval of having low serum cobalamin (< 148 pmol/L) was 2.1 (1.3, 3.5) if the diet contained no meat versus some meat, 5.0 (2.5, 9.8) in the lowest versus the highest tertile of milk intake, and 6.6 (3.1, 12.8) in the lowest versus the highest tertile of any animal source food (McLean et al. 2007). However, factors other than intake can also contribute to cobalamin deficiency in developing countries (see Chapter 6). In a study of over 500 Guatemalan school-children with a mean age of 10 (range 8–12) years, average cobalamin intake was approximately 5.5 μg/d, presumably sufficient to cover daily requirements for this age group (= 1.8 μg/d) in the absence of malabsorption (Rogers et al. 2003a). However, mean plasma cobalamin (285 pmol/L), MMA (479 nmol/L), and homocysteine (9.4 μmol/L) suggested that cobalamin deficiency was common among these children (Rogers et al. 2003b). Elevated serum MMA (> 624 nmol/L) was observed in 32% and 28% of children with low and marginal plasma cobalamin, respectively. Cobalamin intake was correlated with serum cobalamin (Rogers et al. 2003a), but low cobalamin status only partly explained the elevated MMA in the Guatemalan study (Rogers et al. 2003b). Unexpectedly, concentrations of cobalamin were lower in the oldest group of school-children (12 years) compared to the younger groups: 8, 9, or 10 years (Rogers et al. 2003b). These results suggested that cobalamin intake in the older group was not sufficient to meet requirements, leading to depletion over time. In comparison to the Guatemalan study, higher cobalamin (median 436 pmol/L), and lower MMA

(170 nmol/L) and homocysteine (6.5 µmol/L) were reported in Norwegian children (aged 10.5–15 years) (Monsen et al. 2003). Infections with helicobacter pylori or bacterial overgrowth could interfere with cobalamin absorption in low income countries, but this needs further investigation.

In affluent countries, cobalamin deficiency can affect educated people with alternative dietary attitudes, such as strict vegetarians or individuals following a macrobiotic diet. Infants on a macrobiotic diet show metabolic signs of cobalamin deficiency (increased MMA and homocysteine) (Schneede et al. 1994). Dutch children (0–10 years or 4–18 months) who had been on a macrobiotic diet since birth had low intakes of cobalamin (0.3 vs 2.9 µg/day) and several other nutrients (riboflavin, thiamin, and calcium) compared to control children (Dagnelie and van Staveren 1994). Biochemical signs of low cobalamin status have been reported in a group of adolescents (aged 9–15 years) fed a macrobiotic diet up to the age of 6 years then followed by a lacto- or lacto-ovo-vegetarian or omnivorous diet (Van et al. 1999). Therefore, abnormal cobalamin markers due to insufficient intake are likely to continue until early adulthood in formerly macrobiotic infants and children.

7. Clinical Manifestations of Cobalamin Deficiency in Infancy and Childhood

Clinical manifestations covered in this section do not include those caused by genetic disorders of cobalamin absorption and transport, where the symptoms are more severe and are manifested just after birth.

7.1 Clinically manifested severe nutritional cobalamin deficiency

Symptoms of acquired cobalamin deficiency in infancy appear between 3–6 months of age. This postnatal time frame agrees with that in which a decline in cobalamin status is observed in breastfed infants. Clinical symptoms of infantile cobalamin deficiency have been described in numerous case reports, classically of breastfed infants from mothers with low cobalamin intake or cobalamin malabsorption. Physical, haematological and neurological symptoms have been described (Graham et al. 1992) (Table 2).

Table 2. Summary of symptoms reported in cobalamin deficient infants, children or adolescents compared to their non-deficient counterparts.

• Lethargy	• Wasting
• Feeding difficulties	• Anaemia
• Irritation	• Skin hyperpigmentation
• Not smiling	• Low score in fluid intelligence test that measures reasoning, problem solving, learning and abstract thinking
• Muscular hypotonia	
• Low skinfold thickness	
• Low physical activity	• Impaired gross motor function
• Low birth weight	• Bone fracture risk
• Stunting, low weight-for-age	

Magnetic resonance imaging techniques revealed structural changes in the brains of cobalamin deficient infants from strict vegetarian women. Enlargements of cortical sulci and subarachnoid spaces (Kocaoglu et al. 2014), diffuse delayed myelination, and mild dilatation of the lateral ventricles have been reported (Guez et al. 2012). Delayed myelination or demyelination, resembling similar effects in older patients with cobalamin deficiency, are well-documented (Hyland et al. 1988; Miller et al. 2005). The majority of symptoms are reversed after treatment if diagnosed and treated during the early stages. However, long-term neurological complications beyond school age have been reported (Graham et al. 1992). Table 3 shows a few examples of studies reporting clinical conditions associated with biochemical signs of cobalamin deficiency.

Mechanisms of neurodevelopmental manifestations of cobalamin deficiency in infants are not well explained [reviewed in (Dror and Allen 2008) and in Chapter 8]. Table 4 briefly describes the main mechanisms that have been suggested for cobalamin deficiency related neurodevelopmental disorders in infants and children.

7.2 Clinical manifestations of mild cobalamin deficiency in children

Subtle cobalamin deficiency is common in children and it may not lead to urgent hospitalisation, but if it remains untreated it can lead to chronic conditions and long-term consequences. Causes and health consequences of mild cobalamin deficiency differ between different age groups and country settings. Low animal food intake in children negatively affects their physical growth, cognitive development and behaviour even after controlling for confounders (Allen 1993).

Diagnosis of cobalamin deficiency should be considered when exploring causes of childhood anaemia, stunting, wasting, low birth weight, underweight and unexplained neurodevelopmental delay in childhood. Moreover, cobalamin deficiency should be suspected in children with clinical conditions that are risk factors for deficiency such as food intolerance, coeliac disease, and inflammatory bowel disease.

7.2.1 Cobalamin deficiency and physical growth in children

The physical growth of children and adolescents on macrobiotic, vegan, or poor cobalamin diets has been intensively studied. Clinical manifestations of low cobalamin can start at birth and extend beyond weaning, because children whose mothers have a low intake during pregnancy and lactation have low cobalamin stores and are very likely to continue with an insufficient intake after weaning.

In a study of Dutch infants born to women on a macrobiotic diet, the average birth weight was 200 g less than that of a comparable Dutch infant population (p < 0.05) (Dagnelie and van Staveren 1994). In a study of a

Table 3. Symptoms and conditions related to cobalamin deficiency in infants ≤ 1 year.

Reference	Time of manifestation	Child nutrition	Biochemical signs	Clinical symptoms	Maternal nutrition	Treatment	Treatment outcome
(Weiss et al. 2004)	Age 6 months	exclusively breastfed until 6 months	Serum B12 = 75 pg/mL, Hb = 6.8 g/dl, MCV = 102 fl	Mild cerebral atrophy, vomiting, intermittent diarrhoea, slow growth, hypotonia, somnolence, delayed age-expected developmental milestones and communication	Vegan mother for the last 6 years, maternal B12 = 130 pg/mL	1 mg B12/d for 2 wks, then once/month for 6 months plus iron and folate	Rapid improvement in weight, height and developmental milestones; at 5 yrs there was still a delay in the development of gross motor and delay in fine motor function
(Monagle and Tauro 1997)	6 Patients, mean age 7.5 months (range 3.5–10)	All exclusively breastfed	Megaloblastic anaemia with mean Hb = 7.2 g/dl	Symptoms reported in some of the 6 children: failure to thrive, hypotonia, poor response, irritability, regression milestones, apnoea, vomiting, anorexia, poor feeding, diarrhoea, pallor, cardiac failure	3 patients with previously undiagnosed pernicious anaemia, 1 undefined malabsorption, 1 short gut syndrome, 1 vegan	Start B12 therapy, stop breastfeeding or introduce solid foods or formula containing B12	Normalisation of haematological parameters in all children; normal neurodevelopment in 3 children; Significant long term neurological sequelae in 2 children
(Kocaoglu et al. 2014)	Presented at 12 months, following the initiation of symptoms after 6th month	Exclusively breastfed	Megaloblastic anaemia, MCV = 97.3 fl, Hb = 8.8 g/dl, serum B12 = 117 pg/mL	Cerebral atrophy, enlargement of cortical sulci and subarachnoid spaces, developmental regression, growth retardation, lethargy, hypotonia	Vegetarian diet for years, maternal B12 = 232 pg/mL Treatment continues up to 18 months	1 mg B12 injections	Cerebral atrophy corrected; recovery of neurological symptoms; smiling within three days

(Codazzi et al. 2005)	Symptoms started at 6th month, child hospitalised at 10 months	Exclusively breastfed for 10 months	Hb = 5.9 g/dl, MCV = 115 fl, uMMA = 851 mmol/mol creatinine, serum B12 = 30 pg/mL	Diffuse cortical and subcortical atrophy without focal lesions or myelination disorders (on MRI and CT). Coma and extreme muscular weakness, respiratory failure, psychomotor regression, apathic, unable to sit	Strict vegetarian diet for the last 10 years	Cobalamin (1 mg/day), folate (7.5 mg/day), and iron (50 mg/day)	Fast haematological recovery; limited neurological recovery by 3-yrs (normal diet plus B12 supplementation); moderate hypotonia, mild growth delay, psychomotor delay, ataxia and involuntary movements when agitated, impaired speaking at 4 yrs
(Guez et al. 2012)	Symptoms started at 3 months, hospital admission at 5 months	Exclusively breastfed until 5 months	Hb = 4.7 g/dl, MCV = 84.2 fl, B12 = 57 pg/mL, Hcy = 11 μmol/L	Diffuse delayed myelination and mild dilatation of the lateral ventricles (MRI), poor weight gain, feeding difficulties, severe pallor, muscle hypotonia and somnolence, enlarged liver and spleen	Vegan mother, used vitamins during pregnancy (2.5 μg B12), supplements stopped after delivery, maternal B12 = 155 pg/mL	I.M. injections of 1 mg B12/d for 2 weeks then weekly for 6 months plus oral iron for 3 months	After 7 months: persistence of mild hypotonia, child able to remain seated, delay in gross motor function and language; Griffith's scale 86 (normal values, 92–100); B12 increased in plasma and homocysteine decreased

CT, computer tomography; Hb, haemoglobin; Hcy, homocysteine; I.M., intramuscular; MCV, mean corpuscular volume; MRI, magnetic resonance imagining; uMMA, urinary methylmalonic acid. 1 pg/ml vitamin B12 = 0.74 pmol/L.

Table 4. Mechanisms of neurodevelopmental disorders in cobalamin deficient children.

Biochemical and clinical conditions	Mechanisms relevant to brain functions
Reduced methylation potential (i.e., S-adenosylmethionine)	- Epigenetic modifications (DNA and histone methylation) - Hypomethylation of specific methylated positions on myelin
Hyperhomocysteinemia	Possible stimulation of N-methyl-D-aspartate receptors by neurotoxic Hcy
Depletion of alternative methyl donors (choline, betaine)	Reduced intracerebral choline and derivatives in cobalamin deficiency (Horstmann et al. 2003)
Higher cytokines and growth factors	Cause imbalance between tumor necrosis factor-α (TNF-a) and epidermal growth factor. TNF-α acts as a neurotoxin in CNS diseases characterized by demyelination
Accumulation of neurotoxic metabolites	Example: lactate accumulation in the brain and depletion of choline-containing molecules (Horstmann et al. 2003)
Changes in brain energy metabolism	Reduction of N-acetylaspartate, creatine, choline-containing compounds, myo-inositol, glutamate and glutamine in white and gray matter of the brain at 6 and 12 months of age in a case report of a deficient infant
Diarrhoea	Cause of malabsorption of other nutrients (Horstmann et al. 2003)
Low plasma amino acids	Low methionine (Grattan-Smith et al. 1997)
Feeding problems and low appetite	Low intake of other essential nutrients necessary for growth and development such as iron, folate and minerals

macrobiotic community, infants younger than 6 months showed comparable age-specific anthropometrical measures (length, weight, sitting height, and arm circumference) to the Dutch standards (Dagnelie and van Staveren 1994). However, a marked decline of height, weight, and other physical measures was observed from the age of 6 months onwards (Dagnelie and van Staveren 1994), showing that depleted children can manifest deficiency-related developmental delay after weaning. Rickets in children were also associated with macrobiotic diet during early life (Dagnelie and van Staveren 1994) and might be caused by deficiency in other relevant nutrients such as vitamin D and calcium. Lower serum cobalamin and higher MMA in children on macrobiotic diets were also associated with lower bone mineral density and content even after adjusting for other factors such as age, body size, bone area, lean body mass, puberty and calcium intake (Dhonukshe-Rutten et al. 2005).

In many populations, a causal role for cobalamin deficiency in anaemia and stunting cannot be established due to other interfering factors such as micronutrient deficiencies, elevated C-reactive protein, intestinal parasites or malaria infection (McLean et al. 2007; Siekmann et al. 2003). In Guatemalan school children with mild to moderate cobalamin depletion, 83% of the children had Helicobacter pylori infection (Rogers et al. 2003b), a condition that is less common in Western countries. However, the prevalence of infection with Helicobacter pylori did not differ between children with low and marginal cobalamin status (Rogers et al. 2003a), suggesting that the infection was not causally related to cobalamin deficiency. Moreover, height, weight, haemoglobin, and haematocrit did not differ with cobalamin status in this study (Rogers et al. 2003b).

7.2.2 Cobalamin deficiency and other developmental domains in children

Recently, the role of nutritional insufficiency during early life on child brain development has gained more attention (Nyaradi et al. 2013). Indian children born to mothers with adequate versus very low plasma cobalamin at 28 weeks of gestation (> 224 pmol/L versus < 77 pmol/L, respectively) performed better in some psychological tests (color trial test and digit span backward test) at nine years (Bhate et al. 2008). In a follow up study, lower child psychomotor and mental development scores were observed in Indian infants (age 12–18 months) with biochemical signs of cobalamin deficiency compared to infants with better cobalamin status (Strand et al. 2013).

Psychomotor development in areas such as gross motor (sitting, locomotion), speech and language was better in omnivorous children compared to those on a macrobiotic diet (Dagnelie and van Staveren 1994). Cognitive abilities in adolescents (aged 10–16 years) were compared between those that had been on a macrobiotic diet from birth until 6 years and those that had never been on a macrobiotic diet (Louwman et al. 2000). Cobalamin intake and status were lower in adolescents on macrobiotic diets during early life versus never (approximately 1.5 vs 3.6 µg/d) (Louwman et al. 2000). Serum MMA concentrations were negatively associated with fluid intelligence test scores measuring reasoning, problem solving, learning and abstract thinking (Louwman et al. 2000).

Therefore, studies from different settings suggest that low cobalamin status and the resulting biochemical abnormalities are causally associated with impaired development in several neurological domains. Increasing cobalamin status in newborns of depleted women can improve neurodevelopment, growth and other health outcomes in the child (discussed below).

8. Prevention and Treatment of Cobalamin Deficiency in Children

8.1 Early intervention during pregnancy and lactation

Sufficient intake of cobalamin should be ensured from early pregnancy throughout lactation. Supplements and nutrient preparations made to prevent pregnancy anaemia or neural tube defects should also contain cobalamin. Supplementing lactating women is an efficient way to prevent maternal depletion and deficiency in infants.

A recent randomised placebo controlled study of pregnant women from Bangladesh provided 250 µg/d of oral cobalamin from 11–14 weeks of gestation throughout 3 months postpartum (Siddiqua et al. 2015). Compared with the placebo, cyanocobalamin supplementation increased serum cobalamin and reduced MMA and homocysteine at delivery and after 3 months. Supplementation with cyanocobalamin also increased milk cobalamin content (Siddiqua et al. 2015). Furthermore, higher cobalamin status was observed in infants at birth and thereafter (Siddiqua et al. 2015).

Similar results were reported from a study in India. Dietary cobalamin intake in pregnant women (i.e., 1.2–1.4 µg/d) was unlikely to provide sufficient vitamin to the foetus (Duggan et al. 2014). However, cobalamin supplementation in women during pregnancy and lactation improved infant cobalamin status. Milk cobalamin content was higher in women supplemented with 50 µg/d cyanocobalamin during pregnancy and up to 6 weeks postpartum compared with the placebo group (Duggan et al. 2014) (Figure 8). However, when supplement use was discontinued, milk cobalamin declined and by 3 to 6 months after birth, it no longer differed between the placebo and supplemented groups (Duggan et al. 2014; Srikantia and Reddy 1967). Infants (6 weeks old) from supplemented women had higher cobalamin status, confirming the transport of the recently ingested cobalamin to the child (Figure 9).

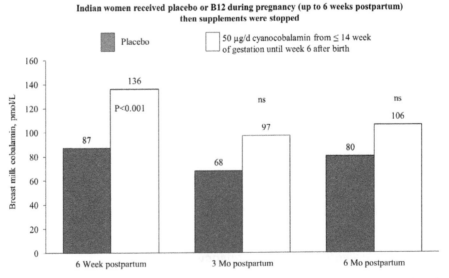

Figure 8. Indian pregnant women were supplemented with a placebo or 50 µg/d oral cyanocobalamin up to 6 weeks after birth. Milk cobalamin declined after stopping the supplements (Duggan et al. 2014).

Indian women – Time: 6 weeks after birth

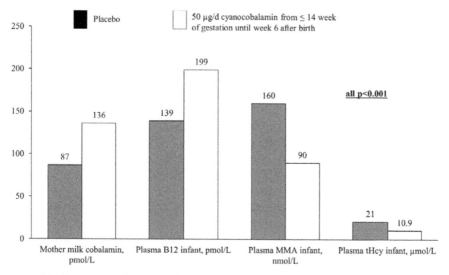

Figure 9. Indian women who received 50 µg/d oral cyanocobalamin until 6 weeks after birth had higher milk cobalamin than those who received a placebo. Infants from cobalamin supplemented pregnant women had a better status after birth (Duggan et al. 2014).

8.2 Cobalamin intervention in children

There is no consensus or evidence-based recommendations regarding the dose and frequency of cobalamin that should be used to treat children with clinically manifested deficiency. Available knowledge is based on trial and error and treatment options need to be personalised according to causes, clinical conditions, and metabolic response. The duration of treatment of acquired cobalamin deficiency is from a few months up to 2 years after manifestation. In contrast, infants with genetic cobalamin disorders should be treated for life. Genetic cobalamin disorders are traditionally treated with hydroxycobalamin injections (1 mg i.m.) at different intervals (1–4 weeks). However, cyanocobalamin has also been used if the genetic anomaly does not affect its conversion into active co-enzymes. For example, patients with methylmalonic aciduria and homocystinuria type C protein (MMACHC, *cblC* gene) defect require hydroxycobalamin that has advantages over cyanocobalamin because of its pharmacokinetic features and potential to circumvent the MMACHC defect, particularly after parenteral application of therapeutic doses (1–2 mg) (Oladipo et al. 2011; Waggoner et al. 1998). Defects in the *cblC* gene lead to inability to synthesise cobalamin coenzyme from supplemental cyanocobalamin (Banerjee 2006), but patients respond to 1–2 mg of hydroxycobalamin (Andersson and Shapira 1998). The high affinity of hydroxycobalamin to plasma proteins causes a prolonged effect after injection because the elimination of this form via the kidney is slower than that of cyanocobalamin (Gimsing et al. 1982; Hertz et al. 1964). However, in patients without genetic defects in the *cblC* gene, it remains questionable whether hydroxycobalamin is advantageous over cyanocobalamin [reviewed in (Obeid et al. 2015)]. In genetic disorders affecting cobalamin absorption, doses less than 1 mg have been successfully used to correct cobalamin deficiency (Bor et al. 2008). Oral treatment has proven effective in many, but not all cases of inherited malabsorption (Bor et al. 2008). In manifested cases or genetic cobalamin disorders, treatment starts with hydroxycobalamin injections. Oral cobalamin treatment can then be tested in certain cases. Monitoring of metabolic markers (i.e., plasma MMA and homocysteine) can give better insights into the efficacy of the treatment route, form, frequency, and dose.

In a study of 6–12 month old infants with cobalamin deficiency due to nutritional factors (i.e., vegan mother), cobalamin injections were initiated as a first step of treatment (1 mg i.m. of any cobalamin form). Injections were continued for up to several years depending on the clinical response (Dror and Allen 2008) (Table 3). The treatment regime is often switched to oral cobalamin after several injections. The responses of clinical symptoms vary according to the severity of the disease, the time of detection, and the start of treatment. Haematological signs (i.e., haemoglobin and mean cell volume) improve within a few days of treatment initiation. Co-supplementation with folate and iron appears to enhance recovery from anaemia. Rapid reversal of muscle hypotonia, communication signs, apathy, vomiting, and irritation within a few days has been reported. Infants start to gain weight and growth rate increases after treatment. However, some neurological damage may persist until school age or even longer. Cerebral atrophy and nerve demyelination (if present) are improved within a few months (examples in Table 3).

Few small and large scale intervention studies have been conducted on infants, pre-school or school children from developing countries. Non-therapeutic doses of cobalamin (close to the RDA) alone or in combination with other micronutrients have been used. Intervention studies using either supplementation or dietary sources of the vitamin have shown normalisation of cobalamin status and improvements in health outcomes. In low socioeconomic settings, improving cobalamin status is not always accompanied by recovery from anaemia that may be related to deficiency in other nutrients.

In school age children from rural Kenya, serum concentrations of cobalamin increased (mean change + 47 to 66 pmol/L) with increasing intake of either milk (200–250 ml/day) or meat (60–85 g/day) in one school meal provided daily for one school year (Siekmann et al. 2003). The meal contained between 0.75–1.16 µg cobalamin leading to higher serum cobalamin in the group receiving meat (mean baseline cobalamin = 132 pmol/L vs after one year 189 pmol/L). Milk consumption also significantly increased serum cobalamin (from 164 pmol/L at baseline to 236 pmol/L after one year) compared with the control group in the isocaloric energy diet (serum cobalamin declined from 195 pmol/L to 151 pmol/L after one year) (Siekmann et al. 2003). Anaemia, elevated C-reactive protein, malaria infection, and stunting were very common in school children from rural Kenya (McLean et al. 2007) and were explained by low animal food intake. Despite increasing serum cobalamin after one year of introducing milk or meat at school in rural Kenya, no effects were seen on hemoglobin, ferritin or iron. This was explained by the high prevalence of infections and the unchanged low status of other nutrients after the intervention (retinol, iron, Zn, vitamin B2, and folate) (Siekmann et al. 2003).

Similar studies in Colombian school children providing a snack containing approximately 9 µg/day cobalamin for a shorter period (3 months) also reported increased serum cobalamin after the intervention (Arsenault et al. 2009). Increased height-for-age z-scores as well as less sick days and doctor visits were also observed in the intervention arm (Arsenault et al. 2009).

Moderately elevated preconception tHcy (≥ 9.0 µmol/L) has been associated with lower psychomotor and cognitive performance scores in children 4 months and 6 years, respectively after birth (Murphy et al. 2016). Few studies have investigated the effect of cobalamin on child neurodevelopmental domains and gross motor development. A placebo-controlled randomised study of Indian children aged 6–30 months tested cyanocobalamin (0.9 µg/d for infants up to 1 year of age and 1.8 µg for those > 1 year), folic acid (75 or 150 µg/d), or a combination of those vitamins for 6 months (Kvestad et al. 2015). Cobalamin increased from 269 to 362 pmol/L (mean difference 118 pmol/L) in the B12 group and from 307 to 446 pmol/L (mean difference 102 pmol/L) in the B12 plus folic acid group (Kvestad et al. 2015). Stunted children had reduced odds ratio [OR (95% confidence intervals, CI) = 0.26 (0.09, 0.78)] of being in the lowest quartile of the neurodevelopmental ASQ-3 scores (Kvestad et al. 2015). Children supplemented with vitamin B12 plus folic acid were less likely to be in the lowest quartile of total ASQ-3 score or scores related to gross motor function and problem solving (Kvestad et al. 2015). Children who were younger than 18 months at the beginning of the study and those who had baseline tHcy > 10 µmol/L were more likely to benefit than those who were older than 18 months, or with tHcy <10 µmol/L, respectively (Kvestad et al. 2015). Strand et al. reported a growth supportive effect of cobalamin and folic acid supplementation in Indian children. Stunted, wasted, or underweight children showed increases in weight-for-age and height-for-age z-scores after oral supplementation with cobalamin or cobalamin plus folic acid (Strand et al. 2015).

Prevention and treatment studies from developed countries are also available. Introducing sources of cobalamin such as dairy products and fish to children on a macrobiotic diet led to slight but significant improvements in several anthropometric markers (Dagnelie and van Staveren 1994). In infants (age < 8 months) with elevated plasma homocysteine (> 6.5 µmol/L), intramuscular cobalamin injection of 400 hydroxycobalamin as a single dose improved motor function and regurgitations (Torsvik et al. 2013).

Summary and Conclusions

1. Concentrations of cobalamin markers at birth (i.e., in cord blood) are predicted by maternal blood markers.
2. After birth, breastfed infants depend on maternal milk cobalamin as a sole source of the nutrient up to the age of 4 months.
3. Exclusively breastfed infants show a dramatic decline in serum cobalamin and holotranscobalamin associated with increases in plasma MMA and homocysteine between 4 and 6 months after birth.
4. The decline in cobalamin status in infants at around 4–6 months appears to coincide with a decrease in cobalamin content in the mother's milk.
5. Extended breastfeeding beyond the age of 6 months is associated with cobalamin depletion in the infant if complementary animal foods are not introduced or not adequately supplied.
6. Low cobalamin status in infancy and early childhood is strongly related to low maternal status during pregnancy, breastfeeding, prolonged lactation and to the introduction of complementary foods.
7. Human milk cobalamin content is predicted by recent intake of the vitamin rather than by maternal vitamin stores. The recent intake of cobalamin is thought to be readily available for transportation to the milk.
8. Low infant cobalamin status can be associated with stunting, low weight-for-length, inflammation, anaemia, and neurodevelopmental disorders. Long term consequences of marginal cobalamin status on toddler and adolescents are not well studied.
9. In situations of poverty and malnutrition, cobalamin deficiency occurs simultaneously with deficiencies in other micronutrients. Therefore, in many studies, it is not possible to conclude that symptoms are solely due to cobalamin deficiency.
10. Supplementing depleted women with 50–250 µg/d cobalamin from pregnancy throughout lactation causes higher levels of cobalamin and lower MMA in cord blood and in infants after birth, and increases breast milk cobalamin. This effect is not maintained after discontinuing supplement use.
11. Supplementation with cobalamin or dietary interventions that include more animal food sources can correct cobalamin status in children and many adverse effects of cobalamin deficiency on health such as anaemia, stunting, and neurodevelopmental disorders.

Keywords: Birth, children, infants, intake, human milk, treatment, neurodevelopmental, supplementation, prevention, lactation, growth, weaning, animal foods

Abbreviations

AI	:	adequate intake
CV	:	coefficient of variation
HC	:	haptocorrin
HoloTC	:	holotranscobalamin
HoloHC	:	holohaptocorrin
IOM	:	Institute of Medicine
RDA	:	Recommended Dietary Allowance
MMA	:	methylmalonic acid
TC	:	transcobalamin

References

Allen LH. 1993. The nutrition CRSP: what is marginal malnutrition, and does it affect human function? Nutr Rev. 51: 255–267.

Andersson HC and Shapira E. 1998. Biochemical and clinical response to hydroxocobalamin versus cyanocobalamin treatment in patients with methylmalonic acidemia and homocystinuria (cblC). J Pediatr. 132: 121–124.

Arsenault JE, Mora-Plazas M, Forero Y, Lopez-Arana S, Marin C, Baylin A and Villamor E. 2009. Provision of a school snack is associated with vitamin B12 status, linear growth, and morbidity in children from Bogota, Colombia. J Nutr. 139: 1744–1750.

Banerjee R. 2006. B12 trafficking in mammals: A for coenzyme escort service. ACS Chem Biol. 1: 149–159.

Berry CG. 1955. Anaemia of pregnancy in Africans of Lagos. Br Med J. 2: 819–823.

Bhate V, Deshpande S, Bhat D, Joshi N, Ladkat R, Watve S, Fall C, de Jager CA, Refsum H and Yajnik C. 2008. Vitamin B12 status of pregnant Indian women and cognitive function in their 9-year-old children. Food Nutr Bull. 29: 249–254.

Bijur AM and Desai AG. 1985. Composition of breast milk with reference to vitamin B12 and folic acid in Indian mothers. Indian J Pediatr. 52: 147–150.

Bjorke Monsen AL, Ueland PM, Vollset SE, Guttormsen AB, Markestad T, Solheim E and Refsum H. 2001. Determinants of cobalamin status in newborns. Pediatrics. 108: 624–630.

Bjorke-Monsen AL, Torsvik I, Saetran H, Markestad T and Ueland PM. 2008. Common metabolic profile in infants indicating impaired cobalamin status responds to cobalamin supplementation. Pediatrics. 122: 83–91.

Black AK, Allen LH, Pelto GH, de Mata MP and Chavez A. 1994. Iron, vitamin B12 and folate status in Mexico: associated factors in men and women and during pregnancy and lactation. J Nutr. 124: 1179–1188.

Bondevik GT, Schneede J, Refsum H, Lie RT, Ulstein M and Kvale G. 2001. Homocysteine and methylmalonic acid levels in pregnant Nepali women. Should cobalamin supplementation be considered? Eur J Clin Nutr. 55: 856–864.

Bor MV, Cetin M, Aytac S, Altay C, Ueland PM and Nexo E. 2008. Long term biweekly 1 mg oral vitamin B12 ensures normal hematological parameters, but does not correct all other markers of vitamin B12 deficiency. A study in patients with inherited vitamin B12 deficiency. Haematologica. 93: 1755–1758.

Carmel R, Mallidi PV, Vinarskiy S, Brar S and Frouhar Z. 2002. Hyperhomocysteinemia and cobalamin deficiency in young Asian Indians in the United States. Am J Hematol. 70: 107–114.

Casterline JE, Allen LH and Ruel MT. 1997. Vitamin B12 deficiency is very prevalent in lactating Guatemalan women and their infants at three months postpartum. J Nutr. 127: 1966–1972.

Chambers JC, Obeid OA, Refsum H, Ueland P, Hackett D, Hooper J, Turner RM, Thompson SG and Kooner JS. 2000. Plasma homocysteine concentrations and risk of coronary heart disease in UK Indian Asian and European men. Lancet. 355: 523–527.

Chanarin I, Malkowska V, O'Hea AM, Rinsler MG and Price AB. 1985. Megaloblastic anaemia in a vegetarian Hindu community. Lancet. 2: 1168–1172.

Codazzi D, Sala F, Parini R and Langer M. 2005. Coma and respiratory failure in a child with severe vitamin B(12) deficiency. Pediatr Crit Care Med. 6: 483–485.

Dagnelie PC and van Staveren WA. 1994. Macrobiotic nutrition and child health: results of a population-based, mixed-longitudinal cohort study in The Netherlands. Am J Clin Nutr. 59: 1187S–1196S.

Dagnelie PC, van Staveren WA, Vergote FJ, Dingjan PG, van den BH and Hautvast JG. 1989. Increased risk of vitamin B12 and iron deficiency in infants on macrobiotic diets. Am J Clin Nutr. 50: 818–824.

Deegan KL, Jones KM, Zuleta C, Ramirez-Zea M, Lildballe DL, Nexo E and Allen LH. 2012. Breast milk vitamin B12 concentrations in guatemalan women are correlated with maternal but not infant vitamin B12 status at 12 months postpartum. J Nutr. 142: 112–116.

Dhonukshe-Rutten RA, van Dusseldorp M, Schneede J, de Groot LC and van Staveren WA. 2005. Low bone mineral density and bone mineral content are associated with low cobalamin status in adolescents. Eur J Nutr. 44: 341–347.

Donangelo CM, Trugo NM, Koury JC, Barreto Silva MI, Freitas LA, Feldheim W and Barth C. 1989. Iron, zinc, folate and vitamin B12 nutritional status and milk composition of low-income Brazilian mothers. Eur J Clin Nutr. 43: 253–266.

Dror DK and Allen LH. 2008. Effect of vitamin B12 deficiency on neurodevelopment in infants: current knowledge and possible mechanisms. Nutr Rev. 66: 250–255.

Duggan C, Srinivasan K, Thomas T, Samuel T, Rajendran R, Muthayya S, Finkelstein JL, Lukose A, Fawzi W, Allen LH, Bosch RJ and Kurpad AV. 2014. Vitamin B12 supplementation during pregnancy and early lactation increases maternal, breast milk, and infant measures of vitamin B12 status. J Nutr. 144: 758–764.

Gimsing P, Hippe E, Helleberg-Rasmussen I, Moesgaard M, Nielsen JL, Bastrup-Madsen P, Berlin R and Hansen T. 1982. Cobalamin forms in plasma and tissue during treatment of vitamin B12 deficiency. Scand J Haematol. 29: 311–318.

Giugliani ER, Jorge SM and Goncalves AL. 1985. Serum vitamin B12 levels in parturients, in the intervillous space of the placenta and in full-term newborns and their interrelationships with folate levels. Am J Clin Nutr. 41: 330–335.

Gomber S, Kumar S, Rusia U, Gupta P, Agarwal KN and Sharma S. 1998. Prevalence & etiology of nutritional anaemias in early childhood in an urban slum. Indian J Med Res. 107: 269–273.

Graber SE, Scheffel U, Hodkinson B and McIntyre PA. 1971. Placental transport of vitamin B12 in the pregnant rat. J Clin Invest. 50: 1000–1004.

Graham SM, Arvela OM and Wise GA. 1992. Long-term neurologic consequences of nutritional vitamin B12 deficiency in infants. J Pediatr. 121: 710–714.

Grattan-Smith PJ, Wilcken B, Procopis PG and Wise GA. 1997. The neurological syndrome of infantile cobalamin deficiency: developmental regression and involuntary movements. Mov Disord. 12: 39–46.

Greibe E, Lildballe DL, Streym S, Vestergaard P, Rejnmark L, Mosekilde L and Nexo E. 2013a. Cobalamin and haptocorrin in human milk and cobalamin-related variables in mother and child: a 9-mo longitudinal study. Am J Clin Nutr. 98: 389–395.

Greibe E, Trolle B, Bor MV, Lauszus FF and Nexo E. 2013b. Metformin lowers serum cobalamin without changing other markers of cobalamin status: A study on women with polycystic ovary syndrome. Nutrients. 5: 2475–2482.

Guerra-Shinohara EM, Morita OE, Peres S, Pagliusi RA, Sampaio Neto LF, D'Almeida V, Irazusta SP, Allen RH and Stabler SP. 2004. Low ratio of S-adenosylmethionine to S-adenosylhomocysteine is associated with vitamin deficiency in Brazilian pregnant women and newborns. Am J Clin Nutr. 80: 1312–1321.

Guez S, Chiarelli G, Menni F, Salera S, Principi N and Esposito S. 2012. Severe vitamin B12 deficiency in an exclusively breastfed 5-month-old Italian infant born to a mother receiving multivitamin supplementation during pregnancy. BMC Pediatr. 12: 85.

Hampel D and Allen LH. 2016. Analyzing B-vitamins in human milk: Methodological approaches. Crit Rev Food Sci Nutr. 56: 494–511.

Hay G, Clausen T, Whitelaw A, Trygg K, Johnston C, Henriksen T and Refsum H. 2010. Maternal folate and cobalamin status predicts vitamin status in newborns and 6-month-old infants. J Nutr. 140: 557–564.

Hay G, Johnston C, Whitelaw A, Trygg K and Refsum H. 2008. Folate and cobalamin status in relation to breastfeeding and weaning in healthy infants. Am J Clin Nutr. 88: 105–114.

Hellegers A, Okuda K, Nesbitt RE Jr, Smith DW and Chow BF. 1957. Vitamin B12 absorption in pregnancy and in the newborn. Am J Clin Nutr. 5: 327–331.

Herrmann W, Isber S, Obeid R, Herrmann M and Jouma M. 2005. Concentrations of homocysteine, related metabolites and asymmetric dimethylarginine in preeclamptic women with poor nutritional status [abstract]. Clin Chem Lab Med. 43: A7.

Hertz H, Kristensen HP and Hoff-Jorgensen E. 1964. Studies on vitamin B12 retention. Composition of retention following intramuscular injection of cyanocobalamin and hydroxocobalamin. Scand J Haematol. 1: 5–15.

Hoey H, Linnell JC, Oberholzer VG and Laurance BM. 1982. Vitamin B12 deficiency in a breastfed infant of a mother with pernicious anaemia. J R Soc Med. 75: 656–658.

Horstmann M, Neumaier-Probst E, Lukacs Z, Steinfeld R, Ullrich K and Kohlschutter A. 2003. Infantile cobalamin deficiency with cerebral lactate accumulation and sustained choline depletion. Neuropediatrics. 34: 261–264.

Hyland K, Smith I, Bottiglieri T, Perry J, Wendel U, Clayton PT and Leonard JV. 1988. Demyelination and decreased S-adenosylmethionine in 5,10-methylenetetrahydrofolate reductase deficiency. Neurology. 38: 459–462.

Institute of Medicine. 1998. Dietary Reference Intakes for Thiamin, Riboflavin, Niacin, Vitamin B6, Folate, Vitamin B12, Pantothenic Acid, Biotin, and Choline. In: USA: Washington, DC, National Academy Press. pp. 390–422.

Institute of Medicine. 2000. Vitamin B12. *In*: Dietary Reference Intakes for Thiamin, Riboflavin, Niacin, Vitamin B6, Folate, Vitamin B12, Pantothenic Acid, Biotin, and Choline. USA: Washington, DC, National Academy Press. pp. 306–356.

Johnson PR Jr. and Roloff JS. 1982. Vitamin B12 deficiency in an infant strictly breast-fed by a mother with latent pernicious anemia. J Pediatr. 100: 917–919.

Karademir F, Suleymanoglu S, Ersen A, Aydinoz S, Gultepe M, Meral C, Ozkaya H and Gocmen I. 2007. Vitamin B12, folate, homocysteine and urinary methylmalonic acid levels in infants. J Int Med Res. 35: 384–388.

Koc A, Kocyigit A, Soran M, Demir N, Sevinc E, Erel O and Mil Z. 2006. High frequency of maternal vitamin B12 deficiency as an important cause of infantile vitamin B12 deficiency in Sanliurfa province of Turkey. Eur J Nutr. 45: 291–297.

Kocaoglu C, Akin F, Caksen H, Boke SB, Arslan S and Aygun S. 2014. Cerebral atrophy in a vitamin B12-deficient infant of a vegetarian mother. J Health Popul Nutr. 32: 367–371.

Kvestad I, Taneja S, Kumar T, Hysing M, Refsum H, Yajnik CS, Bhandari N and Strand TA. 2015. Vitamin B12 and Folic Acid Improve Gross Motor and Problem-Solving Skills in Young North Indian Children: A Randomized Placebo-Controlled Trial. PLoS One. 10: e0129915.

Lildballe DL, Hardlei TF, Allen LH and Nexo E. 2009. High concentrations of haptocorrin interfere with routine measurement of cobalamins in human serum and milk. A problem and its solution. Clin Chem Lab Med. 47: 182–187.

Louwman MW, Van DM, van d V, Thomas CM, Schneede J, Ueland PM, Refsum H and van Staveren WA. 2000. Signs of impaired cognitive function in adolescents with marginal cobalamin status. Am J Clin Nutr. 72: 762–769.

McLean ED, Allen LH, Neumann CG, Peerson JM, Siekmann JH, Murphy SP, Bwibo NO and Demment MW. 2007. Low plasma vitamin B12 in Kenyan school children is highly prevalent and improved by supplemental animal source foods. J Nutr. 137: 676–682.

Miller A, Korem M, Almog R and Galboiz Y. 2005. Vitamin B12, demyelination, remyelination and repair in multiple sclerosis. J Neurol Sci. 233: 93–97.

Miller DR, Specker BL, Ho ML and Norman EJ. 1991. Vitamin B12 status in a macrobiotic community. Am J Clin Nutr. 53: 524–529.

Monagle PT and Tauro GP. 1997. Infantile megaloblastosis secondary to maternal vitamin B12 deficiency. Clin Lab Haematol. 19: 23–25.

Monsen AL, Refsum H, Markestad T and Ueland PM. 2003. Cobalamin status and its biochemical markers methylmalonic acid and homocysteine in different age groups from 4 days to 19 years. Clin Chem. 49: 2067–2075.

Murphy MM, Fernandez-Ballart JD, Molloy AM and Canals J. 2016. Moderately elevated maternal homocysteine at preconception is inversely associated with cognitive performance in children 4 months and 6 years after birth. Matern Child Nutr. in press. DOI: 10.1111/mcn.12289.

Murphy MM, Molloy AM, Ueland PM, Fernandez-Ballart JD, Schneede J, Arija V and Scott JM. 2007. Longitudinal study of the effect of pregnancy on maternal and fetal cobalamin status in healthy women and their offspring. J Nutr. 137: 1863–1867.

Nyaradi A, Li J, Hickling S, Foster J and Oddy WH. 2013. The role of nutrition in children's neurocognitive development, from pregnancy through childhood. Front Hum Neurosci. 7: 97.

Obeid R, Fedosov SN and Nexo E. 2015. Cobalamin coenzyme forms are not likely to be superior to cyano- and hydroxyl-cobalamin in prevention or treatment of cobalamin deficiency. Mol Nutr Food Res. 59: 1364–1372.

Obeid R, Jouma M and Herrmann W. 2002. Cobalamin status (holo-transcobalamin, methylmalonic acid) and folate as determinants of homocysteine concentration. Clin Chem. 48: 2064–2065.

Obeid R, Morkbak AL, Munz W, Nexo E and Herrmann W. 2006. The cobalamin-binding proteins transcobalamin and haptocorrin in maternal and cord blood sera at birth. Clin Chem. 52: 263–269.

Oladipo O, Rosenblatt DS, Watkins D, Miousse IR, Sprietsma L, Dietzen DJ and Shinawi M. 2011. Cobalamin F disease detected by newborn screening and follow-up on a 14-year-old patient. Pediatrics. 128: e1636–e1640.

Pasricha SR, Shet AS, Black JF, Sudarshan H, Prashanth NS and Biggs BA. 2011. Vitamin B12, folate, iron, and vitamin A concentrations in rural Indian children are associated with continued breastfeeding, complementary diet, and maternal nutrition. Am J Clin Nutr. 94: 1358–1370.

Reghu A, Hosdurga S, Sandhu B and Spray C. 2005. Vitamin B12 deficiency presenting as oedema in infants of vegetarian mothers. Eur J Pediatr. 164: 257–258.

Rogers LM, Boy E, Miller JW, Green R, Rodriguez M, Chew F and Allen LH. 2003a. Predictors of cobalamin deficiency in Guatemalan school children: diet, Helicobacter pylori, or bacterial overgrowth? J Pediatr Gastroenterol Nutr. 36: 27–36.

Rogers LM, Boy E, Miller JW, Green R, Sabel JC and Allen LH. 2003b. High prevalence of cobalamin deficiency in Guatemalan school children: associations with low plasma holotranscobalamin II and elevated serum methylmalonic acid and plasma homocysteine concentrations. Am J Clin Nutr. 77: 433–440.

Rosenthal J, Lopez-Pazos E, Dowling NF, Pfeiffer CM, Mulinare J, Vellozzi C, Zhang M, Lavoie DJ, Molina R, Ramirez N and Reeve ME. 2015. Folate and Vitamin B12 Deficiency Among Non-pregnant Women of Childbearing-Age in Guatemala 2009–2010: Prevalence and Identification of Vulnerable Populations. Matern Child Health J. 19: 2272–2285.

Sakurai T, Furukawa M, Asoh M, Kanno T, Kojima T and Yonekubo A. 2005. Fat-soluble and water-soluble vitamin contents of breast milk from Japanese women. J Nutr Sci Vitaminol (Tokyo). 51: 239–247.

Sandberg DP, Begley JA and Hall CA. 1981. The content, binding, and forms of vitamin B12 in milk. Am J Clin Nutr. 34: 1717–1724.

Schneede J, Dagnelie PC, van Staveren WA, Vollset SE, Refsum H and Ueland PM. 1994. Methylmalonic acid and homocysteine in plasma as indicators of functional cobalamin deficiency in infants on macrobiotic diets. Pediatr Res. 36: 194–201.

Siddiqua TJ, Ahmad SM, Ahsan KB, Rashid M, Roy A, Rahman SM, Shahab-Ferdows S, Hampel D, Ahmed T, Allen LH and Raqib R. 2015. Vitamin B12 supplementation during pregnancy and postpartum improves B12 status of both mothers and infants but vaccine response in mothers only: a randomized clinical trial in Bangladesh. Eur J Nutr. 55: 281–293.

Siekmann JH, Allen LH, Bwibo NO, Demment MW, Murphy SP and Neumann CG. 2003. Kenyan school children have multiple micronutrient deficiencies, but increased plasma vitamin B12 is the only detectable micronutrient response to meat or milk supplementation. J Nutr. 133: 3972S–3980S.

Specker BL. 1994. Nutritional concerns of lactating women consuming vegetarian diets. Am J Clin Nutr. 59: 1182S–1186S.

Specker BL, Black A, Allen L and Morrow F. 1990. Vitamin B12: low milk concentrations are related to low serum concentrations in vegetarian women and to methylmalonic aciduria in their infants. Am J Clin Nutr. 52: 1073–1076.

Specker BL, Miller D, Norman EJ, Greene H and Hayes KC. 1988a. Increased urinary methylmalonic acid excretion in breast-fed infants of vegetarian mothers and identification of an acceptable dietary source of vitamin B12. Am J Clin Nutr. 47: 89–92.

Specker BL, Miller D, Norman EJ, Greene H and Hayes KC. 1988b. Increased urinary methylmalonic acid excretion in breast-fed infants of vegetarian mothers and identification of an acceptable dietary source of vitamin B12. Am J Clin Nutr. 47: 89–92.

Srikantia SG and Reddy V. 1967. Megaloblastic anaemia of infancy and vitamin B12. Br J Haematol. 13: 949–953.

Strand TA, Taneja S, Kumar T, Manger MS, Refsum H, Yajnik CS and Bhandari N. 2015. Vitamin B12, folic acid, and growth in 6- to 30-month-old children: a randomized controlled trial. Pediatrics. 135: e918–e926.

Strand TA, Taneja S, Ueland PM, Refsum H, Bahl R, Schneede J, Sommerfelt H and Bhandari N. 2013. Cobalamin and folate status predicts mental development scores in North Indian children 12–18 mo of age. Am J Clin Nutr. 97: 310–317.

Taneja S, Bhandari N, Strand TA, Sommerfelt H, Refsum H, Ueland PM, Schneede J, Bahl R and Bhan MK. 2007. Cobalamin and folate status in infants and young children in a low-to-middle income community in India. Am J Clin Nutr. 86: 1302–1309.

Thompson GN, Chalmers RA, Walter JH, Bresson JL, Lyonnet SL, Reed PJ, Saudubray JM, Leonard JV and Halliday D. 1990. The use of metronidazole in management of methylmalonic and propionic acidaemias. Eur J Pediatr. 149: 792–796.

Torsvik I, Ueland PM, Markestad T and Bjorke-Monsen AL. 2013. Cobalamin supplementation improves motor development and regurgitations in infants: results from a randomized intervention study. Am J Clin Nutr. 98: 1233–1240.

Trugo NM and Sardinha F. 1994. Cobalamin and cobalamin binding capacity in human milk. In: p. 23–33.

Ueland PM and Monsen AL. 2003. Hyperhomocysteinemia and B-vitamin deficiencies in infants and children. Clin Chem Lab Med. 41: 1418–1426.

Ulak M, Chandyo RK, Adhikari RK, Sharma PR, Sommerfelt H, Refsum H and Strand TA. 2014. Cobalamin and folate status in 6 to 35 months old children presenting with acute diarrhea in Bhaktapur, Nepal. PLoS One. 9: e90079.

Ullberg S, Kristoffersson H, Flodh H and Hanngren A. 1967. Placental passage and fetal accumulation of labelled vitamin B12 in the mouse. Arch Int Pharmacodyn Ther. 167: 431–449.

van Beynum I, den Heijer M, Thomas CM, Afman L, Oppenraay-van ED and Blom HJ. 2005. Total homocysteine and its predictors in Dutch children. Am J Clin Nutr. 81: 1110–1116.

Van DM, Schneede J, Refsum H, Ueland PM, Thomas CM, de BE and van Staveren WA. 1999. Risk of persistent cobalamin deficiency in adolescents fed a macrobiotic diet in early life. Am J Clin Nutr. 69: 664–671.

Waggoner DJ, Ueda K, Mantia C and Dowton SB. 1998. Methylmalonic aciduria (cblF): case report and response to therapy. Am J Med Genet. 79: 373–375.

Wardinsky TD, Montes RG, Friederich RL, Broadhurst RB, Sinnhuber V and Bartholomew D. 1995. Vitamin B12 deficiency associated with low breast-milk vitamin B12 concentration in an infant following maternal gastric bypass surgery. Arch Pediatr Adolesc Med. 149: 1281–1284.

Weiss R, Fogelman Y and Bennett M. 2004. Severe vitamin B12 deficiency in an infant associated with a maternal deficiency and a strict vegetarian diet. J Pediatr Hematol Oncol. 26: 270–271.

12

Cobalamin—Folate Interactions

Pol Solé-Navais,[1,2] *Rima Obeid*[3] and *Michelle M Murphy*[*,1,2]

1. Introduction

Folate and cobalamin are essential components of a key reaction in methionine homoeostasis and methyl cycling, ultimately affecting the supplies of tetrahydrofolate (THF) destined for nucleotide synthesis and methionine for methylation processes. Folate and cobalamin play interdependent roles in the cytosolic reaction catalysed by the cobalamin dependent 5-methyltetrahydrofolate-homocysteine methyltransferase (MTR, EC 2.1.1.13) in which the substrate, 5-methyltetrahydrofolate (5-CH3-THF), provides the methyl group that binds to the cobalamin cofactor to form methylcobalamin and THF. Subsequently the methyl group is transferred from methylcobalamin to homocysteine to form methionine and cobalamin. Apart from delivering methyl groups, this reaction also provides THF for nucleotide synthesis and DNA repair. Deficiencies in either folate or cobalamin lead to similar alterations in 1-C metabolism causing elevated plasma total homocysteine (tHcy) and also to macrocytosis (megaloblastic anaemia) resulting from impaired DNA-synthesis and thus cell division. Cobalamin deficiency leads to a phenomenon known as the "folate trap", in which intracellular 5-CH3-THF cannot be converted to THF by MTR, eventually limiting the supply of THF destined for nucleotide synthesis and of 5-CH3-THF for homocysteine metabolism.

---

[1] Area of Preventive Medicine and Public Health, Department of Basic medical Sciences, Faculty of Medicine and Health Sciences, Universitat Rovira i Virgili, IISPV, Spain.

[2] Centro de Investigación Biomédica en Red (CIBER; CB06/03), Instituto de Salud Carlos III (ISCIII).

[3] Aarhus Institute of Advanced Studies, University of Aarhus, Høegh-Guldbergs Gade 6B, building 1632, Dk-8000, Aarhus C, Denmark.

* Corresponding author: michelle.murphy@urv.cat

Folic acid, the fully oxidised synthetic form of the vitamin, used in vitamin supplements and fortified foods, can be reduced to THF and support nucleotide synthesis without passing through the MTR reaction and therefore independently of cobalamin status. Both cobalamin and folate deficiency can lead to a similar megaloblastic anaemia, though anaemia is not confined to deficient subjects. The anaemia usually precedes cobalamin neuropathy. Since the reduction of folic acid to bioactive folate is cobalamin independent, supplementation with folic acid may lead to correction of the anaemia leading to the masking of the haematological symptoms of cobalamin deficiency. Thus detection of cobalamin deficiency before the progression of the associated irreversible neuropathy might be deterred. This chapter will focus on the evidence for metabolic and clinical effects of cobalamin and folate interactions in the light of improved folate status on a global level due to the widespread fortification of cereal based products with folic acid and to folic acid supplement use.

2. Folate—Cobalamin Metabolic Interactions

2.1 Folate—cobalamin interactions in cobalamin deficiency

Herbert and Zalusky reported that intravenous injection of folic acid in cobalamin deficient patients led to the accumulation of plasma 5-CH3-THF, due to its impaired conversion to THF via MTR (Herbert and Zalusky 1962). They proposed that insufficient cobalamin for MTR to carry out its function led to a folate trap, reducing available 5-CH3-THF for other 1-C pathways such as purine and thymidylate synthesis. The resulting impairment in DNA synthesis would lead to megaloblastosis. Scott and Weir later reported that deficiency in folate, cobalamin or methionine leads to folate being preferentially supplied to the methionine cycle and to restricted DNA synthesis and reduced cell division. They proposed that in the face of imminent methyl group deficiency, the folate trap is a physiological mechanism that acts to prioritise methionine use for vital methylation reactions in the nervous system over its use for protein synthesis (Scott and Weir 1981). Evidence for occurrence of the methyl trap in humans has been reported in a patient with profound cobalamin deficiency and normal folate status. The intracellular fraction of 5-CH3-THF was increased with respect to the other forms of folates in the total folate pool (Smulders et al. 2006). The biochemical consequences of cobalamin deficiency are twofold; impaired nucleotide synthesis leading mainly to megaloblastic anaemia, and hypomethylation leading to neuropathy. Impaired methylation of myelin basic protein when methionine synthesis is low (Scott et al. 1981) and altered structural integrity of myelin as a result of odd carbon or branch chain fatty acid synthesis (Frenkel 1973), have been proposed as mechanisms for the development of neuropathy due to cobalamin deficiency. However, each of these hypotheses has numerous limitations (Allen et al. 1993). A more recent study suggested that altered intracellular cobalamin metabolism affects neuroplasticity. It showed that cellular cobalamin deficiency was associated with increased intracellular s-adenosylhomocysteine (SAH) and slower proliferation but faster differentiation of neuroblastoma mammalian cells (Battaglia-Hsu et al. 2009). Nevertheless, megaloblastic anaemia often precedes the neuropathy and therefore, though not haematologically distinct from the anaemia caused by folate deficiency, if it leads to the timely diagnosis of cobalamin deficiency, secondary preventive intervention measures can be implemented to prevent its progression to the irreversible neuropathy. The neurological symptoms can be expressed in the absence of anaemia. Therefore, blood count is not recommended as a screening test for cobalamin or folate deficiency, though it is commonly used in many countries where advanced biomarkers are not available. When megaloblastic cells are detected, further tests should be performed to explain the cause.

THF is a substrate for folylpoly-g-glutamate synthetase that produces long chain folylpolyglutamate forms to retain folate in the cell. *In vitro* studies have shown that 5-CH3-THF has a relatively low affinity for this enzyme and so its conversion to long chain polyglutamate folate forms is very slow (Cook et al. 1987). It was proposed that in the case of being unable to convert to other folate forms, as would happen if MTR activity is low due to cobalamin deficiency, that this form of folate would be lost from the cell and cause a reduction in intracellular folate (Cook et al. 1987). In fact, since this would mean that cobalamin deficient cells cannot retain folate but rather lose it, the term "methyl trap" has been criticised (Savage and Lindenbaum 1995).

Folic acid is readily absorbed and converted into active folate. The bioavailability of naturally occurring folates is considerably lower than that of folic acid (Cuskelly et al. 1996; Hannon-Fletcher et al. 2004). The site of folic acid conversion is not clear, but the liver is thought to be the main folic acid metabolizing organ. Once in the cell, folic acid is reduced to dihydrofolate and further to THF by dihydrofolate reductase (DHFR, EC 1.5.1.3). The conversion of folic acid by DHFR is slow in humans and it has been proposed that the enzyme may become saturated in situations of high folic acid intake, thus leading to circulating unmetabolised folic acid (UMFA) (Bailey and Ayling 2009). The same authors also reported that DHFR activity varies considerably between people. UMFA has been reported to appear in the circulation within hours of a single dose of folic acid above 200 µg (Kelly et al. 1997). This would mean that studies testing folate forms in subjects who were not fasting and using cereals fortified with folic acid are likely to detect UMFA in blood (Pfeiffer et al. 2015). Whether free folic acid may have harmful effects or not, is still unclear. UMFA was detected in 94% of an elderly Irish population unexposed to mandatory folic acid fortification, but UMFA only accounted for 1.3% of total plasma folate (Boilson et al. 2012). Other studies have reported that plasma UMFA concentrations increase with folic acid supplementation (Obeid et al. 2011) and recently UMFA has been reported to be detectable across all age ranges of the USA population (Pfeiffer et al. 2015). Two randomised controlled trials reported that supplementation with 400 µg/d of folic acid during pregnancy or lactation did not lead to a higher concentration of circulating UMFA in blood (Pentieva et al. 2016) or breastmilk (Houghton et al. 2009) in the supplemented versus placebo groups. However, UMFA was detected in a higher proportion of supplemented versus non-supplemented women in the study that was carried out in Northern Ireland where voluntary, but not mandatory, fortification with folic acid occurs (Pentieva et al. 2016). Obeid et al. (2010) reported that maternal folic acid supplementation was not associated with UMFA accumulation in cord blood samples. UMFA was detected in 96% of breastmilk samples, representing approximately 8% of total milk folate concentration, in the Canadian study (Houghton et al. 2009).

Elevated tHcy due to impaired remethylation of homocysteine to methionine is a consequence of either folate or cobalamin deficiency. In cobalamin replete populations, folate status is a major determinant of tHcy (Jacques et al. 2001) and evidence for interaction between folate and cobalamin status in their effect on tHcy has been reported. Cobalamin status and renal function became the most important determinants of mild hyperhomocysteinemia in coronary artery disease patients following the introduction of mandatory fortification with folic acid in the USA (Liaugaudas et al. 2001). An Irish study of the effect of gradually increasing doses of folic acid supplements over 26 weeks from 100 to 400 µg/d showed that initially there was a strong inverse correlation between plasma folate and tHcy and the weak inverse correlation between plasma cobalamin and tHcy was not significant. However, the correlation between plasma folate and tHcy became weaker and lost significance, while the inverse correlation between plasma cobalamin and tHcy became stronger and became significant during the final phase of the intervention with 400 µg folic acid/d (Quinlivan et al. 2002). These findings suggest that cobalamin status becomes an important determinant of plasma tHcy in situations of replete folate status.

The role of cobalamin in tHcy and folate metabolism has received more attention in countries where cobalamin deficiency is endemic. Cobalamin deficiency has been reported to be common in Pune in India (Yaknik et al. 2006) and this has been confirmed in numerous other parts of India where the diet is typically vegetarian. A randomised controlled trial in Pune of parents and children from the Pune Maternal Nutritional Study cohort and in which 27% of the children, 72% of the fathers and 48% of the mothers had plasma cobalamin < 150 pmol/L at baseline, showed that supplementation with 2 µg/d of cyanocobalamin for 12 months led to an average reduction in tHcy of 5.9 (95% confidence interval: –7.8 –4.1) µmol/L. The same trial reported no additional tHcy-lowering effect of low dose folic acid supplementation (200 µg/d) (Deshmukh et al. 2010).

Cobalamin deficiency is associated with elevated methylmalonic acid (MMA) and an elevated blood concentration of this metabolite is considered more sensitive and specific for cobalamin deficiency than that of tHcy. MMA is an end product of methylmalonyl-CoA that increases in cobalamin deficient subjects. Methylmalonyl-CoA mutase (EC 5.4.99.2) requires adenosylcobalamin as a cofactor and catalyses the conversion of methylmalonyl-CoA to succinyl-CoA in the mitochondria. This cobalamin-dependent mitochondrial reaction and the folate pathway do not interact so folate deficiency is not expected to affect MMA. However, high folate status has been reported to be associated with MMA in cobalamin deficient subjects (Selhub et al. 2007; Miller et al. 2009) and this will be covered later in the chapter.

Plasma concentrations of tHcy cannot be exclusively used to discern between cobalamin and folate deficiencies. In a comparative study of cobalamin deficient and folate deficient patients, tHcy was elevated in both sets of patients. Elevated MMA occurred in 98% of cobalamin deficient patients. Its occurrence in 12% of the folate deficient patients was attributed to impaired renal function (Savage et al. 1994).

2.2 Folate and cobalamin: nutrient-nutrient and gene-nutrient interactions

Cobalamin and vitamin B6 play a central role in folate metabolism which explains why co-supplementation of these 2 vitamins leads to less UMFA compared with folic acid alone (Obeid et al. 2015). Folate metabolism is largely dependent on other interacting nutrients and provides an example of nutrient-nutrient interactions whose role in health and disease is still largely underestimated.

The betaine homocysteine-methyltransferase (BHMT) pathway is folate and cobalamin independent, but it helps to maintain folate and methionine homeostasis under conditions of deficiency or excess of folate or cobalamin. Both vitamins interact with choline and betaine metabolism via the BHMT pathway. Experimental studies have shown that folate restriction in rats leads to depletion of liver choline and phosphatidylcholine (Kim et al. 1994). When folate is deficient, the BHMT pathway is upregulated following a methionine load (Holm et al. 2004; McGregor et al. 2002) thus using more betaine and choline. In contrast, choline-restriction depletes hepatic folate (Horne et al. 1989) and decreases hepatic methionine synthesis (Zeisel et al. 1989). A switch between folate and betaine as methyl donors has been described in African women with strong fluctuations in nutrient intake over the year (Dominguez-Salas et al. 2013). There is also evidence that the contribution of the BHMT pathway to homocysteine remethylation is enhanced during late human pregnancy when maternal folate reserves are reduced (Fernàndez-Roig et al. 2013).

The effects on tHcy of 2 common polymorphisms that affect folate status (methylenetetrahydrofolate reductase, MTHFR 677C > T) (Bueno et al. 2016) and MTRR activity (methionine synthase transferase, MTRR 66A > G) (Olteanu et al. 2002) have recently been shown to depend on riboflavin status (Garcia-Minguillan et al. 2014). Details of the nature of the role of riboflavin in MTHFR and MTRR functions are beyond the scope of this chapter but it is noted that MTHFR has a FAD cofactor and that MTRR is a flavoprotein and that riboflavin interacts with the substrates of each of these enzymes (folate and cobalamin respectively), and can modify their effect on tHcy.

3. Effects of Folic Acid Supplementation and Fortification

3.1 Worldwide folic acid fortification policies and fortification levels

One of the most important public health discoveries of the twentieth century was the prevention of neural tube defects (NTDs) with periconceptional folic acid supplementation. This finding led to the implementation of mandatory fortification of cereal grain products/flour with folic acid in numerous countries worldwide and food products voluntarily fortified with folic acid are also widely available in many countries (Flour Fortification Initiative 2015). Two studies in separate countries reported that 906 nmol/L is the optimal red blood cell folate threshold for reducing the risk of NTD-affected pregnancy in women of reproductive age (Daly et al. 1995; Crider et al. 2014). This cut-off for red blood cell folate concentration has been recently recommended by the World Health Organization to protect against NTDs (World Health Organization 2015). Since the maximum effect was observed at considerably replete folate status levels, it appears that merely eliminating folate deficiency is insufficient to reach the optimal folate status for NTD prevention. Therefore, whether folic acid from fortified foods is enough to achieve optimal blood folate concentrations remains unclear.

The levels of mandatory fortification with folic acid differ between countries but are designed in all cases to reduce the incidence of NTDs by increasing folate status in young women (Table 1). In South-America, fortification levels range between 35 µg to 330 µg per 100 g of food (Rosenthal et al. 2014).

Globally, reports from countries where mandatory fortification has been implemented show that the policy successfully increases folate intake, improves folate status and reduces hyperhomocysteinaemia (Table 2). However, some reports show that not all of the population is reached and a recent report showed that red blood cell folate was still below the optimum

Table 1. Folic acid mandatory fortification levels of wheat flour in different countries.

Country	Fortification levels[1]	Date of implementation
United States	95–308 µg	1998
Canada	150 µg	1998
Costa Rica	180 µg	1998
Chile	100–260 µg	2000
Argentina	220 µg	2004
South Africa	200 µg	2003

[1] Data from the Food Fortification Initiative (accessed November 2015). Amount per 100 grams.

Table 2. Folate intake and status following the implementation of mandatory fortification of flour with folic acid in the USA since January 1998.

Authors (Study)	Time frame	Aim	Observations
Bentley et al. 2006 (NHANES)	1988–1994 versus 1999–2000	To compare changes in average daily folate intake between pre- and post-fortification	Folate intake increased by 100 µg/d an intake > 400 µg/d only reached in 50% of fertile women
Jacques et al. 1999 (Framingham)	1995–1996 versus 1997–1998	To evaluate pre- and post-fortification changes in the prevalences of folate deficiency (plasma folate < 7 nmol/L) and hyperhomocysteinemia (tHcy > 13 µmol/L)	Folate deficiency and hyperhomocysteinemia were reduced from 22% to 1.7% and 18.7 to 9.8%, respectively
Pfeiffer et al. 2008 (NHANES)	1991–1994 versus 1999–2004	To evaluate pre- and post-fortification changes in plasma total homocysteine (tHcy)	Average tHcy was reduced by 10%
Pfeiffer et al. 2012 (NHANES)	1998–2010	To evaluate pre- and post-fortification changes in serum and red blood cell folate	Prevalence of low serum folate (< 10 nmol/L) was reduced from 24% to ≤ 1% Prevalence of red blood cell folate < 340 nmol/L was reduced from 3.5% to 1%
Yang et al. 2010 (NHANES)	2003–2006	To assess average daily folate intake according to reported dietary patterns	Average folate intake from fortified foods only: 117 µg/d from supplements and breakfast cereals: 621 µg/d
Tinker et al. 2015 (NHANES)	2007–2012	To assess the prevalence of suboptimal red blood cell folate status in women of pregnancy age	22.8% had sub-optimal red cell folate despite fortification

NHANES: National Health and Nutrition Examination Survey.

for NTD prevention in 22.8% of fertile women from the National Health and Nutrition Examination Survey (NHANES) (Tinker et al. 2015). Women not consuming folic acid supplements, smokers and non-Hispanic black or Hispanics were at increased risk of suboptimal red blood cell folate in countries where fortification is in place. Improved post-fortification folate intake and/or status have also been reported in Canada (Ray et al. 2004), Chile (Hertrampf et al. 2003; Hertrampf and Cortés 2004) and South Africa (Papathakis and Pearson 2012).

Despite the recognised health benefits of improved folate status, some concerns have emerged regarding metabolic abnormalities associated with elevated folate status that might result from the combined intake of folic acid from supplements and fortified foods. It was considered that the level of added folic acid at 140 µg/100 g of enriched cereal grain product in mandatory fortification policies such as that in the USA would not lead to the masking of underlying cobalamin deficiency by enhanced folate status (US FDA 1996). The policy was expected to deliver 100 to 200 µg/d of the vitamin to women of fertile age. By 2015, mandatory fortification of at least industrialised milled wheat flour with folic acid was in place in 79 countries (Flour Fortification Initiative 2015). While policies vary among the European countries, mandatory fortification is not implemented in Europe. The United Kingdom Standing Advisory Committee on Nutrition recommended that mandatory fortification of flour with folic acid should be implemented to reduce the incidence of NTD affected births (UK Standing Advisory Committee on Nutrition 2006) but so far this policy has not been implemented. Some experts have advocated for prudence and cautious assessment of all beneficial and detrimental effects (Smith et al. 2008).

The Food Safety Authority of Ireland has concluded that there would be no additional benefits for mandatory fortification of bread in Ireland, given the widespread availability of voluntarily fortified foods on the market (Food Safety Authority of Ireland 2008). In Denmark and in The Netherlands, fortification with folic acid is prohibited (Bailey et al. 2015).

Some European studies have studied the effect of voluntary fortification of foods with folic acid on daily folate intake or on folate status. In Spain, it is estimated that voluntary fortification with folic acid provides 35% of the daily reference intakes per serving (Samaniego-Vaesken et al. 2016). Two Irish studies have reported a positive association between folic acid intake from fortified foods and optimal red cell folate status but have highlighted that the folic acid intake threshold to achieve this is not reached by a large proportion of women of fertile age (Hoey et al. 2007; Hopkins et al. 2015). The folic acid intake threshold from fortified foods to reach the optimum red cell folate concentration (>906 nmol/L) was reported to be 99 µg/d (Hoey et al. 2007). Folic acid from fortified foods or supplements was reported not to reach 18% of the population (Hopkins et al. 2015). The same study reported folate intakes of 582 µg/d in consumers of supplements and fortified foods and 445 µg/d in consumers of folic acid from fortified foods, only, compared to 205 µg/d in nonconsumers of either of these sources of folic acid. Overall, 66% of women of childbearing age had suboptimal red blood cell folate concentrations ($\leq$ 906 nmol/L), and only 16% of non-consumers of folic acid had optimal red blood cell folate (Hopkins et al. 2015). On the other hand, 70% of the women had optimal cobalamin status for NTD prevention (>221 pmol/L) and this was observed across the folic acid intake range.

A meta-analysis from six studies using the microbiological assay for the determination of plasma and red blood cell folate reported that for every 10% increase in natural food folate intake, red blood cell and serum/plasma folate increased by 6 and 7% respectively (Marchetta et al. 2015). An intake of more than 450 µg/d folate equivalents could achieve optimal red blood cell folate concentrations.

A randomised dose-finding trial reported that low-dose folic acid supplements (0.2 mg/d) over a six-month period had the same tHcy reducing effect as normal-high doses (0.4 or 0.8 mg/d) (Tighe et al. 2011). Importantly, compared to previous trials, the duration of supplementation was longer and showed that long term supplementation with low-dose folic acid was as effective as higher doses. The observed reduction in tHcy in participants with baseline tHcy below the median (the cut off concentration is unspecified in the paper but mean baseline tHcy in this group was 10.1 ± 1.9 µmol/L) was on par with that reported between pre- and post-fortification with folic acid in the USA (Pfeiffer et al. 2008). Red blood cell folate change was not reported so it is not known whether optimal folate status was reached with this low dose supplemental regime.

Cobalamin is one of the greatest predictors of homocysteine in folate-replete populations. It seems plausible that NTD risk reduction after mandatory fortification can be further improved by cobalamin supplementation (or fortification). This reduction would be expected to be even greater in countries with concominant high prevalence of cobalamin deficiency (Refsum et al. 2001) and high rates of NTDs (Cherian et al. 2005; Jaikrishan et al. 2013) such as India. The prevalence of cobalamin deficiency (< 148 pmol/L) or borderline deficiency (< 185 pmol/L) in a study of Chilean women before and after mandatory folic acid fortification was considerably greater than that of folate deficiency (Hertrampf et al. 2003).

Molloy and collaborators suggested that women should aim for cobalamin concentrations > 300 ng/L (221 pmol/L) at preconception in order to reduce the incidence of NTDs, as they observed that cobalamin concentrations < 250 ng/L (185 pmol/L) increased the risk of NTD by 2.5- to 3 fold (Molloy et al. 2009).

Cobalamin deficiency is widespread in parts of the world where low amounts of animal-foods are consumed for religious, cultural, or economic reasons. Folic acid fortification and periconception supplement use may not be sufficient to prevent NTD-affected pregnancies in the presence of cobalamin deficiency. Apart from the fact that cobalamin deficiency is a risk factor for NTDs, there is also the issue of supplementing cobalamin deficient mothers with folic acid. Increasing folic acid intake will maintain purine and pyrimidine synthesis but would not correct the methylation abnormalities associated with cobalamin deficiency. Mandatory fortification of flour with folic acid has been in place in Oman, where many Indians and Asians emigrate to in search of work, since 1996. Among Asian Indian men living in Oman, vegetarians had lower plasma cobalamin but higher serum folate as well as markedly higher MMA and tHcy compared with omnivores and with Omani men (Table 3). It is possible that high serum folate in vegetarians is further enhanced by the extra folic acid provided by fortification or it might reflect folate "trapping" due to very low cobalamin status. As mentioned previously, in subjects with high folate intake and cobalamin deficiency, cobalamin becomes the main determinant of plasma homocysteine (Figure 1). Therefore, countries planning to apply fortification programs need to consider the population structure and dietary habits in minorities or population subgroups that may not benefit from the fortification or require co-fortification with small amounts of cobalamin.

Table 3. Concentrations of vitamin B12 and methylmalonic acid in Asian Indian immigrant men in Oman compared to Omani men.

	Asian Indians Omnivorous N=82	Asian Indians Vegetarians N=52	Omnivorous Omanis N=63
Age, years	43 (40–59)	43 (34–60)	46 (40–56)*
Total B12, pg/ml	315 (113–894)	160 (63–514)*	476 (204–906)*
Folate, ng/ml	10.0 (2.0–44.8)	14.3 (3.0–33.6)	9.0 (2.0–40.0)
MMA, nmol/L	212 (84–2066)	650 (134–7935)*	153 (83–645)*
tHcy, μmol/L	8.9 (5.3–78.7)	13.4 (6.3–49.8)*	8.2 (4.3–14.1)

Median (range); *$p < 0.05$ compared to omnivorous Asian Indians (Mann-Whitney test) (Obeid et al., unpublished data). Mandatory fortification is in place in Oman since 1996. The study was conducted in 2004 and some of the results have been published (Herrmann et al. 2009).

4. Potential Adverse Effects of Elevated Folate Status in Cobalamin Deficient Subjects

4.1 Haematological and metabolic abnormalities

There is conflicting evidence in the literature regarding the haematological and metabolic effects of elevated folate status combined with low cobalamin status. Many reports are based on results from population studies with measurements of haematological or metabolic outcomes such as anaemia, macrocytosis, hyperhomocysteinaemia, or elevated methylmalonic acid concentrations in situations of mandatory or voluntary fortification with folic acid. Nevertheless, there is considerable interest in this topic, given the extent of mandatory folic acid fortification and availability of voluntarily fortified foods worldwide. Early case reports suggested that treating pernicious anaemia patients with folic acid, either aggravated or precipitated neurological complications (Chodos and Ross 1951; Reynolds 2006) because the underlying cause

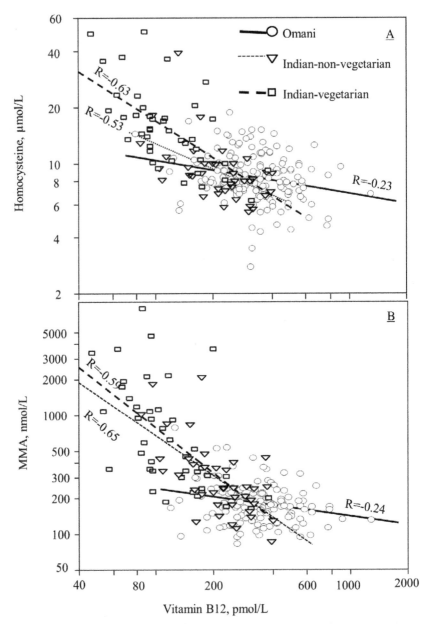

Figure 1. Scatter plot of vitamin B12 with methylmalonic acid (MMA) and homocysteine in 3 different cobalamin status groups. The study was conducted in Oman in 2004. Mandatory fortification of flour with folic acid has been in place there since 1996. In vegetarian Asian Indians, plasma cobalamin is a more important determinant of plasma homocysteine. R; is the correlation coefficient according to Pearson test. The x- and y axises are shown as log-transformed data.

(cobalamin deficiency) remained undetected. A hospital-based study reported that the prevalence of patients with low cobalamin status without macrocytosis had increased from 70% before fortification to 87% post-fortification (Wyckoff and Ganji 2007). However, cobalamin deficiency (or even mild deficiency) in the absence of anaemia or macrocytosis has not increased after mandatory fortification with folic acid in the general population in the United States (Qi et al. 2014). This observation was upheld even after categorising the participants by groups exposed to more or less folic acid intake, depending on their dietary and supplement use habits.

Other studies have investigated whether elevated folate status aggravates haematological or metabolic anomalies associated with cobalamin deficiency. Four percent of participants aged 60 or over in the NHANES 1992–2002 survey had cobalamin deficiency (serum cobalamin < 148 pmol/L or serum MMA > 210 nmol/L) combined with high folate status (plasma folate > 59 nmol/L). Anaemia was more prevalent in these than in participants with cobalamin deficiency combined with normal plasma folate (Morris et al. 2007). When the effects of the 5-CH3-THF and folic acid (UMFA) components of total serum folate were assessed separately, 6% of the participants had cobalamin deficiency combined with circulating UMFA and 6.9% had cobalamin deficiency combined with high 5-CH3-THF (> 50 nmol/L). The prevalences of anaemia and macrocytosis did not vary between participants with UMFA versus no UMFA or elevated 5-CH3-THF versus normal 5-CH3-THF. However, UMFA versus no UMFA was associated with lower mean cell volume (Morris et al. 2010). In a setting of voluntary fortification with folic acid, a UK study of 2559 adults aged > 65 years, from 2 population cohorts, observed no differences in the prevalence of anaemia between the high folate (≥30 or ≥60 nmol/L) and low cobalamin status (holoTC < 45 pmol/L) versus the normal folate and low cobalamin groups (Clarke et al. 2008). Only 5% of participants had serum folate greater than 60 nmol/L, whereas the 80th percentile was 59 nmol/L in the study by Morris et al. (2007). Moreover, less participants were affected by the high folate and low cobalamin combination and the prevalence of anaemia was lower in this group in the UK compared to the USA study (Morris et al. 2007).

After mandatory fortification with folic acid in the US, high folate status (plasma folate > 44 nmol/L) combined with cobalamin deficiency was associated with a tendency towards higher tHcy and MMA concentrations compared to cobalamin deficiency combined with non-elevated folate status (Selhub et al. 2007). These tendencies were not observed before mandatory fortification with folic acid, nor in situations of normal cobalamin status. Similar observations were reported for tHcy and MMA, and holoTC and holoTC/cobalamin were lower in the low cobalamin/elevated folate versus low cobalamin/non-elevated folate groups, in the Sacramento Area Latino Study on Ageing (SALSA) (Miller et al. 2009). A recent intervention trial in Chile reported that enhancement of cobalamin status in participants with baseline serum folate (< 33.9 nmol/L) by cyanocobalamin injections was more pronounced than in those with serum folate ≥ 33.9 nmol/L (serum cobalamin

< 120 pmol/L) (Brito et al. 2016). Studies performed in settings of voluntary fortification with folic acid have not replicated these results. In Ireland for example, cobalamin deficient subjects with a high folate status (> 30 nmol/L) had lower plasma tHcy and a tendency towards lower MMA than subjects with low-normal folate status (Mills et al. 2011). Consistent with these findings, but using different criteria to classify cobalamin deficiency, a UK study found no deleterious effect of folate status on metabolic markers of cobalamin deficiency (Clarke et al. 2008).

4.2 Cognitive function

Severe cobalamin deficiency is characterised by subacute combined degeneration of the spinal cord, peripheral neuropathy, and psychiatric disorders (Shorvon et al. 1980; Healton et al. 1991; Hemmer et al. 1998). Affective disorders, peripheral neuropathy and neurological changes are found in both folate and cobalamin deficient subjects (Shorvon et al. 1980). Early and more recent case reports have reported that treatment of pernicious anaemia patients with folic acid, either aggravated or precipitated neurological complications (Chodos and Ross 1951; Reynolds 2006). Numerous studies were reviewed by Savage and Lindenbaum (1995). They observed that more cases of neuropathy emerged following supplementation of cobalamin deficient patients with ≥ 10 mg/d of folic acid compared to ≤ 5 mg/d. However, it is not clear whether the neuropathy started at the same time as the treatment was initiated. They also observed that neuropathy appeared 1–12 weeks after initiation of supplementation in pernicious anaemia patients treated with ≥ 10 mg/d of folic acid compared to no cases 8–27 months later in those that were not treated. They also suggested that worsening of cobalamin stores was as important a factor in neuropathy progression as folic acid treatment.

High folate and folic acid intake from supplements greater than 400 µg/day (compared to non-users) were associated with a faster rate of cognitive decline in adults aged over 65 years from the Chicago Health and Aging Project with 6 years of follow-up (Morris et al. 2005). However, Fridman attributed these results to a severe bias in the detection of cognitive changes in the reference group due to the low sensitivity of the cognitive tests in this group (Fridman 2005). In the NHANES study, described above, the odds of cognitive impairment were almost 5-fold greater for the cobalamin deficient combined with high folate status group compared to those with a normal status in both B-vitamins (Morris et al. 2007). However, the prevalence of cognitive impairment was lower (11%) in participants with normal cobalamin/high folate status. UMFA and elevated 5-CH3-THF showed no association with cognition when cobalamin status was normal but UMFA combined with cobalamin deficiency was associated with a lower score in the Digit-Symbol Substitution Test compared to cobalamin deficiency with no UMFA. 5-CH3-THF was positively associated with cognitive score in participants with normal cobalamin status (Morris et al. 2010). In a separate 8 year follow-up study of 549 old adults from the Framingham Heart Study, participants with plasma cobalamin < 258 pmol/L and plasma folate ≥ 21.75 nmol/L, had greater cognitive decline (assessed by the Mini-Mental State Examination Score) over the 8 years compared to participants in the low cobalamin combined with any other folate status category (Morris et al. 2012). Similarly, folic acid supplement users with low cobalamin status had a higher cognitive decline over 8 years than non-consumers with low cobalamin status (Morris et al. 2012). The design of this study overcomes the temporality drawback of cross-sectional designs, but nonetheless given its observational nature, causation cannot be established. These results have not been replicated in other settings of mandatory or voluntary fortification with folic acid. No effects of high folate status (plasma folate > 45.3 nmol/L) were observed on cognitive scores in participants with low cobalamin status in the SALSA study (Miller et al. 2009). In settings of voluntary fortification, low cobalamin status (plasma cobalamin < 274 pmol/L) combined with high folate status (plasma folate > 18.5 nmol/L) was associated with a reduced risk of cognitive impairment in a Norwegian study (Doets et al. 2014). Similarly in a UK study, low cobalamin status combined with non-elevated folate status increased slightly the odds of cognitive impairment but the odds were not significant for low cobalamin status combined with elevated folate status (Clarke et al. 2008).

4.3 Suggested explanations

Different hypotheses have been proposed for the apparent exacerbation of haematological and metabolic abnormalities in situations of combined low cobalamin/high folate status. One is based on the possible adverse effects of excess folic acid on 1-C metabolism. The other is that combined low cobalamin/high folate status actually reflects more severe cobalamin deficiency than the low cobalamin/non-elevated folate status combination. In situations of severe cobalamin deficiency, folates are released from the cell leading to high plasma and low red blood cell folate concentrations (Stokstad et al. 1998). It has been suggested that the low cobalamin/high folate status combination may in fact be due to impaired cobalamin absorption, leading to a more severe deficiency (Berry et al. 2007). The greater use of folic acid and cobalamin containing supplements in participants in the combined low cobalamin/high folate status group compared to low cobalamin/non-elevated folate status group in the SALSA study supported this hypothesis (Miller et al. 2009). This hypothesis was also supported by Carmel, who suggested that reverse causation cannot be ruled out (Carmel 2009). Quinlivan suggested that low cobalamin/high folate status is caused by the methyl trap in situations of greater severity of cobalamin deficiency in which folate that cannot be converted to THF is lost from the cell (Quinlivan 2008). This situation would lead to a rise in plasma folate and a loss in red blood cell folate in the presence of cobalamin deficiency. However, the results of the SALSA study where red blood cell folate was not lower in participants with low cobalamin/high folate status versus low cobalamin/nonelevated folate status do not support this hypothesis (Miller et al. 2009). Much of the available evidence supporting the harmful effects of a high folate status in cobalamin subjects arises from animal studies. High folate status inhibits MTHFR in rats thus leading to lower 5-CH3-THF, S-adenosylmethionine (SAM) and SAM/SAH ratio (Christensen et al. 2015). This coupled with the low activity of MTR due to cobalamin deficiency, could lead to impaired myelin synthesis, and to increased cognitive decline. Reduced haematopoiesis was also observed in rats fed a high folate diet with lower red blood cell counts, haematocrit, haemoglobin and higher mean cell volume (Christensen et al. 2015), suggesting impaired nucleotide synthesis. However, underlying mechanisms for the reduced haematopoiesis have not been described yet. Dihydrofolate has been shown to be an inhibitor of thymidylate synthase (Dolnick and Cheng 1978), and its accumulation could lead to reduced thymidylate synthesis (Smith et al. 2008). But this would not explain the deleterious or beneficial effects of a high folate status on anaemia in cobalamin deficient or non-deficient subjects respectively (Morris et al. 2008). Selhub et al. have suggested that excess folic acid might cause the irreversible oxidation of cobalamin (Selhub et al. 2007), comparable to the effects of nitric oxide (Deacon et al. 1978). This hypothesis could explain the effects on both reduced MTR and methylmalonyl-CoA mutase activity, but still remains to be tested.

A number of unresolved questions and gaps in knowledge in the cobalamin field of work are outstanding and listed in Table 4.

Table 4. Elevated folate in cobalamin deficiency: gaps in knowledge.

- Does cobalamin deficiency contribute to or cause high serum folate in situations of exposure to mandatory fortification and supplementation with folic acid (folate trap)?
- Can high serum folate be lowered if cobalamin status is corrected by providing cobalamin?
- What is the nature of the association between serum and red blood cell folate in situations of cobalamin deficiency?
- What biochemical indicators and which cut-offs should be used to define cobalamin deficiency in epidemiological studies?
- Do high red cell folate concentrations in cobalamin deficient subjects have a comparable effect with that described for plasma or serum folate on metabolic and clinical markers of cobalamin deficiency? How should cut-offs for elevated folate status be defined?
- What is the adequate cobalamin-folate balance?
- What mechanisms might explain a deleterious or beneficial effect of high folate status in cobalamin deficient or non-deficient subjects on haematological, cognitive and metabolic outcomes?
- Could severity or stage of cobalamin deficiency, or an underlying unrecognised medical condition explain the observations attributed to high folate status in cobalamin deficient subjects?
- Are populations with prevalent cobalamin deficiency and exposed to high folic acid intakes (pregnant women) in non-fortified countries at increased risk of anaemia, cognitive or metabolic impairment?
- Does enhancing cobalamin deficiency in subjects with high folate improve haematological, cognitive and metabolic outcomes?

In summary, cobalamin is required for folate metabolism. Both nutrients show interactions with several other nutrients such as methionine, choline, betaine, vitamin B6 and B2. Deficiencies of both nutrients produce similar forms of megaloblastic anaemia. Measurements of plasma methylmalonic acid and homocysteine can help in differential diagnosis. There is insufficient evidence, to date, that folic acid supplement use or exposure to mandatory fortification with folic acid causes adverse neurological effects in cobalamin deficient people. Replication of the results of studies that reported such effects and studies designed to test the mechanisms involved, are required. However, a balance between folate and cobalamin intake appears to be important for maintaining cell methylation and DNA synthesis.

Keywords: cobalamin, folate, folic acid, folate trap, cobalamin neuropathy, unmetabolised folic acid

Abbreviations

DHFR	:	dihydrofolate reductase
dTMP	:	deoxythymidylate
MMA	:	methylmalonic acid;
5-CH3-THF	:	5-methyltetrahydrofolate
MTHFR	:	methylene tetrahydrofolate reductase
MTR	:	5-methyltetrahydrofolate-homocysteine methyltransferase
NHANES	:	National Health and Nutrition Examination Survey
NTD	:	neural tube defect
SAH	:	S-adenosylhomocysteine
SAM	:	S-adenosylmethionine
tHcy	:	plasma total homocysteine
THF	:	tetrahydrofolate
UMFA	:	unmetabolised folic acid

References

Allen RH, Stabler SP, Savage DG and Lindenbaum J. 1993. Metabolic abnormalities in cobalamin (vitamin B12) and folate deficiency. Faseb J. 7: 1344–53.

Bailey LB, Stover PJ, McNulty H, Fenech MF, Gregory III JF, Mills JL, Pfeiffer CM, Fazili Z, Zhang M, Ueland PM, Molloy AM, Caudill MA, Shane B, Berry RJ, Bailey Rl, Hausman DB, Raghavan R and Raiten D. 2015. Biomarkers of nutrition for development—folate review. J Nutr. 145: 1636S–80S.

Bailey SW and Ayling JE. 2009. The extremely slow and variable activity of dihydrofolate reductase in human liver and its implications for high folic acid intake. Proc Natl Acad Sci U S A. 106: 15424–9.

Battaglia-Hsu S, Akchiche N, Noel N, Alberto J-M, Jeannesson E, Orozco-Barrios CE, Martinez-Fong D, Daval J-L and Guéant J-L. 2009. Vitamin B12 deficiency reduces proliferation and promotes differentiation of neuroblastoma cells and up-regulates PP2A, proNGF, and TACE. Proc Natl Acad Sci U S A. 106: 21930–5.

Bentley TGK, Willett WC, Weinstein MC and Kuntz KM. 2006. Population-level changes in folate intake by age, gender, and race/ethnicity after folic acid fortification. Am J Public Health. 96: 2040–7.

Berry RJ, Carter HK and Yang Q. 2007. Cognitive impairment in older Americans in the age of folic acid fortification. Am J Clin Nutr. 86: 265–7; author reply 267–9.

Boilson A, Staines A, Kelleher CC, Daly L, Shirley I, Shrivastava A, Bailey SW, Alverson PB, Ayling JE, Parle-McDermott A, MacCooey A, Scott JM and Sweeney MR. 2012. Unmetabolized folic acid prevalence is widespread in the older Irish population despite the lack of a mandatory fortification program. Am J Clin Nutr. 96: 613–21.

Brito A, Verdugo R, Hertrampf E, Miller JW, Green R, Fedosov SN, Shahab-Ferdows S, Sanchez H, Albala C, Castillo JL, Matamala JM, Uauy R and Allen LH. 2016. Vitamin B12 treatment of asymptomatic, deficient, elderly Chileans improves conductivity in myelinated peripheral nerves, but high serum folate impairs vitamin B12 status response assessed by the combined indicator of vitamin B12 status. Am J Clin Nutr. 103: 250–7.

Bueno O, Molloy AM, Fernandez-Ballart JD, García- Minguillán CJ, Ceruelo S, Ríos L, Ueland PM, Meyer K and Murphy MM. 2016. Common polymorphisms that affect folate transport or metabolism modify the effect of the MTHFR 677C > T polymorphism on folate status. J Nutr. 146: 1–8.

Carmel R. 2009. Does high folic acid intake affect unrecognized cobalamin deficiency, and how will we know it if we see it? Am J Clin Nutr. 90: 1449–50.

Cherian A, Seena S, Bullock RK and Antony AC. 2005. Incidence of neural tube defects in the least-developed area of India: a population-based study. Lancet (London, England). 366: 930–1.

Chodos R and Ross J. 1951. The effects of combined folic acid and liver extract therapy. Blood. 6: 1213–33.

Christensen KE, Mikael LG, Leung K-Y, Lévesque N, Deng L, Wu Q, Malysheva OV, Best A, Caudill MA, Greene NDE and Rozen R. 2015. High folic acid consumption leads to pseudo-MTHFR deficiency, altered lipid metabolism, and liver injury in mice. Am J Clin Nutr. 101: 646–58.

Clarke R, Sherliker P, Hin H, Molloy AM, Nexo E, Ueland PM, Emmens K, Scott JM and Evans JG. 2008. Folate and vitamin B12 status in relation to cognitive impairment and anaemia in the setting of voluntary fortification in the UK. Br J Nutr. 100: 1054–9.

Cook JD, Cichowicz DJ, George S, Lawler A and Shane B. 1987. Mammalian folylpolyl-gamma-glutamate-synthetase. 4. *In vitro* and *in vivo* metabolism of folates and analogues and regulation of folate homeostasis. Biochemistry. 26: 530–9.

Crider KS, Devine O, Hao L, Dowling NF, Li S, Molloy AM, Li Z, Zhu J and Berry RJ. 2014. Population red blood cell folate concentrations for prevention of neural tube defects: Bayesian model. BMJ. 349: g4554.

Cuskelly GC, McNulty H and Scott JM. 1996. Effects of increasing dietary folate on red cell faolte. Implications for the prevention of neural tube defects. Lancet. 347: 657–9.

Daly LE, Kirke PN, Molloy A, Weir DG and Scott JM. 1995. Folate levels and neural tube defects. Implications for prevention. JAMA. 274: 1698–702.

Deacon R, Lumb M, Perry J, Chanarin I, Minty B, Halsey MJ and Nunn JF. 1978. Selective inactivation of vitamin B12 in rats by nitrous oxide. Lancet (London, England). 2: 1023–4.

Deshmukh US, Joglekar CV, Lubree HG, Ramdas LV, Bhat DS, Naik SS, Hardikar PS, Raut DA, Konde TB, Wills AK, Jackson AA, Refsum H, Nanivadekar AS, Fall CH and Yajnik CS. 2010. Effect of physiological doses of oral vitamin B12 on plasma homocysteine: a randomized, placebo-controlled, double-blind trial in India. Eur J Clin Nutr. 64: 495–502.

Doets EL, Ueland PM, Tell GS, Vollset SE, Nygård OK, Van't Veer P, de Groot LCPGM, Nurk E,

Dolnick BJ and Cheng YC. 1978. Human thymidylate synthetase. II. Derivatives of pteroylmono- and -polyglutamates as substrates and inhibitors. J Biol Chem. 253: 3563–7.

Dominguez-Salas P, Moore SE, Cole D, da Costa KA, Cox SE, Dyer RA, Fulford AJ, Innis SM, Waterland RA, Zeisel SH, Prentice AM and Hennig BJ. 2013. DNA methylation potential: dietary intake and blood concentrations of one-carbon metabolites and cofactors in rural African women. Am J Clin Nutr. 97: 1217–27.

Fernandez-Roig S, Cavallé-Busquets P, Fernandez-Ballart JD, Ballesteros M, Berrocal Zaragoza MI, Salat-Batlle J and Murphy MM. 2013. Low folate status enhances pregnancy changes in plasma betaine and dimethylglycine concentrations and the association between betaine and homocysteine. Am J Clin Nutr. 97: 1252–9.

Flour Fortification Initiative. 2015. Flour Fortification Initiative home page. Available from: http://www.ffinetwork.org/about/stay_informed/releases/2014 Review.html.

Food and Drug Administration. 1996. Food standards: amendment of standards of identity for enriched grain products to require addition of folic acid.

Food Safety Authority of Ireland. 2008. Report of the National Committee on Folic Acid Food Fortification.

Frenkel EP. 1973. Abnormal fatty acid metabolism in peripheral nerves of patiens with pernicious anemia. J Clin Invest. 52: 1237–45.

Fridman S. 2005. High folic acid intake is not a risk factor for cognitive decline: misinterpretation of results. Arch Neurol. 62: 1785–6; author reply 1786.

García-Minguillán CJ, Fernandez-Ballart JD, Ceruelo S, Ríos L, Bueno O, Berrocal-Zaragoza MI, Molloy AM, Ueland PM, Meyer K and Murphy MM. 2014. Riboflavin status modifies the effects of methylenetetrahydrofolate reductase (MTHFR) and methionine synthase reductase (MTRR) polymorphisms on homocysteine. Genes Nutr. 9: 435.

Hannon-Fletcher MP, Armstrong NP, Scott JM, Pentieva K, Bradbury I, Ward M, Strain JJ, Dunn AA, Molloy AM, Kerr MA and McNulty H. 2004. Determining bioavailability of food folates in a controlled intervention study. Am J Clin Nutr. 80: 911–8.

Healton EB, Savage DG, Brust JC, Garrett TJ and Lindenbaum J. 1991. Neurologic aspects of vitamin B12 deficiency. Medicine (Baltimore). 70: 229–45.

Hemmer B, Glocker FX, Schumacher M, Deuschl G and Lücking CH. 1998. Subacute combined degeneration: clinical, electrophysiological, and magnetic resonance imaging findings. J Neurol Neurosurg Psychiatry. 65: 822–7.

Herbert V and Zalusky R. 1962. Interrelations of vitamin B12 and folic acid metabolism: folic acid clearance studies. J Clin Invest. 41: 1263–76.

Herrmann W, Obeid R, Schorr H, Hubner U, Geisel J, Sand-Hill M, Ali N and Herrmann M. 2009. Enhanced bone metabolism in vegetarians—the role of vitamin B12 deficiency. Clin Chem Lab Med. 47: 1381–1387.

Hertrampf E, Cortés F, Erickson JD, Cayazzo M, Freire W, Bailey LB, Howson C, Kauwell GPA and Pfeiffer C. 2003. Consumption of folic acid-fortified bread improves folate status in women of reproductive age in Chile. J Nutr. 133: 3166–9.

Hertrampf E and Cortés F. 2004. Folic acid fortification of wheat flour: Chile. Nutr Rev. 62: S44–8; discussion S49.

Hoey L, McNulty H, Askin N, Dunne A, Ward M, Pentieva K, Strain J, Molloy AM, Flynn CA and Scott JM. 2007. Effect of a voluntary food fortification policy on folate, related B vitamin status, and homocysteine in healthy adults. Am J Clin Nutr. 86: 1405–13.

Holm PI, Bleie Ø, Ueland PM, Lien EA, Refsum H, Nordrehaug JE and Nygård O. 2004. Betaine as a determinant of postmethionine load total plasma homocysteine before and after B-vitamin supplementation. Arterioscler Thromb Vasc Biol. 24: 301–7.

Hopkins SM, Gibney MJ, Nugent AP, McNulty H, Molloy AM, Scott JM, Flynn A, Strain JJ, Ward M, Walton J and McNulty BA. 2015. Impact of voluntary fortification and supplement use on dietary intakes and biomarker status of folate and vitamin B12 in Irish adults. Am J Clin Nutr. 101: 1163–72.

Horne DW, Cook RJ and Wagner C. 1989. Effect of dietary methyl group deficiency on folate metabolism in rats. J Nutr. 19: 618–21.

Houghton LA, Yang J and O'Connor DL. 2009. Unmetabolized folic acid and folate concentrations in breast milk are unaffected by low-dose folate supplements. Am J Clin Nutr. 89: 216–20.

Jacques PF, Bostom AG, Wilson PW, Rich S, Rosenberg IH and Selhub J. 2001. Determinants of plasma total homocysteine concentration in the Framingham Offspring cohort. Am J Clin Nutr. 73: 613–21.

Jaikrishan G, Sudheer KR, Andrews VJ, Koya PKM, Madhusoodhanan M, Jagadeesan CK and Seshadri M. 2013. Study of stillbirth and major congenital anomaly among newborns in the high-level natural radiation areas of Kerala, India. J Community Genet. 4: 21–31.

Kelly P, McPartlin J, Goggins M, Weir DG and Scott JM. 1997. Unmetabolized folic acid in serum: acute studies in subjects consuming fortified food and supplements. Am J Clin Nutr. 65: 1790–5.

Kim YI, Miller JW, da Costa KA, Nadeau M, Smith D, Selhub J, Zeisel SH and Mason JB. 1994. Severe folate deficiency causes secondary depletion of choline and phosphocholine in rat liver. J Nutr. 124: 2197–203.

Liaugaudas G, Jacques PF, Selhub J, Rosenberg IH and Bostom AG. 2001. Renal insufficiency, Vitamin B12 status, and population attributable risk for mild hyperhomocysteinemia among coronary artery disease patients in the era of folic acid-fortified cereal grain flour. Arterioscler Thromb Vasc Biol. 21: 849–851.

Marchetta CM, Devine OJ, Crider KS, Tsang BL, Cordero AM, Qi YP, Guo J, Berry RJ, Rosenthal J, Mulinare J, Mersereau P and Hamner HC. 2015. Assessing the association between natural food folate intake and blood folate concentrations: a systematic review and Bayesian meta-analysis of trials and observational studies. Nutrients. 7: 2663–86.

McGregor DO, Dellow WJ, Robson RA, Lever M, George PM and Chambers ST. 2002. Betaine supplementation decreases post-methionine hyperhomocysteinemia in chronic renal failure. Kidney Int. 61: 1040–6.

Miller JW, Garrod MG, Allen LH, Haan MN and Green R. 2009. Metabolic evidence of vitamin B12 deficiency, including high homocysteine and methylmalonic acid and low holotransvitamin B12, is more pronounced in older adults with elevated plasma folate. Am J Clin Nutr. 90: 1586–92.

Mills JL, Carter TC, Scott JM, Troendle JF, Gibney ER, Shane B, Kirke PN, Ueland PM, Brody LC and Molloy AM. 2011. Do high blood folate concentrations exacerbate metabolic abnormalities in people with low vitamin B12 status? Am J Clin Nutr. 94: 495–500.

Molloy AM, Kirke PN, Troendle JF, Burke H, Sutton M, Brody LC, Scott JM and Mills JL. 2009. Maternal vitamin B12 status and risk of neural tube defects in a population with high neural tube defect prevalence and no folic Acid fortification. Pediatrics. 123: 917–23.

Morris MC, Evans DA, Bienias JL, Tangney CC, Hebert LE, Scherr PA and Schneider JA. 2005. Dietary folate and vitamin B12 intake and cognitive decline among community-dwelling older persons. Arch Neurol. 62: 641–5.

Morris MS, Jacques PF, Rosenberg IH and Selhub J. 2007. Folate and vitamin B12 status in relation to anemia, macrocytosis, and cognitive impairment in older Americans in the age of folic acid fortification. Am J Clin Nutr. 85: 193–200.

Morris MS, Jacques PF, Rosenberg IH and Selhub J. 2010. Circulating unmetabolized folic acid and 5-methyltetrahydrofolate in relation to anemia, macrocytosis, and cognitive test performance in American seniors. Am J Clin Nutr. 91: 1733–44.

Morris MS, Selhub J and Jacques PF. 2012. Vitamin B12 and folate status in relation to decline in scores on the mini-mental state examination in the framingham heart study. J Am Geriatr Soc. 60: 1457–64.

Obeid R, Kasoha M, Kirsch SH, Munz W and Herrmann W. 2010. Concentrations of unmetabolized folic acid and primary folate forms in pregnant women at delivery and in umbilical cord blood. Am J Clin Nutr. 92: 1416–22.

Obeid R, Kirsch SH, Kasoha M, Eckert R and Herrmann W. 2011. Concentrations of unmetabolized folic acid and primary folate forms in plasma after folic acid treatment in older adults. Metabolism. 60: 673–80.

Obeid R, Kirsch SH, Dilmann S, Klein C, Eckert R, Geisel J and Herrmann W. 2016. Folic acid causes higher prevalence of detectable unmetabolized folic acid in serum than B-complex: a randomized trial. Eur J Nutr. 55: 1021–1028.

Olteanu H, Munson T and Banerjee R. 2002. Differences in the efficiency of reductive activation of methionine synthase and exogenous electron acceptors between the common polymorphic variants of human methionine synthase reductase. Biochem. 41: 13378–13385.

Papathakis PC and Pearson KE. 2012. Food fortification improves the intake of all fortified nutrients, but fails to meet the estimated dietary requirements for vitamins A and B6, riboflavin and zinc, in lactating South African women. Public Health Nutr. 15: 1810–7.

Pentieva K, Selhub J, Paul L, Molloy AM, McNulty B, Ward M, Marshall B, Dornan J, Reilly R, Parle-McDermott A, Bradbury I, Ozaki M, Scott JM and McNulty H. 2016. Evidence from a randomized trial that exposure to supplemental folic acid at recommended levels during pregnancy does not lead increased unmetabolized folic acid concentrations in maternal or cord blood. J Nutr. 146: 494–500.

Pfeiffer CM, Osterloh JD, Kennedy-Stephenson J, Picciano MF, Yetley EA, Rader JI and Johnson CL. 2008. Trends in circulating concentrations of total homocysteine among US adolescents and adults: findings from the 1991–1994 and 1999–2004 National Health and Nutrition Examination Surveys. Clin Chem. 54: 801–13.

Pfeiffer CM, Hughes JP, Lacher DA, Bailey RL, Berry RJ, Zhang M, Yetley EA, Rader JI, Sempos CT and Johnson CL. 2012. Estimation of trends in serum and RBC folate in the U.S. population from pre- to postfortification using assay-adjusted data from the NHANES 1988–2010. J Nutr. 142: 886–93.

Pfeiffer CM, Sternberg MR, Fazili Z, Lacher DA and Johnson CL. 2015. Unmetabolized folic acid is detected in nearly all serum samples from US children, adolescents and adults. J Nutr. 145: 520–31.

Qi YP, Do AN, Hamner HC, Pfeiffer CM and Berry RJ. 2014. The prevalence of low serum vitamin B12 status in the absence of anemia or macrocytosis did not increase among older U.S. adults after mandatory folic acid fortification. J Nutr. 144: 170–6.

Quinlivan EP. 2008. In vitamin B12 deficiency, higher serum folate is associated with increased homocysteine and methylmalonic acid concentrations. Proc Natl Acad Sci U S A. 105: E7; author reply E8.

Quinlivan EP, McPartlin J, McNulty H, Ward M, Strain JJ, Weir DG and Scott JM. 2002. Importance of both folic acid and vitamin B12 in reduction of risk of vascular disease. Lancet (London, England). 359: 227–8.

Ray JG. 2004. Folic acid food fortification in Canada. Nutr Rev. 62: S35–9.

Refsum H, Yajnik CS, Gadkari M, Schneede J, Vollset SE, Orning L, Guttormsen AB, Joglekar A, Sayyad MG, Ulvik A and Ueland PM. 2001. Hyperhomocysteinemia and elevated methylmalonic acid indicate a high prevalence of vitamin B12 deficiency in Asian Indians. Am J Clin Nutr. 74: 233–41.

Refsum H, Smith AD and Eussen SJPM. 2014. Interactions between plasma concentrations of folate and markers of vitamin B(12) status with cognitive performance in elderly people not exposed to folic acid fortification: the Hordaland Health Study. Br J Nutr. 111: 1085–95.

Reynolds E. 2006. Vitamin B12, folic acid, and the nervous system. Lancet Neurol. 5: 949–960.

Samaniego-Vaesken ML, Alonso-Aperte E and Varela-Moreiras G. 2016. Voluntary fortification with folic acid in Spain: An updated food composition database. Food Chem. 193: 148–53.

Savage DG, Lindenbaum J, Stabler SP and Allen RH. 1994. Sensitivity of serum methylmalonic acid and total homocysteine determinations for diagnosing vitamin B12 and folate deficiencies. Am J Med. 96: 239–46.

Savage DG and Lindenbaum J. 1995. Folate-cobalamin interactions. pp. 237–85. In: LB Bailey (ed.). Folate in Health and Disease. 1st ed. Marcel Dekker Inc NY.

Savage DG and Lindenbaum J. 1995. Neurological complications of acquired vitamin B12 deficiency: clinical aspects. Baillieres Clin Haematol. 8: 657–78.

Scott JM, Dinn JJ, Wilson P and Weir DG. 1981. Pathogenesis of subacute combined degeneration: a result of methyl group deficiency. Lancet (London, England). 2: 334–7.

Scott JM and Weir DG. 1981. The methyl folate trap. A physiological response in man to prevent methyl group deficiency in kwashiorkor (methionine deficiency) and an explanation for folic-acid induced exacerbation of subacute combined degeneration in pernicious anaemia. Lancet. 2: 337–40.

Selhub J, Morris MS and Jacques PF. 2007. In vitamin B12 deficiency, higher serum folate is associated with increased total homocysteine and methylmalonic acid concentrations. Proc Natl Acad Sci U S A. 104: 19995–20000.

Shorvon SD, Carney MW, Chanarin I and Reynolds EH. 1980. The neuropsychiatry of megaloblastic anaemia. Br Med J. 281: 1036–8.

Smith AD, Kim Y-I and Refsum H. 2008. Is folic acid good for everyone? Am J Clin Nutr. 87: 517–33.

Smulders YM, Smith DEC, Kok RM, Teerlink T, Swinkels DW, Stehouwer CDA and Jakobs C. 2006. Cellular folate vitamer distribution during and after correction of vitamin B12 deficiency: a case for the methylfolate trap. Br J Haematol. 132: 623–9.

Stokstad EL, Reisenauer A, Kusano G and Keating JN. 1998. Effect of high levels of dietary folic acid on folate metabolism in vitamin B12 deficiency. Arch Biochem Biophys. 265: 407–14.

Tighe P, Ward M, McNulty H, Finnegan O, Dunne A, Strain J, Molloy AM, Duffy M, Pentieva K and Scott JM. 2011. A dose-finding trial of the effect of long-term folic acid intervention: implications for food fortification policy. Am J Clin Nutr. 93: 11–8.

Tinker SC, Hamner HC, Qi YP and Crider KS. 2015. U.S. women of childbearing age who are at possible increased risk of a neural tube defect-affected pregnancy due to suboptimal red blood cell folate concentrations, National Health and Nutrition Examination Survey 2007 to 2012. Birth Defects Res A Clin Mol Teratol. 103: 517–26.

US Food and Drug Administration, Food standards: amendment of standards of identity for enriched grain products to require addition of folic acid, Fed. Regist. 61(1996) 8781–8797.

UK Standing Advisory Committee on Nutrition. 2006. Folate and Disease Prevention Report. London.

World Health Organization. 2015. Optimal serum and red blood cell folate concentrations in women of reproductive age for prevention of neural tube defects.

Wyckoff KF and Ganji V. 2007. Proportion of individuals with low serum vitamin B12 concentrations without macrocytosis is higher in the post folic acid fortification period than in the pre folic acid fortification period. Am J Clin Nutr. 86: 1187–92.

Yajnik CS, Deshpande SS, Lubree HG, Naik SS, Bhat DS, Uradey BS, Deshpande JA, Rege SS, Refsum H and Yudkin JS. 2006. Vitamin B12 deficiency and hyperhomocysteinemia in rural and urban Indians. J Assoc Physicians India. 54: 775–82.

Zeisel SH, Zola T, daCosta KA and Pomfret EA. 1989. Effect of choline deficiency on S-adenosylmethionine and methionine concentrations in rat liver. Biochem J. 259: 725–729.

13

Extreme Vitamin B12 Concentrations in Clinical Practice in the Absence of Symptoms or B12 Treatment

Rima Obeid

1. Introduction

Despite its low sensitivity and specificity, serum vitamin B12 test is still used as a first line marker to diagnose vitamin B12 deficiency. The improvement in medical care and the attempt to focus on prevention, drive physicians to order vitamin B12 serum test even without clear clinical indications. Vitamin B12 test has been even viewed as a "wellness marker" in some countries. The number of vitamin B12 tests ordered between 2005 and 2012 has shown a dramatic increase (2 fold) in a French hospital (Arlet et al. 2015). The same tendency was observed in Denmark between 2001 and 2013 (Arendt et al. 2015). The clinical conditions that lead physicians to order B12 test are very heterogeneous with a remarkable bias towards over-prescription among the most ill patient's groups (as a last option to help the patient) or the most healthy ones (as a wellness marker). Some hospitals have even included vitamin B12 test as part of a regular 'routine profile' in some patient groups (i.e., dialysis units). Nevertheless, independent on the presence or the absence of clinical symptoms, high test results are generally interpreted as 'non-deficient' and low results are interpreted as 'deficient'.

Aarhus Institute of Advanced Studies, University of Aarhus, Høegh-Guldbergs Gade 6B, building 1632, Dk-8000, Aarhus C, Denmark.
Email: rima.obeid@uks.eu

Vitamin B12 blood test is widely prescribed among hospital residents, patients on haemo- or peritoneal dialyses, and outpatients. The traditional 'indications' for vitamin B12 level assessment are anaemia, unexplained neurological disorders, bariatric surgery, and all malabsorption disorders. A recent study has shown that the most common indicators of B12 status assessment were anemia, cognitive impairment, and undernutrition accounting for 63%, 20%, and 17% of the total tests, respectively (Chiche et al. 2013). All traditional indications together (i.e., macrocytic non-regenerative anemia, isolated macrocytosis, dementia and proprioceptive disorders) accounted for only 34% of all tests ordered (Chiche et al. 2013). The majority of vitamin B12 tests in real health care settings could be theoretically over-prescribed, not-justified by the clinical condition, or performed in the wrong target group.

Surprising enough, high vitamin B12 concentrations can be more frequent than low ones in samples sent to clinical labs for vitamin B12 testing. In the absence of clinical symptoms, vitamin B12 levels below the detection limit are not rare in the lab. However, while low vitamin B12 levels are seriously taken, high serum levels receive less attention from physicians. Extremely high or low serum vitamin B12 levels can be difficult to interpret when clinical symptoms of deficiency are absent or they do not fit in the classical picture of deficiency. On the other hand, serum concentrations of vitamin B12 may disagree with that of methylmalonic acid (i.e., a metabolic marker of B12) and thus, a serum B12 test result alone is not always considered indicative of vitamin B12 status.

This chapter discusses the occurrence and interpretation of very low serum vitamin B12 in the absence of clinical signs of deficiency. The significance and potential clinical meaning of very high serum vitamin B12 in the absence of vitamin B12 treatment are also discussed.

2. Very Low Serum Vitamin B12

Serum concentrations of vitamin B12 below the detection limit of common assay methods (approximately 70 pmol/L) have been reported in many cases in the absence of clinical symptoms of deficiency or elevated concentrations of the metabolic markers, methylmalonic acid and homocysteine. Congenital R-binder (haptocorrin) deficiency is a condition that is associated with very low serum vitamin B12 without symptoms (Hall and Begley 1977). The first patients with very low vitamin B12 were described by Carmel et al. in 1969 (Carmel and Herbert 1969). Later, it was found that R-binder was not detectable in saliva and blood of these patients (Carmel 1982). The main fraction of vitamin B12 is bound to haptocorrin under normal physiological conditions. Only 10–30% of serum vitamin B12 is attached to transcobalamin. The vitamin B12-fraction that is attached to transcobalamin is considered biologically active, since it can be internalized through transcobalamin receptor (CD320). In patients with deficiency of R-binder, the endogenous vitamin B12, as fractionated on Sephadex G-200 gel, was mainly carried by transcobalamin, instead of haptocorrin.

Unlike transcobalamin deficiency, deficiency of R-binder is considered a benign condition. Cells can maintain their normal functions when a sufficient amount of holotranscobalamin (i.e., cobalamin bound to transcobalamin) is available. In subjects with R-binder deficiency, low B12 is not associated with elevated plasma levels of methylmalonic acid or homocysteine.

Most individuals having this condition are misdiagnosed as having pernicious anaemia because of their low serum total vitamin B12 level (Carmel 1982). R-binder deficiency does not require treatment with vitamin B12.

3. High Serum Concentrations of Vitamin B12

In hospital settings, elevated concentrations of serum vitamin B12 are more common than low ones. The definition of high serum B12 is not fixed yet, but various publications referred to B12 > 600 pmol/L or > 800 pmol/L as being 'high'. Elevated serum concentrations of vitamin B12 (> 600 pmol/L) occur in about 8–24% of all samples analyzed for vitamin B12 status in the absence of vitamin supplementation (Carmel et al. 2001). In a population cohort of over 25000 patients with cancer and with B12 measured (between 2001–2013), 14% of the patients had serum vitamin B12 > 600 pmol/L (Arendt et al. 2015).

Several studies reported about clinical conditions or laboratory markers that predict high serum vitamin B12 among patients delivered to internal medicine departments or those suspected to be deficient (Table 1). High B12 levels are commonly associated with renal failure (Carmel et al. 2001). Patients with malignancies or liver diseases also show high concentrations of vitamin B12. The reasons behind elevated vitamin B12 in blood of patients with different clinical conditions are not clear. Several mechanisms have been suggested (Table 2). However, it must be noted that high serum B12 could also be related to the presence of immune complexes with IgM and IgG or to analytical failure due to forming immune complexes or interfering with B12 assay (Bowen et al. 2006; Jeffery et al. 2010).

High vitamin B12 levels have currently no health significance for the patient, but careful interpretation of the results by physicians is needed. This

Table 1. Clinical and biochemical conditions associated with high serum/or plasma vitamin B12.

Clinical conditions that predict high serum vitamin B12 (Brah et al. 2014; Carmel et al. 2001; Halsted et al. 1959)
Chronic kidney failure
Liver diseases
Malignancies of different types
Chronic myelogenous myeloma
Alcoholism
Haematopathy
Diabetes and its complications (i.e., retinopathy, renal insufficiency)
Laboratory results associated with high serum vitamin B12 (Brah et al. 2014)
Hepatic abnormalities (elevated liver enzymes)
Elevated creatinine or reduced eGFR, albuminurea

Table 2. Mechanisms speculated to cause or explain high serum B12.

1. Increased synthesis rate of vitamin B12 binding proteins (haptocorrin and transcobalamin) that can capture more vitamin.
2. Increased saturation of haptocorrin or transcobalamin with vitamin B12.
3. Release of vitamin B12 from body stores or necrotic cells following organ damage (i.e., liver damage).
4. Disturbed renal function can lead to disturbed vitamin B12 homeostasis, altered renal filtration or re-uptake of the vitamin, elevated levels of transcobalamin and/or haptocorrin or their saturation, or change the distribution of the vitamin between its two plasma binding proteins.

chapter will focus on the three most common conditions associated with high serum vitamin B12:

- Diabetes and renal insufficiency;
- Liver disorders;
- Different malignancies.

4. Serum Vitamin B12 in Patients with Diabetes and Renal Diseases

Concentrations of vitamin B12 and holoTC are generally higher in patients with diabetes and renal failure compared with healthy subjects. The prevalence of high vitamin B12 is more common than that of low vitamin B12 in this risk population.

In 1954, Chow et al. reported elevated serum levels of vitamin B12 in patients with diabetes and retinopathy (Chow et al. 1954). Later studies showed contradicted results. Boger et al. reported no association between high vitamin B12 and diabetes (Boger et al. 1957). Whereas, Halsted et al. (Halsted et al. 1959) and Beckett et al. attributed high B12 to renal diseases (Beckett and Matthews 1962). The association between diabetes and high vitamin B12 has been attributed to the presence of renal insufficiency that is common in patients with diabetes (Beckett and Matthews 1962). This was supported by data showing that the clearance of a single dose of radiolabeled vitamin B12 was delayed in patients with renal insufficiency (Beckett and Matthews 1962).

In a study based on hospital archive, Carmel et al. identified renal failure as a main clinical condition associated with high serum vitamin B12 (Carmel et al. 2001). Low serum albumin and elevated serum creatinine were two main significant laboratory markers associated with elevated serum vitamin B12, suggesting renal dysfunction as a main confounding factor that interfere with vitamin B12 homeostasis and interpretation of serum B12. However, low serum albumin or albuminuria does not completely explain high vitamin B12 in patients with chronic kidney disease.

The association between serum vitamin B12, serum homocysteine and albuminuria has been recently investigated in over 4000 samples in a cross sectional and follow up study designs (McMahon et al. 2015). In the cross-sectional study, higher vitamin B12 levels were associated with the presence of albuminuria [Odds Ratio, 95% Confidence Intervals (OR, 95% CI) = 1.57, 1.10–2.26]. The association was stronger in participants with plasma homocysteine concentrations above 9.1 µmol/L (McMahon et al. 2015). There was an association between serum vitamin B12 and reduced renal function (eGFR < 60 ml/min/1.73 m²) in subjects with homocysteine ≥ 9.1 µmol/L (OR = 2.17, 95% CI: 1.44–3.26), but not in those with homocysteine levels below this cut-off (OR = 1.22, 95% CI: 0.62–2.41). Follow up studies showed no associations between serum vitamin B12 and incident albuminuria or reduced kidney function (eGFR < 60 ml/min/1.73 m²), suggesting that elevated vitamin B12 levels cannot predict future decline of renal function. Thus, an elevated level of vitamin B12 is not likely to be a risk factor for future chronic kidney disease (McMahon et al. 2015). Instead, vitamin B12 metabolism is likely to be altered in subjects with lowered eGFR (McMahon et al. 2015). This is in agreement with a higher percentage of holoTC to total vitamin B12 in patients with high creatinine as compared with those with low creatinine (24.4% vs 15.5%) (Table 3).

In addition to folate, vitamin B12 is a significant determinant of serum homocysteine levels that are elevated in renal patients. Elevated serum homocysteine is an independent risk factor for vascular diseases. Indeed, high serum vitamin B12 is not consistent with the metabolic markers, homocysteine and methylmalonic acid, both of which show a marked elevation in renal failure.

Increasing the lower cut-off value for normal serum vitamin B12 in chronic kidney disease is not useful in detecting patients with possible deficiency, because B12 levels are influenced by B12 status and the degree of renal dysfunction thus leading to a large overlap in serum B12 between

Table 3. Serum vitamin B12 markers in relation to renal function.

	Quintiles of serum creatinine					
	Q1 (lowest)	Q2	Q3	Q4	Q5 (highest)	p
Creatinine, mg/dL	0.54 (0.11)	0.76 (0.05)	0.90 (0.01)	1.07 (0.08)	5.00 (3.91)	-
Age, years	52 (18)	55 (17)	57 (17)	62 (16)	67 (15)	< 0.001
Vitamin B12, pmol/L	471 (816)	343 (545)	374 (679)	340 (279)	419 (794)	0.027*
HoloTC, pmol/L	58 (82)	59 (69)	60 (101)	66 (83)	87 (115)	0.001*
MMA, nmol/L	229 (159)	332 (1104)	246 (159)	309 (332)	807 (765)	< 0.001*
tHcy, µmol/L	11.6 (4.2)	13.3 (8.8)	14.3 (7.5)	13.4 (4.4)	20.7 (6.0)	0.001*
(holoTC x 100)/B12	15.5 (8.9)	18.6 (12.2)	18.9 (18.3)	19.3 (11.2)	24.4 (22.3)	< 0.001*

Data are mean (standard deviation). * p adjusted for age using General Linear Model univariate analysis of variance applied on the log-transformed data (age entered as covariate). The study included 1358 samples that were sent to the lab for B12 testing. The original study has been published in (Herrmann and Obeid 2013). tHcy, total homocysteine; MMA, methylmalonic acid; holoTC, holotranscobalamin.

individuals with and those without a deficiency (i.e., low test specificity). It is also not expected that monitoring holotranscobalamin (holoTC), instead of total vitamin B12, can be more useful. Also serum holoTC concentrations show dependency on renal function (positive association with creatinine) (Table 3, and Figures 1 and 2). Serum holoTC correlates with serum creatinine at higher creatinine concentrations (Figure 1). The distribution of holoTC is shifted towards higher values in patients with elevated creatinine (> vs ≤ 97.2 µmol/L) (Figure 2). The mean difference (95% CI) of holoTC between the two creatinine ranges was 21 (10–32) pmol/L; $p < 0.001$ (Figure 2). Therefore, holoTC has a limited value as a vitamin B12 marker in patients with renal diseases.

5. Mechanisms of Hyper-Vitaminemia in Diabetes and Renal Diseases

Mechanisms responsible for altered vitamin B12 blood markers in diabetes and renal diseases are not well investigated. Possible mechanisms could be impaired B12 metabolism, reduced renal filtration, or altered homeostasis.

Cubulin has been suggested as a key determinant of serum vitamin B12 levels in renal diseases. This protein is responsible for uptake of vitamin B12 that is filtered in urine. The same protein is also responsible for reabsorption of albumin and other substrates. Genetic defects in CUBN cause vitamin B12 and

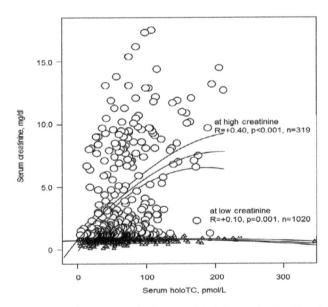

Figure 1. Worsening renal function (i.e., higher creatinine) is associated with a shift in holoTC distribution towards higher values. In the high range of serum creatinine, serum holoTC and creatinine show a moderate positive correlation (Spearman correlation coefficient R = 0.40; p < 0.001). The correlation between holoTC and creatinine was weak at low creatinine level (R = 0.10; p = 0.001). Creatinine ranges were > or ≤ 97.2 µmol/L (Obeid R unpublished data).

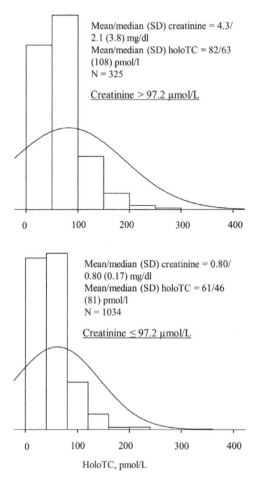

Mean/median (SD) creatinine = 4.3/
2.1 (3.8) mg/dl
Mean/median (SD) holoTC = 82/63
(108) pmol/l
N = 325

Creatinine > 97.2 μmol/L

Mean/median (SD) creatinine = 0.80/
0.80 (0.17) mg/dl
Mean/median (SD) holoTC = 61/46
(81) pmol/l
N = 1034

Creatinine ≤ 97.2 μmol/L

HoloTC, pmol/L

Figure 2. The distribution of serum holoTC levels in random samples sent to the lab for evaluation of vitamin B12 status. The distribution is shown according to serum creatinine levels (> or ≤ 97.2 μmol/L). The mean difference (95% confidence intervals) of holoTC between the creatinine ranges was 21 (10–32) pmol/L; $p < 0.001$; 1 mg/dL creatinine = 88.4 μmol/L.

albumin loss in urine and megaloblastic anaemia (Aminoff et al. 1999; Hauck et al. 2008). Also common SNPs in CUBN show association with albuminuria (Boger et al. 2011).

The kidney plays a significant role in transcobalamin-mediated re-uptake of vitamin B12 in the proximal tubule back to the circulation. Defects in filtration of transcobalamin-bound vitamin B12 to the glomerulus and the increase in the amount of vitamin B12 bound to serum transcobalamin and haptocorrin could explain high serum vitamin B12. Elevated serum concentrations of vitamin B12-binding proteins have been reported in patients with chronic kidney diseases and could probably explain that vitamin B12 is captured in blood (Carmel et al. 2001).

Elevated vitamin B12 concentrations do not fit, however, into increased metabolic markers of B12 (methylmalonic acid and homocysteine). *In-vitro* studies have shown less uptake of vitamin B12 into peripheral blood cells isolated from pre-dialyses patients compared with control subjects (Obeid et al. 2005a). A reduced cellular uptake could indicate low amount of vitamin B12 reaching the intracellular compartments which can lead to accumulation of homocysteine and methylmalonic acid despite high extracellular (i.e., serum) vitamin B12.

We investigated serum concentrations of vitamin B12 markers in relation to renal function (Table 3). The concentrations of vitamin B12, holoTC, methylmalonic acid, and homocysteine show positive associations with serum creatinine. Renal insufficiency was associated with higher total amount of vitamin B12 in blood, altered distribution of vitamin B12 between transcobalamin and haptocorrin (i.e., holoTC/total B12 ratio), and altered intracellular vitamin B12 metabolism as indicated by increased metabolic markers. The positive association between serum creatinine and the percentage of holoTC/to total B12 (Table 3), suggests that more B12 is attached to transcobalamin in people with high creatinine than in people with low creatinine.

Concentrations of holoTC in urine tended to be lower in patients with diabetes compared with those free of the disease ($p = 0.09$). The ratio of holoTC/urinary albumin was lower in patients with diabetes compared with that in the controls ($p = 0.048$) (Figure 3. Obeid et al. unpublished data). Urinary holoTC showed no significant correlation with any vitamin B12 marker in blood (total vitamin B12, holoTC, methylmalonic acid, or homocysteine) (Obeid et al. unpublished data). Thus, less amount of transcobalamin-B12 appears to be filtered into the urine of patients with diabetes compared with the controls, though this difference is unlikely to be the sole explanation for the high serum holoTC or total vitamin B12 in blood.

Taken together, the complex mechanisms behind elevated vitamin B12 in diabetes and renal insufficiency are poorly understood. A reduced ability to filter vitamin B12 from the blood, altered distribution of vitamin B12 between its main transporters, transcobalamin and haptocorrin or a combination of these mechanisms could be involved. It remains unclear whether transcobalamin receptor, CD320, or its soluble form, sCD320, are differently regulated or have different affinities to vitamin B12 in diabetes and renal dysfunction. It is also unclear whether diabetes per see or its later complications (renal, vascular) are responsible for high serum vitamin B12.

6. Impact of High Vitamin B12 in Diabetes and Renal Diseases

High serum vitamin B12 for reasons not related to vitamin B12 supplementation has implications for diagnosis and treatment of patients. The aim of measuring blood vitamin B12 markers is to diagnose a possible deficiency. Keeping this in mind, this aim cannot be reached in most of renal patients who might be

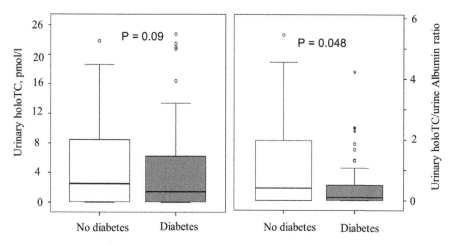

Figure 3. Urinary excretion of holoTC in random urine collected from 81 patients with type 2 diabetes and 50 controls. Urinary holoTC concentrations were measured using the same automated assay used for serum holoTC (AxSYM® Active-B12, Abbott) (Obeid et al. unpublished data).

deficient. The real challenge following a high normal/or high serum vitamin B12 result is to exclude or confirm vitamin B12 deficiency (Herrmann et al. 2005; Hyndman et al. 2003; Rasmussen et al. 1990) (see also Chapter 9). There is currently no general agreement on a strategy to diagnose vitamin B12 deficiency in these cases. In patients with renal dysfunction or diabetes-related renal insufficiency, we strongly recommended measurement of methylmalonic acid and interpreting vitamin B12 markers in the light of renal function.

Concentrations of methylmalonic acid are elevated in the majority of this risk population, but it is not possible to judge how much of this elevation is explained by renal dysfunction and how much is explained by vitamin B12 deficiency. Treatment with pharmacological doses of vitamin B12 (cyanocobalamin >0.5 mg) can help judging whether methylmalonic acid elevation is reversible (or partly reversible). A significant decrease in serum methylmalonic acid (by 150–250 nmol/L) indicates a pre-treatment deficiency. Normalisation of methylmalonic acid after B12 treatment is unlikely in patients with renal insufficiency. Moelby et al. (Moelby et al. 2000), suggested an approximate threshold of 800 nmol/L for methylmalonic acid that can be reached in end stage renal diseases after treatment with B12. Our studies on patients with end stage renal disease confirm this threshold (Obeid et al. 2005b).

Another unsettled issue is whether vitamin B12 treatment should be started in patients with renal dysfunction and elevated methylmalonic acid despite high serum vitamin B12. High serum vitamin B12 has also implications for prevention of vitamin B12 deficiency. Treatment with vitamin B12 is known to lower serum methylmalonic acid and homocysteine in patients with renal failure in the absence of low serum vitamin B12 or holoTC at baseline (Elian and Hoffer 2002; Hyndman et al. 2003; Obeid et al. 2005b). The reduction of methylmalonic acid after cyanocobalamin treatment was stronger in patients with lower holoTC at baseline, despite that none of the patients had holoTC below 35 pmol/L (the lower cut-off of the reference range) (Figure 4).

A concern has been raised by a study using B-vitamins including cyanocobalamin in patients with diabetic nephropathy where the vitamin arm showed a decline in renal function after 36 months of intervention compared with the placebo (House et al. 2010). High serum vitamin B12 was not confirmed as a predictor of future decline in renal function in a recent follow up study (McMahon et al. 2015). There is currently no evidence that vitamin B12 treatment is causally related to the prognoses of renal dysfunction. Comparisons between the effect of cyano- and methylcobalamin are not available yet, but needed before a conclusion can be made about a causal link between cyanocobalamin form and declining renal function.

7. High Serum Vitamin B12 in Patients with Liver Diseases

The liver is the principal storage organ of vitamin B12 and the main site of B12-dependent metabolism. Early studies found higher serum vitamin B12 in patients with certain liver diseases compared to control subjects (Aronovitch et al. 1956; Halsted et al. 1959; Rachmilewitz et al. 1958; Rachmilewitz et al.

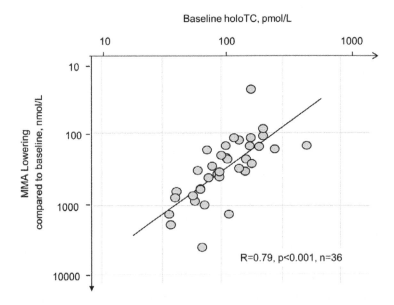

Figure 4. The correlation between baseline holoTC levels and the change of methylmalonic acid after B12 treatment in 36 patients with end stage renal disease who had hyperhomocysteinemia at baseline (homocysteine > 18 µmol/L). All patients had holoTC > 50 nmol/L at baseline. The y-axis represents baseline MMA minus MMA after treatment. The treatment was with 0.7 mg cyanocobalamin/i.v. 3 times per week for 4 weeks. MMA, methylmalonic acid. The original study was published in (Obeid et al. 2005b).

1959). Liver cell necrosis is associated with high serum B12. Liver damage can be related to any reason such as virus infection, hepatic cirrhosis, drugs, toxins or cancer (Fremont et al. 1991). Hepatic necrosis due to gall-stone obstruction of the bile duct has been also associated with elevated serum vitamin B12 (Mackay et al. 1957). Whereas conditions associated with mild to moderate secondary liver diseases such as extrahepatic biliary obstruction are associated with normal B12 levels (Holdsworth et al. 1964; Stevenson and Beard 1959). Patients with hepatitis show high serum B12 during the active phase of the disease, but B12 levels decline upon recovery (Stevenson and Beard 1959).

Alcoholic liver disease is a well described clinical condition associated with high serum vitamin B12 (Baker et al. 1987; Baker et al. 1998; Medici et al. 2010). In hospitalised patients, concentrations of the liver enzymes (ALT, AST and γ-GT) were positively related to serum vitamin B12 (Brah et al. 2014). Hyperhomocysteinemia shows a dose-response relationship with liver cell damage. Higher homocysteine levels are associated with severe liver cirrhosis as measured by the Child–Pugh score (Ferre et al. 2002; Garcia-Tevijano et al. 2001). Mean plasma homocysteine values increased from non-alcoholic cirrhosis (10.3 µmol/L) to abstaining cirrhosis (12.5 µmol/L, 21% higher than non-alcoholic) and to non-abstaining cirrhosis (14.9 µmol/L, 45% higher than non-alcoholic) (Ferre et al. 2002). Polyzos et al. reported that levels of homocysteine may not be elevated in patients with non-alcoholic steatohepatitis compared with patients with non-alcoholic fatty liver (Polyzos et al. 2012). A U-shape relationship between plasma homocysteine and liver steatosis or fibrosis was reported and lower homocysteine in liver patients with inflammation (Polyzos et al. 2012).

In a series of studies, Stevenson et al. observed elevated serum concentrations of vitamin B12 that were reasonably correlated to the severity of liver diseases (Stevenson and Beard 1959). Improvement of liver disease was associated with a decline in serum B12 upon serial testing (Stevenson and Beard 1959). Cobalt[60]-labelled B12 was used to study the absorption and elimination of the vitamin from the plasma in patients with liver diseases (Stevenson and Beard 1959). The studies have shown normal absorption of radioactive vitamin B12 in patients with various liver disorders. Also the urinary (in 24 hours urine) and fecal excretions of radioactive vitamin B12 were within the reference range in the majority of the patients. Less than 20% of the radioactive cobalamin remained in plasma after 24 hours (Stevenson and Beard 1959), which is in contrast to results in patients with leukemia, where higher plasma binding capacity and prolonged clearance time of the radioactive B12 dose were observed (Mollin et al. 1956).

The mechanisms of elevated serum vitamin B12 in liver diseases are not well studied, but they could be different in different liver pathologies. In general, high vitamin B12 could be related to a reduced uptake or storage or to the release of vitamin B12 from the damaged liver tissues. A lower content of vitamin B12 has been reported in liver biopsies from patients with liver cirrhosis compared with those without liver cirrhosis (mean = 0.263 vs 0.614

μg/mg wet weight) (Halsted et al. 1959). This may reflect the inability of diseased liver cells to store vitamin B12 that is captured through binding to the vitamin B12-dependent enzymes, methionine synthase and methylmalonyl-CoA mutase. High serum vitamin B12 may be related to releasing the vitamin from the damaged hepatocytes (Halsted et al. 1959; Stevenson and Beard 1959). Accordingly, patients recovering from liver damage show a decline in serum vitamin B12.

Concentrations of vitamin B12 binding proteins (haptocorrin, and transcobalamin) were significantly higher in patients with liver diseases (cancer or chronic liver diseases) compared with the controls (Simonsen et al. 2014). The same was found for serum holoTC, percentage of transcobalamin saturation, sCD320 (soluble transcobalamin receptor), and total vitamin B12 (all higher in liver patients independent on the cause) (Simonsen et al. 2014). These observations confirm earlier ones, though they do not explain whether elevated vitamin B12 or its binding proteins are related to their release from the liver, reduced uptake, or increased production.

Elevated serum levels of vitamin B12 in patients with liver disorders challenge the diagnosis of vitamin B12 deficiency. There is currently no consensus on diagnosis or treatment strategy. Hyperhomocysteinemia is common in liver disorders and can be prevented by B-vitamin supplementation. Steatohepatitis (induced by alcohol and high fat diet) can be attenuated when feeding a methyl donor–enriched diet containing folic acid, and vitamin B12 (Powell et al. 2010). Using B-vitamins to reduce alcohol-induced liver injury and generally enhance liver function has been addressed by several studies (Halsted and Medici 2012; Portugal et al. 1995; Portugal et al. 1996).

8. High Serum Vitamin B12 in Patients with Cancer

Elevated serum (or plasma) concentrations of vitamin B12 disagree with elevated concentrations of homocysteine and methylmalonic acid in patients with cancer and can thus challenge the diagnosis of deficiency conditions in this risk group.

Elevated serum concentrations of vitamin B12 are common in myeloproliferative diseases such as in chronic myeloid leukemia (Carmel and Hollander 1978; Mendelsohn et al. 1958), as well as in hepatocellular carcinoma, multiple myeloma and Waldenstrom macroglobulinema and inflammatory diseases. By using radioactive vitamin B12, it has been shown that elevated serum vitamin B12 in patients with chronic myelocytic leukemia is associated with increased vitamin B12-plasma binding capacity and extended retention of B12 in plasma (Mendelsohn et al. 1958; Miller 1958). Increased unsaturated vitamin B12-binding capacity in patients with cancer has been recognized in the late 1950th. Further studies demonstrated that in particular, patients with myelocytic leukemia show a delay in the disappearance of labelled cobalamin (Co[58]B12) from the blood after oral or intravenous administration (Corbus et al. 1957; Mollin et al. 1956; Weinstein and Watkin 1960). Studies using oral labelled cobalamin (doses 1–500 μg) did not show an upregulation of intestinal absorption of the vitamin in patients with myelocytic leukemia as compared with other non-neoplastic diseases or with patients with solid tumors (Heinrich and Erdmann-Oehlecker 1956; Weinstein and Watkin 1960). Despite technical limitations, these early studies suggested that high serum vitamin B12 in patients with myelocytic leukemia was not explained by increased absorption or decreased urinary excretion of the vitamin, but was likely to be related to increased B12-binding proteins and a shift of vitamin B12 from tissue sites to plasma (Weinstein and Watkin 1960). Several studies reported an elevation of a specific B12-binding glycoprotein in those patients (Mendelsohn et al. 1958; Weinstein and Watkin 1960), but the identity of this protein was not known at that time.

The highest transcobalamin values are found in patients with lymphoma involving the liver, those with chronic lymphocytic leukemia and infection, myeloma, cirrhosis, lupus erythematosus and myelofibrosis, and esophageal cancer and hypercalcemia (Carmel and Hollander 1978), suggesting that transcobalamin elevation may resemble an acute phase response to cancer infection and inflammation. A recent study in patients with liver cancer who were followed for up to 12 weeks during the course of treatment has shown that only holoTC increased in the first week after starting the treatment (Simonsen et al. 2014). Other vitamin B12 binders showed slight, but insignificant changes that could be related to the small sample size or the heterogeneity of the treatments. Long term changes of vitamin B12 and its binders in patients with cancer are currently not known.

Carmel et al. recognized that concentrations of serum total vitamin B12 and haptocorrin were elevated in approximately 6% and lowered in approximately 10% of patients with cancer and metastatic disease (Carmel and Eisenberg 1977). Elevated vitamin B12 was explained by increased serum haptocorrin. Importantly, vitamin B12 deficiency was common among patients with cancer (Carmel and Eisenberg 1977), though this phenomenon receives less attention in seriously ill patients.

Tumor tissues are thought to store vitamin B12-binding enzymes and are speculated to produce more vitamin B12 binding proteins (Burger et al. 1975; Nexo et al. 1975). Further explanations discussed to cause high total vitamin B12 and haptocorrin are a reduced elimination of haptocorrin from the circulation (Burger et al. 1975), but this hypothesis has not been verified. Haptocorrin elevation appears to be sensitive for liver cancer. An elevated serum vitamin B12 level in a patient with cancer was associated with a poor prognosis (Arendt et al. 2015; Carmel and Eisenberg 1977). The median survival for patients with cancer and high vitamin B12 was 1 month compared with 4 months in other patients with cancer, but with normal serum vitamin B12 (Carmel and Eisenberg 1977).

Elevated concentrations of serum vitamin B12 were reported in a significant numbers of patients with different types of solid tumors or haematological malignancies (6.6% with serum vitamin B12 > 800 pmol/L) (Arendt et al. 2015).

In a health register-based cohort study, the association between increased serum vitamin B12 (> 800 pmol/L) and cancer mortality was significant and not explained by cancer type, age, or co-morbidities. The study showed that the number of the vitamin B12 tests that were ordered in cancer patients was increased by > 200% in the years between 2001 to 2013, but the percentage of patient with increased vitamin B12 remain rather stable (Arendt et al. 2015) (Figures 5 and 6). As mentioned before in this chapter, vitamin B12 test has been considered as a wellness marker and has been over-prescribed for seriously ill patients, even without clear clinical indications. Physicians tend to order vitamin B12 test more frequently in more seriously ill patients (selection bias), but they rarely follow high vitamin B12 test results with more specific markers, such as methylmalonic acid.

The association between high concentrations of serum B12 and advanced cancer is not sensitive or specific. Independent on the type of morbidities, high serum vitamin B12 predicts short term mortality in elderly or hospital residents (see below) and thus it cannot be considered as a markers with prognostic or diagnostic value for a specific illness (Simonsen et al. 2014). Keeping this mind, high vitamin B12 is likely to be a sign of cell apoptosis or necrosis, for whatever reason.

On the other hand, high serum B12 can lead to underestimating vitamin B12 deficiency in patients with cancer. Normal or high B12 levels can be associated with elevated homocysteine and methylmalonic acid (Chang et al.

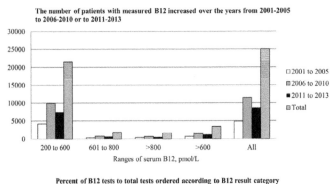

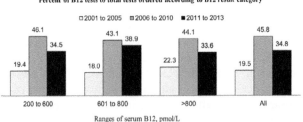

Figure 5. The increase in the number (upper panel) and percentage (lower panel) of patients with measured serum vitamin B12 over the years from 2001–2005 to 2006–2010 or to 2011–2013. Data are from 25.017 patients with cancer and measured B12 (Arendt et al. 2015).

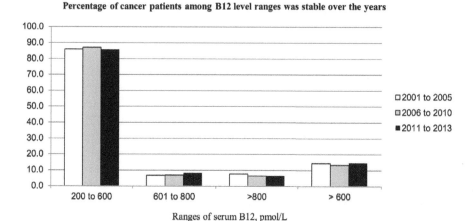

Figure 6. Out of all 25.017 cancer patients studied, 6.6% had vitamin B12 levels above 800 pmol/L and 86.2% had levels in the reference range (200–600 pmol/L) (Arendt et al. 2015).

2015; Cinemre et al. 2015; Gauchan et al. 2012; Vashi et al. 2016). Intracellular vitamin B12 deficiency in seriously ill patients or elderly people is associated with neuropathy, dementia, depression and several other common neurological co-morbidities.

9. Serum Vitamin B12 and the Risk of Cancer

In 2005, a nested case control study in Swedish patients with prostate cancer was published where blood samples were obtained before diagnosis of cases (median, 4.9 years; 25–75th percentile, 2.9–7.0) (Hultdin et al. 2005). Cases with prostate cancer showed higher plasma vitamin B12 at baseline (mean = 329 vs 300 pmol/L), and lower homocysteine (12.2 vs 13.2 µmol/L) compared with the controls. The risk of cancer was higher in subjects with plasma vitamin B12 > 370 pmol/L compared with those < 238 pmol/L. The association between prostate cancer and high vitamin B12 or low homocysteine was significant in older patients (age ≥ 59 year) (Hultdin et al. 2005). These results were confirmed by later studies (Collin et al. 2010; Price et al. 2016). However, supplementation studies did not confirm an increased risk of any cancer after relatively high doses of B-vitamins (containing cyanocobalamin, folic acid and B6) (Vollset et al. 2013).

Animal foods (i.e., red meat) are rich sources of vitamin B12. Therefore, serum levels of vitamin B12 could also reflect certain dietary patterns, instead of a primary difference in B12 intakes or statuses. Higher serum vitamin B12 was associated with the risk of prostate cancer occurring after a shorter follow up period of < 5 years [1.19 (1.02–1.38)], but not that occurring after a longer follow-up (> 5 years) (Price et al. 2016). This observation could be related to the well described elevation in serum B12 in patients with established cancer or probably existing, but undiagnosed tumor in some follow up studies.

It remains unsettled whether patients with cancer should be treated with vitamin B12 if they have elevated serum vitamin B12, but also elevated serum methylmalonic acid levels.

10. Serum Vitamin B12 and General Mortality

Studies investigating baseline vitamin B12 levels and 10-year follow up outcomes have shown no association between high vitamin B12 and future mortality in the general population (Waters et al. 1971). In contrast, low vitamin B12 was associated with a higher mortality which was explained by the prevalence of low vitamin B12 in elderly people (Waters et al. 1971).

Elevated serum vitamin B12 has been discussed as a predictor for short term mortality irrespective of the cause of death. However, a causal role for high B12 in future mortality cannot be assumed. A serum vitamin B12 level above 400 pmol/L in hospitalised elderly patients was predictive of higher risk of mortality within 90 days, though all patients were free of cancer (Salles et al. 2008). A prospective study on 161 patients with different cancers investigated serum vitamin B12 concentrations and the time of death (Geissbuhler et al. 2000). The global median survival time was 45 days (CI 95%: 32–56 days). The highest mortality corresponded to the highest vitamin B12 levels. A significant link was found between elevated vitamin B12 (> 600 pmol/l) and the presence of metastasis, a tumor or liver problems (Geissbuhler et al. 2000).

Therefore, serum vitamin B12 has been suggested to have a predictive value for mortality in general. However, careful interpretation of the results is necessary to avoid wrong conclusions on causality. Confounding by indication is very likely in observational studies on measuring B12 in hospital or health register setting. Because the allocation for vitamin B12 test in observational studies is not randomized and the indication for its measurement can be related to the risk of future health outcomes, the imbalance in the risk profile between patients with measured B12 and those without measured B12 can generate biased results. It is well known that physicians over-prescribing vitamin B12 tests in seriously ill patients. Data on vitamin B12 markers in asymptomatic people and future disease outcome can be more informative and less affected by selection bias.

11. Extreme Vitamin B12 Concentrations in Blood: Gaps in Knowledge

The issue of high serum vitamin B12 level constitutes a serious challenge for setting a reliable diagnosis and making a meaningful clinical decision. Even when combining vitamin B12 test with other metabolic markers, it remains open whether vitamin B12 treatment should be recommended to lower methylmalonic acid. Table 4 provides insights into some open questions related to this research area.

Table 4. Summary of open questions related to extreme serum vitamin B12 concentrations.

Low serum B12 without deficiency symptoms or elevated serum functional markers

- What is the biological function of haptocorrin that binds ~70% of serum cobalamin?
- Levels and biological variations of transcobalamin and its saturation in patients with haptocorrin deficiency are not known.

High serum vitamin B12

- What are the mechanisms behind elevated serum B12 in different diseases?
- Are there changes in the B12-binders or their saturation with the vitamin in different conditions (i.e., renal diseases, alcohol liver disorders)?
- Is there a change in the affinity of B12 to its binding proteins, haptocorrin and transcobalamin, at higher serum B12?
- What is the source of this B12 (i.e., increased absorption, tissue release, less filtration in the kidney, etc.)?
- Why are serum vitamin B12 and holoTC increased in renal patients?
- Is there any diagnostic or prognostic value for serum B12 or its binders in different conditions?
- Can vitamin B12 treatment promote existing or pre-clinical cancer?
- Should regular vitamin B12 supplementation be recommended for patients with high vitamin B12 levels, if serum homocysteine or methylmalonic acid are elevated?
- What criteria should be used for diagnosing B12 deficiency in patients with cancer?

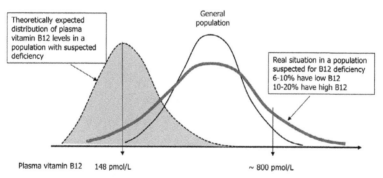

Figure 7. The distribution of vitamin B12 concentrations in the general population is expected to take a Gaussian form. Cobalamin test intends at detecting 'deficient cases'. In a group of subjects suspected for deficiency, the distribution of vitamin B12 is in theory expected to be shifted to the left of the population distribution (towards lower B12 levels). However, in practice, we observe a flattened distribution in a population suspected for deficiency. In a group with suspected vitamin B12 deficiency in a hospital setting, the percentage of subjects with elevated vitamin B12 (> 800 pmol/L) may exceed the number of those with low B12. This is related to selection bias, or the decision of the health care providers to measure B12 in more seriously ill patients who are likely to have conditions associated with high B12.

Summary and Conclusions

Serum or plasma vitamin B12 test is used as a tool to diagnose deficiency of the vitamin. Therefore, low concentrations of this marker are commonly interpreted as deficiency and values above the low cut-off are classically considered as "normal". However, blood samples with extreme values (very low, or high) are common, despite that low values are more expected when samples are sent to the lab with the aim of exploring a 'deficiency' condition (Figure 7).

Very low concentrations of vitamin B12 in blood may occur in asymptomatic people without any clinical relevance, if they have sufficient vitamin carried by transcobalamin (i.e., in case of haptocorrin deficiency). On the other hand, subjects who are not treated with vitamin B12 can have very high serum vitamin B12, but also metabolic signs that suggest a deficiency condition. Interpretation of these extreme, and often contradicted, results and judgment of the need for testing further blood markers or starting supplementation are challenging questions in clinical practice.

Currently, extremely high serum vitamin B12 values in the absence of treatment with cyanocobalamin and extremely low levels in the absence of deficiency symptoms have no direct implications for diagnosing certain clinical conditions in the patient. However, physicians should be aware that extreme serum vitamin B12 can be associated with established or yet undiagnosed clinical conditions.

Keywords: Hypervitaminemia, very low vitamin B12, diagnostic value, liver, cancer, renal function, high B12, confounding by indication

Abbreviation

HoloTC	:	holotranscobalamin
MMA	:	methylmalonic acid
tHcy	:	total homocysteine

References

Aminoff M, Carter JE, Chadwick RB, Johnson C, Grasbeck R, Abdelaal MA, Broch H, Jenner LB, Verroust PJ, Moestrup SK, de la CA and Krahe R. 1999. Mutations in CUBN, encoding the intrinsic factor-vitamin B12 receptor, cubilin, cause hereditary megaloblastic anaemia 1. Nat Genet. 21: 309–313.

Arendt JF, Farkas DK, Pedersen L, Nexo E and Sorensen HT. 2015. Elevated plasma vitamin B12 levels and cancer prognosis: A population-based cohort study. Cancer Epidemiol. 40: 158–165.

Arlet JB, Rachas A, Colombet I, Pouchot J and Chiche L. 2015. [Is cobalamin measurement overprescribed by physicians? Results of an eight-year single academic centre survey]. Rev Med Interne. 36: 495–497.

Aronovitch J, Grossowicz N and Rachmilewitz M. 1956. Serum concentrations of vitamin B12 in acute and chronic liver disease. J Lab Clin Med. 48: 339–344.

Baker H, Frank O and DeAngelis B. 1987. Plasma vitamin B12 titres as indicators of disease severity and mortality of patients with alcoholic hepatitis. Alcohol Alcohol. 22: 1–5.

Baker H, Leevy CB, DeAngelis B, Frank O and Baker ER. 1998. Cobalamin (vitamin B12) and holotranscobalamin changes in plasma and liver tissue in alcoholics with liver disease. J Am Coll Nutr. 17: 235–238.

Beckett AG and Matthews DM. 1962. Vitamin B12 in diabetes mellitus. Clin Sci. 23: 361–370.

Boger CA, Chen MH, Tin A, Olden M, Kottgen A, de Boer IH, Fuchsberger C, O'Seaghdha CM, Pattaro C, Teumer A, Liu CT, Glazer NL, Li M, O'Connell JR, Tanaka T, Peralta CA, Kutalik Z, Luan J, Zhao JH, Hwang SJ, Akylbekova E, Kramer H, van der Harst P, Smith AV, Lohman

K, de AM, Hayward C, Kollerits B, Tonjes A, Aspelund T, Ingelsson E, Eiriksdottir G, Launer LJ, Harris TB, Shuldiner AR, Mitchell BD, Arking DE, Franceschini N, Boerwinkle E, Egan J, Hernandez D, Reilly M, Townsend RR, Lumley T, Siscovick DS, Psaty BM, Kestenbaum B, Haritunians T, Bergmann S, Vollenweider P, Waeber G, Mooser V, Waterworth D, Johnson AD, Florez JC, Meigs JB, Lu X, Turner ST, Atkinson EJ, Leak TS, Aasarod K, Skorpen F, Syvanen AC, Illig T, Baumert J, Koenig W, Kramer BK, Devuyst O, Mychaleckyj JC, Minelli C, Bakker SJ, Kedenko L, Paulweber B, Coassin S, Endlich K, Kroemer HK, Biffar R, Stracke S, Volzke H, Stumvoll M, Magi R, Campbell H, Vitart V, Hastie ND, Gudnason V, Kardia SL, Liu Y, Polasek O, Curhan G, Kronenberg F, Prokopenko I, Rudan I, Arnlov J, Hallan S, Navis G, Parsa A, Ferrucci L, Coresh J, Shlipak MG, Bull SB, Paterson NJ, Wichmann HE, Wareham NJ, Loos RJ, Rotter JI, Pramstaller PP, Cupples LA, Beckmann JS, Yang Q, Heid IM, Rettig R, Dreisbach AW, Bochud M, Fox CS and Kao WH. 2011. CUBN is a gene locus for albuminuria. J Am Soc Nephrol. 22: 555–570.

Boger WP, Strickland SC, Wright LD and Ciminera JL. 1957. Diabetes mellitus and serum vitamin B12 concentrations: 333 patients. Proc Soc Exp Biol Med. 96: 316–319.

Bowen RA, Drake SK, Vanjani R, Huey ED, Grafman J and Horne MK III. 2006. Markedly increased vitamin B12 concentrations attributable to IgG-IgM-vitamin B12 immune complexes. Clin Chem. 52: 2107–2114.

Brah S, Chiche L, Mancini J, Meunier B and Arlet JB. 2014. Characteristics of patients admitted to internal medicine departments with high serum cobalamin levels: results from a prospective cohort study. Eur J Intern Med. 25: e57–e58.

Burger RL, Waxman S, Gilbert HS, Mehlman CS and Allen RH. 1975. Isolation and characterization of a novel vitamin B12-binding protein associated with hepatocellular carcinoma. J Clin Invest. 56: 1262–1270.

Carmel R. 1982. A new case of deficiency of the R binder for cobalamin, with observations on minor cobalamin-binding proteins in serum and saliva. Blood. 59: 152–156.

Carmel R and Eisenberg L. 1977. Serum vitamin B12 and transcobalamin abnormalities in patients with cancer. Cancer. 40: 1348–1353.

Carmel R and Herbert V. 1969. Deficiency of vitamin B12-binding alpha globulin in two brothers. Blood. 33: 1–12.

Carmel R and Hollander D. 1978. Extreme elevation of transcobalamin II levels in multiple myeloma and other disorders. Blood. 51: 1057–1063.

Carmel R, Vasireddy H, Aurangzeb I and George K. 2001. High serum cobalamin levels in the clinical setting—clinical associations and holo-transcobalamin changes. Clin Lab Haematol. 23: 365–371.

Chang SC, Goldstein BY, Mu L, Cai L, You NC, He N, Ding BG, Zhao JK, Yu SZ, Heber D, Zhang ZF and Lu QY. 2015. Plasma folate, vitamin B12, and homocysteine and cancers of the esophagus, stomach, and liver in a Chinese population. Nutr Cancer. 67: 212–223.

Chiche L, Mancini J and Arlet JB. 2013. Indications for cobalamin level assessment in departments of internal medicine: a prospective practice survey. Postgrad Med J. 89: 560–565.

Chow BF, Rosen DA and Lang CA. 1954. Vitamin B12 serum levels and diabetic retinopathy. Proc Soc Exp Biol Med. 87: 38–39.

Cinemre H, Serinkan Cinemre BF, Cekdemir D, Aydemir B, Tamer A and Yazar H. 2015. Diagnosis of vitamin B12 deficiency in patients with myeloproliferative disorders. J Investig Med. 63: 636–640.

Collin SM, Metcalfe C, Refsum H, Lewis SJ, Zuccolo L, Smith GD, Chen L, Harris R, Davis M, Marsden G, Johnston C, Lane JA, Ebbing M, Bonaa KH, Nygard O, Ueland PM, Grau MV, Baron JA, Donovan JL, Neal DE, Hamdy FC, Smith AD and Martin RM. 2010. Circulating folate, vitamin B12, homocysteine, vitamin B12 transport proteins, and risk of prostate cancer: a case-control study, systematic review, and meta-analysis. Cancer Epidemiol Biomarkers Prev. 19: 1632–1642.

Corbus HF, Miller A and Sullivan JF. 1957. The plasma disappearance, excretion, and tissue distribution of cobalt 60 labelled vitamin B12 in normal subjects and patients with chronic myelogenous leukemia. J Clin Invest. 36: 18–24.

Elian KM and Hoffer LJ. 2002. Hydroxocobalamin reduces hyperhomocysteinemia in end-stage renal disease. Metabolism. 51: 881–886.

Ferre N, Gomez F, Camps J, Simo JM, Murphy MM, Fernandez-Ballart J and Joven J. 2002. Plasma homocysteine concentrations in patients with liver cirrhosis. Clin Chem. 48: 183–185.

Fremont S, Champigneulle B, Gerard P, Felden F, Lambert D, Gueant JL and Nicolas JP. 1991. Blood transcobalamin levels in malignant hepatoma. Tumour Biol. 12: 353–359.

Garcia-Tevijano ER, Berasain C, Rodriguez JA, Corrales FJ, Arias R, Martin-Duce A, Caballeria J, Mato JM and Avila MA. 2001. Hyperhomocysteinemia in liver cirrhosis: mechanisms and role in vascular and hepatic fibrosis. Hypertension. 38: 1217–1221.

Gauchan D, Joshi N, Gill AS, Patel V, Debari VA, Guron G and Maroules M. 2012. Does an elevated serum vitamin B(12) level mask actual vitamin B(12) deficiency in myeloproliferative disorders? Clin Lymphoma Myeloma Leuk. 12: 269–273.

Geissbuhler P, Mermillod B and Rapin CH. 2000. Elevated serum vitamin B12 levels associated with CRP as a predictive factor of mortality in palliative care cancer patients: a prospective study over five years. J Pain Symptom Manage. 20: 93–103.

Hall CA and Begley JA. 1977. Congenital deficiency of human R-type binding proteins of cobalamin. Am J Hum Genet. 29: 619–626.

Halsted CH and Medici V. 2012. Aberrant hepatic methionine metabolism and gene methylation in the pathogenesis and treatment of alcoholic steatohepatitis. Int J Hepatol. 2012: 959746.

Halsted JA, Carroll J and Rubert S. 1959. Serum and tissue concentration of vitamin B12 in certain pathologic states. N Engl J Med. 260: 575–580.

Hauck FH, Tanner SM, Henker J and Laass MW. 2008. Imerslund-Grasbeck syndrome in a 15-year-old German girl caused by compound heterozygous mutations in CUBN. Eur J Pediatr. 167: 671–675.

Heinrich HC and Erdmann-Oehlecker S. 1956. Der Vitamin B12-Stoffwechsel bei Haemolastosen. III. Resorption, blutverteilung, serumproteinbindung, etention und Exkretion der B12-Vitamine bei Haemoblastosen nach oraler und parenteraler B12-Applikation. In: p 326.

Herrmann W and Obeid R. 2013. Utility and limitations of biochemical markers of vitamin B12 deficiency. Eur J Clin Invest. 43: 231–237.

Herrmann W, Obeid R, Schorr H and Geisel J. 2005. The usefulness of holotranscobalamin in predicting vitamin B12 status in different clinical settings. Curr Drug Metab. 6: 47–53.

Holdsworth CD, Atkinson M, Dossett JA and Hall R. 1964. An assessment of the diagnostic and prognostic value of serum vitamin B12 levels in liver disease. Gut. 5: 601–606.

House AA, Eliasziw M, Cattran DC, Churchill DN, Oliver MJ, Fine A, Dresser GK and Spence JD. 2010. Effect of B-vitamin therapy on progression of diabetic nephropathy: a randomized controlled trial. JAMA. 303: 1603–1609.

Hultdin J, Van GB, Bergh A, Hallmans G and Stattin P. 2005. Plasma folate, vitamin B12, and homocysteine and prostate cancer risk: a prospective study. Int J Cancer. 113: 819–824.

Hyndman ME, Manns BJ, Snyder FF, Bridge PJ, Scott-Douglas NW, Fung E and Parsons HG. 2003. Vitamin B12 decreases, but does not normalize, homocysteine and methylmalonic acid in end-stage renal disease: a link with glycine metabolism and possible explanation of hyperhomocysteinemia in end-stage renal disease. Metabolism. 52: 168–172.

Jeffery J, Millar H, Mackenzie P, Fahie-Wilson M, Hamilton M and Ayling RM. 2010. An IgG complexed form of vitamin B12 is a common cause of elevated serum concentrations. Clin Biochem. 43: 82–88.

Mackay IR, Cowling DC and Gray A. 1957. Highly raised serum vitamin B12 levels in obstructive hepatic necrosis. Br Med J. 2: 800–801.

McMahon GM, Hwang SJ, Tanner RM, Jacques PF, Selhub J, Muntner P and Fox CS. 2015. The association between vitamin B12, albuminuria and reduced kidney function: an observational cohort study. BMC Nephrol. 16: 7.

Medici V, Peerson JM, Stabler SP, French SW, Gregory JF, III, Virata MC, Albanese A, Bowlus CL, Devaraj S, Panacek EA, Rahim N, Richards JR, Rossaro L and Halsted CH. 2010. Impaired homocysteine transsulfuration is an indicator of alcoholic liver disease. J Hepatol. 53: 551–557.

Mendelsohn RS, Watkin DM, Horbett AP and Fahey JL. 1958. Identification of the vitamin B12-binding protein in the serum of normals and of patients with chronic myelocytic leukemia. Blood. 13: 740–747.

Miller A. 1958. The *in vitro* binding of cobalt 60 labeled vitamin B12 by normal and leukemic sera. J Clin Invest. 37: 556–566.

Moelby L, Rasmussen K, Ring T and Nielsen G. 2000. Relationship between methylmalonic acid and cobalamin in uremia. Kidney Int. 57: 265–273.

Mollin DL, PITNEY WR, Baker SJ and BRADLEY JE. 1956. The plasma clearance and urinary excretion of parenterally administered 58Co B12. Blood. 11: 31–43.

Nexo E, Olesen H, Norredam K and Schwartz M. 1975. A rare case of megaloblastic anaemia caused by disturbances in the plasma cobalamin binding proteins in a patient with hepatocellular carcinoma. Scand J Haematol. 14: 320–327.

Obeid R, Kuhlmann M, Kirsch CM and Herrmann W. 2005a. Cellular uptake of vitamin B12 in patients with chronic renal failure. Nephron Clin Pract. 99: c42–c48.

Obeid R, Kuhlmann MK, Kohler H and Herrmann W. 2005b. Response of homocysteine, cystathionine, and methylmalonic acid to vitamin treatment in dialysis patients. Clin Chem. 51: 196–201.

Polyzos SA, Kountouras J, Patsiaoura K, Katsiki E, Zafeiriadou E, Deretzi G, Zavos C, Gavalas E, Katsinelos P, Mane V and Slavakis A. 2012. Serum homocysteine levels in patients with nonalcoholic fatty liver disease. Ann Hepatol. 11: 68–76.

Portugal V, Garcia-Alonso I, Barcelo P and Mendez J. 1995. Effect of allopurinol, folinic acid, SOD and cyclosporine A on ischemic liver regeneration. Eur Surg Res. 27: 69–76.

Portugal V, Garcia-Alonso I and Mendez J. 1996. Hepatotrophic effect of folinic acid in rats. J Surg Res. 61: 527–530.

Powell CL, Bradford BU, Craig CP, Tsuchiya M, Uehara T, O'Connell TM, Pogribny IP, Melnyk S, Koop DR, Bleyle L, Threadgill DW and Rusyn I. 2010. Mechanism for prevention of alcohol-induced liver injury by dietary methyl donors. Toxicol Sci. 115: 131–139.

Price AJ, Travis RC, Appleby PN, Albanes D, Barricarte GA, Bjorge T, Bueno-de-Mesquita HB, Chen C, Donovan J, Gislefoss R, Goodman G, Gunter M, Hamdy FC, Johansson M, King IB, Kuhn T, Mannisto S, Martin RM, Meyer K, Neal DE, Neuhouser ML, Nygard O, Stattin P, Tell GS, Trichopoulou A, Tumino R, Ueland PM, Ulvik A, de VS, Vollset SE, Weinstein SJ, Key TJ and Allen NE. 2016. Circulating Folate and Vitamin B and Risk of Prostate Cancer: A Collaborative Analysis of Individual Participant Data from Six Cohorts Including 6875 Cases and 8104 Controls. Eur Urol. (doi:10.1016/j.eururo.2016.03.029).

Rachmilewitz M, Stein Y, Aronovitch J and Grossowicz N. 1959. Serum cyanocobalamin (vitamin B12) as an index of hepatic damage in chronic congestive heart failure. Arch Intern Med. 104: 406–410.

Rachmilewitz M, Stein Y, Aronovitch J and Grossowicz N. 1958. The clinical significance of serum cyanocobalamin (vitamin B12) in liver disease. AMA Arch Intern Med. 101: 1118–1125.

Rasmussen K, Vyberg B, Pedersen KO and Brochner-Mortensen J. 1990. Methylmalonic acid in renal insufficiency: evidence of accumulation and implications for diagnosis of cobalamin deficiency. Clin Chem. 36: 1523–1524.

Salles N, Herrmann F, Sieber C and Rapin C. 2008. High vitamin B12 level and mortality in elderly inpatients. J Nutr Health Aging. 12: 219–221.

Simonsen K, Rode A, Nicoll A, Villadsen G, Espelund U, Lim L, Angus P, Arachchi N, Vilstrup H, Nexo E and Gronbaek H. 2014. Vitamin B(1)(2) and its binding proteins in hepatocellular carcinoma and chronic liver diseases. Scand J Gastroenterol. 49: 1096–1102.

Stevenson TD and Beard MF. 1959. Serum vitamin B12 content in liver disease. N Engl J Med. 260: 206–210.

Vashi P, Edwin P, Popiel B, Lammersfeld C and Gupta D. 2016. Methylmalonic Acid and Homocysteine as Indicators of Vitamin B12 Deficiency in Cancer. PLoS One. 11: e0147843.

Vollset SE, Clarke R, Lewington S, Ebbing M, Halsey J, Lonn E, Armitage J, Manson JE, Hankey GJ, Spence JD, Galan P, Bonaa KH, Jamison R, Gaziano JM, Guarino P, Baron JA, Logan RF, Giovannucci EL, den HM, Ueland PM, Bennett D, Collins R and Peto R. 2013. Effects of folic acid supplementation on overall and site-specific cancer incidence during the randomised trials: meta-analyses of data on 50000 individuals. Lancet. 381: 1029–1036.

Waters WE, Withey JL, Kilpatrick GS and Wood PH. 1971. Serum vitamin B 12 concentrations in the general population: a ten-year follow-up. Br J Haematol. 20: 521–526.

Weinstein IB and Watkin DM. 1960. Co58B12 absorption, plasma transport and excretion in patients with myeloproliferative disorders, solid tumors and non-neoplastic diseases. J Clin Invest. 39: 1667–1674.

14

Vitamin B12 and Drug Development

Jayme L Workinger and *Robert P Doyle**

1. Background and Scope on Utilizing the Vitamin B12 Dietary Pathway for Drug Development

1.1 Vitamin B12 and its dietary uptake

The consumption of vitamin B12 (B12), also known as cobalamin (Cbl), is essential for humans. B12 is produced naturally by select bacteria (and likely certain archea) (Doxey et al. 2015) and organisms must acquire the vitamin through their diet (about 2.5 µg per day for humans) (Martens et al. 2002; Nielsen et al. 2012). There are two primary biologically active forms of B12: methylcobalamin and adenosylcobalamin. Methionine synthase uses methylcobalamin to produce the amino acid methionine from homocysteine, and methylmalonyl-CoA mutase uses adenosylcobalamin as a cofactor to produce succinyl CoA (Nielsen et al. 2012; Kräutler 2005). Mammals have developed a complex dietary uptake pathway for B12 involving a series of transport proteins and specific receptors across various tissues and organs (vide infra) (Nielsen et al. 2012; Gherasim et al. 2013). It is the understanding and exploitation of this uptake pathway that offers considerable scope for drug development.

B12 is a water-soluble vitamin with a highly complex structure, comprising a midplanar corrin ring composed of four pyrrole rings linked to a central cobalt (III) atom (Hodgkin et al. 1955). The corrin ring is similar to the more commonly known porphyrin structure, but with key differences in terms of degree of saturation, symmetry and planarity (see Figure 1).

Department of Chemistry, Syracuse University, Syracuse NY 13244, USA.
* Corresponding author: rpdoyle@syr.edu

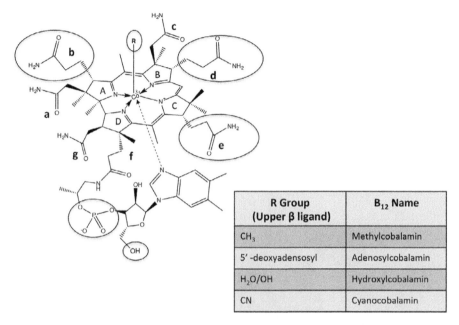

R Group (Upper β ligand)	B$_{12}$ Name
CH$_3$	Methylcobalamin
5′-deoxyadensosyl	Adenosylcobalamin
H$_2$O/OH	Hydroxylcobalamin
CN	Cyanocobalamin

Figure 1. Structure of vitamin B12. R is –methyl or -adenosyl in active cofactors. In center is the corrin ring, which has a greater number of sp3 carbons than a porphyrin rendering it less planar and less conjugated, and with one less carbon (19 rather than 20) due to the lack of a methylene spacer unit between the 'C' and 'D' rings as is found in a porphyrin. Listed are the common variable R groups found on B12. H$_2$O/OH are the same group but is dependent on pH (OH at alkaline pH and H$_2$O at neutral and acidic). Also highlighted are the major modifiable sites on B12 for conjugation to small molecules or peptides/proteins, whereby binding by the dietary uptake proteins can be minimally affected or selected for, based on drug development requirements.

Several functional groups are synthetically available for modification on B12 (see Figure 1). Only select modification sites, however, maintain the recognition needed to utilize the full B12 uptake pathway (see Figures 1 and 2). Modifications can, however, be made to target specific proteins while reducing affinity for others, a fact recently exploited to target haptocorrin-positive tumors (Waibel et al. 2008). An in-depth discussion of modification sites for either complete, or targeted pathway access, can be found in Section 1.2.

Transport and delivery of B12 through the gastrointestinal tract is dependent on three primary carrier proteins: haptocorrin (HC; K_d = 0.01 pM), intrinsic factor (IF; K_d = 1 pM), and transcobalamin II (TCII; K_d = 0.005 pM), each responsible for carrying a single B12 molecule (Fedosov et al. 2002). B12 is initially released from food by the action of peptic enzymes and the acidic environment of the gastrointestinal system and bound by HC (also known as R-binder or transcobalamin I (TCI)) (Nielsen et al. 2012). HC is a glycoprotein with an apparent molecular mass of between 60–70 kDa and is secreted by the salivary glands (Furger et al. 2013). HC has a high affinity for B12 under acidic conditions (pH < 3), allowing it to protect B12 (Holo-HC) from acid

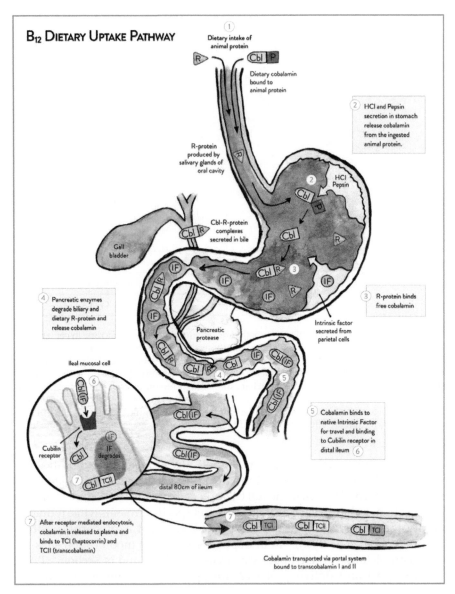

Figure 2. Dietary uptake pathway for B12 in humans. Abbreviations used: R-protein/TCI: Haptocorrin (HC); IF: Intrinsic factor; TCII: Transcobalamin II; Cbl: cobalamin/B12. Image produced and used with permission of Xeragenx LLC (St. Louis, MO, USA).

hydrolysis. Holo-HC travels from the stomach to the duodenum, where the increase in pH (> 5) decreases the affinity of HC for B12 and, combined with pancreatic digestion of HC, causes B12 release, whereupon it is bound by gastric intrinsic factor (Glass 1963).

IF is a ~50 kDa glycosylated protein that is secreted from the gastric mucosa (Mathews et al. 2007). Once B12 is bound to IF, it facilitates transport to the ileum and passage across intestinal enterocytes. This occurs by receptor-mediated endocytosis via an IF-B12 receptor cubilin (CUB) (Christensen et al. 2013). CUB transports holo-IF in concert with a transmembrane protein amnionless (AM), creating a CUBAM receptor for holo-IF (Fyfe et al. 2004). Following internalization, IF is degraded by lysosomal proteases, such as cathepsin L, and B12 is released into the blood stream, either as free B12 or pre-bound to transcobalamin II (TCII) (Nielsen et al. 2012; Beedholm-Ebsen et al. 2010). There is some controversy in this area as to whether both methods occur or one dominates over the other, and indeed whether there is a third mechanism at play also. Cells that require B12 express the holo-TCII receptor, CD320. Upon internalization, TCII is degraded and B12 is transported from the lysosome for cellular use (Kräutler 2005). Kidney cells also express the megalin receptor, which in part reabsorbs filtered holo-TCII from urine (Moestrup et al. 1996).

Knowledge of the binding between B12 and is various transport proteins is critical if the system is to be successfully exploited from bench-top to bedside. In the last 10 years there has been a huge advance in our critical understanding of protein structure as it relates to the B12 uptake pathway, with the publication of HC, IF, TCII, and cubilin-IF-B12 structures (Furger et al. 2013; Mathews et al. 2007; Wuerges et al. 2006; Andersen et al. 2010). The first solution structure of a B12 conjugate, that of B12 coupled to the anorectic peptide PYY_{3-36}, was also recently reported (using NMR methodology) (Henry et al. 2016).

Researchers have a better understanding now of how B12 interacts with its transport proteins, and how these transport proteins interact with their receptors. The implications this can have on drug delivery and sites of potential conjugation can then be better predicted, detailed, rationalized and hence optimized. These endeavors need fundamental knowledge of properties and behavior of B12 in biological systems but also require new or improved synthetic routes to introduce the exact desired modification into the vitamin necessary for specific exploitation.

1.2 Modifying B12: the crossing of synthetic and end-goal considerations

There are multiple sites for chemical modification on the B12 molecule, depending on whether retention of recognition by all transport proteins, or the selective recognition of a subset is required. Therefore, it is critical to consider solvent-accessible surfaces of B12 transport proteins and how such proteins physically bind B12. For IF, this exposure is ~13% (~163 Å^2), with TCII at ~6.5% (~80 Å^2) and HC, having the least accessible area at 3.2% (~40 Å^2) (Figure 3) (Wuerges et al. 2007). These exposures allow for a wide range of modification at select sites on B12 conjugates while retaining (A)

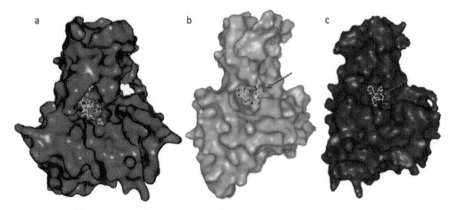

Figure 3. Crystal Structures of the Three Transport Proteins Bound to B12. The ribose 5'OH is solvent accessible in all three transport proteins as shown by the arrows. (a) TCII, (b) IF, (c) HC. The structure of TCII, IF, and HC was obtained from www.pdb.org; (PDB code 2BB5, 2PMV, 4KK1 respectfully). The program Protein Workshop Version 3 was used to display the image.

general pathway acceptance or (B) for selecting specific parts thereof (both (A) and (B) are discussed separately below in Section 1.2).

B12 and the molecule of interest ('drug') can be: (1) coupled directly together; (2) held apart by "spacer" units to give distance between B12 and the molecule; or (3) have the desired drug contained within the carrier, unconjugated, but with the carrier covalently bound to B12. Several functional groups are available for modification on B12, including propionamides, acetamides, hydroxyl groups, the cobalt(III) ion and the phosphate moiety (Proinsias et al. 2013). However, the sites for modification capable of maintaining the recognition of all three transport proteins is limited due to the manner in which B12 is bound by each protein.

All three transport proteins (HC, IF and TCII) bind to B12 with high affinities, but the specificity varies. IF shows the highest specificity for B12, followed closely by TCII, with HC have a broad substrate base including B12 analogs such as cobinamides (Fedosov et al. 2002; Fedosov et al. 2007). It is thought because of the affinity of HC for many inactive B12 analogs that it acts as a scavenger, removing such from the blood and partially digested B12 from the intestine, preventing bacterial access (thus suggesting a role for B12 in bacteriostasis). Viewed synthetically, this implies of course that B12 can be readily modified and retain recognition by HC, whereas IF and TCII offer significantly less range for modification.

1.2.1 General pathway acceptance

Conjugation of molecules to B12 resulting in the recognition of HC, IF, and TCII have been successful with five major sites to date: (1) the peripheral corrin ring e-propionamide (Alsenz et al. 2000); (2) the peripheral corrin ring b-propionamide (Waibel et al. 2008); (3) the 5'-hydroxyl group of the ribose ring on the dimethylbenzimidazole 'tail' (Petrus et al. 2007); (4) the 2'-hydroxyl group of the ribose ring (Wang et al. 2007); and (5) the cobalt cation (Tran et al. 2016).

The crystal structure of holo-TCII provides a rationale for why these positions are favored for modification. The phosphate moiety, 2' hydroxyl group, and a, c, d, and g-propionamides have various hydrogen bonds between multiple TCII residues and the solvent molecules indicating any modifications would disrupt that bonding and stability of the TCII-B12 complex (Wuerges et al. 2006). In addition, TCII does not completely encompass B12 and leaves a 1.4 nm solvent-accessible pocket of B12. This pocket shows the phosphate and the ribose moieties protruding and both have been exploited in conjugate design whereby TCII and IF binding has been maintained (Figure 3).

The 2'-hydroxyl group has been modified through activation by diglycolic anhydride but this has not been extensively used in the B12 conjugation field, given the ease of 5'-OH coupling (Wang et al. 2007). Conjugation to the b- and e-propionamides have been a popular choice for chemists for conjugation (Alsenz et al. 2000; Waibel et al. 2008). Such a route requires acid hydrolysis of the amides, typically using 1 N HCl (Pathare et al. 1996). This synthetic route creates multiple mono-acids at the b, d and e-positions, which makes access to targeted specific acids low yielding ($\leq 15\%$) and complex to purify (Pathare et al. 1996). A recent result exploiting this approach however, is that of Schubiger et al., who showed that, based on the tether length off of the b-acid side chain, selectively towards specific transport proteins (IF over TCII, for example) could be achieved (see Section 2.5) (Waibel et al. 2008).

The most common site for modification has, however, became the 5' hydroxyl group, for three main reasons: (1) molecules conjugated here still allow binding retention of the transport proteins (Bonaccorso et al. 2015; McEwan et al. 1999; Fowler et al. 2013); and (2) conjugation to this site is highly facile and selective (Clardy-James et al. 2012; Chromiński and Gryko 2013); and (3) a wide range of modifications have been developed for this site, expanding scope for substrate conjugation (Clardy-James et al. 2012; Wierzba et al. 2016; Chromiński and Gryko 2013; McEwan et al. 1999). The "classic" activation with 1,1'-carbonyldiimidazole or 1,1'-carbonyldi(1,2,4-triazole) with an addition of a primary amine, producing a carbamate linked conjugation, allows for a wide range of molecules to be used (McEwan et al. 1999). Doyle et al. directly modified this position, using 2-iodoxybenzoic acid and 2-hydroxypyridine, to create a carboxylic acid at this position, which could then be readily used to produce amide linked conjugates (Clardy-James et al. 2012; Bonaccorso et al. 2015; Henry et al. 2016; Henry et al. 2015). Later, Gryko et al. developed a "clickable" B12 conjugate, replacing the 5'hydroxyl with an azide, which allows for a high yielding Huisgen-Sharpless copper-azide-alkyl reaction, creating a stable triazole linker with alkyne containing molecules (Chromiński and Gryko 2013). Most recently, Gryko et al. also developed a reactive pyridyl disulfide group at this site with moderate yield (~60%) (Wierzba et al. 2016). A reactive thiol group creates the possibility of direct disulfide bonds to proteins and molecules, opening up a new area for conjugation that readily exploits redox for the first time.

Another increasingly popular site for modification is the cobalt atom. Transport proteins accommodate significant change at the cobalt β-ligand site, a feature exploited in the biochemistry of B12 (Kräutler 2005). Comparing the binding constants of different biological axial ligands such as methyl, hydroxyl, 5'-adenosylcobalamin, and cyano-cobalamin show no significant difference (Fedosov et al. 2002). In synthetic approaches the Co(III) is typically reduced to Co(I) and then reacted with electrophiles, similar to the biological enzymatic process by methionine synthases (Ruetz et al. 2013; Kräutler 2005). Modifications at the cobalt ion have been limited by the fact that most products are extremely light sensitive (Ruetz et al. 2013; Ruiz-Sánchez et al. 2007). However, in 2013, two groups published light stable phenylethynylcobalamin in two separate syntheses: (1) radical reduction chemistry (Ruetz et al. 2013) and (2) reduction free synthesis (Chromiński et al. 2013). These syntheses allowed for a wide tolerance to functional groups, allowing for more complex ligands on this site to be conjugated or subsequently reacted forward. Another way of modifying the cobalt ion is by metalating the cyano axial group. Fluorophores, radionuclides, cisplatin, vanadate, chlorambucil, and colchicine have been attached to this site (see Table 1).

In 2016, Gryko et al. published moderate-to-high yielding (~40–80%) modifications to the phosphate moiety, the first complete investigation of this group to also include binding studies (Proinsias et al. 2016). The phosphate modification showed preferential binding to IF based on the length of the linker attached to the phosphate. Interestingly, these conjugates are acid, heat, and UV light sensitive possibly allowing for a future new class of cleavable B12 conjugates for drug delivery.

A more recent approach in B12 modifications is creating a dual functionalization of the (a) cobalt ion and (b) 5' hydroxyl group. These conjugates are used for detection and delivery by designing a detectable component with a drug on the same molecule. The detectable component, such as a radionuclide or fluorophore, is conjugated to the 5' hydroxyl group and a drug is either added directly to the cobalt atom or attached via a linker (Tran et al. 2016; Shell et al. 2014). In 2014, Lawrence et al. used this idea to target erythrocytes with the 5'-hydroxyl moiety and delivered a cleavable drug via the cobalt linker (see Section 2.4) (Smith et al. 2014).

While understanding the maintenance of transport protein binding and modified B12 conjugate design and synthesis has made significant strides, it is important to note that little is known about the effects a modified holo-protein complex had on recognition and binding to its receptor. In 2010, Andersen et al. published the crystal structure of cubilin$_{(5-8)}$-IF-B12 (Andersen et al. 2010). This structure provides an outline of how the holo-IF interacts with cubilin,

Table 1. Vitamin B12-Small Molecule Conjugates.

Molecule	Conjugation Site	Linker	Application	Ref
b-nido-carborane	b-acid	diaminobutane	Antitumor therapy	(Hogenkamp et al. 2000)
d-nida-carborane	d-acid	diaminobutane	Antitumor therapy	
Bis-nido-carbarane	b- and d-acid	diaminobutane	Antitumor therapy	
NO	cobalt atom	direct	Antitumor therapy	(Bauer et al. 2002)
SulfoCy5	Ribose 5'-OH	1,6-diaminohexane	Imaging	(McGreevy et al. 2003)
Colchicine	Cobalt atom	4-chlorobutyricacid chloride	Delivery	
	Ribose 5'-OH*	Octadecylamine	Delivery	(Smith et al. 2014)
[Re(OH$_2$)(CO)$_3$]	Cyano ligand	Imidazolecarboxylic acid	Linker for biomolecules	(Kunze et al. 2004)
	Cyano ligand	2,4-dipicolinic acid	Linker for biomolecules	
	Cyano ligand	Serine	Linker for biomolecules	
	Cyano ligand	N,N-dimethylglycine	Linker for biomolecules	
Cisplatin	Cobalt atom	Cyano ligand	Oral anticancer	(Mundwiler et al. 2005)
Cisplatin-2'deoxyguanosine	Cobalt atom	Cyano ligand	Proof of principle	
Cisplatin-methylguanosine	Cobalt atom	Cyano ligand	Proof of principle	
[trans-PtCl(NH$_3$)$_2$]	Cobalt atom	Cyano ligand	Delivery	(Ruiz-Sánchez et al. 2007)
[trans-PtCl$_2$(NH$_3$)]	Cobalt atom	Cyano ligand	Delivery	
[Cis-PtCl$_2$(NH$_3$)]	Cobalt atom	Cyano ligand	Delivery	
PtCl$_3$	Cobalt atom	Cyano ligand	Delivery	
VO$_2$(OH/H)	Cobalt atom	3-hydroxy-2-methyl-1-propyl-IH-pyridin-4-one	Diabetes treatment	(Mukherjee et al. 2008)
VO$_2$	Cobalt atom		Diabetes treatment	
Khodamine 6G	Ribose 5'-OH	Trans-1,4-diaminocyclohexane	Imaging	(Lee and Grissom 2009)
Gd^{3+}	Ribose 5'-OH	Anhydride of DTPA	Imaging	(Siega et al. 2009)
		Anhydride of TTHA	Imaging	
Rhodamine isothiocyanate	NA	HPMA	Imaging	(Russell-Jones et al. 2011)
Re(CO)$_3$ – 1,1 -bisthiazole	Ribose 5'-OH	1,4- diaminobutane	Imaging	(Vortherms et al. 2011)
[{trans-Pt(NH$_3$)$_2$}-{Cyt}]$^{+2}$	Cobalt atom	Cyano ligand	Delivery, IC$_{50}$ of 230 nM*	(Tran et al. 2013)

Table 1. contd.....

Table 1. contd.

Molecule	Conjugation Site	Linker	Application	Ref
[{trans-Pt(NH$_3$)$_2$}-{DTIC}]$^{+2}$	Cobalt atom	Cyano ligand	Delivery	
[{trans-Pt(NH$_3$)$_2$}-{Ana}]$^{+2}$	Cobalt atom	Cyano ligand	Delivery	
Bodipy650	Cobalt atom	3-aminopropyl	Delivery	(Shell et al. 2014)
cAMP	Cobalt atom	3-aminopropyl	Delivery	
Doxorubicin	Cobalt atom	3-aminopropyl	Delivery^F	
Methotrexate	Cobalt atom / Ribose 5'-OH*	3-aminopropyl / Octadecylamine	Delivery	(Smith et al. 2014)
Dexamethasone	Cobalt atom / Ribose 5'-OH*	3-aminopropyl / Octadecylamine	Delivery	
[{Re}-{cis-PtCl(NH$_3$)$_2$}]$^+$	Cobalt and 5'-OH	Cobalt and Ribose-5'-OH	Delivery	(Tran et al. 2016)

NA: Not available; DTPA: diethylenetriamine-N,N;N',N'', N'''-pentaacetic acid; DCC: N,N'-dicyclohexylcarbodiimide; TTHA: Triethylenetetramine-N,N,N',N',N'',N''',N''''-hexacetic acid; HPMA: Lysine-modified-hydroxypropyl-methacrylamide; Cyt: cytarabine; DTIC: dacarbazine; Ana: anastrozole. *compared to IC$_{50}$ of B12-[trans-PtCl(NH$_{3/2}$)] of 8 μM. *A two functional B12 conjugate. ^{F}Showed similar cell death as the control doxorubicin. Imaging defined as fluorescence/confocal microscopy.

allowing researchers a new tool to determine the implications of modification on receptor binding (a study not yet conducted to our knowledge).

1.2.2 Targeting specific transport proteins

A B12 conjugate injected in the systemic circulation can be bound by HC or TCII. Initially, it was hypothesized that cancer therapy/imaging using a B12 based delivery mechanism would work based on a projected increase in the TCII receptor, CD320, in a variety of cancer types such as breast, ovarian, thyroid, uterine, testis, and brain cancer (Mundwiler et al. 2005; Collins and Hogenkamp 1997). This overexpression of CD320 would provide sufficient uptake of a tracer bound to endogenous TCII vs uptake in healthy tissue. Such studies, however, suffered from high background (Section 2.5).

The use of HC binding was not investigated until 2008, when Schubiger et al. made a series of B12 conjugates that would selectively bind HC (and IF), but not TCII (Waibel et al. 2008). The hypothesis here was that, given the presence of TCII and HC in serum, and assigning the high background to TCII mediated cell entry, targeting only HC would offer improved results. Membrane associated HC, expressed *de novo*, in certain cancer cell lines offered a possible route to selectivity, absent from the approach to CD320 uptake (Carmel 1975). In 2014, a select conjugate, ^{99m}TC-PAMA-cobalamin capable of selectively binding HC in blood serum, used in the detection of breast, colon, lung, and pharyngeal cancers in human patients, showed greater tumor

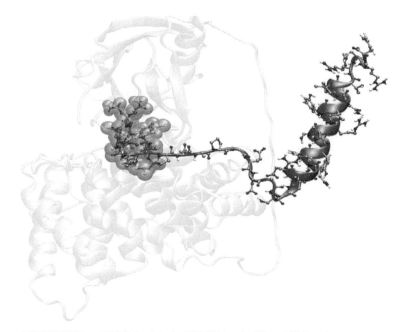

Figure 4. IF-B12-PYY$_{3-36}$. Highlighted is the B12-PYY$_{3-36}$ (gold) and IF (gray).

uptake and reduced TCII-based background (see Figure 4) (Sah et al. 2014). This publication is highly significant for B12 drug development, especially since it was performed in a human patient.

1.2.3 Structural investigations of B12 conjugates

As mentioned above there are three ways in which B12 and a molecule can be connected (elaborated upon in Section 1.2). Connecting through a "linker" to create length is favored in conjugates mainly for ease of synthesis (a result of favored modification techniques). In considering modification a few questions arise; (a) are all linker lengths created equal, and (b) do you need a certain "space" to better allow binding to transport proteins as well as allow the peptide to function properly?

In 2016, Doyle et al. published a structural study, using NMR and molecular dynamics, to predict agonism at the peptide receptor for a B12-peptide (Henry et al. 2016). They used two B12-peptides (both based on B12 conjugates of the anorectic Y2-receptor agonist Peptide YY (PYY_{3-36})) with a difference of one methylene unit length and assessed the B12-peptide's agonism at its receptor as well as performed constrained molecular dynamics (MD) with each. The data (collected by NMR and used to constrain MD) showed that the conjugate with a longer tether had more hydrogen bonding to B12 and predicted lower agonism at the Y2-R target receptor, a result confirmed by *in vitro* cell assay (Henry et al. 2016). Such work offers the possibility of using constrained MD to predict function a priori and minimize the need to synthesize extensive libraries of compounds for screening. This data also showed for the first time in a solution structure, that a B12-peptide/protein can be made without prominently affecting the peptides secondary structure (and hence function).

2. Recent Highlights in B12 Drug Development

There are several excellent reviews in this area that the reader is referred to here (Zelder 2015; Clardy et al. 2011). What is noted below are recent specific highlights and general overview considerations that endeavor to ask where the field is going and what the big hurdles/goals are in the field of B12 drug development.

2.1 Oral delivery

Few peptide/protein-based drugs have the ability to survive the gastrointestinal tract and/or cross the intestinal wall to make it to the systemic circulation. The B12 pathway has naturally developed a complex mechanism for this uptake. Researchers can "hijack" this pathway to deliver B12-drugs in an oral manner. Early research in B12-peptide/protein oral drug delivery was conducted by Russell-Jones and co-workers in the 1990's focusing on B12 conjugates of

Table 2. Vitamin B12-Peptide Bioconjugates.

Peptides	Conjugation Site	Linker	Application	Ref
BS Albumin	Phosphate	Phosphaye-amine	NA	(Hippe et al. 1971)
YG-globulin	Phosphate	Phosphaye-amine	NA	
HS-albumin	e-acid	GABA	Antibody response	(Ahrenstedt and Thorell 1979)
IFN-con	Ribose 5'-OH	Glutaroyl	24 – 28% activity*	(Habberfield et al. 1996)
G-CSF	e-acid	Disulfide	61-66% activity[F]	(Russell-Jones, Westwood, and Habberfield 1995)
		Amide	29-85% activity[F]	
		Hydrazine	ND-100% activity[F]	
EPO	e-acid	Amide	ND-<4% activity[F]	
		Hydrazide	17-22% activity[F]	
ANTIDE-1	e-acid	EGS	ND	(Russell-Jones et al. 1995)
		Amide	0% IF recognition	
		disulfide	65% IF recognition	
		Hindered thiol	54% IF recognition	
		thioester	81% IF recognition	
		Transglutamase cleavable tetrapeptide	60% IF recognition	
ANTIDE-3	e-acid	EGS	ND	
		Amide	ND	
		disulfide	ND	
		Hindered thiol	7% IF recognition	
		thioester	65% IF recognition	
		Transglutamase cleavable tetrapeptide	48% IF recognition	
LHRH	e-acid	Amide	45% absorbed	(Alsenz et al. 2000)
DP3	e-acid	Amide	21% absorbed	
		Hexyl	42% absorbed	
Insulin	Ribose 5'-OH	Amide	26% drop is glucose	(Petrus et al. 2007)
PYY (3-36)	Ribose 5'-OH	Amide	Oral delivery, 200 times higher plasma levels[Æ]	(Fazen et al. 2011)

Table 2. contd.....

Table 2. contd.

Peptides	Conjugation Site	Linker	Application	Ref
	Ribose 5'-OH	1-amino-3-butyne	11% decrease in food intake[θ]	(Henry et al. 2015)
Exendin-4	Ribose 5'-OH	1-amino-□-butyne	4 fold increase in stability[ψ]	(Bonaccorso et al. 2015)
Encapsulated insulin delivery				
B12 coated-dextran nanoparticles	Ribose 5'-OH	Amide	70 – 75% drop in plasma glucose	(Chalasani et al. 2007)
B12-phthaloyl chitosan carboxymethyl nanoparticles	Ribose 5'OH	dihydrazine	□5% drug release	(Fowler et al. 2013)
B12-trimethyl chitosan nanoparticles	Ribose 5'-OH	dihydrazine	29% drug release	
B12-chitosan layered CaPO4 nanoparticles	Ribose 5'-OH	Amide	12 fold increase in effective duration[Δ]	(Verma et al. 2016)

*compared with native IFN-con. [F]compared with unconjugated G-CSF and EPO. [Æ]compared to native PYY levels are meal ingestion. [Φ]compared to unconjugated PYY$_{(3-36)}$. [ψ]when bound to IF vs unbound to IF. [Δ]compared to injected insulin. ND = not determined. BS, HS: Bovine and human serum; IFN-con: Consensus interferon; G-CSF: Granulocyte colony stimulating factor; EPO: Erythropoietin; ANTIDE: N-Ac-D-Nal(2)D, D-Phe (pCl), D-Pal(3), Ser, Lys (Nic), D-Lys(Nic), Leu, Lys(IPr), Pro, D-Ala-NH2; LHRH: Luteinizing hormone-releasing hormone; DP3: Octapeptide (Glu-Ala-Ser-Ala-Ser-Tyr-Ser-Ala); GABA: γ amino butyric acid; EGS: Ethylene glycol bis(succinimidyl succinate).

granulocyte colony stimulating factor, erythropoietin, luteinizing hormone-releasing hormone, ANTIDE-1, and ANTIDE-3 (see Table 2) (Russell-Jones et al. 1995; Russell-Jones, Westwood and Habberfield 1995). Since then other groups have shown B12-molecules being transported via the B12 pathway across intestine cell lines *in vitro* and *in vivo* (Petrus et al. 2009; Dix et al. 1990; Verma et al. 2016). More recent highlights include that of a B12-PYY$_{3-36}$ conjugate, which, when administered orally achieved clinically relevant levels of PYY$_{3-36}$ in blood of a rat model (~200 pg/mL after 1 h) (see Table 2 and Figure 4) (Fazen et al. 2011). This conjugate was not, however, shown to have any functional effect. In 2016, Mishra et al. showed a B12-chitosan layered nanoparticle that encapsulated insulin had a 10-fold increase in effective insulin duration *in vivo* when administered orally, achieving a maximum drop in glucose of ~40% (Verma et al. 2016).

Although data suggests drugs can be delivered orally through this pathway there are some questions left unanswered such as; (a) would oral administration be feasible because of the limit of uptake due to the expression of CUB, (b) amount of drug that gets into the blood serum and (2) would pre-binding of B12-drug to IF allow more efficient uptake? It is known that there is a limited pool of CUB expressed in the terminal ilium, which limits IF-mediated absorption to around 1.5 µg per meal (1 nmole/dose) (Schjonsby and Andersen 1974). Survival of enterocyte passage by a peptide bound to B12 is also unknown, as is whether such a conjugate would arrive in serum bound or unbound to TCII (with implications for subsequent function) (see Section 3).

2.2 Subcutaneous delivery

As mentioned above a "hijacking" of the B12 pathway can be exploited for oral delivery. However, this "hijacking" is not limited to the use of oral administration. In 2015, Doyle et al. published on a subcutaneously administered B12-PYY$_{3-36}$ (Henry et al. 2015). Peptide YY$_{3-36}$ (PYY$_{3-36}$) is an endogenous appetite suppressing peptide that is an agonist for the NPY2 receptor in the intestines and arcuate nucleus of the hypothalamus. Food intake (FI) was significantly reduced over a five-day course for B12-PYY$_{3-36}$ (24%) compared to PYY$_{3-36}$ (13%). In addition, reduction of FI was more consistent after each dose through the course of a rat feeding cycle for B12-PYY$_{3-36}$ (26%, 29%, 27%) compared with PYY$_{3-36}$ treatment (3%, 21%, 16%)[ref]. These findings demonstrate significant pharmacodynamic (PD) improvement upon simple conjugation of B12 to PYY$_{3-36}$ for subcutaneous delivery. Of interest also was the fact that, when looking at the pharmacokinetic (PK) parameters of the B12 conjugate compared to free PYY$_{3-36}$, it is clear there is minimal improvement in terms of serum half-life, clearance, volume of distribution. Of note, in PK terms, was the observed increased Cmax for B12-PYY$_{3-36}$ compared to PYY$_{3-36}$, but at the same T$_{1/2}$ (suggesting B12 conjugation did not increase subcutaneous uptake rate, but did improve amount that was passaged).

2.3 Peptide and protein protection

One of the major open questions in the field has been whether binding a peptide/protein to B12, with or without subsequent B12 binding protein interaction, could offer any protection against proteolysis. In 2015, Doyle et al. attempted to address this focusing on the stability (as defined by retention of peptide/protein agonist receptor function) of a B12 conjugate of the glucose controlling (GLP-1 receptor agonist) incretin exendin-4 (Ex-4) (Bonaccorso et al. 2015). Either as the straight B12-conjugate, or bound by IF, function at the GLP1-R relative to undigested controls was investigated using proteases from both the gastrointestinal tract (trypsin and chymotrypsin) and kidney (meprin β). The addition of IF produced up to a four-fold increase in function compared to Ex-4 alone, when digested by trypsin, and no statistical decrease in function when challenged by meprin β (Bonaccorso et al. 2015). These results offer a significant opportunity for exploitation. Increase in gastric stability, even on a small percentage scale, could provide a route to achieving the desired effect orally. This work also suggests the possibility of utilizing an IF-B12-drug complex in serum, thus expanding use of IF beyond oral administration. If the fact that IF is not found in serum produces antigenicity, it is likely that switching IF for HC would achieve similar improvements in protease protection, and mitigate the antigenicity (thus, suggesting here using the pathway for PK improvements).

2.4 Photo-cleavable conjugates

In 2014, Lawrence et al. published a series of B12-flurophores that were modified on the cobalt atom. This series was designed, and proven, to be selectively photocleavable at different wavelengths, tissue-penetrating light (600–900 nm), in a mixture (Shell et al. 2014). The photocleavable-B12 system was then used as a platform to selectively deliver drugs. Initially, B12-cAMP and B12-doxorubicin (B12-Dox) activity was shown *in vitro*. B12-Dox showed cell viability equal to that of the control doxorubicin and B12-cAMP showed a light only induced cell morphology change typical of cAMP-dependent protein kinase activity (Shell et al. 2014).

A follow up paper, also published in 2014 by Lawrence et al., used the photocleavable-B12 platform to deliver three anti-inflammatories: methotrexate, colchicine, and dexamethasone (see Table 2) (Smith et al. 2014). They used the 5′-hydroxyl group to target erythrocytes by using a C_{18} hydrophobic linker and functionalized the cobalt atom with each drug. After loading the erythrocyte with the B12-drug and a modified fluorophore (added as a separate molecule) they were able to photocleave the drug into the surrounding media and observe cell morphology changes (Smith et al. 2014). This approach then was able to selectively deliver drugs to targeted cell lines.

2.5 Imaging

Table 3 lists the B12-imaging agents reported to date. As mentioned above in Section 1.2.2, historically, imaging using B12 targeted the CD320 receptor, based on the premise that overexpression of CD320 on rapidly proliferating tumor cells would provide necessary tumor to background ratio's (Collins and Hogenkamp 1997; Mundwiler et al. 2005). However, this technique proved highly limited due to observed high background uptake across tissues. In 2014, Burger et al. published a human *in vivo* study using a ^{99m}Tc probe, based on a B12-conjugate modified at the b-acid with a 4-carbon linker (B12-PAMA) in patients with five types of cancer: lung, colon, hypopharyngeal, prostate, and breast (Sah et al. 2014). The conjugate was shown initially to be selectivity bound by HC, but not to be bound by TCII (Waibel et al. 2008). Initial results showed uptake in select cancers but with moderate background (see Figure 5). After pre-dosing with excess B12, background was further reduced with an average uptake of ~4.5% was observed (Sah et al. 2014).

In 2014, Ikotun et al. published a B12-PET imaging probe based on a B12-NOTA conjugate with ^{64}Cu (Ikotun et al. 2014). Small animals tumor studies were conducted with four cancer cell lines: pancreatic, ovarian, colorectal, and murine melanoma. However, as was observed with ^{99m}Tc studies, the same high background trend was seen. The tumor % ID/g achieved was ~4%.

2.6 Anti-vitamins

For a full comprehensive review of anti-vitamins, we refer you to two recent reviews by Zelder and Kräutler (Kräutler 2015; Zelder et al. 2015).

B12 based anti-vitamins are a new class of B12 conjugates. Anti-vitamins are defined as molecules that counteract biological action of vitamins. B12 based conjugates interfere with B12's ability during the enzymatic process by preventing the redox potential of the cobalt atom. There are two types of antivitamins: ones they prevent B12 from (a) forming into methylcobalamin and adenosylcobalamin (Ruetz et al. 2013), and (b) converting into a base "off" form (Zhou and Zelder 2010). These conjugates are designed to "lock" the cobalt atom and therefore remove its capability to act as a B12 vitamin.

Zelder et al. has focused on modifying the dimethylbenzimidazole (DMB) group within B12 (see Figure 1) (Zhou et al. 2012). By altering DMB's linker to resemble peptides, creating a rigid backbone, the B12 is locked into a "base on" form. These conjugates have been shown *in vitro* to inhibit L. Delbruekii growth in a concentration dependent manner (Zhou et al. 2012). Kräutler et al. have modified the R axial group, shifting the cobalt redox potential more negative and therefore making reduction very difficult (Ruetz et al. 2013). The inert 'aryl-cobalamins' thus produced have been shown to bind to transport proteins but are not functional in B12 dependent enzymes (Mutti et al. 2013).

Table 3. Vitamin B_{12} Used in Radio Imaging.

Molecule	Conjugation Site	Linker	Use	Ref
^{99m}Tc-DTPA	b-acid	1,4diaminobutane	Radio Imaging	(Collins and Hogenkamp 1997)
^{111}In-DTPA	b-acid	1,4diaminobutane	Radio Imaging	
[^{99m}TcO$_4$]	cyano ligand	Imidazolecarboxylic acid	Radiodiagnosis	(Kunze et al. 2004)
		Picolinic acid	Radiodiagnosis	
		2,4-dipicolinic acid	Radiodiagnosis	
		Serine	Radiodiagnosis	
		N,N-dimethylglycine	Radiodiagnosis	
^{131}I	Cobalt atom	Cisplatin	Radio Imaging	(Ruiz-Sánchez et al. 2007)
[^{99m}Tc(CO)$_3$(OH$_2$)$_3$]	b-acid	Propyl-PAMA-OEt	Radio Imaging	(Waibel et al. 2008)
		Ethyl-PAMA-OEt	Radio Imaging	
		Butyl-PAMA-OEt	Radio Imaging	(Sah et al. 2014)
		Pentyl-PAMA-OEt	Radio Imaging	
		Hexyl-PAMA-OEt	Radio Imaging	
^{64}Cu-SCN-Bn-NOTA	Ribose 5'-OH	Ethylenediamine	Radio Imaging	(Ikotun et al. 2014)

DTPA: diethylenetriamine-N,N,N″, N‴- pentaacetic acid; PAMA: [pyridine-2-ylmethyl-amino]-acetic acid; SCN-Bn-NOTA; isothiocyanatobenzyl-1,4,7- triazacyclononane-N,N′,N″-triacetic acid. * Conjugate was selected for human studies targeting tumors.

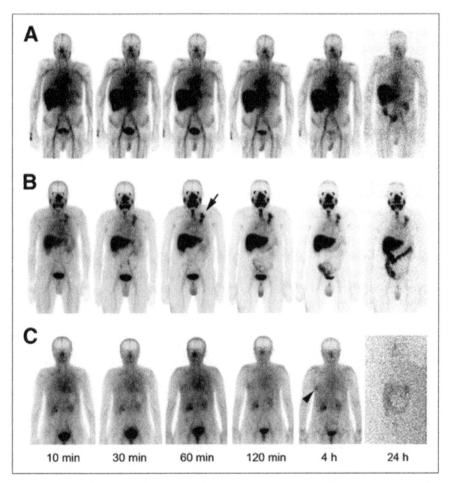

Figure 5. SPECT-CT Scans Targeting Haptocorrin in Human Patients. Partial-body scans after distribution of ^{99m}Tc-PAMA-cobalamin at 10, 30, 60, 120, 240 min, and 24 h. (A) Without Cbl pre-dose has high blood-pool uptake over 24 h and no tracer accumulation in the tumor (hypopharynx), (B) After 20-mg Cbl pre-dose has reduced blood-pool activity and high uptake in bronchial carcinoma (arrow) stable over 24 h, (C) After 1,000-mg Cbl pre-dose has reduced liver uptake and only faint uptake in metastatic right axillary lymph node (arrowhead). This research was originally published in the Journal of Nuclear Medicine by I. A. Burger et al. 2014; 55: 43–49. © by the Society of Nuclear Medicine and Molecular Imaging, Inc.

B12 anti-vitamins could be used to attempt to effectively starve cancer cells of B12. The overall effect on the patient would most likely be concomitant pernicious anemia. Looked at another way, these antivitamins could also be used in animal models, to study the effects of B12 deficiency in a variety of diseases.

2.7 Considerations for murine models of B12 conjugates in In vivo studies

Upon careful consideration of structural design and *in vitro* target validation, the next obvious step is to move into animal models, and in particular that of rodent models. What needs to be made clear here is the fact that there are several major concerns about using murine models for extrapolation to humans.

The first major issue in the use of murine models lies in the fact that, humans, as described in Section 1.1, have two B12 binding proteins in serum, namely TCII and HC (Nielsen et al. 2012). As demonstrated in a 2011 paper by Nexo et al., mouse TCII has a single serum protein with features of both TCII and HC (Hygum et al. 2011). This work can be extrapolated to rats and other common small animal models such as guinea pigs by BLAST analysis. Developing systems to prevent TCII binding (to lower background uptake in imaging studies or to prevent loss of function upon serum delivery in oral studies, for example) by modifying the B12 structure (as discussed in Section 1.2.2 and 2.5) are significantly hampered then, since the broader specificity of binding inherent in the murine TCII prevents the desired effect from being manifest. In such situations, it is likely that models, such as the rabbit (documented to contain both the serum TCII and HC proteins as in humans) would be a more appropriate choice (Nexø and Olesen 1981). Cow (Polak et al. 1979), monkey, pig, and dog (Hygum et al. 2011) have also been documented to contain each of the two serum proteins, although these are not typically first pass *in vivo* screen models.

Another issue with the choice of murine models is the variation in unsaturated TCII concentrations both within model (depending on diet) and then (in terms of nmol/L) between the mouse (> 20 nmol/L), rat (2 nmol/L) and human (0.6–1.5 nmol/L). The concern here has to be with knowing the concentration of unsaturated TCII in the actual model as measured on the specific diet that model is being provided. Factoring this information into the concentration range difference inherent across the species noted, the fact that the murine models have a single serum TCII protein with HC-type properties and understanding that typically 80% of total bound B12 in human serum is bound to HC, not TCII, all must be considered when looking to use a murine model in *in vivo* B12 conjugate drug development studies (Nielsen et al. 2012).

When considering model choice for oral studies, the B12 uptake capacity must also be noted. In humans the uptake capacity is in the range of 1 nmoles per dose (Schjonsby and Andersen 1974), whereas in rats, for example, it is in the order of 10 pmoles per dose (Nexø et al. 1985). Combining such an uptake capacity with the increased serum TCII levels noted above, makes extrapolating anything observed (or not observed) from murine model to human relevance exceedingly difficult.

3. Open Questions and Possible Directions

The use of B12 supplementation in the treatment of B12 deficiencies or in tandem with other drugs to treat or mitigate disease is well documented. A recent review of clinical trials data through the US National Institutes of Health (www.clinicaltrials.gov; accessed April 7th 2016; search term: vitamin B12) reveals 79 open studies (of 346 total studies), all of which are focusing on B12 supplementation in the methyl-, hydroxo-, or cyano-B12 forms primarily, and all aiming to investigate such for improved uptake responses in the elderly, in children, those with feeding issues, etc., or looking at whether such can lower homocysteine levels in vegetarians, or positively effect neurological development, etc. While there has been considerable progress and developments made in exploring and pushing pharmaceutical development based on exploiting the B12 dietary pathway and its various components through use of B12 conjugation (or in general terms, covalent attachment to B12 or B12 analogs such as cobinamides), there remains no FDA approved drug based on such a system, or indeed any open trials noted under the criteria above).

Several questions that remain to be addressed then are:

1. Whether a drug (particularly a nucleic acid or peptide/protein), can be successfully and reproducibly delivered orally upon conjugation to B12 and produce a clinically relevant response, especially beyond the rodent model. Empirical studies of function upon oral dosing would answer this directly but interesting side questions that remain are (a) whether a bound nucleic acid or peptide/protein, can survive the lysosome upon cubilin mediated uptake, and (b) if the conjugate arrives in the blood pre-bound to TCII or not (or both).

    Question 1a is important because partial destruction upon enterocyte passage would undermine, or change, PK function, but also may provide false positive data for successful delivery of certain levels of drug, if part of the surviving B12-peptide conjugate for example, retained the target epitope (e.g., using ELISA). Validating, *in vitro*, IF protection against lysosomal degradation would be a missing link between the gastrointestinal track and blood for a B12-conjugate.

    Question 1b is critical when one considers that, if/when bound to TCII, there is a strong likelihood that the delivered pharmaceutical will be rapidly endocytosed into proliferating cells. This rapid clearance and non-specific targeting would be expected to have a detrimental effect on drug function where it is necessary to target receptors on a particular organ, for example. Establishing the surety of conjugate arrival in the blood, but loss of function due to TCII binding, would then warrant an investigation of oral uptake using B12 modified to maintain IF recognition but not that by TCII. Even if it is shown that the primary mechanism of B12 delivery into blood is bound to TCII, it is clear from the work of Nexo et al. that at least 50% of the B12 dose enters the blood when TCII is removed (Beedholm-Ebsen et al. 2010). This reduced uptake would then need to be factored into uptake capacity calculations.

2. Whether the B12 pathway can be used to (a) increase or decrease drug delivery across the blood-brain barrier (BBB), and (b) whether upon BBB passage, it can change localization within the brain. These questions are of interest because, in general, it is difficult to passage drugs to the brain and any effect on such can be a positive. In some cases, removing brain uptake, while maintaining general systemic effects, is warranted. A B12-peptide conjugate that maintained glucoregulatory control via the pancreas while removing CNS activation that triggers food intake reduction (and under-wanted malaise/nausea) would be of interest for example. Along a similar line, modifying localization within specific brain architectures by, for example, targeting a B12-peptide to the paraventricular neurons and keeping them from the arcuate nucleus would make for a platform technology with, probably inherent and definable, characteristics that could be applied to a particular drug subset.

3. Whether the pathway can be used to develop a new pharmacokinetic (PK) platform technology along the lines of targeting serum albumin to improve serum half-life *in vivo*. A recent patent (R. P. Doyle; Syracuse University, 62/323,013), describes HC targeting substrates such as dicyano-cobinamide peptide conjugates to achieve this half-life improvement. With a half-life in blood of ~9 hours and no known receptors in healthy cells when fully glycosylated, HC is an exciting avenue for PK improvement. The unsaturated binding concentration for HC in serum is 0.3 nmol/L (compared to ~1 nmol/L for TCII in humans with 80% of B12 and B12 analogs bound up by HC and the remaining 20% by TCII) so, while B12 itself would be expected to be bound up by both HC and TCII, some of the administered drug would be lost to TCII if such were used (Sheppard et al. 1984). Exploiting this area of the dietary pathway remains mostly unexplored, although it worth noting again here that Alberto et al. did attempt to target *de novo* expressed membrane associated HC in cancer cells for imaging (Waibel et al. 2008; Sah et al. 2014).

4. Whether prolonged administration of a B12-drug would have detrimental effects on healthy B12 dependent physiology. Nexo et al. produced a study over 27 days in mice administered high doses of dicyano-cobinamide (4.25 nmol/h) by osmotic pump and followed B12 bio-markers such

as the plasma levels of cysteine, total homocysteine, methionine and methylmalonic acid (Lildballe et al. 2012). This study showed no significant changes in plasma levels for the markers in question over the time-period under study. The production of so-called 'antivitamin B12' will allow further elaboration of this area, as would the incorporation of the B12-conjugate under investigation into a B12 dependent assay to gauge if such B12 remained functional (or to what degree it remained functional, at least).

The future of the field lies in expanding and exploiting on the successes of the past several years. What is evident from the work to date is that there is considerable potential in the use of B12 and/or its transport proteins, be it for delivery, targeting, improved PK/PD, etc. The full potential of the B12 dietary uptake pathway has not been realized and the authors believe that with such realization, will come clinical development.

Acknowledgments

This research was supported by the National Institute of Diabetes and Digestive and Kidney Diseases of the US National Institutes of Health under award number R15K097675-01A1. This work was also supported by Xeragenx LLC (St. Louis, MO, USA). The content is solely the responsibility of the authors and does not necessarily represent the official views of the US NIH or Xeragenx LLC.

Keywords: Vitamin B12, drug delivery, dietary pathway, targeting, synthesis, modification, cubilin, CD320, blood-brain barrier, imaging, cancer, pharmacokinetics, pharmacodynamics, oral uptake

Abbreviations

HoloTC	:	holotranscobalamin
HC	:	haptocorrin
TCII	:	transcobalamin
holoHC	:	holohaptocorrin
IF	:	intrinsic factor
B12/Cbl/ cobalamin	:	vitamin B12
CUB	:	cubilin
AMN	:	amnionless
CUBAM	:	cubilin/amnionless receptor
BBB	:	blood brain barrier
NMR	:	nuclear magnetic resonance
PYY	:	peptide YY
FI	:	food intake

PD	:	pharmacodynamics
PK	:	pharmacokinetics
Cmax	:	maximum serum concentration
$T_{1/2}$	:	half-life
Ex4	:	exdendin-4
^{99m}Tc	:	99m-technetium
PAMA	:	pyridine-2-ylmethyl-amino]-acetic acid
NOTA	:	1,4,7- triazacyclononane-N,N′,N″-triacetic acid
PET	:	positron emission tomography
SPEC-CT	:	single-photon emission computed tomography
^{64}Cu	:	64-copper
DMB	:	dimethylbenzimidazole.

References

Ahrenstedt S Stefan and Jan I Thorell. 1979. The production of antibodies to vitamin B12. Clinica Chimica Acta. 95(3): 419–423.

Alsenz J, Russell-Jones GJ, Westwood S, Levet-Trafit B and de Smidt PC. 2000. Oral absorption of peptides through the cobalamin (vitamin B12) pathway in the rat intestine. Pharmaceutical Research. 17(7): 825–832.

Andersen Christian Brix Folsted, Mette Madsen, Tina Storm, Søren K Moestrup and Gregers R Andersen. 2010. Structural basis for receptor recognition of vitamin-B12–intrinsic factor complexes. Nature. 464(7287): 445–448.

Bagnato Joshua D, Alanna L Eilers, Robert A Horton and Charles B Grissom. 2004. Synthesis and characterization of a cobalamin-colchicine conjugate as a novel tumor-targeted cytotoxin. The Journal of Organic Chemistry. 69(26): 8987–8996.

Bauer Joseph A, Bei H Morrison, Ronald W Grane et al. 2002. Effects of interferon beta on transcobalamin II-receptor expression and antitumor activity of nitrosylcobalamin. Journal of the National Cancer Institute. 94(13): 1010–1019.

Beedholm-Ebsen Rasmus, Koen van de Wetering, Tore Hardlei et al. 2010. Identification of multidrug resistance protein 1 (MRP1/ABCC1) as a molecular gate for cellular export of cobalamin. Blood. 115(8): 1632–1639.

Bonaccorso Ron L, Oleg G Chepurny, Christoph Becker-Pauly, George G Holz and Robert P Doyle. 2015. Enhanced peptide stability against protease digestion induced by intrinsic factor binding of a vitamin B12 conjugate of exendin-4. Molecular Pharmaceutics. 12(9): 3502–3506.

Carmel Ralph. 1975. Extreme elevation of serum transcobalamin I in patients with metastatic cancer. New England Journal of Medicine. 292(6): 282–284.

Chalasani Kishore B, Gregory J Russell-Jones, Akhlesh K Jain, Prakash V Diwan and Sanjay K Jain. 2007. Effective oral delivery of insulin in animal models using vitamin B12-coated dextran nanoparticles. Journal of Controlled Release. 122(2): 141–150.

Christensen Erik I, Rikke Nielsen and Henrik Birn. 2013. From bowel to kidneys: The role of cubilin in physiology and disease. Nephrology Dialysis Transplantation. 28(2): 274–281.

Chromiński Mikołaj and Dorota Gryko. 2013. "Clickable" vitamin B12 derivative. Chemistry—A European Journal. 19(16): 5141–5148.

Chromiński Mikołaj, Agnieszka Lewalska and Dorota Gryko. 2013. Reduction-free synthesis of stable acetylide cobalamins. Chemical Communications. 49(97): 11406–11408.

Clardy-James Susan, Jaime Bernstein, Deborah Kerwood and Robert Doyle. 2012. Site-selective oxidation of vitamin B12 using 2-iodoxybenzoic acid. Synlett. 23(16): 2363–2366.

Clardy Susan M, Damian G Allis, Timothy J Fairchild and Robert P Doyle. 2011. Vitamin B12 in drug delivery: Breaking through the barriers to a B12 bioconjugate pharmaceutical. Expert Opinion on Drug Delivery. 8(1): 127–140.

Collins Douglas A, Harry PC Hogenkamp and Mark W Gebhard. 1999. Tumor imaging via indium 111–labeled DTPA–adenosylcobalamin. Mayo Clinic Proceedings. 74(7): 687–691.

Collins Douglas A and Hogenkamp H.P.C. 1997. Transcobalamin II receptor imaging via radiolabeled diethylene-triaminepentaacetate cobalamin analogs. Journal of Nuclear Medicine. 38(5): 717–723.

Dix CJ, Hassan IF, Obray HY, Shah R and Wilson G. 1990. The transport of vitamin B12 through polarized monolayers of Caco-2 cells. Gastroenterology 98(5 PART 1). Scopus. 1272–1279.

Doxey Andrew C, Daniel A Kurtz, Michael DJ Lynch, Laura A Sauder and Josh D Neufeld. 2015. Aquatic metagenomes implicate thaumarchaeota in global cobalamin production. The ISME Journal. 9(2): 461–471.

Fazen Christopher H, Debbie Valentin, Timothy J Fairchild and Robert P Doyle. 2011. Oral delivery of the appetite suppressing peptide hPYY(3–36) through the vitamin B12 uptake pathway. Journal of Medicinal Chemistry. 54(24): 8707–8711.

Fedosov Sergey N, Lars Berglund, Natalya U Fedosova, Ebba Nexø and Torben E Petersen. 2002. Comparative analysis of cobalamin binding kinetics and ligand protection for intrinsic factor, transcobalamin, and haptocorrin. Journal of Biological Chemistry. 277(12): 9989–9996.

Fedosov Sergey N, Natalya U Fedosova, Bernhard Kräutler, Ebba Nexø and Torben E Petersen. 2007. Mechanisms of discrimination between cobalamins and their natural analogues during their binding to the specific B12-transporting proteins. Biochemistry. 46(21): 6446–6458.

Fowler Robyn, Driton Vllasaliu, Franco H Falcone et al. 2013. Uptake and transport of B12-conjugated nanoparticles in airway epithelium. Journal of Controlled Release. 172(1): 374–381.

Furger Evelyne, Dominik C Frei, Roger Schibli, Eliane Fischer and Andrea E Prota. 2013. Structural basis for Universal corrinoid recognition by the cobalamin transport protein haptocorrin. Journal of Biological Chemistry. 288(35): 25466–25476.

Fyfe John C, Mette Madsen, Peter Højrup et al. 2004. The functional cobalamin (vitamin B12)–intrinsic factor receptor is a novel complex of Cubilin and Amnionless. Blood. 103(5): 1573–1579.

Gherasim Carmen, Michael Lofgren and Ruma Banerjee. 2013. Navigating the B12 road: Assimilation, delivery, and disorders of cobalamin. Journal of Biological Chemistry. 288(19): 13186–13193.

Glass GBJ. 1963. Gastric intrinsic factor and its function in the metabolism of vitamin B12. Physiological Reviews 43. CAB Direct. 2: 529–849.

Habberfield Alan David, Olaf Boris Kinstler and Colin Geoffrey Pitt. 1996. Conjugates of vitamin b12 and proteins. http://www.google.com/patents/WO1996004016A1, accessed April 17, 2016.

Henry Kelly E, Clinton T Elfers, Rachael M Burke et al. 2015. Vitamin B12 conjugation of peptide-YY3–36 decreases food intake compared to native peptide-YY3–36 upon subcutaneous administration in male rats. Endocrinology. 156(5): 1739–1749.

Henry Kelly E, Deborah J Kerwood, Damian G Allis et al. 2016. Solution structure and constrained molecular dynamics study of Vitamin B12 conjugates of the anorectic peptide PYY(3–36). Chem Med Chem. DOI: 10.1002/cmdc.201600073.

Hippe Erik, Edgar Haber and Henrik Olesen. 1971. Nature of vitamin B12 binding. Biochimica et Biophysica Acta (BBA)—Protein Structure. 243(1): 75–82.

Hodgkin Dorothy Crowfoot, Jenny Pickworth, John H Robertson et al. 1955. Structure of vitamin B12 : The crystal structure of the hexacarboxylic acid derived from B12 and the molecular structure of the vitamin. Nature. 176(4477): 325–328.

Hogenkamp HP, Collins DA, Live D, Benson LM and Naylor S. 2000. Synthesis and characterization of Nido-Carborane-Cobalamin conjugates. Nuclear Medicine and Biology. 27(1): 89–92.

Hygum Katrine, Dorte L Lildballe, Eva H Greibe et al. 2011. Mouse Transcobalamin Has Features Resembling Both Human Transcobalamin and Haptocorrin. PLOS ONe. 6(5): e20638.

Ikotun Oluwatayo F, Bernadette V Marquez, Christopher H Fazen et al. 2014. Investigation of a vitamin B12 conjugate as a PET imaging probe. Chem Med Chem. 9(6): 1244–1251.

Kräutler B. 2005. Vitamin B12: Chemistry and biochemistry. Biochemical Society Transactions. 33(4): 806–810.

Kräutler B. 2015. Antivitamins B12—A structure- and reactivity-based concept. Chemistry—A European Journal. 21(32): 11280–11287.

Kunze Susanne, Fabio Zobi, Philipp Kurz, Bernhard Spingler and Roger Alberto. 2004. Vitamin B12 as a Ligand for Technetium and Rhenium Complexes. Angewandte Chemie (International Ed. in English). 43(38): 5025–5029.

Lee Manfai and Charles B Grissom. 2009. Design, synthesis, and characterization of fluorescent cobalamin analogues with high quantum efficiencies. Organic Letters. 11(12): 2499–2502.

Lildballe Dorte L, Elena Mutti, Henrik Birn and Ebba Nexo. 2012. Maximal load of the vitamin B12 transport system: A study on mice treated for four weeks with high-dose vitamin B12 or cobinamide. PLOS ONE. 7(10): e46657.

Martens J-H, Barg H, Warren M and Jahn D. 2002. Microbial production of vitamin B12. Applied Microbiology and Biotechnology. 58(3): 275–285.

Mathews FS, Gordon MM, Chen Z et al. 2007. Crystal structure of human intrinsic factor: Cobalamin complex at 2.6-Å resolution. Proceedings of the National Academy of Sciences. 104(44): 17311–17316.

McEwan JF, Veitch HS and Russell-Jones GJ. 1999. Synthesis and biological activity of ribose-5'-carbamate derivatives of vitamin B12. Bioconjugate Chemistry. 10(6): 1131–1136.

McGreevy James M, Michelle J Cannon and Charles B Grissom. 2003. Minimally invasive lymphatic mapping using fluorescently labeled vitamin B12. The Journal of Surgical Research. 111(1): 38–44.

Moestrup SK, Birn H, Fischer PB et al. 1996. Megalin-mediated endocytosis of transcobalamin-vitamin-B12 complexes suggests a role of the receptor in vitamin-B12 homeostasis. Proceedings of the National Academy of Sciences. 93(16): 8612–8617.

Mukherjee Riya, Edward G Donnay, Michal A Radomski et al. 2008. Vanadium-vitamin B12 bioconjugates as potential therapeutics for treating diabetes. Chemical Communications (Cambridge, England). (32): 3783–3785.

Mundwiler Stefan, Bernhard Spingler, Philipp Kurz, Susanne Kunze and Roger Alberto. 2005. Cyanide-bridged vitamin B12-cisplatin conjugates. Chemistry (Weinheim an Der Bergstrasse, Germany). 11(14): 4089–4095.

Mundwiler Stefan, Robert Waibel, Bernhard Spingler, Susanne Kunze and Roger Alberto. 2005. Picolylamine-methylphosphonic acid esters as tridentate ligands for the labeling of alcohols with the Fac-[M(CO)3]+ core (M=99mTc, Re): Synthesis and biodistribution of model compounds and of a 99mTc-Labeled cobinamide. Nuclear Medicine and Biology. 32(5): 473–484.

Mutti Elena, Markus Ruetz, Henrik Birn, Bernhard Kräutler and Ebba Nexo. 2013. 4-Ethylphenyl-cobalamin impairs tissue uptake of vitamin B 12 and causes vitamin B 12 deficiency in mice. PLOS ONE. 8(9): e75312.

Nexø Ebba and Jørgen Andersen. 1977. Unsaturated and cobalamin saturated transcobalamin I and II in normal human plasma. Scandinavian Journal of Clinical and Laboratory Investigation. 37(8): 723–728.

Nexø Ebba and Henrik Olesen. 1981. Purification and characterization of rabbit haptocorrin. Biochimica et Biophysica Acta (BBA)—Protein Structure. 667(2): 370–376.

Nexø Ebba, Mads Hansen, Steen Seier Poulsen and Peter Skov Olsen. 1985. Characterization and immunohistochemical localization of rat salivary cobalamin-binding protein and comparison with human salivary haptocorrin. Biochimica et Biophysica Acta (BBA)—General Subjects. 838(2): 264–269.

Nielsen Marianne J, Mie R Rasmussen, Christian BF Andersen, Ebba Nexø and Søren K Moestrup. 2012. Vitamin B12 transport from food to the body's cells—a sophisticated, multistep pathway. Nature Reviews Gastroenterology and Hepatology. 9(6): 345–354.

Pathare Pradip M, D Scott Wilbur, Shannon Heusser et al. 1996. Synthesis of cobalamin–Botin conjugates that vary in the position of cobalamin coupling. Evaluation of cobalamin derivative binding to transcobalamin II. Bioconjugate Chemistry. 7(2): 217–232.

Petrus Amanda K, Timothy J Fairchild and Robert P Doyle. 2009. Traveling the vitamin B12 pathway: Oral delivery of protein and peptide drugs. Angewandte Chemie International Edition. 48(6): 1022–1028.

Petrus Amanda K, Anthony R Vortherms, Timothy J Fairchild and Robert P Doyle. 2007. Vitamin B12 as a carrier for the oral delivery of insulin. Chem Med Chem. 2(12): 1717–1721.

Polak DM, Elliot JM and Haluska M. 1979. Vitamin B12 binding proteins in bovine serum1. Journal of Dairy Science. 62(5): 697–701.

Proinsias Keith ó, Maciej Giedyk and Dorota Gryko. 2013. Vitamin B12: Chemical modifications. Chemical Society Reviews. 42(16): 6605–6619.

Proinsias Keith ó, Michal Ociepa, Katarzyna Pluta, Mikolaj Chrominski, Ebba Nexo and Dorota Gryko. 2016. Vitamin B12 phosphate conjugation and its effect on binding to the human B12 binding proteins intrinsic factor and haptocorrin. Chemistry—A European Journal. IN PRESS.

Ruetz Markus, Carmen Gherasim, Karl Gruber et al. 2013. Access to organometallic arylcobaltcorrins through radical synthesis: 4-Ethylphenylcobalamin, a potential "Antivitamin B12." Angewandte Chemie International Edition. 52(9): 2606–2610.

Ruetz Markus, Robert Salchner, Klaus Wurst, Sergey Fedosov and Bernhard Kräutler. 2013. Phenylethynylcobalamin: A light-stable and thermolysis-resistant organometallic vitamin B12 derivative prepared by radical synthesis. Angewandte Chemie International Edition. 52(43): 11406–11409.

Ruiz-Sánchez Pilar, Stefan Mundwiler, Alfredo Medina-Molner, Bernhard Spingler and Roger Alberto. 2007. Iodination of Cisplatin adduct of vitamin B12 [{B12}-CN-{cis-PtCl(NH3)2}]+. Journal of Organometallic Chemistry 692(6). Third International Symposium on Bioorganometallic Chemistry(ISBOMC'06)Third International Symposium on Bioorganometallic Chemistry. 1358–1362.

Ruiz-Sánchez Pilar, Stefan Mundwiler, Bernhard Spingler et al. 2007. Syntheses and characterization of vitamin B12–Pt(II) conjugates and their Adenosylation in an enzymatic assay. JBIC Journal of Biological Inorganic Chemistry. 13(3): 335–347.

Russell-Jones GJ, Westwood SW and Habberfield AD. 1995. Vitamin B12 mediated oral delivery systems for granulocyte-colony stimulating factor and erythropoietin. Bioconjugate Chemistry. 6(4): 459–465.

Russell-Jones Gregory, Kirsten McTavish and John McEwan. 2011. Preliminary studies on the selective accumulation of vitamin-targeted polymers within tumors. Journal of Drug Targeting. 19(2): 133–139.

Russell-Jones G, Westwood S, Farnworth P, Findlay J and Burger H. 1995. Synthesis of LHRH antagonists suitable for oral administration via the vitamin B12 uptake system. Bioconjugate Chemistry. 6(1): 34–42.

Sah Bert-Ram, Roger Schibli, Robert Waibel et al. 2014. Tumor imaging in patients with advanced tumors using a new 99mTc-radiolabeled vitamin B12 derivative. Journal of Nuclear Medicine. 55(1): 43–49.

Schjonsby H and Andersen KJ. 1974. The intestinal absorption of vitamin B12. Scan J Gastro Suppl. 29: 7–11.

Seetharam Bellur and David H Alpers. 1994. Cobalamin binding proteins and their receptors. In Vitamin Receptors. Intercellular and Intracellular Communication. Cambridge University Press. dx.doi.org/10.1017/CBO9780511525391.006.

Shell Thomas A, Jennifer R Shell, Zachary L Rodgers and David S Lawrence. 2014. Tunable visible and near-IR photoactivation of light-responsive compounds by using fluorophores as light-capturing antennas. Angewandte Chemie International Edition. 53(3): 875–878.

Sheppard K, Bradbury DA, Davies JM and Ryrie DR. 1984. Cobalamin and folate binding proteins in human tumour tissue. Journal of Clinical Pathology. 37(12): 1336–1338.

Siega Patrizia, Jochen Wuerges, Francesca Arena et al. 2009. Release of toxic Gd3+ ions to tumour cells by vitamin B12 bioconjugates. Chemistry (Weinheim an Der Bergstrasse, Germany). 15(32): 7980–7989.

Smith Weston J, Nathan P Oien, Robert M Hughes et al. 2014. Cell-mediated assembly of phototherapeutics. Angewandte Chemie International Edition. 53(41): 10945–10948.

Takahashi-Iñiguez Tóshiko, Enrique García-Hernandez, Roberto Arreguín-Espinosa and María Elena Flores. 2012. Role of vitamin B12 on methylmalonyl-CoA mutase activity. Journal of Zhejiang University. Science. B. 13(6): 423–437.

Tran Mai Thanh Quynh, Evelyne Furger and Roger Alberto. 2013. Two-step activation prodrugs: Transplatin mediated binding of chemotherapeutic agents to vitamin B12. Organic & Biomolecular Chemistry. 11(19): 3247–3254.

Tran Mai Thanh Quynh, Stefan Stürup, Ian Henry Lambert et al. 2016. Cellular uptake of metallated cobalamins. Metallomics. 8(3): 298–304.

Verma Ashwni, Shweta Sharma, Pramod Kumar Gupta et al. 2016. Vitamin B12 functionalized layer by layer calcium phosphate nanoparticles: A mucoadhesive and pH responsive carrier for improved oral delivery of insulin. Acta Biomaterialia. 31: 288–300.

Vortherms Anthony R, Anna R Kahkoska, Amy E Rabideau et al. 2011. A water soluble vitamin B12-Re(I) fluorescent conjugate for cell uptake screens: Use in the confirmation of Cubilin in the lung cancer line A549. Chemical Communications. 47(35): 9792–9794.

Waibel Robert, Hansjörg Treichler, Niklaus G Schaefer et al. 2008. New derivatives of vitamin B12 show preferential targeting of tumors. Cancer Research. 68(8): 2904–2911.

Wang Xiaoyang, Lianhu Wei and Lakshmi P Kotra. 2007. Cyanocobalamin (vitamin B12) conjugates with enhanced solubility. Bioorganic & Medicinal Chemistry. 15(4): 1780–1787.

Wierzba Aleksandra, Monika Wojciechowska, Joanna Trylska and Dorota Gryko. 2016. Vitamin B12 suitably tailored for disulfide-based conjugation. Bioconjugate Chemistry. 27(1): 189–197.

Wuerges, Jochen, Gianpiero Garau, Silvano Geremia et al. 2006. Structural basis for mammalian vitamin B12 transport by transcobalamin. Proceedings of the National Academy of Sciences of the United States of America. 103(12): 4386–4391.

Wuerges Jochen, Silvano Geremia, Sergey N Fedosov and Lucio Randaccio. 2007. Vitamin B12 transport proteins: Crystallographic analysis of β-axial ligand substitutions in cobalamin bound to transcobalamin. IUBMB Life. 59(11): 722–729.

Zelder Felix. 2015. Recent trends in the development of vitamin B12 derivatives for medicinal applications. Chemical Communications. 51(74): 14004–14017.

Zelder Felix, Marjorie Sonnay and Lucas Prieto. 2015. Antivitamins for medicinal applications. Chem Bio Chem. 16(9): 1264–1278.

Zhou Kai, René M Oetterli, Helmut Brandl et al. 2012. Chemistry and bioactivity of an artificial adenosylpeptide B12 cofactor. Chem Bio Chem. 13(14): 2052–2055.

Zhou Kai and Felix Zelder. 2010. Vitamin B12 mimics having a peptide backbone and tuneable coordination and redox properties. Angewandte Chemie International Edition. 49(30): 5178–5180.

Index

Printed and bound by CPI Group (UK) Ltd, Croydon, CR0 4YY

01/11/2024

01782622-0009